Traumatic Brain Injury/Chronic Traumatic Encephalopathy as Cause of Alzheimer's Disease: Physics and Molecular Biology in the Genesis of Neurodegeneration?

Traumatic Brain Injury/Chronic Traumatic Encephalopathy as Cause of Alzheimer's Disease: Physics and Molecular Biology in the Genesis of Neurodegeneration?

Guest Editor

Dimitrios Kanakis

MDPI

Basel • Beijing • Wuhan • Barcelona • Belgrade • Novi Sad • Cluj • Manchester

Guest Editor
Dimitrios Kanakis
Pathology
University of Nicosia Medical
School
Nicosia
Cyprus

Editorial Office
MDPI AG
Grosspeteranlage 5
4052 Basel, Switzerland

This is a reprint of the Special Issue, published open access by the journal *International Journal of Molecular Sciences* (ISSN 1422-0067), freely accessible at: https://www.mdpi.com/journal/ijms/special_issues/1C14W2152Y.

For citation purposes, cite each article independently as indicated on the article page online and as indicated below:

Lastname, A.A.; Lastname, B.B. Article Title. *Journal Name* **Year**, *Volume Number*, Page Range.

ISBN 978-3-7258-7378-4 (Hbk)
ISBN 978-3-7258-7379-1 (PDF)
https://doi.org/10.3390/books978-3-7258-7379-1

Contents

About the Editor

Dimitrios Kanakis

Dimitrios Kanakis is a Professor of Pathology and Neuropathology at the University of Nicosia Medical School, Cyprus, and an Honorary Senior Lecturer at St George's, University of London, UK. He earned an MD from the University of Novi Sad (Federal Republic of Yugoslavia) and a PhD, magna cum laude, from Otto von Guericke University Magdeburg, Germany. He holds a specialty certification in neuropathology from the Medical Association of North Rhine, Dusseldorf, Germany (September 2011), and in general pathology from the General Directorate of Public Health and Social Welfare of the Region of Thessaly in Larissa, Greece (January 2013). His academic work spans general pathology and neuropathology. He edited *Pituitary Adenoma: Pathophysiology, Diagnosis and Treatment Options* (NOVA, 2016), *Biomedicines* Special Issue Establishment of State-of-the-Art Methods/Techniques and Identification of Novel Biomarkers in Brain Tumor Diagnosis (later published as a Reprint volume, 2026), and co-edited *Frontiers in Neurology* Research Topic: Recent Advances in Diagnosis and Treatment of Brain Tumors (published as an e-book, 2025). He has also served as a Guest Editor for seven more Special Issues, two in *Biomedicines*, four in the *International Journal of Molecular Sciences* (both MDPI journals), and one in *Frontiers in Immunology*. He has authored four book chapters, including three for the Hellenic Society of Pathology Manuals and one for Springer's *Biomarkers in Cancer*, 2015. He has published peer-reviewed papers and serves as a Reviewer and Editorial Board Member for several journals.

International Journal of
Molecular Sciences

Editorial

Special Issue: "Traumatic Brain Injury/Chronic Traumatic Encephalopathy as Cause of Alzheimer's Disease: Physics and Molecular Biology in the Genesis of Neurodegeneration?"

Dimitrios Kanakis [1,2]

[1] Laboratory of Pathology, Department of Basic and Clinical Sciences, University of Nicosia Medical School, Nicosia 2408, Cyprus; kanakis.d@unic.ac.cy; Tel.: +357-22471902

[2] Centre of Neuroscience and Integrative Brain Research (CENIBRE), University of Nicosia Medical School, Nicosia 2408, Cyprus

In recent years, interest has grown in clarifying the relationship between brain trauma and subsequent neurodegeneration. Current research increasingly aims to define pathogenetic mechanisms that could explain how traumatic brain injury (TBI) and/or chronic traumatic encephalopathy (CTE) may influence later neurodegenerative processes, potentially culminating in a clinical diagnosis of Alzheimer's disease. Importantly, TBI/CTE and Alzheimer's disease share clinical features and histopathological changes that warrant careful consideration. This Special Issue was conceived to examine this possible link and to contribute to a clearer interpretation of any potential cause-and-effect relationship between TBI/CTE and Alzheimer's disease. The articles included here advance this goal by addressing these entities from multiple perspectives and methodological approaches, thereby enriching our understanding of their intersections and distinctions.

Katramadou and colleagues provide evidence connecting the physics of head trauma, especially rotational acceleration and tissue deformation, to diffuse axonal injury (DAI) and white matter vulnerability. These primary insults trigger secondary cascades related to AD-relevant biology: blood–brain barrier (BBB) disruption, neuroinflammation, neurovascular and mitochondrial dysfunction, and protein aggregation, with dementia risk rising in a dose-responsive manner with injury severity and repetition and moderated by factors such as ApoE4. The review underscores stage specific therapeutic interventions, spanning stem cell-based methods for BBB and neural repair, use of neuroprotective agents, mitochondrial stabilization and immunomodulation [Contribution 1].

Blaschke and colleagues introduce a closed-head concussive brain injury (CBI) mouse model designed to evaluate the integrity of the blood–brain barrier (BBB) after single and repetitive CBI. The disruption of BBB, which becomes evident as a result of the administered dextran and IgG extravasation, correlates with functional deficits and increases with injury frequency. In this study, multimodal assessment (MRI, PET with [18F]DPA-714, histology) links "mild" injury phenotypes to measurable vascular compromise [Contribution 2].

Joseph's focused review argues that physiology-sensitive imaging, including late-phase perfusion analysis by 3D arterial spin labeling (ASL) MRI and complementary modalities, may provide empirical evidence of subtle BBB dysfunction and white matter shear injury. Using physiologic testing measures could improve predictions about when it is safe for athletes with mild traumatic brain injury (mTBI) to return to sport. Better understanding of BBB repair could further enable new therapies and outcome tracking, and detecting BBB dysfunction may also support earlier intervention in preclinical neurodegenerative disease [Contribution 3].

Neuschmid and colleagues focus on calcium homeostasis as a shared vulnerability across TBI and AD. Their review highlights calcium-driven enzyme networks (kinases, phosphatases, proteases) that promote synaptic dysfunction, axonal degeneration, amyloid-β accumulation, and tau hyperphosphorylation, with calpain positioned as a key regulator. The editorial claims for more precise therapeutic strategies (including the combined inhibition of selected calcium-dependent enzymes or downstream targets) along with improved TBI phenotyping, long-term observational studies, and polygenic risk scoring to better define dementia risk and guide precision interventions [Contribution 4].

Mitroshina and Vedunova underline the importance of hypoxia as a contributor to neurodegeneration, while emphasizing that HIF-1α involvement can be complex and context-dependent. An imbalanced HIF-1-mediated response to hypoxia may contribute to neurodegeneration. In Alzheimer's disease (AD) and Parkinson's disease (PD), HIF-1 can be protective in some contexts yet proinflammatory and neurotoxic in others. Its effects likely depend on the severity and duration of hypoxia and related conditions. For this reason, therapeutic strategies should aim to modulate HIF-1 carefully: boosting HIF-1 in early disease or mild hypoxia to support neuronal survival and reduce inflammation, but potentially suppressing it in chronic hypoxia or advanced AD where harmful effects may dominate [Contribution 5].

Egoraeva and colleagues provide a concrete example of multi-process repair in a mouse model of mild traumatic brain injury (mTBI) using N-stearidonoylethanolamine (SDEA), an ω-3 ethanolamide. SDEA reduces astrocytic reactivity, restores Arc expression, improves CA1 dendritic spine density and morphology, and partially rescues dentate gyrus proliferative activity (Ki-67). These molecular and cellular changes coincide with improvements in anxiety-like behavior and working-memory performance. Overall, these findings suggest that ω-3 ethanolamides could be promising candidates for multi-target therapeutic approaches in mTBI [Contribution 6].

Pszczołowska and colleagues explicitly explore CTE in relation to AD, highlighting overlapping neuropathological features (notably tau- and amyloid-related themes) while emphasizing differences, including BBB-related damage patterns. The authors synthesize the currently available evidence on potential therapeutic options for chronic traumatic encephalopathy (CTE) and conclude that future progress will likely depend on two complementary priorities: improved prevention of traumatic brain injuries (TBIs) and the development of more effective approaches to limit neuroinflammation. An optimal therapeutic strategy would target the cellular and molecular drivers of secondary injury and cell death, including immune cells, astrocytes, and key inflammatory mediators such as cytokines and chemokines [Contribution 7].

Volloch and Rits-Volloch push toward theory revision. Their perspective traces the evolution of an "Amyloid Cascade Hypothesis 2.0" toward a model centered on sustained neuronal integrated stress response (ISR) biology. They present C99 and intraneuronal amyloid-related species as key drivers and propose that proteolytic A-beta production is suppressed in AD-affected neurons under ISR conditions. They argue this should reshape the interpretation of model systems and therapeutic strategy, including skepticism that extracellular A-beta-focused approaches will reverse symptomatic disease, and propose alternatives aimed at ISR modulation and targeted degradation pathways [Contribution 8].

Funding: This research received no external funding.

Conflicts of Interest: The author declares no conflicts of interest.

https://doi.org/10.3390/ijms27052266

List of Contributions

1. Katramadou, A.; Bender, E.S.; Kanakis, D. From Traumatic Brain Injury to Alzheimer's Disease: Multilevel Biomechanical, Neurovascular, and Molecular Mechanisms with Emerging Therapeutic Directions. *Int. J. Mol. Sci.* **2026**, *27*, 1570. https://doi.org/10.3390/ijms27031570.
2. Blaschke, S.J.; Rautenberg, N.; Endepols, H.; Jendro, A.; Konrad, J.; Vlachakis, S.; Wiedermann, D.; Schroeter, M.; Hoffmann, B.; Merkel, R.; et al. Early Blood–Brain Barrier Impairment as a Pathological Hallmark in a Novel Model of Closed-Head Concussive Brain Injury (CBI) in Mice. *Int. J. Mol. Sci.* **2024**, *25*, 4837. https://doi.org/10.3390/ijms25094837.
3. Joseph, C.R. Assessing Mild Traumatic Brain Injury-Associated Blood–Brain Barrier (BBB) Damage and Restoration Using Late-Phase Perfusion Analysis by 3D ASL MRI: Implications for Predicting Progressive Brain Injury in a Focused Review. *Int. J. Mol. Sci.* **2024**, *25*, 11522. https://doi.org/10.3390/ijms252111522.
4. Neuschmid, S.; Schallerer, C.; Ehrlich, B.E.; McGuone, D. Pathological Calcium Signaling in Traumatic Brain Injury and Alzheimer's Disease: From Acute Neuronal Injury to Chronic Neurodegeneration. *Int. J. Mol. Sci.* **2025**, *26*, 9245. https://doi.org/10.3390/ijms26189245.
5. Mitroshina, E.V.; Vedunova, M.V. The Role of Oxygen Homeostasis and the HIF-1 Factor in the Development of Neurodegeneration. *Int. J. Mol. Sci.* **2024**, *25*, 4581. https://doi.org/10.3390/ijms25094581.
6. Egoraeva, A.; Manzhulo, I.; Ivashkevich, D.; Tyrtyshnaia, A. N-Stearidonoylethanolamine Restores CA1 Synaptic Integrity and Reduces Astrocytic Reactivity After Mild Traumatic Brain Injury. *Int. J. Mol. Sci.* **2026**, *27*, 471. https://doi.org/10.3390/ijms27010471.
7. Pszczołowska, M.; Walczak, K.; Mi'sków, W.; Antosz, K.; Batko, J.; Kurpas, D.; Leszek, J. Chronic Traumatic Encephalopathy as the Course of Alzheimer's Disease. *Int. J. Mol. Sci.* **2024**, *25*, 4639. https://doi.org/10.3390/ijms25094639.
8. Volloch, V.; Rits-Volloch, S. Alzheimer's Is a Multiform Disease of Sustained Neuronal Integrated Stress Response Driven by the C99 Fragment Generated Independently of AβPP; Proteolytic Production of Aβ Is Suppressed in AD-Affected Neurons: Evolution of a Theory. *Int. J. Mol. Sci.* **2025**, *26*, 4252. https://doi.org/10.3390/ijms26094252.

International Journal of
Molecular Sciences

MDPI

Review

From Traumatic Brain Injury to Alzheimer's Disease: Multilevel Biomechanical, Neurovascular, and Molecular Mechanisms with Emerging Therapeutic Directions

Aikaterini Katramadou [1], Eva Sonja Bender [1] and Dimitrios Kanakis [1,2,*]

1 Laboratory of Pathology, Department of Basic and Clinical Sciences, University of Nicosia Medical School, Nicosia 2408, Cyprus; katramadou.a@live.unic.ac.cy (A.K.); bender.e@live.unic.ac.cy (E.S.B.)
2 Centre of Neuroscience and Integrative Brain Research (CENIBRE), University of Nicosia Medical School, Nicosia 2408, Cyprus
* Correspondence: kanakis.d@unic.ac.cy; Tel.: +357-22471902

Abstract

Traumatic brain injury (TBI) is being increasingly recognized as a major risk factor for chronic neurodegenerative disease, including chronic traumatic encephalopathy (CTE) and Alzheimer's disease (AD). Biomechanical forces during head trauma, particularly rotational acceleration and angular deformation, produce diffuse axonal injury (DAI) and widespread white matter damage that trigger persistent neurobiological cascades. These include axonal transport failure, blood–brain barrier (BBB) disruption, neuroinflammation, neurovascular and mitochondrial dysfunction, and pathological protein aggregation, closely paralleling core AD features. Epidemiological data support a dose–response relationship between TBI severity or repetition and subsequent dementia risk, moderated by genetic factors such as apolipoprotein E4 (ApoE4). Converging experimental and early clinical studies have begun to target shared injury and neurodegenerative pathways through acute neuroprotection, stem cell-based strategies for BBB restoration and neural repair, transcriptional and hormonal modulation, mitochondrial stabilization, and immunomodulation of chronic inflammation. This review synthesizes evidence linking biomechanical injury to molecular and neurovascular pathways of neurodegeneration and summarizes emerging temporally targeted interventions. By integrating mechanistic and therapeutic perspectives, we aim to narrow the translational gap between TBI and AD, refine identification of at-risk populations, and inform priorities for prevention and development of disease-modifying therapies.

Keywords: traumatic brain injury (TBI); Alzheimer's disease (AD); chronic traumatic encephalopathy (CTE); diffuse axonal injury (DAI); neurovascular changes; therapeutic interventions

1. Introduction

1.1. Traumatic Brain Injury (TBI): Definition, Prevalence, and Biomechanics

Traumatic brain injury (TBI) is a significant global public health concern substantially contributing to mortality and disability with major medical, social, and economic impacts. TBI results from an external force causing temporary or permanent cognitive, physical, or psychosocial impairment [1]. Mechanical energy is transferred to the head by events such as direct impact, penetrating injury, or rapid acceleration and deceleration of the brain within the skull. In 2021, there were 20.84 million new and 37.93 million existing TBI

 https://doi.org/10.3390/ijms27031570

cases globally [2]. Estimates vary, but TBI is widely recognized as affecting over 50 million people annually and accounting for 30–40% of all injury-related deaths worldwide [3,4].

Epidemiological patterns indicate that TBI varies significantly based on geography, socioeconomic context, age, and sex. In low- and middle-income countries, road traffic collisions account for most cases. However, in higher-income countries, most TBIs are caused by falls, particularly in the older population with other comorbidities [3]. Sports-related concussions and military blast exposures notably contribute to TBI incidence [5]. In most settings, young males experience a disproportionately high incidence of TBI due to high-risk-taking activities, whereas older adults are particularly susceptible due to age-related vulnerability and increased likelihood of falls [6].

At the biomechanical level, TBI results from the interplay between external forces and individual brain tissue properties. When the head undergoes normal movements such as rotation, the brain is protected from potential damage by the cushioning effect of the meninges and the cerebrospinal fluid (CSF) [7]. The brain exhibits viscoelastic properties, suggesting that its deformation under stress conditions is both time-dependent and non-linear [8]. Under high strain rates, during high-energy impacts or blast exposures, brain tissue demonstrates marked stiffness and mechanically distinct behavior from that observed under gradual loading [9,10]. The types of forces developed during head trauma, namely shear, tensile, and compressive, produce heterogeneous patterns of brain tissue damage that affect cortical and subcortical junctions, the corpus callosum, and deep white matter pathways, and rely on impact direction, magnitude, and duration [11].

Among the manifestations of these biomechanical forces, diffuse axonal injury (DAI) is considered one of the most critical. Unlike focal lesions or hematomas, DAI is a hallmark of rotational acceleration and deceleration injuries, where the shear and torsional forces that develop disrupt the axonal integrity, especially in deeper structures such as the corpus callosum, brainstem, and white matter tracts [12]. At the microscopic level, axonal injury is initiated by mechano-transduction pathways, including microtubule disruption, axolemmal damage, and axonal transport failure, which leads to secondary swelling and disconnection [13]. These pathological cellular processes are accompanied by accumulated proteins such as amyloid precursor protein and tau, which may lead to neurodegenerative diseases [14].

1.2. TBI Classification (Severity, Type, and Temporal Phases)

There are several ways to classify a TBI, based on injury severity, neuroimaging, pathophysiology, prognostic models, or the duration of retrograde amnesia [15]. One of the most common methods for determining injury severity is the Glasgow Coma Scale (GCS), first introduced in 1974, used as a clinical tool to evaluate a patient's level of consciousness through eye, motor, and verbal responses [2,16]. Based on GCS, traumas are categorized as mild (scores 13–15), moderate (scores 9–12), or severe (scores 3–8). However, this classification system has several limitations, since it does not account for the underlying pathology, as patients with divergent imaging findings may present with smaller scores. To address this issue, an updated framework definition was proposed by Manley et al., 2025 [17]. This recent revision promotes a four-pillar system, including clinical examination (GCS with pupillary reactivity), biomarkers (blood-based assays), imaging (structural and functional measures), and modifiers (patient-specific factors that influence outcomes). The Clinical, Biomarker, Imaging Modifiers (CBI-M) framework provides a more holistic approach to TBI classification.

TBI can also be classified based on injury type, which is divided into focal and diffuse lesions. Focal injuries are typically characterized by localized structural damage such as contusions, hematomas, and penetrating lesions, most often confined to cortical regions [18].

 https://doi.org/10.3390/ijms27031570

In contrast, diffuse injuries present with widespread disruption of axonal networks and microvascular structures across multiple brain areas [19]. Diffuse axonal injury (DAI) is a characteristic pathology resulting from the shear and tensile forces generated during rapid acceleration and deceleration or rotational head movements [20]. In clinical presentation, both focal and diffuse widespread injury mechanisms often co-exist.

Another significant classification is based on the temporal evolution. The primary injury occurs at the moment of impact and reflects the immediate mechanical disruption of neurons, glia, and vasculature [21]. As a result, our research focuses on the secondary injury cascade, which starts from minutes to years post-trauma. While primary injury can be mitigated through preventative measures, such as helmets and seatbelts, it is not amenable to direct medical intervention [22]. Acute secondary injury events unfold with blood–brain barrier (BBB) breakdown, edema, hemorrhage, excitotoxicity, and hypoperfusion [23]. Over subsequent days to weeks, progressive mechanisms such as oxidative stress, mitochondrial dysfunction, axonal degeneration, and neuroinflammation lead to cellular loss [24]. In the chronic phase, these mechanisms converge on persistent neurodegeneration characterized by abnormal protein aggregation, synaptic dysfunction, and long-term cognitive and behavioral decline, features that overlap with neuropathological hallmarks of Alzheimer's disease (AD) and chronic traumatic encephalopathy (CTE) [25]. Beyond neuropathology, TBI represents the most prevalent neurological disorder worldwide and is associated with a broad spectrum of persistent symptoms. Survivors experience enduring cognitive deficits, emotional and behavioral disturbances, and somatomotor impairments that persist for years after the initial affliction [26]. Sleep disturbances are more common in patients experiencing insomnia, hypersomnia, and poor sleep quality than in the general population [27]. Understanding these temporal and pathological dimensions of TBI is crucial for mapping its overlap with chronic neurodegenerative diseases such as AD.

1.3. Alzheimer's Disease (AD): Overview and Pathology

Alzheimer's disease (AD) is the most common form of dementia, accounting for 60–80% of all dementia cases worldwide [28,29]. Clinically, AD is characterized by progressive cognitive decline, memory impairment, and behavioral disturbances that compromise daily functioning significantly [30]. From a neuropathological perspective, AD is defined by the accumulation of extracellular amyloid-beta (Aβ) plaques and intracellular neurofibrillary tangles (NFTs) composed of hyperphosphorylated tau protein [31]. These pathological features contribute to synaptic dysfunction, neuronal death, and cortical atrophy in the hippocampus [32]. Although Aβ and tau are considered the central pathological features of AD, its pathophysiology is more complex. Evidence implicates additional molecular mechanisms, including brain insulin resistance, chronic oxidative stress, mitochondrial dysfunction, and persistent neuroinflammation [33]. Dysregulated insulin signaling impairs neuronal glucose utilization, while oxidative stress and mitochondrial damage compromise cellular energy metabolism, leading to synaptic failure [34]. Neuroinflammation processes, mediated by activated microglia and astrocytes, contribute to neuronal injury through the release of pro-inflammatory cytokines, chemokines, and reactive oxygen species (ROS) [35]. Furthermore, vascular abnormalities have a critical role in driving neurodegenerative progression. Disrupted cerebral blood flow compromises oxygen and glucose delivery to neurons, while BBB breakdown allows the infiltration of peripheral immune cells and circulating inflammatory mediators into the central nervous system (CNS). These neurovascular processes increase neuroinflammation, impair the clearance of neurotoxic proteins, and promote neuronal dysfunction, gradually leading to cognitive decline.

 https://doi.org/10.3390/ijms27031570

1.4. Epidemiological Link Between TBI and Neurodegenerative Diseases

Epidemiological studies have demonstrated a strong connection between TBI and the subsequent development of neurodegenerative diseases such as Alzheimer's disease [36]. A large-scale, propensity-matched cohort study of over 170,000 U.S. military veterans demonstrated that dementia risk increased proportionally with TBI severity, with adjusted hazard ratios of 2.36 for mild TBI without loss of consciousness, 2.51 for mild TBI with loss of consciousness, and 3.77 for moderate-to-severe TBI [37].

A meta-analysis further confirmed that TBI almost doubles the risk of all-cause dementia (OR = 1.79, 95% CI: 1.66–1.92), while also increasing the risk of AD (OR = 1.60, CI: 1.44–1.77), vascular dementia (VaD OR = 2.03, 95% CI: 1.79–2.30), and frontotemporal lobar degeneration (FTLD OR = 3.99, 95% CI: 2.20–7.20) [38]. These findings highlight that TBI could be considered a modifiable risk factor for late-life cognitive decline. Human neuropathological studies suggest that TBI can directly initiate or accelerate AD-like processes, including neuroinflammation, excitotoxity, cerebrovascular dysfunction, and BBB breakdown [39]. As stated in the literature, TBI exacerbates the upregulation of amyloid precursor protein (APP) and tau hyperphosphorylation, which are core characteristics of AD [40]. These epidemiological and neuropathological findings support the hypothesis that TBI may serve not only as a trigger but also as a potential accelerator of Alzheimer's disease pathology.

1.5. Study Aim

This review endeavors to provide a comprehensive mechanistic and clinical perspective of the association between traumatic brain injury (TBI) and subsequent neurodegeneration. Increasing evidence suggests a causal link between TBI and the initiation or acceleration of Alzheimer's disease. The synthesis of epidemiological, neuropathological, and molecular evidence aims to show how acute and repetitive head injuries lead to long-term cognitive decline and the development of chronic neurodegenerative diseases. A crucial objective of the review is to highlight the role of biomarkers in facilitating early detection of TBI-related neurodegenerative risk, while also presenting emerging therapeutic solutions that may mitigate or prevent the progression from acute injury to chronic disease.

2. Traumatic Brain Injury vs. Chronic Traumatic Encephalopathy vs. Alzheimer's Disease

2.1. Physical and Neuropathological Consequences of TBI

Traumatic brain injury represents a highly heterogeneous condition characterized by diverse patterns of neuropathological events, with its severity and long-term outcomes determined by the initial biomechanical impact, the magnitude of secondary injury mechanisms, and the brain's intrinsic capacity for tissue repair. Immediately after trauma, both diffuse and focal structural damage, often accompanied by vascular disruption, predominate. Over time, these acute processes evolve and can give rise to progressive neurodegenerative cascades that unfold over years to decades. As a result, the neuropathological sequelae of TBI are best understood by considering both the injury severity and the temporal trajectory from the acute to chronic phase.

TBI is increasingly recognized as a trigger for progressive neurodegeneration and increases susceptibility to Alzheimer's disease (AD), Parkinson's disease (PD), motor neuron disease (MND), and chronic traumatic encephalopathy (CTE) [41,42]. Diffuse axonal injury (DAI) emerges as a common pathological feature of TBI, particularly affecting white matter due to mechanical loading and shear forces generated during impact [14]. DAI pathology begins with primary cytoskeletal disruption, impaired axonal transport, and axonal swelling, which progress to secondary axonal disconnection [43]. These processes

 https://doi.org/10.3390/ijms27031570

can ultimately lead to Wallerian degeneration and proteinopathies involving amyloid-beta and tau [44]. In humans, establishing direct causal evidence linking DAI to downstream neurodegeneration remains challenging, as conventional imaging techniques cannot monitor axonal injury at the cellular level [45]. However, diffusion magnetic resonance imaging (MRI) facilitates in vivo quantification of axonal injury and longitudinal atrophy [46].

DAI Pathophysiology

Mechanical indicators such as von Mises stress, which describes brain tissue distortion, are also used to model the likelihood of DAI in TBI [11]. Primary mechanical injury following trauma involves the rapid deformation of white matter tracts, resulting in cytoskeletal failure and the formation of axonal varicosities and bulbs, i.e., "retraction balls," which show disconnection. Stretching these axonal fibers to their limit at a rate of 10 s^{-1} can lead to mechanical damage and secondary injury, resulting in neuronal cell death [11]. This secondary phase involves calcium influx and calpain activation, leading to spectrin and microtubule degradation, mitochondrial swelling, and progressive cytoskeletal breakdown over several hours [47]. Experimental injury models suggest that DAI may serve as a central initiating factor for neurodegeneration after TBI. Disrupted axonal transport also leads to intra-axonal accumulation of amyloid precursor protein (APP), along with β-secretase (BACE1) and presenilin-1, promoting intra-axonal amyloid-beta generation [48,49]. Tau further dissociates from microtubules, undergoes hyperphosphorylation, and propagates into a prion-like manner across the neuronal networks [50].

From Axonal Injury to Neurodegeneration

DAI is observed across the full spectrum of TBI severity, from mild concussion to severe injuries, and manifests clinically as transient loss of consciousness or confusion to persistent coma and cognitive dysfunction [14]. Histopathological evidence shows that APP accumulation in damaged axons can be detected within hours upon TBI and can often be more sensitive than silver staining, identifying DAI [51,52]. However, axonal degeneration and white matter loss persist for weeks to years following injury in certain patients [53]. As a result, two non-exclusive mechanistic pathways are proposed to underlie chronic degeneration following TBI: (i) Wallerian degeneration of axons characterized by transport failure; (ii) proteinopathies, particularly amyloid-beta and tau, originating from traumatic axonal pathology and propagating across interconnected brain networks.

Neuroimaging Evidence Linking DAI to Neurodegeneration

Graham et al. 2023 [54] investigated whether the severity and spatial distribution of diffuse axonal injury (DAI) predict progressive neurodegeneration in a cohort of 55 patients with moderate-to-severe TBI. Using diffusion MRI, the study demonstrated that the DAI burden is the leading factor of chronic white degeneration, promoting a causal relationship between axonal injury and subsequent tissue loss. Fractional anisotropy (FA), a diffusion-derived metric used to calculate the white matter structure, has emerged as an important marker of axonal integrity in the chronic phase of TBI [55]. Reduced FA reflects axonal disruption and associated neuroinflammation and has been correlated with large-scale network dysfunction and cognitive impairment [56]. Serial T1-weighted volumetric imaging further enables accurate quantification [57]. Diffusion abnormalities have also been linked to gray matter volume changes through network diffusion modeling, indicating the relationship between white matter disconnection and cortical degeneration [58]. Conventional imaging markers of DAI, including T2, fluid-attenuated inversion recovery (FLAIR), and T2*-weighted sequences, have been associated with region-specific atrophy pathways early after injury [45].

Overall, DAI has emerged as a strong predictor of progressive neurodegeneration after moderate-to-severe TBI, explaining most of the variability in white matter atrophy and long-term neurodegeneration. Evidence from both pathology and neuroimaging supports

https://doi.org/10.3390/ijms27031570

the central role of axonal injury in progressive tissue loss, functional impairment, and finally increased dementia risk.

2.1.1. TBI Types and Severity

The clinical spectrum of TBI is classified as mild, moderate, or severe based on the Glasgow Coma Scale (GCS), the duration of loss of consciousness (LOC), and the extent of post-traumatic amnesia [59]. Mild TBI is often described as a concussion and typically presents with transient LOC, confusion, and subtle cognitive changes. While symptoms usually resolve, a subset of patients develop persistent post-concussive complaints [60]. Moderate-to-severe injuries are characterized by extended loss of consciousness, radiologically visible lesions such as contusions or hematomas, and a greater risk of long-term disability [61,62]. At the pathological level, severe TBI is associated with diffuse axonal injury (DAI), hemorrhage, and necrotic cavitation [12]. Even in mild cases, microscopic axonal disruption and glial reactivity may occur, changes that are not readily detectable on conventional imaging but have functional consequences [63].

2.1.2. Acute Phase of TBI

The acute phase of TBI progresses within minutes to days after the initial affliction and involves a cascade of cellular, vascular, and metabolic disturbances [64]. Mechanical deformation and shearing forces disrupt neuronal and glial membranes, causing further ionic disequilibrium, glutamate release, and excitotoxic activation of N-methyl-D-aspartate (NMDA) and α-amino-3-hydroxy-5-methyl-4-isoxazolepropionic acid (AMPA) receptors [65]. This cascade leads to calcium overload, mitochondrial dysfunction, and excessive production of reactive oxygen and nitrogen species, driving oxidative stress and energy depletion. Cytotoxic edema develops as astrocytic and neuronal swelling accumulate, exacerbating intracranial pressure and perfusion deficits [66]. At the same time, the blood–brain barrier becomes compromised and permeable as tight-junction proteins degrade, endothelial adhesion molecules are upregulated, and peripheral immune cells infiltrate the parenchyma, amplifying neuroinflammatory responses [67,68]. Microvascular dysfunction, microthrombosis, and early hypoperfusion enhance the secondary injury cascade, indicating chronic pathology.

2.1.3. Chronic Phase of TBI

Beyond the acute phase of TBI, many survivors undergo a chronic phase characterized by long-lasting neuropathological changes that all contribute to progressive cognitive, behavioral, and motor deficits. Diffuse axonal injury evolves into chronic axonopathy associated with disconnected white matter tracts, demyelination, and gliosis. Persistent microglial activation and astrocytosis maintain a pro-inflammatory environment for years following even a single moderate-to-severe TBI, as shown by both human postmortem studies and in vivo imaging [69]. These inflammatory processes converge with vascular injury, blood–brain barrier disruption, and problematic clearance pathways to drive abnormal protein accumulation, particularly hyperphosphorylated tau and amyloid-beta deposits [19]. This results in a neuropathological process that links TBI with later neurodegenerative diseases, including chronic traumatic encephalopathy and Alzheimer's disease. Clinically, the chronic phase presents with enduring cognitive dysfunction, psychiatric disturbances, namely depression and anxiety, and increased risk of dementia.

2.1.4. Diffuse Axonal Injury (DAI)

Diffuse axonal injury (DAI) is recognized as one of the most characteristic substrates of TBI, particularly in acceleration-deceleration and rotational injuries, and primarily affects the axons of cerebral white matter, which are vulnerable to mechanical strain during

https://doi.org/10.3390/ijms27031570

trauma [43]. Under a pathological spectrum, DAI represents not a single lesion but a spectrum of axonal damage ranging from cytoskeletal disruption and impaired axonal transport to swelling and axonal disconnection, further leading to Wallerian degeneration. The hallmark histological findings include axonal swellings and bulbs, with amyloid precursor protein (APP) accumulating within hours of injury. Although the definitive diagnosis still requires post-mortem histopathology, innovative neuroimaging methods such as diffusion tensor imaging (DTI) have been presented as sensitive tools used in vivo to detect white matter abnormalities, with fractional anisotropy (FA) reduction being a sensitive marker of axonal injury [70].

Experimental evidence suggests that DAI is a trigger of progressive neurodegeneration and not only an acute injury phenomenon. Cytoskeletal disruption initiates axonal swelling and bulb formation in sites where APP accumulates with β-secretase (BACE1) and presenilin-1, producing amyloid-β [71]. Dissociation from microtubules and abnormal phosphorylation following injury were also seen in young athletes and experimental mouse models, showing that traumatic axonal injury can induce early tauopathy [72]. Recent studies showed prion-like propagation of tau pathology in animal models of TBI, suggesting a potential mechanistic link between DAI and progressive neurodegeneration. A single moderate-to-severe TBI could accelerate dementia onset and increase the risk of AD through the same axonal pathways of protein accumulation and degeneration. Imaging studies show that DAI predicts progressive brain atrophy after TBI. Using diffusion MRI and volumetric T1 imaging, Graham and colleagues demonstrated that fractional anisotropy in the chronic phase following TBI was a strong predictor of subsequent atrophy in white matter tracts such as the corpus callosum [44]. In addition, calcium influx and calpain activation following axonal stretch injury trigger proteolysis of structural proteins, disruption of microtubules, and mitochondrial dysfunction, which all lead to axonal failure [73]. These inflammatory processes may persist for years following trauma in the corpus callosum [53]. DAI presents as a critical mechanical and imaging marker of progressive post-traumatic neurodegeneration, since it establishes a link between axonal disruption and chronic degeneration.

2.1.5. Chronic Traumatic Encephalopathy (Neuropathological Features, Clinical Manifestation, Risk Factors, and Diagnosis)

Chronic traumatic encephalopathy (CTE) is a progressive neurodegenerative tauopathy associated with repetitive head impacts (RHIs) [74]. Originally, it was described in boxers as 'punch drunk' syndrome; this condition is now well-recognized across a range of contact sports and military personnel [75]. It is neuropathologically unique as it consists of perivascular deposits of hyperphosphorylated tau found at the depths of cortical sulci [76]. Clinically, CTE manifests with cognitive, mood, and behavioral alterations that can manifest years after exposure. The first descriptions of long-term neurological impairment following repeated head trauma were reported by Martland in 1928, who described boxers' symptomatology with gait disturbance, slowed speech, and cognitive decline [77]. Millspaugh later described the punk drunk syndrome as dementia pugilistica [78]. In 1973, Corsellis and colleagues published a neuropathological study of 15 retired boxers identified as having cortical atrophy, neurofibrillary tangles, and cerebellar scarring, further enhancing the theory of dementia pugilistica [79]. These early observations set the basis for later recognition of a neurodegenerative disease. In the 2000s, the work of McKee and Colleagues established modern diagnostic criteria and a four-stage neuropathological staging scheme [80,81]. Early disease presents with focal perivascular lesions, while advanced stages show widespread cortical, limbic, and brainstem involvement [82].

The central lesion in CTE is the perivascular deposition of hyperphosphorylated tau (p-tau) in neurons and astrocytes, located deep in the cortical sulci [83]. This is the most im-

https://doi.org/10.3390/ijms27031570

portant difference compared to AD, because although both are tauopathies, CTE lesions are perivascular, sulcal, and not associated with neuritic amyloid plaques. Molecular studies, using cryo-electron microscopy, showed that tau filaments in CTE adopt a distinct conformation compared to those in AD, proposing CTE as a unique tauopathy [84]. Although tau remains the defining pathology, other findings include transactive response DNA-binding protein 43 (TDP-43) inclusions, axonal degeneration, and white matter degradation [85]. These changes overlap with other neurodegenerative diseases, further highlighting the need for specificity in biomarker CTE-tailored development. Systematic comparisons between CTE and normal aging, as well as other tauopathies, validate the specificity of perivascular sulcal tau lesions [86]. Specifically, the presence of TDP-43 and amyloid-related pathology in both CTE and non-CTE brains further highlights the importance of distinct pathological interpretation. A recent study, which examined 185 athletes, demonstrated cortical thinning, neuronal loss, and synaptic alterations, particularly in sulcal regions, further enhancing the hypothesis that regional biomechanical vulnerability determines lesion localization [87]. Increased astrocytic activity and aquaporin-4 upregulation around the lesions showed BBB dysfunction; however, microgliosis was less prominent [88].

Clinical Manifestations

The clinical presentation of CTE is varied. Symptoms present years after injury and cluster into four main domains: (i) cognitive with executive dysfunction and memory loss, (ii) behavioral with impulsivity, (iii) mood with psychiatric disorders such as depression and suicidality, (iv) motor with Parkinsonian characteristics and dysarthria [89–91]. Clinical studies of former football players showed a high prevalence of behavioral and mood changes before cognitive decline [92,93]. To explain the gap between neuropathology and clinical presentation, Montenigro et al. 2014 [94] proposed the concept of traumatic encephalopathy syndrome (TES).

Risk Factors

The principal risk factor for CTE is exposure to repetitive head trauma. Mez et al. 2017 [95] reported CTE pathology in 87% of 202 deceased former American football players, with disease severity being directly correlated to their athletic career duration. Alosco et al. 2018 [96] further demonstrated a dose–response relationship between years of football and risk of CTE, which was thought to be independent from concussion history. Other modifiers include genetic risk factors such as APOE and TMEM106B, although these associations are still under investigation [97,98].

Diagnosis

At present, CTE can only be diagnosed definitively at autopsy. The DIAGNOSE CTE Research project, a prospective study of 240 men with different exposures to American football, aims to develop clinical criteria and biomarkers [99]. Fluid biomarkers, namely plasma tau, neurofilament light chain, and markers of microglial activation (soluble triggering receptor expressed on myeloid cells 2 [sTREM 2]), are also being evaluated [100].

2.2. *Molecular and Cellular Mechanisms Linking TBI to AD*

2.2.1. Blood–Brain Barrier Dysfunction

The blood–brain barrier (BBB) is a specialized neurovascular unit formed by endothelial cells, pericytes, and astroglial cells, which together maintain central nervous system homeostasis [101]. Following TBI, the BBB undergoes rapid and sustained disruption, which exacerbates secondary injury cascades. Mechanical shear stress and vascular rupture comprise barrier integrity, allowing plasma proteins and immune cells to enter the parenchyma [39]. Endothelial cell activation upregulates adhesion molecules such as intercellular adhesion molecule (ICAM), vascular cell adhesion molecule-1 (VCAM-1), and selectins, facilitating leukocyte adhesion and transmigration, which further amplifies

 https://doi.org/10.3390/ijms27031570

neuroinflammation and oxidative stress cascade [102]. Elevated P-selectin levels promote multicellular aggregation and microvascular occlusion, aggravating ischemic vulnerability. Vascular endothelial growth factor (VEGF) rises sharply after TBI, inducing angiogenesis but simultaneously downregulating tight junction proteins such as occludin and claudin-5, weakening endothelial cohesions and increasing vascular leakage [103,104]. These alterations result in vasogenic edema, neuroinflammation, and impaired tissue repair capacity. BBB disruption is not transient, since chronic dysfunction is characterized by pericyte loss, endothelial degeneration, and tight junction protein disorganization [105]. Chronic mural cell degeneration and reduced expression of Caveolin-1 (Cav-1) and low-density lipoprotein receptor-related protein 1 (LRP1) impair Aβ and tau clearance, which establishes neurodegeneration [106].

Findings from both human and animal studies suggest that the BBB disruption is a central link between TBI and Alzheimer's disease. Evidence indicates BBB breakdown in AD precedes cognitive decline, manifesting as vascular leakage, reduced glucose transport, and leukocyte infiltration [107]. In TBI, the "two-hit" hypothesis model proposes that the initial mechanical disruption is compounded by secondary inflammatory and metabolic cascades, altering vascular and cellular permeability [108,109]. Pharmacological interventions targeting BBB repair, even one year post-injury, reversed axonal degeneration and restored cognition, highlighting barrier restoration as a promising therapeutic strategy [110].

At the molecular level, BBB vulnerability is shaped by genetic and signaling factors. In APOε4 carriers, increased BBB permeability has been linked to cyclophilin A/matrix metalloproteinase 9 (CypA/MMP-9) pathway activation and LRP1 degradation. Concurrently, reduced expression of the major facilitator superfamily domain-containing protein 2A (Mfsd2a) promotes caveolae-mediated transcytosis, further compromising barrier integrity [111]. Impairment of Wingless/Integrated (Wnt)/β-catenin signaling further impairs tight junction integrity and endothelial-pericyte communication, linking genetic and molecular predisposition to blood–brain barrier fragility [112].

2.2.2. Chronic Vascular Changes

Beyond the parenchymal injury, TBI triggers early alterations in leptomeningeal arterial vascular smooth muscle cells (VSMCs) that disrupt cerebrovascular homeostasis and promote amyloidogenic processes. Experimental and post-mortem studies have demonstrated pathological changes in cerebral vessels, including altered distribution of NOTCH3 and alpha-smooth muscle actin (α-SMA) within the tunica media, indicating vascular smooth muscle cell dysfunction. These structural abnormalities are accompanied by the early aggregation of Aβ peptides (Aβ1-40/42, Aβ1-16, beta C-terminal fragment of APP (β-CTF)/C99) around VSMCs, even in young individuals following TBI, whereas controls show strong uniform marker expression and minimal vascular amyloid deposition [113]. These vascular changes are thought to arise from local hypoxia and oxidative stress, which lead to beta-secretase-mediated (BACE-1) APP cleavage with relative downregulation of the non-amyloidogenic a disintegrin and metalloproteinase domain-containing protein 10 (ADAM10) pathway. Therefore, VSMC dysfunction impairs cerebrovascular reactivity, compromises cerebral perfusion, and might lead to secondary ischemia and long-term neurodegeneration [114].

2.2.3. Excitotoxicity

Mechanical forces generated upon head impact strain the neuronal and glial membranes, leading to widespread depolarization of axonal networks and excessive glutamate release. Elevated extracellular glutamate will then activate synaptic and extrasynaptic NMDA receptors [115,116]. The inability of the astrocytic transporter to sustain

 https://doi.org/10.3390/ijms27031570

the extracellular glutamate drives Ca^{2+} influx and downstream apoptotic and necrotic signaling [117]. The rapid entry of cations depolarizes neurons, which leads to the indiscriminate release of neurotransmitters such as glutamate and disrupts synaptic integrity and network stability, further contributing to both acute neuronal injury and long-term neurodegenerative vulnerability.

2.2.4. Oxidative Stress

TBI promotes the generation of reactive oxygen species (ROS) and reactive nitrogen species (RNS) through mitochondrial dysfunction, altered cellular metabolism, and inflammatory oxidase activation [118]. The resulting oxidative and nitrosative stress induces lipid peroxidation, protein nitration, and DNA damage. These alterations contribute to neuronal loss, glial reactivity, and neuroinflammation, accelerating neurodegenerative processes [119].

2.2.5. Mitochondrial Dysfunction and Ferroptosis

Mitochondria regulate cellular energy metabolism, redox homeostasis, and Ca^{2+} regulation, all of which are disrupted following TBI [120]. Impaired electron transport and oxidative phosphorylation (OXPHOS) lead to adenosine triphosphate (ATP) depletion, ROS overproduction, and maladaptive mitochondrial dynamics that compromise repair and synaptic plasticity, which are phenomena observed in early stages of Alzheimer's disease. A recent study implicated the role of ferroptosis, which is an iron-dependent lipid peroxidation pathway, in TBI-associated neuronal death [121]. In experimental models, pharmacological or genetic inhibition of mitoNEET (CDGSH iron–sulfur domain-containing protein 1 [CISD1]) reduces ferroptosis and preserves cognition by stabilizing dihydroorotate dehydrogenase (DHODH)-mediated mitochondrial defenses, indicating that the mitoNEET/DHODH axis could act as a potential therapeutic target [122]. Region-specific proteomic signatures further reveal that repetitive as opposed to single, mild TBI persistently dysregulates mitochondrial and synaptic proteins, especially within the hippocampus, correlating with cognitive decline [123].

2.2.6. Coagulopathy and Microvascular Hemorrhage

TBI initiates an early hypercoagulable state through BBB disruption, extrinsic pathway activation, endothelial glycocalyx degradation, and platelet leukocyte endothelium interaction, followed by coagulopathy and hyperfibrinolysis [124–127].

2.2.7. Cerebral Blood Flow Changes

Altered cerebral blood flow is common following TBI and strongly influence prognosis. Hypoperfusion correlates with cognitive impairment and poor outcomes, whereas timely cerebral blood flow (CBF) restoration is associated with functional improvement [128,129]. Loss of vascular muscle cell contractility and impaired pericyte-mediated regulation disrupt neurovascular coupling, since they limit oxygen and glucose delivery in periods of increased metabolic demand [130–132].

2.2.8. Genetic Susceptibility Linking TBI and Alzheimer's Disease

Inherited genetic factors and dysregulated molecular pathways are recognized as important contributors in determining how traumatic brain injury (TBI) and diffuse axonal injury (DAI) progress toward Alzheimer's disease (AD) and chronic traumatic encephalopathy (CTE) [133]. Recent studies have expanded the understanding of how somatic mutations, trauma-related alterations in lipid metabolism, protein clearance pathways, and blood–brain barrier (BBB) regulators may accelerate neurodegeneration after head injury [134]. Somatic mosaicism has recently been identified as a potential mechanism

https://doi.org/10.3390/ijms27031570

linking acute trauma to chronic neurodegeneration. Single-cell sequencing of TBI and AD brain tissue revealed overlap in somatic mutations across molecular pathways, which suggests that trauma may accelerate genomic instability in neurons and glia. This analysis highlighted that trauma is not only a mechanical insult but also a mutagenic trigger that may exacerbate age-related genetic vulnerabilities [135].

Among germline susceptibility factors, the APOE locus remains the most extensively studied. The ε4 allele was associated with worse outcomes following TBI and an increased risk of AD [136]. Previous studies have shown a tenfold increased risk of AD diagnosis among ε4 carriers with a TBI history [137]. A meta-analysis of 2600 patients suggested that ε4 carriers have poorer global outcomes following TBI [138]. The structural and vascular consequences of APOE isoforms show that ε4 exacerbates cerebral amyloid angiopathy, BBB leakage, and perivascular tau deposition. The human apolipoprotein E gene, APOE, encodes three isoforms—APOEε2, APOEε3, and APOEε4—and is predominantly expressed by astrocytes in the brain. APOEε4 has been linked with being the strongest genetic risk factor for early- and later-onset AD, as well as poor outcomes following TBI, since it disrupts the BBB [139]. APOE ε4 interacts with lipid metabolism and neuroinflammatory cascades, accelerating both the traumatic and degenerative pathology, whereas ε2 appears protective in certain cases but is a risk factor in severe vascular injury. The age of AD onset is three years lower among ε4 carriers with TBI compared to non-carriers with TBI [140]. APOE's role in oxidative stress and excitotoxic process after trauma enhances the hypothesis that isoform-specific effects extend to acute injury physiology and long-term degeneration.

New molecular mechanisms within the APOE locus have also been identified. The translocase of outer mitochondrial membrane 40 (TOMM40)-APOE (T9A2), a chimeric fusion transcript, reportedly modified mitochondrial trafficking and accelerated amyloidogenic processing in AD models [141]. Given the importance of mitochondrial dysfunction in TBI pathology, the TOMM40-APOE fusion may represent an important molecular link between acute axonal injury and late amyloid-driven degeneration.

Autophagy and protein clearance pathways also have a central role in linking TBI to subsequent neurodegenerative pathology. BCL2-associated athanogene 3 (BAG3), a co-chaperone that regulates autophagy–lysosome flux, plays a key role in tau clearance. Reduced BAG3 function disrupts tau regulation and accelerates neurofibrillary pathology, which enhances the hypothesis that repetitive TBI, through impaired proteostatic recovery, may increase vulnerability to CTE and AD tauopathy [142]. Another study showed that genetic variants affecting neprilysin activity have been shown to reduce enzymatic clearance of Aβ, thereby potentiating plaque accumulation. These findings suggest that genetic regulation of proteostasis is important in determining whether traumatic brain injury progresses towards AD pathology [143,144]. Additional susceptibility loci have also been identified through genome-wide association studies (GWASs), including variants in Triggering Receptor Expressed on Myeloid Cells 2 (TREM2). In particular, the R47H variant of TREM2 has been associated with a fivefold increased risk of Alzheimer's disease (OR = 5.05, 95% CI: 2.77–9.16, $p < 0.001$) [145].

Signaling pathways that regulate BBB integrity and neurovascular stability present as essential regulators of TBI to AD progression. Transforming growth factor beta (TGF-β) signaling, which normally has a neuroprotective role since it promotes the expression of tight junction proteins and mediates the recruitment of astrocytes to the BBB, can become maladaptive under chronic trauma conditions, while promoting glial scar formation and impaired repair [146]. Mfsd2a, a lipid transporter crucial for BBB integrity, is downregulated following a TBI, potentially facilitating amyloid influx and vascular inflammation [147]. In particular, this transporter downregulates with age, with a stronger reduction in 5xFAD Alzheimer's disease (AD) model mice compared to wild-type mice. This downregulation is

 https://doi.org/10.3390/ijms27031570

amplified further in the presence of APOE4, a major AD risk allele, and a similar reduction is also observed in human APOE4, suggesting that enhancing Mfsd2a expression could potentially restore BBB integrity [148]. The Wnt/β-catenin signaling pathway, which regulates endothelial tight junctions and cellular proliferation, is disrupted in both AD and TBI, contributing to vascular dysfunction. As a result, the upregulation of Wnt/β-catenin signaling is important to vascular repair, as shown with the intranasal application of recombinant Wnt3a, which promoted BBB repair after TBI [149]. Matrix metalloproteinase-9 (MMP-9) is upregulated following trauma and contributes to BBB disruption and neuronal apoptosis. In TBI, MMP-9 expression is significantly elevated in neutrophils after the initial insult [150]. Sonic the Hedgehog (Shh) signaling is pivotal in neural development and normal brain function. However, in TBI, it has been implicated that endogenous Shh expression is downregulated after TBI, leading to impaired signaling in the acute phase and suppressed neural stem cell renewal in chronic TBI [151]. Finally, mechanistic (mammalian) targeting of rapamycin (mTOR) dysregulation has been observed in TBI, with hyperactivation linked to impaired autophagy and accumulated toxic protein aggregates. Evidence indicates that mTOR activation occurs early in Alzheimer's disease pathogenesis, with upregulated mTOR signaling promoting Aβ and tau aggregation, neuroinflammation, and oxidative stress [106].

2.3. *Clinical Presentation of TBI-Related Neurodegeneration*

The clinical spectrum of traumatic brain injury-related neurodegeneration (TBI-ND) includes a plethora of cognitive, behavioral, and psychiatric manifestations that can impair the quality of life of the individual. Memory impairment is frequently reported in the literature, with patients experiencing difficulties in learning, retaining, and retrieving new information [152,153]. Both anterograde and retrograde amnesia were observed, disrupting the hippocampal cortical network [154]. Executive dysfunction with impaired decision-making processes and organization was also recorded [155]. Beyond cognitive decline, mood disturbances and behavioral dysregulation were common clinical features [156]. Irritability, agitation, depression, and anxiety emerge, which disrupt the interpersonal relationships of the individual [157].

2.4. *Risk Factors*

TBI does not operate as an isolated affliction but instead as a contributor that interacts with multiple biological and demographic risk factors to accelerate the progression towards Alzheimer's disease and other neurodegenerative diseases [158], including age, sex, vascular and systemic comorbidities, sleep disturbances, and neuroinflammation.

2.4.1. Age

Age is a crucial determinant of dementia risk in the general population and modifies the effects of TBI on long-term outcomes. Epidemiological data suggest that younger individuals experience a disproportionately higher relative risk. A meta-analysis of 32 observational studies found that TBI was associated with a pooled risk ratio (RR) of 1.51 (95% CI: 1.26–1.80) for developing Alzheimer's disease [159]. Gardner et al. (2014) [160] similarly reported that moderate-to-severe TBI reported nearly double the risk of dementia in older adults, while mild TBI showed weaker associations. Interestingly, age was also found to suppress the association between traumatic brain injury severity and functional outcomes. Adding age at injury to regression models significantly increased the association between TBI severity and functioning at all follow-up points (up to 10 years) [161]. These findings support the hypothesis that TBI accelerates brain aging by lowering the threshold for clinical dementia onset. Recent neuroimaging evidence supports this biological aging model, as mild TBI patients demonstrate significantly increased brain-predicted age

 https://doi.org/10.3390/ijms27031570

gaps (brain-PAGs) correlated with cognitive impairment and neurofilament light chain levels [162].

2.4.2. Sex and Gender

Sex differences in TBI incidence are increasingly being investigated in the literature, reflecting interest in potential demographic influences on injury patterns. Men sustain more TBIs overall, with peaks in young adulthood and late life, but women experience worse functional and cognitive consequences [163]. A meta-analysis of 19 studies showed that female patients demonstrated poorer outcomes in 85% of examined domains, with a small-to moderate effect size (d = −0.15) [164].

Another study found that female athletes with sports-related concussion reported more severe post-concussive symptoms and slower recovery trajectories than males [165]. Although some studies indicate better psychosocial adjustment in women, most of the evidence points towards sex-specific vulnerabilities, reflecting hormonal influences and immune regulation.

2.4.3. Vascular and Systemic Comorbidities

TBI shares common vascular and systemic risk factors with AD, including hypertension, diabetes, and hyperlipidemia [166]. Hypertension exacerbates blood–brain barrier (BBB) breakdown, a hallmark of both TBI and AD, and is associated with worse cognitive outcomes after head trauma [167]. Diabetes mellitus promotes amyloidosis and oxidative stress, while it exacerbates the TBI-induced cortical damage and is associated with increased mortality and poorer recovery [168]. Hyperlipidemia and atherosclerosis promote amyloid deposition and tau hyperphosphorylation, while high-fat diets in experimental TBI models delay BBB repair and worsen cognitive outcome [169].

2.4.4. Lifestyle Exposures

Modifiable lifestyle factors increase the risk of dementia and worsen post-TBI recovery. Smoking is associated with vascular dysfunction, oxidative stress, and increased tau phosphorylation in AD models, while it is linked with poorer neurocognitive outcomes after TBI [170,171]. Alcohol use increases BBB permeability, exacerbates amyloid β-aggregation, and contributes to higher mortality after TBI.

2.4.5. Sleep and Neuroinflammation

Sleep disturbance and neuroinflammation are important mechanisms through which TBI predisposes towards AD. Sleep issues are observed in most AD patients and frequently precede cognitive decline [172]. In TBI, sleep disturbances are highly prevalent and linked with increased neuroinflammation [173]. Green et al.'s 2020 [174] study highlighted, among others, how cytokines such as interleukins, IL-1β and IL-6, and TNF-α, which normally act as sleep regulators, upregulate after trauma, increase microglial activation, and ultimately exacerbate secondary brain injury.

2.4.6. Neurovascular Dysfunction

The neurovascular unit (NVU) provides a convergent mechanism linking these diverse risk factors. Genetic susceptibility, vascular changes, systemic exposures, and TBI resulting in mechanical damage are linked with BBB integrity disruption and cerebral flow regulation. This dysfunction allows the entry of plasma proteins, namely fibrinogen, into the CNS, triggers chronic neuroinflammation, and accumulates amyloid and tau [175]. Hippocampal CA1 pyramidal neurons exhibit selective neuronal vulnerability in the setting of NVU dysfunction. TBI-related disruption of NVU integrity explains why survivors show increased susceptibility to hippocampal-dependent memory decline characteristic of AD.

 https://doi.org/10.3390/ijms27031570

2.5. Biomarker Evidence

2.5.1. Fluid Biomarkers

Fluid-based biomarkers in cerebrospinal fluid (CSF) and blood are critical in traumatic brain injury pathophysiology. Among the most investigated are amyloid-β (Aβ), total tau (t-tau), phosphorylated tau (p-tau), neurofilament light chain (NfL), glial fibrillary acidic protein (GFAP), ubiquitin carboxy-terminal hydrolase L1 (UCH-L1), and S100B.

These markers reflect axonal degeneration, astroglial damage, amyloid metabolism, and cytoskeletal disruption [176,177]. Acute TBI triggers a cascade of biomarker alterations with distinct temporal patterns. GFAP becomes detectable within the first hour after injury and peaks between 20 and 24 h, and declines over the following 72 h [178]. S100B concentrations are detectable in the peripheral circulation as early as 15 min and return to baseline within 1–2 h [179]. UCH-L1 is detectable in peripheral blood shortly post-injury and increases, reaching a peak 8–12 h after initial trauma [180]. Findings from the large CENTER-TBI cohort of 2869 patients showed that serum levels of GFAP, NfL, NSE, S100B, t-tau, and UCH-L1 are strongly correlated with injury severity and lesion extension with diffuse axonal injury (Marshall III/IV), producing higher concentrations than focal mass lesions [181]. In another study, serum NfL outperformed other biomarkers, namely tau, S100B, and neuron-specific enolase (NSE), in characterizing sports-related concussion (SRC) [182]. Proteomics analyses emphasize the biological aspect of TBI. A study identified 16 plasma proteins altered within 10 days following moderate-to-severe TBI, including neuronal (UCH-L1, visinin protein 1), astroglial (GFAP, S100B), neurodegenerative (tau, pTau231, presenilin-1 (PSEN1), Aβ42), and inflammatory (IL-16, C-C motif chemokine ligand 2 [CCL2]) markers. These molecular alterations correlated with lesion volumes and white matter integrity, underscoring the interplay between biochemical signals and pathology [183]. Recent evidence also demonstrated long-term biomarker changes in athletes exposed to repetitive head impacts. Elevated plasma p-tau217 and NfL were noted in former rugby players compared to controls, with p-tau217 levels associated with traumatic encephalopathy syndrome and smaller hippocampal volume, while NfL was linked to anxiety symptoms [184].

2.5.2. Imaging Biomarkers

Neuroimaging is essential to understand the neuropathological consequences of TBI and RHI, particularly in relation to neurodegenerative diseases such as Alzheimer's disease and chronic traumatic encephalopathy (CTE). Advances in MRI, positron emission tomography (PET), and fluid biomarker imaging have drastically evolved the ability to detect both acute injuries and long-term neurological changes.

PET imaging facilitates the in vivo detection of hallmark AD pathologies, especially beta-amyloid (Aβ) and tau aggregates. In AD, PET findings showed that plaques emerge in a predictable spatial and temporal sequence [185]. Amyloid PET has been proposed as the earliest biomarker of AD, appearing years before clinical manifestation [186,187]. However, Eagle et al. 2020 [188] reported that middle-aged individuals with a history of repeated TBIs showed no significant elevations in amyloid or tau PET uptake compared with controls, despite a higher symptom burden. These results highlight that conventional AD biomarkers may not reliably differentiate risk groups in trauma-related neurodegeneration; thus, alternative TBI-targeted biomarkers may be needed. MRI techniques provide sensitive markers for traumatic and degenerative processes. Diffusion tensor imaging (DTI) is particularly designed to provide information on axonal injury.

Preclinical models demonstrated DTI's sensitivity in detecting traumatic white matter damage [189]. Zimmerman et al. [190] identified signs of axonal and vascular injury in elite rugby players, with abnormal trajectories of white matter changes. Volumetric MRI changes

 https://doi.org/10.3390/ijms27031570

reveal reduced whole-brain and hippocampal volume in professional rugby players [191]. Larger studies, such as the BRAIN Health Cohort, are now examining whether volumetric brain loss, white matter changes, and fluid biomarkers (NfL, GFAP, phosphorylated tau) correlate with exposure extent and clinical outcome [192].

Raji et al. 2024 [193] reported that MRI volumetry strongly distinguishes TBI from early- and late-onset Alzheimer's and behavioral variant frontotemporal dementia (bvFTD) by distinct atrophy patterns. Atrophy is identified in the hippocampus in AD, the frontal lobe in the bvFTD, and white matter loss in TBI.

Shetty et al. 2015 [194] reviewed the established role of CT and MRI in detecting acute trauma, including cortical contusions, DAI, and secondary complications such as hydrocephalus. While CT remains the first-line modality for acute injuries, MRI provides increased sensitivity for smaller contusions and chronic conditions such as encephalomalacia. Recent advances highlight the important role of imaging in correlation with blood-based biomarkers. Lagramante et al. [195] demonstrated that GFAP and UCH-L1 assays could exclude with accuracy, CT-detectable lesions in suspected mild TBI. Similarly, this is also shown in multimodal studies assessing the relationship between volumetric MRI and white matter integrity as well as plasma biomarkers in retired athletes.

2.5.3. Comparison to Alzheimer's Disease Biomarker Profiles

An important question in TBI research is whether the biomarker alterations observed following injury resemble those characteristics of Alzheimer's disease (AD). Both conditions share certain common markers, namely amyloid-β, total tau (t-tau), phosphorylated tau (p-tau), neurofilament light chain (NfL), and glial fibrillary acidic protein (GFAP), but they differ in timing of appearance, magnitude, and long-term pathways [196,197]. In AD, the combined presence of decreased CSF Aβ42, elevated p-tau181 or p-tau217, and increased total tau constitutes a validated biomarker indicator, with detection established in CSF and now increasingly detectable in blood through highly sensitive single-molecular array (Simoa) assays [198,199]. In TBI, acute elevations in GFAP, NfL, and tau reflect immediate structural and axonal injury, while p-tau 217 elevations suggest a convergence with AD pathology [200,201]. However, biomarker pathways appear to differ in magnitude and temporal relationships.

Amyloid pathology emerges early in TBI, where increased APP cleavage and impaired clearance due to lymphatic and blood–brain barrier dysfunction contribute to acute Aβ accumulation [202]. While AD is defined by a consistent reduction in CSF Aβ42 and plasma Aβ42/40 ratio, TBI produces more heterogeneous Aβ dynamics, with acute elevations that may transition into chronic deposition [203]. Recent work by Friberg et al. 2024 [204] demonstrated overlapping CSF profiles between chronic TBI (cTBI) patients and those with AD, indicating a partial shared pathway with amyloid-driven pathology.

Tau-related changes also present an evident overlap between the two conditions. In AD, elevated CSF and plasma p-tau, especially subcategories such as p-tau217 and p-tau231, are highly specific for AD pathology and closely linked to neurofibrillary tangle accumulation [205]. TBI is characterized by acute elevations in total tau (t-tau) reflecting axonal damage, while chronic increases in phosphorylated tau (p-tau217) have been reported in athletes with traumatic encephalopathy syndrome (TES) [206].

Axonal injury markers, particularly NfL, present strong prognostic utility in TBI. Andersson et al. (2024) [207] demonstrated that cerebrospinal fluid (CSF) NfL levels continue to rise for up to 2 weeks following severe TBI and independently predict unfavorable outcomes at both one year and 10–15 years post-injury. Elevated plasma NfL has also been associated with depressive and anxiety symptoms, as well as frontal cortical atrophy, in retired rugby players [205]. Astroglial injury markers, most notably glial

 https://doi.org/10.3390/ijms27031570

fibrillary acidic protein (GFAP), show marked divergence between TBI and AD. In severe TBI, CSF GFAP peaks on days 3–4 and remains elevated for months or years, with higher levels predicting poor long-term outcomes [207]. Synaptic injury markers, including synaptosomal-associated protein 25 (SNAP-25) and visinin-like protein 1 (VILIP-1), are emerging as potential informative biomarkers. Evidence from sTBI indicates a biphasic trajectory, with an early rise reflecting primary synaptic damage and a delayed peak linked to neuroinflammatory responses. Importantly, SNAP-25 and VILIP-1 demonstrated superior predictive performance for six-month outcomes compared to NfL or GFAP [208]. These biomarkers were also elevated in AD and other neurological conditions, highlighting their role as indicators of synaptic dysfunction [209].

Neuroinflammatory markers differentiate TBI from AD. Acute TBI triggers robust increases in cytokines (IL-1β, IL-6, TNF, and IL-17A) and persistent immune activation, with autoantibodies against GFAP, S100B, and Myelin-associated glycoprotein (MAG) detected more than a decade post-injury [210,211]. In AD, inflammation is a prominent feature but tends to follow a slower and more chronic pathway without the acute peaks that are characteristic of TBI.

2.5.4. Neuroinflammation as a Mediator of Tau Pathology

The relationship between traumatic brain injury (TBI) and tau pathology is now being increasingly recognized as a key pathway connecting acute trauma to chronic neurodegeneration. Neuroinflammation driven by microglial activation, cytokine release, and inflammasome signaling is an important mediator of tau phosphorylation and subsequent expansion. This mechanistic link provides biological plausibility to the epidemiological association between TBI and tauopathies such as chronic traumatic encephalopathy and Alzheimer's disease. Human post-mortem studies demonstrate that even a single moderate-to-severe TBI can lead to widespread phosphorylated tau (p-tau) accumulation, involving cortical layers, the hippocampus, thalamus, and substantia nigra [212]. The density and distribution of tau depositions differ between age-matched controls and CTE, with CTE characterized by perivascular tau deposits localized at the depths of cortical sulci [74]. Experimental studies have further shown that brain homogenates derived from TBI mice, when inoculated into naïve animals, induce widespread tau pathology, synaptic loss, and persistent memory deficits, supporting a prion-like transmissible mechanism of tau propagation [213]. Evidence also suggests that temporality is an important factor since P-tau deposition emerges years to decades after injury; however, in TBI survivors, it often appears at a younger age than in controls, suggesting that trauma accelerates disease onset [214]. In wild-type mice, severe TBI initiates progressive tau accumulation, initially localized to the injured hemisphere and later spreading to contralateral regions over 12 months [213,215]. Tau protein aggregates in these models resemble neurofibrillary tangles and glial inclusions, frequently with perivascular localization [216]. These results establish the view that trauma-induced tau pathology is self-propagating, with neuroinflammation creating an environment for amplification and spread.

At the mechanistic level, microglial activation has a crucial impact. Activated microglia release pro-inflammatory cytokines, namely IL-1β, IL-6, and tumor necrosis factor (TNF)-α, which exacerbate tau hyperphosphorylation through kinase-dependent cascades [217]. In vivo PET studies show sustained glial reactivity in athletes with repetitive concussion [10]. These inflammatory responses correlate with subsequent neurodegeneration. Another study reported a threefold increase in IL-1α–positive microglia after head injury, which correlated with β-amyloid precursor protein-positive neurites ($R = 0.78$, $p < 0.05$), indicating that inflammation directly promotes neurotoxic protein accumulation [218].

https://doi.org/10.3390/ijms27031570

The inflammasome pathway further links inflammation with tau pathology. NOD-, LRR-, and pyrin domain-containing protein 3 (NLRP3) inflammasome was shown to regulate pathological tau accumulation, with its prolonged activation driving IL-1β release and caspase-mediated pyroptosis [219]. Experimental interventions targeting this pathway, such as inhibiting AKT/nuclear factor kappa-light-chain-enhancer of activated B cells (NFκB)/NLRP3 signaling, demonstrated a potential therapeutic role in attenuating microglial-mediated injury [220]. In addition, pericytes act as amplifiers of inflammation since friend leukemia integration 1 (Fli-1) after TBI increased pericyte apoptosis, blood–brain barrier (BBB) disruption, and pro-inflammatory cytokine release, exacerbating microglial activation [221]. Repetitive injuries are also associated with more persistent and severe inflammation than single injuries, confirming the neuropathological progression of CTE [222].

2.5.5. Evidence from Animal Models Linking TBI to Neurodegeneration

Preclinical models have provided important mechanistic insights into the way that TBI promotes chronic neurodegeneration, highlighting the combination of immune responses, neuronal death pathways, amyloid and tau pathology, and biomarker signatures. Ayerra et al. (2025) [223] demonstrated that adaptive immunity contributes substantially to the long-term progression of TBI. By using T-cell-deficient mice (T cell receptor beta (TCRβ)−/−δ−/−) subjected to controlled cortical impact, it was found that the absence of T-cells heavily reduced monocyte infiltration, attenuated microglial proliferation, and enhanced an anti-inflammatory phenotype in myeloid cells [223]. At the cellular level, evidence indicates that zipper-interacting protein kinase (ZIPK) has been implicated in mediating neuronal apoptotic pathways following TBI. In both controlled cortical impact models and in vitro neuronal injury cases, ZIPK was upregulated in neurons, reaching a peak at two days post-injury and persisting for a week [224]. Partial deletion of ZIPK (Zipk+/− mice) reduced neuronal death, while knockdown experiments in SH-SY5Y cells confirmed a neuroprotective effect under oxidative stress. Biomarkers from animal models have also increasingly been used since they align with preclinical findings in human TBI. In a systematic review of 74 studies, Lisi et al. (2025) [225] reported that glial fibrillary acidic protein (GFAP), ubiquitin terminal hydrolase L1 (UCH-L1), neurofilament light (NfL), and tau isoforms show specific temporal behavior post-injury. GFAP and UCH-L1 rose sharply within 24 h, while NfL elevations persisted for months, demonstrating chronic axonal injury, similarly seen in humans. However, tau biomarkers rose progressively, peaking over weeks, reflecting a long-term neurodegenerative process.

Amyloid pathology has also been researched robustly in animal models, linking DAI and impaired axonal transport to amyloid-beta accumulation. Rodent and swine studies showed that intra-axonal amyloid precursor protein (APP) accumulates as an early and consistent response to TBI, with amyloid-beta formation presenting more slowly. In APP-transgenic mouse models, TBI induced elevations in Aβ levels, peaking within hours; on the other hand, in swine rotational acceleration models, intra-axonal Aβ and diffuse plaque deposition were observed within 3–10 days, persisting for months [203,226]. Additionally, ApoE genotype studies indicated that APOε4-expresssing mice showed earlier and more pronounced Aβ deposition than those expressing ApoE3 [227]. Uryu et al. (2002) [228] demonstrated that repetitive but not single mild TBI in Tg2576 mice accelerated Aβ deposition, lipid peroxidation, and cognitive dysfunction. At 16 weeks post-trauma, repetitive TBI animals exhibited higher isoprostane levels, increased Aβ plaques, and memory impairment, whereas wild-type mice did not develop amyloidosis.

https://doi.org/10.3390/ijms27031570

As a result, animal model evidence supports a temporal sequence, extending from acute biomarker release to chronic Aβ and tau accumulation, thereby providing biological plausibility mediated by immune, apoptotic, and metabolic pathways.

2.6. Causal Relationship Between TBI and AD

The link between traumatic brain injury (TBI) and Alzheimer's disease (AD) is increasingly being conceptualized within a causal relationship, supported by epidemiological evidence, biological plausibility, and temporality. Evidence suggests that TBI, when moderate to severe, may accelerate processes leading to AD.

2.6.1. Epidemiological Evidence Linking TBI to Dementia and Alzheimer's Disease

Epidemiological studies consistently indicate that a history of a previous traumatic brain injury (TBI) confers an increased risk of late-life neurodegeneration, though the association with Alzheimer's disease remains uncertain. A comprehensive population-based analysis found that individuals with a history of TBI exhibited a significantly elevated risk of Alzheimer's disease-related neurodegeneration compared to age- and sex-adjusted referents. A large-scale cohort study of 1418 confirmed TBI cases and 2836 matched controls found that TBI was associated with 32% increased risk of Alzheimer's disease and related dementias (HR = 1.32, 95% CI: 1.11–1.58, $p = 0.002$). Risk was significant for "probable" (HR = 1.42, 95% CI: 1.05–1.92) and "possible" (HR = 1.29, 95% CI: 1.02–1.62) TBI, but not for "definite" TBI (HR = 1.22, 95% CI: 0.68–2.18) [229]. Study populations are heterogeneous, ranging from general population cohorts to at-risk groups such as active-duty military personnel, veterans, and professional rugby and football athletes, most of whom were males and exposed to repetitive head impacts (RHIs). Epidemiology shows that rugby players are more likely to manifest a neurodegenerative disease compared with matched community controls [230].

Mild symptomatic TBI is the most common injury in rugby, and repetitive head impact, which may be symptomatic, has been linked to poorer cognitive function, mental health issues, and poor sleep quality [231]. A systematic review and meta-analysis by Snowden et al. (2020) [232] pooled data from 21 studies and reported that individuals with a history of mTBI had nearly twice the risk of developing dementia compared to those without prior head trauma (OR = 1.96, 95% CI 1.698–2.263).

Similarly, an umbrella systematic review confirmed that TBI across all severities is significantly associated with dementia risk (OR = 1.81, 95% CI: 1.53–2.14). Stratification by severity indicated that both mild (OR = 1.96, 95% CI: 1.70–2.26) and moderate-to-severe TBI (OR = 1.95, 95% CI: 1.55–2.45) carried comparable increased risks. However, the association between TBI and Alzheimer's disease was weaker and inconsistent across studies, with some analyses reporting modest associations (RR = 1.18, 95% CI: 1.11–1.25) and others detecting no significant relationship (OR = 1.02, 95% CI: 0.91–1.15) [233]. Barnes et al. (2014) [234] found that older U.S. Veterans with TBI history had a 60% increased risk of dementia (HR = 1.57, 95% CI 1.35–1.83), with risk persisting after a mean follow-up of 9 years. Graham et al. (2025) [184] demonstrated biomarker evidence of neurodegeneration in mid-life former rugby players, while Saltiel et al. (2024) [235] ighlighted the frequent coexistence of mixed proteinopathies (amyloid-β, tau, TDP-43, α-synuclein) in post-mortem tissue from individuals exposed to RHI. Post-mortem studies demonstrated a wide range of proteinopathies after RHI/TBI, including Aβ, TDP-43, alpha-synuclein, and tau, in addition to other neuropathologies such as white matter rarefaction and vascular injury [184]. The TRACK-TBI longitudinal study further analyzed the chronic outcomes by showing that the mTBI and mild severity TBI group survivors presented higher rates of cognitive, psychiatric, and functional decline observed years after initial injury, when compared

https://doi.org/10.3390/ijms27031570

with orthopedic trauma controls (OTCs) [236]. Hammond et al. (2004) [237] showed progressive communicational, cognitive, and social functioning problems between 1 and 5 years following the injury.

However, risk prediction, exposure assessment, and follow-up remain a limitation as indicated by Hanrahan et al. (2023) [238], who stated that 40% of studies reported no significant association, while 60% found increased risk or earlier dementia onset. Notably, acute severe TBI has been associated with widespread amyloid plaque pathology, but this is considered distinct from Alzheimer's pathology [239]. Dams O'Connor et al. 2013 [240], who followed 4200 dementia-free individuals aged 65 years or older who were evaluated biannually, with a mean follow-up of 7.4 years, and a history of TBI with loss of consciousness, did not find an increase in the risk for development of dementia.

2.6.2. Relationship Temporality

One of Bradford Hill's requirements for causation is temporality, namely that the exposure precedes the outcome. The temporal association linking TBI or RHI and the later emergence of neurodegenerative disease has been demonstrated across human and preclinical studies. Epidemiological cohorts suggest that exposure to TBI, especially when accompanied by loss of consciousness (LOC), predicts later-life dementia [241]. The MIRAGE study, a large family-based case–control investigation involving more than 16,000 relatives, reported that head injury with LOC was associated with an odds ratio (OR) of up to 9.9 for Alzheimer's disease (AD), a risk that persisted for decades after the initial trauma [242]. Neuropathological and imaging studies further enhance this temporal association by demonstrating progressive structural and functional brain alterations that persist long after injury. Longitudinal morphometric analysis from the DIAGNOSE CTE research in former American football players showed cortical thinning of the hippocampus, amygdala, and frontal and temporal cortices decades after exposure, with greater atrophy in professional athletes compared to amateurs, highlighting a dose–response relationship between RHI and neurodegeneration [243]. Similar findings were observed in a study of 185 athletes, in which longer duration in athletic career predicted gyral thinning and neuronal loss, suggesting a cumulative temporal burden of head impact exposures that leads to gyral vulnerability. Preclinical data focused on the mechanistic pathway in CD1 mice showed that a single severe TBI induced sustained brain atrophy, gliosis, reduced cerebral perfusion, and accumulation of beta-amyloid and hyperphosphorylated tau, with cognitive decline presenting over an 11-month follow-up period [244]. These data confirm that the neurodegenerative process evolves slowly after the inciting trauma, proving that temporality is a key element in this cascade. However, some prospective studies have not identified accelerated cognitive decline among TBI survivors when compared to controls. Long-term follow-up for over a decade, incorporating amyloid and tau PET imaging in chronic TBI survivors, revealed no elevation of β-amyloid or tau compared to healthy individuals [245]. These data showcase heterogeneity in post-TBI trajectory; nevertheless, when considered collectively, epidemiological and neuroimaging evidence support a temporal sequence establishing that injury precedes and can initiate neurodegenerative change.

2.6.3. Biological Plausibility

Biological plausibility constitutes an important criterion within Bradford Hill's framework, emphasizing the necessity for further mechanistic and translational analyses to clarify underlying causal pathways [246]. Evidence that stems from pathological, imaging, and preclinical studies indicates strong biological plausibility for a causal relationship between TBI and dementia. Human post-mortem studies of individuals with prior head trauma reveal a polypathology mechanism, including amyloid-β plaques, hyperphosphorylated

https://doi.org/10.3390/ijms27031570

tau, TDP-43 inclusions, and Lewy bodies, which are consistent with different neurodegenerative processes rather than a single disease pathway [247]. In addition, former athletes with white matter rarefaction and hippocampal sclerosis were independently associated with dementia, suggesting that axonal and vascular injury are central mechanisms irrespective of tau pathology [248]. Oxidative stress and inflammatory cascades have also been investigated from cellular and molecular perspectives. TBI-induced oxidative stress accelerates tau phosphorylation and amyloid-β aggregation, two core hallmarks of AD [249]. Chronic neuroinflammation mediated through NLRP3 inflammasome activation leads to sustained release of pro-inflammatory cytokines and neuronal death, processes also shared with AD and CTE [250,251]. Neuroimaging findings suggest in vivo evidence of these mechanisms. Enlarged perivascular spaces in mTBI survivors demonstrated impaired lymphatic clearance and the accumulation of toxic proteins [252]. Axonal injury within the anterior internal capsulae was identified as a predictor of sleep disturbance and memory decline, indicating tract-specific vulnerability [253]. Smaller hippocampal volumes after TBI were also reported, particularly within the CA1 region, with a stronger association in older adults, highlighting how injury interacts with aging processes to exacerbate neurodegeneration [254]. In conclusion, TBI initiates a cascade of vascular, axonal, inflammatory, and proteinopathic processes that lead to progressive neurodegenerative disease. Table 1 provides a comparative overview of the initiating events, neuroanatomical distribution, pathology and imaging characteristics distinguishing post-traumatic TBI, CTE and AD.

Table 1. Comparison of post-traumatic neurodegeneration, chronic traumatic encephalopathy, and Alzheimer's disease.

	Post-TBI Neurodegeneration	**Chronic Traumatic Encephalopathy (CTE)**	**Alzheimer's Disease (AD)**
Initiating Events	Single or repetitive TBI across the severity spectrum (mild-severe), including blast exposure	Repetitive head impacts (RHI), often sub-concussive, associated with contact sports or military blast exposure	Multifactorial etiology involving advanced age, genetic susceptibility (APOε4, TREM2 variants)
Neuroanatomical Topography	Selective vulnerability of long association and white matter tracts, including corpus callosum, internal capsule, with diffuse involvement of deep and periventricular white matter networks	Predominant involvement of cortical sulcal depths, especially in frontal and temporal cortices, with a characteristic perivascular distribution, and extends to limbic structures and brainstem in advanced stages	Early involvement of medial temporal lobe structures (hippocampus and entorhinal cortex), followed by spread to temporoparietal association cortices
Predominant Pathology	Diffuse axonal injury and microvascular disruption. Chronic cases may demonstrate mixed proteinopathy and accumulation of tau and β- amyloid	Pathognomonic perivascular accumulation of hyperphosphorylated tau at sulcal depths, involving both neurons and astrocytes, accompanied by frequent TDP-43 co-pathology in the medial temporal lobe, diencephalon, and progressive degeneration of white matter	Accumulation of extracellular neuritic β-amyloid plaques and intracellular neurofibrillary tangles composed of hyperphosphorylated tau, beginning in the entorhinal cortex and hippocampus, and spreading to temporoparietal and frontal association
Imaging	White matter atrophy, reduced fractional anisotropy on diffusion MRI, and microbleeds on susceptibility-weighted imaging, patterns of cortical atrophy	Regional cortical thinning and sulcal atrophy have been described, alongside tau PET signal in characteristic sulcal and perivascular regions, while standard MRI may be non-specific	Hippocampal and temporoparietal cortical atrophy on structural MRI, positive amyloid and tau PET, imaging consistent with established AD biomarker patterns

These entities share overlapping features such as tau pathology and cortical atrophy; however, they differ in the primary initiating event, the anatomical distribution of the lesion, and the imaging findings. Post-traumatic neurodegeneration is characterized by diffuse axonal and microvascular injury, CTE by perivascular sulcal tau pathology following the repetitive head impacts, and Alzheimer's disease by hippocampal and temporoparietal amyloid and tau deposition.

https://doi.org/10.3390/ijms27031570

2.7. Therapeutic Interventions

Traumatic brain injury (TBI) remains a complex neuropathological condition with disease-specific therapy yet to be established. The available literature showed a range of experimental and preclinical approaches with the main aim of mitigating the pathological mechanisms of TBI, such as neuronal degeneration, excitotoxicity, blood–brain barrier (BBB) disruption, mitochondrial dysfunction, and oxidative stress. Current therapeutic solutions could be further categorized into acute care and neuroprotection, stem-cell-based therapies for BBB repair, mitochondrial stabilization, transcriptional and hormonal modulation, and immunomodulatory approaches aimed at mitigating neuroinflammation. Figure 1 summarizes the proposed pathophysiological mechanisms and therapeutic models of action following TBI.

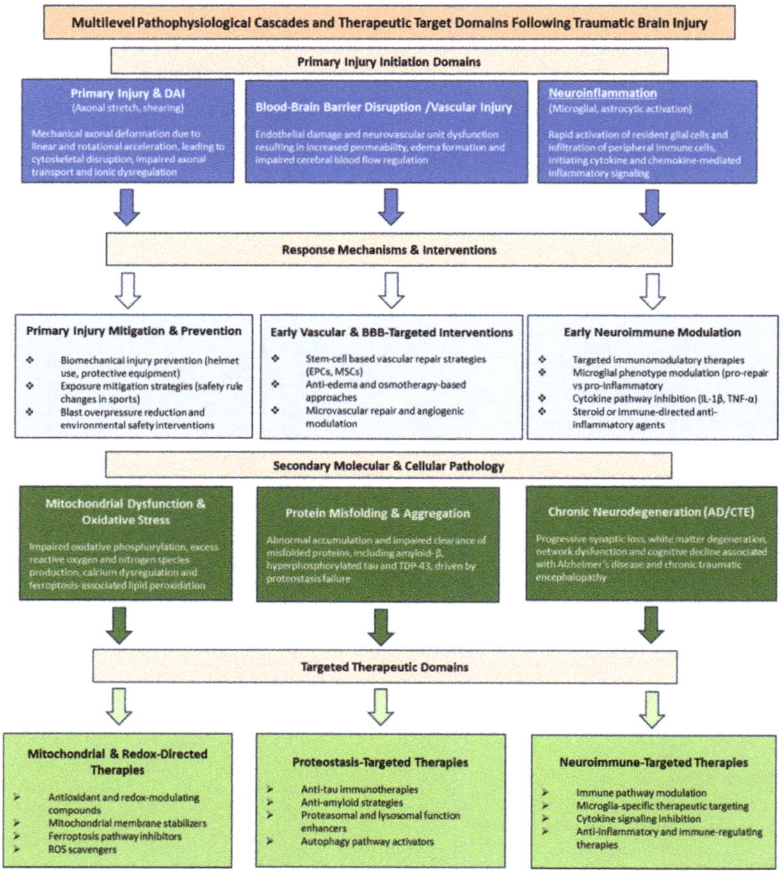

Figure 1. Multilevel pathophysiological cascades and therapeutic target domains following TBI. The figure illustrates the hierarchical organization of injury mechanisms, secondary biological processes, and therapeutic target domains following TBI. Primary injury domains include mechanical axonal injury and DAI, BBB disruption with neurovascular unit dysfunction, and early neuroinflammatory activation involving microglia and astrocytes. These processes trigger coordinated response mechanisms and intervention strategies aimed at mitigating injury severity, preserving vascular integrity, and modulating early immune response. Secondary molecular and cellular pathologies such as mitochondrial dysfunction and oxidative stress, protein misfolding and aggregation, and progressive neurodegeneration follow. The lower tier highlights targeted therapeutic domains aligned with these pathological pathways.

 https://doi.org/10.3390/ijms27031570

2.7.1. Stem Cell-Based Approaches for BBB and Neural Repair

Several preclinical studies demonstrated the capacity of stem cell-based therapies to restore BBB integrity and mitigate neurodegenerative processes. There are different classes of stem cells, namely neural stem cells (NSCs), endothelial progenitor cells (EPCs), hematopoietic stem cells (HSCs), and mesenchymal stem cells (MSCs), which can preserve and promote BBB repair [255,256]. Neural stem cells (NSCs), which self-renew and differentiate into neurons, astrocytes, and oligodendrocytes, demonstrated potential to establish disrupted neural circuits at the site of injury. Preclinical studies by Xia et al. 2025 [256] revealed that transplantation of neural stem cells or premature neurons to the injury site introduces new functional neurons, potentially re-establishing neural circuits and promoting brain tissue repair. They engineered NSCs with micro-patches via micro-contact printing, enabling sustained release of retinoic acid, which increased neuronal differentiation from 28% to 54% in vitro without affecting the viability [256]. Intranasal administration of human NSCs restored pericyte populations and reduced amyloid beta pathology in APP/Presenilin-1 protein (PS1) transgenic mice, suggesting translational potential for Alzheimer's disease [257]. EPCs are bone marrow-derived cells that can migrate to injury sites to proliferate and differentiate into mature endothelial cells, enhancing angiogenesis and vasculogenesis [258]. In TBI models, intracerebroventricular transplantation of EPCs upregulated tight junction proteins, reduced BBB leakage due to disruption, and improved neurological outcomes after TBI [259]. Complementary findings in APP/PS1 mice showed that stereotactic injection of EPCs into the hippocampus enhanced BBB integrity and facilitated Aβ clearance, ultimately improving cognitive outcomes [260]. HSCs are undifferentiated multipotent progenitors for both myeloid and lymphoid lineages, which have also been implicated in BBB repair and modulation of neuroinflammation [261]. In a recent study, Mishra et al. 2023 [262] demonstrated that transplantation of HSCs from wild-type donors into 5x FAD mice reduced amyloid plaque deposition, mitigated neuroinflammation, reinforced BBB integrity, and improved cognitive outcomes. MSCs are multipotent adult stroma stem cells derived from the bone marrow and have been shown to act primarily through paracrine mechanisms [263]. Intravenous administration of MSCs preserved endothelial adherent and tight junctions and protected against BBB dysfunction following TBI [264]. These studies underscore the importance of the therapeutic potential of different stem cell classes in addressing both vascular and neuronal dysfunction after TBI and AD-related pathology. To date, these findings remain preclinical, and no large-scale clinical trials have established efficacy or safety in human TBI patients.

2.7.2. Natural Compounds as Neuroprotective Agents

Secondary injury cascades in TBI involve oxidative stress, excitotoxicity, and BBB disruption. Brooshgahalan (2024) [265] reported that fucoidan, a sulfated polysaccharide extracted from seaweed, exerts potent neuroprotective effects when administered acutely post-injury. It was shown to improve neurological scores, restore BBB integrity, and reduce oxidative damage, while preserving the hippocampal long-term potentiation, a synaptic mechanism essential for cognitive recovery, highlighting its potential role as an antioxidant compound in adjuvant therapy. Melatonin has also emerged as a promising neuroprotective agent. Across multiple preclinical studies, melatonin acts through various mechanisms, including modulation of caspase-dependent apoptosis, activation of the Nrf2 oxidative stress pathway, and stimulation of mitophagy [266–268]. More recent work by Gunara et al. 2025 [269] revealed an additional effect in reducing demyelination, suggesting a potential broader role in preserving glial and myelin integrity. These effects were dose-dependent, with optimal benefit observed at 10 mg/kg. Clinical validation of these compounds in TBI patients requires further optimal dosing, timing, and safety profiles to be established.

 https://doi.org/10.3390/ijms27031570

2.7.3. Targeting Mitochondrial Dysfunction

Mitochondrial dysfunction was identified as a central driver of neuronal injury. Carteri et al.'s 2025 [270] study proposed that accumulated reactive oxygen species (ROS) within mitochondria contribute to neuronal damage [270]. Beyond pharmacological stabilization, mitochondrial transplantation was also investigated as a potential therapy for TBI, supporting its capacity to restore energy metabolism and prevent cell death [271]. The broader literature has also linked defective mitophagy and ROS accumulation to neurodegenerative diseases such as Alzheimer's disease, Huntington's disease (HD), and amyotrophic lateral sclerosis (ALS) [272,273]. As a result, these findings suggest that interventions aimed at mitochondrial stabilization could represent a therapeutic solution across TBI subtypes and subsequent neurodegeneration.

2.7.4. Transcriptional and Hormonal Modulation

Transcriptomic analyses have highlighted the potential for reprogramming injured neurons. Alkalasi et al. 2025 [274] revealed that mild TBI activates the transcription factor Atf3, a stress-responsive transcription factor previously associated with peripheral nerve regeneration. Interestingly, cortical neurons exhibited transcriptional responses distinct from those observed in peripheral neurons, suggesting that targeted reprogramming of CNS transcriptional pathways may require targeted modulation to restore synaptic plasticity and functional recovery. Endocrine modulation has also been investigated as a therapeutic strategy. Zhang et al. 2024 [275] showed that thyroxine (T4) treatment resulted in restoring the disrupted astrocyte and microglia transcriptional networks following TBI, particularly in the hippocampus area. Normally, thyroid hormones such as thyroxine (T4) and triiodothyronine (T3) are important for physiological brain development and maturation [276]. This intervention not only promoted the acute injury repair but also reversed dysregulated metabolic and immune pathways. T4-regulated genes overlapped with human genome-wide association study (GWAS) loci for cognition and psychiatric disorders, providing translational importance. Other studies have also shown that post-injury administration of triiodothyronine (T3) improves motor and cognitive recovery, reduces lesion volume and neuroinflammation, and mediates the upregulation of neurotrophic factors such as brain-derived neurotrophic factor (BDNF) and glial cell line-derived neurotrophic factor (GDNF) [277,278].

2.7.5. Immunomodulation and Neuroinflammation

Primary mechanical injury following TBI initiates a complex neuroinflammatory cascade involving cytokines and chemokines that organize the function of brain cells, such as microglia, astrocytes, and infiltration of peripheral immune cells [279]. While acute inflammation contributes to repair, chronic activation exacerbates secondary injury by producing pro-inflammatory cytokines, including tumor necrosis factor-alpha (TNF-α), interleukin 6 (IL-6), interleukin 1 beta (IL-1β), and interferon-gamma (IFN-γ) [280]. Inflammation has an equivocal role, being initially protective yet chronically maladaptive. Single TBI events may induce transient activation of repair-associated immune pathways, whereas repetitive TBI is linked to persistent neuroinflammatory signaling, synaptic dysfunction, and dysregulation of proteins (e.g., ApoA1, ApoE, Cox6a1, and Snca) [145]. These chronic inflammatory states are being increasingly linked to progressive neurodegenerative changes. Accordingly, immunomodulatory therapies targeting inflammation have been explored to mitigate chronic inflammation. Czaerpaniak et al. (2024) [281] investigated delayed administration (4–6 weeks post-injury) of low-dose interleukin-2 after mild TBI, which reversed chronic headache symptoms and restored cognitive performance. In parallel, monoclonal antibodies (mAbs) targeting inflammatory cytokines (IL-6, IL-1β, and TNF-α) and

 https://doi.org/10.3390/ijms27031570

neurodegenerative proteins (Aβ, tau, α-synuclein) demonstrated efficacy in preclinical TBI models [282]. Kondo et al. 2015 [283] reported that antibodies targeting cis-phosphorylated tau reduced tau accumulation and restored neuronal function, supporting immunotherapy as a potential disease-modifying strategy.

2.7.6. Neurovascular-Linked Therapies

Neurovascular-linked therapies have emerged as a potential intervention based on the shared neurovascular hypothesis of TBI and Alzheimer's disease. Drugs initially developed for cardiovascular intervention, namely telmisartan, simvastatin, dabigatran, and 3K3A-activated protein C (3K3A-APC), have shown anti-inflammatory, antioxidant, and neuroprotective effects in mitigating BBB disruption, proteinopathies, and microvascular injury in preclinical TBI and AD animal models [284].

Collectively, the experimental and preclinical studies show that traumatic brain injury could be characterized as a biologically modifiable condition at multiple levels, including blood–brain barrier integrity, mitochondrial function, transcriptional regulation, and neuronal immune signaling. Findings derived from in vitro studies and animal models indicate that interventions such as stem cell-based strategies, naturally derived neuroprotective agents, mitochondrial stabilization, and hormonal modulation can attenuate secondary injury mechanisms and further limit neurodegenerative progression, improving functional recovery. However, despite strong mechanistic support and reproducible benefits in controlled experimental settings, these approaches remain largely preclinical. These findings underscore both the therapeutic potential of targeting shared neurovascular and neuroinflammatory pathways in TBI and the critical need for clinical trials to enable the progression from experimental efficacy to clinical impact.

2.7.7. Pharmacological Interventions in Acute and Chronic TBI

Unlike the preceding experimental approaches, the interventions discussed in this section represent therapies currently used or evaluated in human TBI patients, with evidence derived from clinical trials and practice guidelines. TBI management is planned according to the post-injury timeline. In the acute phase (0–3 h), interventions focus on hemostasis, intracranial pressure (ICP) control, seizure prophylaxis, and prevention of secondary injury cascades [285]. Tranexamic acid is administered within three hours to limit hemorrhagic expansion in mild to moderate TBI [286], while reversal of anticoagulants (Vitamin K, fresh frozen plasma [FFP], and direct oral anticoagulants [DOACs]) and correction of thrombocytopenia are important to prevent hemorrhage progression. Hyperosmolar agents such as mannitol and hypertonic saline are used to handle intracranial hypertension and cerebral edema [287]. For refractory ICP management and reduction in cerebral metabolism, sedatives such as barbiturates are used [288]. Anticonvulsants are used to prevent early post-traumatic seizures, and antipyretics are used to enhance cerebral oxygenation [289]. Investigational pharmacological therapies in this phase involve corticosteroids, citicoline, and progesterone, although clinical trials show mixed results [290]. The CRASH trial for corticosteroids demonstrated no improvement in the long-term outcomes after TBI, and citicoline failed to improve recovery in the COBRIT study, while progesterone showed efficacy in TBI models [291–293].

In the post-acute and chronic phases, focus is on neurobehavioral, cognitive, and neuropsychiatric support [290]. Selective serotonin reuptake inhibitors (SSRIs) are used to treat depression and attention deficits, while tricyclic antidepressants (TCAs) target mood and sleep disturbances [294,295].

https://doi.org/10.3390/ijms27031570

3. Conclusions

Substantial progress has been made in linking traumatic brain injury (TBI) and diffuse axonal injury (DAI) to long-term neurodegenerative outcomes. However, several critical gaps still require focused investigation, necessary to prove causality, develop innovative therapeutic approaches, and refine strategies for early detection and prevention of Alzheimer's disease and other dementias.

One major ongoing challenge lies not in the absence of appropriate experimental models but in the heterogeneity and translational limitations of existing preclinical TBI models. A wide range of injury models, including controlled cortical impact, blast exposure, rotational acceleration, and repetitive mild TBI, have been developed to reproduce focal and diffuse axonal injury with varying degrees of success.

To enhance translational relevance, future studies should incorporate broader biomechanical diversity, including combinations of focal and diffuse injury, repetitive exposure paradigms, and stratification by age and sex.

In parallel, there is a critical need to standardize and validate fluid and imaging biomarkers across cohorts and time periods. Current evidence focuses on GFAP, UCH-L1, NfL, S100B, and tau as chosen biomarkers. Nevertheless, longitudinal cohort studies with standardized sampling approaches are crucial to characterize acute, subacute, and chronic biomarker pathways and to determine whether these trajectories enhance the risk of Alzheimer's pathology. Risk stratification by demographic and genetic factors is also required for future investigations. Age and sex both affect TBI outcomes, as well as modulate the vulnerability to progressive neurodegeneration. Women tend to present different biomarker profiles and recovery patterns compared to men, while older patients exhibit slower clearance of injury-related proteins and a higher risk of cognitive decline. The APOE ε4 allele, which is already established as a genetic risk factor for sporadic AD, has been proposed to exacerbate the impact of TBI on amyloid deposition and clinical progression. For this, large-scale, genotype-categorized studies will be necessary to fully clarify this interaction. Biomechanical studies highlight the need for more accurate modeling of brain tissue responses to rotational and linear acceleration, as current viscoelastic approximation does not fully resemble the human neuroanatomical complexity.

Therapeutically, intervention temporality plays a crucial role. As highlighted, TBI pathophysiology implicates acute excitotoxicity, subacute neuroinflammation, and delayed tauopathy, indicating the need for multiple therapeutic interventions. However, it appears essential to temporally combine therapies that target excitotoxic cascades and neurovascular unit stabilization, followed by the subacute modulation of inflammation and chronic strategies to reduce tau aggregation.

Ultimately, a multimodal framework that combines TBI models, standardized biomarkers, stratification by sex, age, and genotype, and temporally adapted therapeutic approaches is necessary. Consequently, we could resolve the major question of whether TBI increases the risk of Alzheimer-type pathology or whether it produces spatial vulnerability across brain regions.

In conclusion, traumatic brain injury represents not only an acute neurological event but also a potential initiator of chronic neurodegenerative processes, which overlap with Alzheimer's disease and chronic traumatic encephalopathy. Across epidemiological, clinical, and biomechanical analyses, evidence suggests that TBI is associated with long-term cognitive decline, psychiatric comorbidities, and pathological hallmarks, namely amyloid-β deposition, tau hyperphosphorylation, and neuroinflammation. Nevertheless, the relationship remains complex, influenced by injury severity, frequency, anatomic location, age, sex, and genetic susceptibility. In sum, TBI plays a pivotal role at the intersection of acute injury and chronic neurodegeneration. For this reason, discovering the connection

https://doi.org/10.3390/ijms27031570

between preclinical studies, biomarker validation, and clinical interventions is important for enhanced understanding of TBI-related neurodegeneration in developing strategies for diagnosis, prevention, and appropriate treatment.

Author Contributions: Conceptualization, D.K.; methodology, A.K. and E.S.B.; formal analysis, A.K. and E.S.B.; data curation, A.K. and E.S.B.; writing—original draft preparation, A.K.; writing—review and editing, A.K., E.S.B. and D.K.; supervision, D.K.; project administration, D.K.; funding acquisition, D.K. All authors have read and agreed to the published version of the manuscript.

Funding: This research received no external funding.

Data Availability Statement: No new data were created or analyzed in this study.

Acknowledgments: The authors acknowledge the use of Rayyan (Qatar Computing Research Institute, Doha, Qatar) for facilitating the screening and study selection process. During the preparation of this manuscript, the authors used ChatGPT (OpenAI, GPT-5 family) and Gemini 3 (Google, large language model, 2025) for language editing purposes, including paraphrasing, improving clarity, and enhancing academic vocabulary. The authors have critically reviewed and edited the output and take full responsibility for the content of this publication.

Conflicts of Interest: The authors declare no conflicts of interest.

Abbreviations

The following abbreviations are used in this manuscript:

3K3A-APC	3K3A-activated protein C
AD	Alzheimer's disease
ADAM10	A disintegrin and metalloproteinase domain-containing protein 10
ALS	Amyotrophic lateral sclerosis
AMPA	α-amino-3-hydroxy-5-methyl-4-isoxazolepropionic acid
ApoE	Apolipoprotein E
APP	Amyloid precursor protein
ATP	Adenosine triphosphate
Aβ	Amyloid-beta
BACE1	β-secretase
BAG3	BCL2-associated athanogene 3
BBB	Blood–brain barrier
BDNF	Brain-derived neurotrophic factor
Brain-PAG	Brain-predicted age gaps
bvFTD	Behavioral variant frontotemporal dementia
Cav-1	Caveolin-1
CBF	Cerebral blood flow
CBI-M	Clinical, Biomarker, Imaging Modifiers
CCL2	C-C motif chemokine ligand 2
CISD1	CDGSH iron–sulfur domain-containing protein 1
CNS	Central nervous system
CSF	Cerebrospinal fluid
CTE	Chronic traumatic encephalopathy
CypA	Cyclophilin A
DAI	Diffuse axonal injury
DHODH	Dihydroorotate dehydrogenase
DOACs	Direct oral anticoagulants
DTI	Diffusion tensor imaging
EPCs	Endothelial progenitor cells
FA	Fractional anisotropy
FFP	Fresh frozen plasma

https://doi.org/10.3390/ijms27031570

FLAIR	Fluid-attenuated inversion recovery
Fli-1	Friend leukemia integration 1
FTLD	Frontotemporal lobar degeneration
GCS	Glasgow Coma Scale
GDNF	Glial cell line-derived neurotrophic factor
GFAP	Glial fibrillary acidic protein
GWAS	Genome-wide association studies
HD	Huntington's disease
HSCs	Hematopoietic stem cells
ICAM	Intercellular adhesion molecule
ICP	Intracranial pressure
IFN-γ	Interferon-gamma
IL	Interleukin
LOC	Loss of consciousness
LRP1	Low-density lipoprotein receptor-related protein 1
mAbs	Monoclonal antibodies
MAG	Myelin-associated glycoprotein
MeSH	Medical Subject Headings
Mfsd2a	Major facilitator superfamily domain-containing protein 2A
MMP-9	Matrix metalloproteinase-9
MND	Motor neuron disease
MRI	Magnetic resonance imaging
MSCs	Mesenchymal stem cells
mTOR	Mechanistic (mammalian) target of rapamycin
NfL	Neurofilament light chain
NFTs	Neurofibrillary tangles
NFκB	Nuclear factor kappa-light-chain-enhancer of activated B cells
NLRP3	NOD-, LRR-, and pyrin domain-containing protein 3
NMDA	N-methyl-D-aspartate
NSCs	Neural stem cells
NSE	Neuron-specific enolase
NVU	Neurovascular unit
OR	Odds ratio
OTCs	Orthopedic trauma controls
OXPHOS	Oxidative phosphorylation
PCC	Population-Concept-Context
PD	Parkinson's disease
PET	Positron emission tomography
PS1	Presenilin-1 protein
PSEN1	Presenilin-1 gene
P-tau	Phosphorylated tau
RHI	Repetitive head impacts
RNS	Reactive nitrogen species
ROS	Reactive oxygen species
RR	Risk ratio
Shh	Sonic hedgehog
SNAP-25	Synaptosomal-associated protein 25
SRC	Sports-related concussion
SSRIs	Selective serotonin reuptake inhibitors
sTREM2	Soluble triggering receptor expressed on myeloid cells 2
T3	Triiodothyronine
T4	Thyroxine
TBI	Traumatic brain injury
TBI-ND	Traumatic brain injury-related neurodegeneration

https://doi.org/10.3390/ijms27031570

TCAs	Tricyclic antidepressants
TCR	T-cell receptor
TDP-43	Transactive response DNA-binding protein 43
TES	Traumatic encephalopathy syndrome
TGF-β	Transforming growth factor beta
TNF	Tumor necrosis factor
TOMM40	Translocase of outer mitochondrial membrane 40
T-tau	Total tau
UCH-L1	Ubiquitin carboxy-terminal hydrolase L1
VaD	Vascular dementia
VCAM-1	Vascular cell adhesion molecule-1
VEGF	Vascular endothelial growth factor
VILIP-1	Visinin-like protein 1
VSMCs	Vascular smooth muscle cells
Wnt	Wingless/Integrated
ZIPK	Zipper-interacting protein kinase
α-SMA	Alpha-smooth muscle actin
β-CTF	Beta C-terminal fragment of APP

References

1. National Academies of Sciences, Engineering, and Medicine; Health and Medicine Division; Board on Health Care Services; Committee on the Review of the Department of Veterans Affairs Examinations for Traumatic Brain Injury. *Diagnosis and Assessment of Traumatic Brain Injury*; National Academies Press: Washington, DC, USA, 2019. [CrossRef]
2. Zhong, H.; Feng, Y.; Shen, J.; Rao, T.; Dai, H.; Zhong, W.; Zhao, G. Global Burden of Traumatic Brain Injury in 204 Countries and Territories from 1990 to 2021. *Am. J. Prev. Med.* **2025**, *68*, 754–763. [CrossRef]
3. Gu, L.; Zhang, L.; Li, C.; Jiang, L.; Zhou, J.; Xie, Y.; Yang, J.; Jiang, C.; Zhang, L.; Jiang, Y.; et al. Global, Regional, and National Burden of Traumatic Brain Injury, 1990–2021: A Systematic Analysis for the Global Burden of Disease Study 2021. *J. Neurotrauma* **2025**, *42*, 1805–1815. [CrossRef]
4. Ayele, A.; Anteneh, S.; Degu, F.S.; Dessie, G.; Lonsako, A.A.; Anley, A.; Beyene, G. Time to Death and Its Predictors among Traumatic Brain Injury Patients Admitted to East Amhara Comprehensive Specialized Hospitals, Ethiopia: Retrospective Cohort Study. *BMC Neurol.* **2024**, *24*, 370. [CrossRef]
5. McKee, A.C.; Robinson, M.E. Military-Related Traumatic Brain Injury and Neurodegeneration. *Alzheimer's Dement.* **2014**, *10*, S242–S253. [CrossRef] [PubMed]
6. Leo, P.; McCrea, M. Epidemiology. Available online: https://www.ncbi.nlm.nih.gov/books/NBK326730/ (accessed on 5 August 2025).
7. Jha, R.M.; Kochanek, P.M.; Simard, J.M. Pathophysiology and Treatment of Cerebral Edema in Traumatic Brain Injury. *Neuropharmacology* **2019**, *145*, 230–246. [CrossRef] [PubMed]
8. Reddipaga, M.; Kannan, K. On the Construction of a Viscoelastic Constitutive Model for Brain Tissue Maximizing Tension–Compression Asymmetry. *Int. J. Eng. Sci.* **2025**, *218*, 104415. [CrossRef]
9. Keating, C.E.; Cullen, D.K. Mechanosensation in Traumatic Brain Injury. *Neurobiol. Dis.* **2021**, *148*, 105210. [CrossRef]
10. Cullen, D.M.; Vernekar, V.N.; LaPlaca, M.C. Trauma-Induced Plasmalemma Disruptions in Three-Dimensional Neural Cultures Are Dependent on Strain Modality and Rate. *J. Neurotrauma* **2011**, *28*, 2219–2233. [CrossRef]
11. Du, Z.; Wang, P.; Luo, P.; Fei, Z.; Zhuang, Z.; Liu, Z. Mechanical Mechanism and Indicator of Diffuse Axonal Injury under Blast-Type Acceleration. *J. Biomech.* **2023**, *156*, 111674. [CrossRef]
12. Mesfin, F.B.; Taylor, R.S. Diffuse Axonal Injury (DAI). Available online: https://www.ncbi.nlm.nih.gov/books/NBK448102/ (accessed on 27 July 2025).
13. Smith, D.H.; Meaney, D.F.; Shull, W.H. Diffuse Axonal Injury in Head Trauma. *J. Head Trauma Rehabil.* **2003**, *18*, 307–316. [CrossRef]
14. Johnson, V.E.; Stewart, W.; Smith, D.H. Axonal Pathology in Traumatic Brain Injury. *Exp. Neurol.* **2013**, *246*, 35–43. [CrossRef] [PubMed]
15. Zima, L.; Moore, A.N.; Smolen, P.; Kobori, N.; Noble, B.; Robinson, D.; Hood, K.N.; Homma, R.; Al Mamun, A.; Redell, J.B.; et al. The Evolving Pathophysiology of TBI and the Advantages of Temporally-Guided Combination Therapies. *Neurochem. Int.* **2024**, *180*, 105874. [CrossRef] [PubMed]

 https://doi.org/10.3390/ijms27031570

16. Mena, J.H.; Sanchez, A.I.; Rubiano, A.M.; Peitzman, A.B.; Sperry, J.L.; Gutierrez, M.I.; Puyana, J.C. Effect of the Modified Glasgow Coma Scale Score Criteria for Mild Traumatic Brain Injury on Mortality Prediction: Comparing Classic and Modified Glasgow Coma Scale Score Model Scores of 13. *J. Trauma Inj. Infect. Crit. Care* **2011**, *71*, 1185–1193. [CrossRef] [PubMed]
17. Manley, G.; Dams-O'Connor, K.; Alosco, M.L.; Awwad, H.O.; Bazarian, J.J.; Bragge, P.; Corrigan, J.D.; Doperalski, A.; Ferguson, A.R.; Mac Donald, C.L.; et al. A new characterisation of acute traumatic brain injury: The NIH-NINDS TBI Classification and Nomenclature Initiative. *Lancet Neurol.* **2025**, *24*, 512–523. Correction in *Lancet Neurol.* **2025**, *24*, E14. [CrossRef] [PubMed]
18. Ng, S.Y.; Lee, A.Y.W. Traumatic Brain Injuries: Pathophysiology and Potential Therapeutic Targets. *Front. Cell. Neurosci.* **2020**, *13*, 528. [CrossRef]
19. Su, E.; Bell, M. Diffuse Axonal Injury. Available online: https://www.ncbi.nlm.nih.gov/books/NBK326722/ (accessed on 5 August 2025).
20. Palacios, E.M.; Owen, J.P.; Yuh, E.L.; Wang, M.; Vassar, M.J.; Ferguson, A.R.; Diaz-Arrastia, R.; Giacino, J.T.; Okonkwo, D.O.; Robertson, C.; et al. The Evolution of White Matter Microstructural Changes after Mild Traumatic Brain Injury: A Longitudinal DTI and NODDI Study. *Sci. Adv.* **2020**, *6*, eaaz6892. [CrossRef]
21. Aurelio, M.; Sousa, G.; Bittencourt, L.O.; Falcao, D.; Lima, R.R.; Lopes, R. Cellular and Molecular Pathophysiology of Traumatic Brain Injury: What Have We Learned so Far? *Biology* **2023**, *12*, 1139. [CrossRef]
22. Ganti, L.; Bodhit, A.N.; Daneshvar, Y.; Hatchitt, K.; Kuchibhotla, S.; Pulvino, C.; Ayala, S.W.; Peters, K.R. Effectiveness of Seatbelts in Mitigating Traumatic Brain Injury Severity. *World J. Emerg. Med.* **2021**, *12*, 68–72. [CrossRef]
23. Hinson, H.E.; Rowell, S.; Schreiber, M. Clinical Evidence of Inflammation Driving Secondary Brain Injury. *J. Trauma Acute Care Surg.* **2015**, *78*, 184–191. [CrossRef]
24. Tesco, G.; Lomoio, S. Pathophysiology of Neurodegenerative Diseases: An Interplay among Axonal Transport Failure, Oxidative Stress, and Inflammation? *Semin. Immunol.* **2022**, *59*, 101628. [CrossRef]
25. Wilson, L.; Stewart, W.; Dams-O'Connor, K.; Diaz-Arrastia, R.; Horton, L.; Menon, D.K.; Polinder, S. The Chronic and Evolving Neurological Consequences of Traumatic Brain Injury. *Lancet Neurol.* **2017**, *16*, 813–825. [CrossRef]
26. Devi, Y.; Khan, S.; Rana, P.; Dhandapani, M.; Ghai, S.; Gopichandran, L.; Dhandapani, S. Cognitive, Behavioral, and Functional Impairments among Traumatic Brain Injury Survivors: Impact on Caregiver Burden. *J. Neurosci. Rural Pract.* **2020**, *11*, 629–635. [CrossRef]
27. Aoun, R.; Rawal, H.; Attarian, H.; Sahni, A. Impact of Traumatic Brain Injury on Sleep: An Overview. *Nat. Sci. Sleep* **2019**, *11*, 131–140. [CrossRef] [PubMed]
28. Scheltens, P. Alzheimer's Disease. *Lancet* **2021**, *397*, 1577–1590. [CrossRef] [PubMed]
29. Alzheimer's Association. 2024 Alzheimer's Disease Facts and Figures. *Alzheimer's Dement.* **2024**, *20*, 3708–3821. [CrossRef]
30. Tryphena, K.P.; Shukla, R.; Khatri, D.K.; Vora, L.K. Pathogenesis, Diagnostics, and Therapeutics for Alzheimer's Disease: Breaking the Memory Barrier. *Ageing Res. Rev.* **2024**, *101*, 102481. [CrossRef]
31. Iaccarino, L.; Burnham, S.C.; Dell'Agnello, G.; Dowsett, S.A.; Epelbaum, S. Diagnostic Biomarkers of Amyloid and Tau Pathology in Alzheimer's Disease: An Overview of Tests for Clinical Practice in the United States and Europe. *J. Prev. Alzheimer's Dis.* **2023**, *10*, 426–442. [CrossRef]
32. Weerasinghe-Mudiyanselage, P.D.E.; Ang, M.J.; Kang, S.; Kim, J.-S.; Moon, C. Structural Plasticity of the Hippocampus in Neurodegenerative Diseases. *Int. J. Mol. Sci.* **2022**, *23*, 3349. [CrossRef]
33. Barone, E.; Di Domenico, F.; Perluigi, M.; Butterfield, D.A. The Interplay among Oxidative Stress, Brain Insulin Resistance and AMPK Dysfunction Contribute to Neurodegeneration in Type 2 Diabetes and Alzheimer Disease. *Free Radic. Biol. Med.* **2021**, *176*, 16–33. [CrossRef]
34. Yuan, Y.; Zhao, G.; Zhao, Y. Dysregulation of Energy Metabolism in Alzheimer's Disease. *J. Neurol.* **2024**, *272*, 2. [CrossRef]
35. He, L.; Zhang, R.; Yang, M.; Lu, M. The Role of Astrocyte in Neuroinflammation in Traumatic Brain Injury. *Biochim. Biophys. Acta Mol. Basis Dis.* **2024**, *1870*, 166992. [CrossRef] [PubMed]
36. Brett, B.L.; Gardner, R.C.; Godbout, J.; Dams-O'Connor, K.; Keene, C.D. Traumatic Brain Injury and Risk of Neurodegenerative Disorder. *Biol. Psychiatry* **2022**, *91*, 498–507. [CrossRef] [PubMed]
37. Barnes, D.E.; Byers, A.L.; Gardner, R.C.; Seal, K.H.; Boscardin, W.J.; Yaffe, K. Association of Mild Traumatic Brain Injury with and without Loss of Consciousness with Dementia in US Military Veterans. *JAMA Neurol.* **2018**, *75*, 1055. [CrossRef] [PubMed]
38. Thompson, J.; Vasefi, M. Increased Risk of Dementia and Its Subtypes Following Various Forms of Acquired Brain Injury: A Meta-Analysis and Systematic Review. *Neuroscience* **2025**, *580*, 332–350. [CrossRef]
39. Werner, C.; Engelhard, K. Pathophysiology of Traumatic Brain Injury. *Br. J. Anaesth.* **2017**, *99*, 4–9. [CrossRef]
40. Acosta, S.A.; Tajiri, N.; Sanberg, P.R.; Kaneko, Y.; Borlongan, C.V. Increased Amyloid Precursor Protein and Tau Expression Manifests as Key Secondary Cell Death in Chronic Traumatic Brain Injury. *J. Cell. Physiol.* **2016**, *232*, 665–677. [CrossRef]
41. Delic, V.; Beck, K.D.; Pang, K.C.H.; Citron, B.A. Biological Links between Traumatic Brain Injury and Parkinson's Disease. *Acta Neuropathol. Commun.* **2020**, *8*, 45. [CrossRef]

https://doi.org/10.3390/ijms27031570

42. Evans, T.M.; Jaramillo, C.A.; Sataranatarajan, K.; Watts, L.; Sabia, M.; Qi, W.; Van Remmen, H. The Effect of Mild Traumatic Brain Injury on Peripheral Nervous System Pathology in Wild-Type Mice and the G93A Mutant Mouse Model of Motor Neuron Disease. *Neuroscience* **2015**, *298*, 410–423. [CrossRef]
43. Siedler, D.G.; Chuah, M.I.; Kirkcaldie, M.T.K.; Vickers, J.C.; King, A.E. Diffuse Axonal Injury in Brain Trauma: Insights from Alterations in Neurofilaments. *Front. Cell. Neurosci.* **2014**, *8*, 429. [CrossRef]
44. Graham, N.S.N.; Jolly, A.; Zimmerman, K.; Bourke, N.J.; Scott, G.; Cole, J.H.; Schott, J.M.; Sharp, D.J. Diffuse Axonal Injury Predicts Neurodegeneration after Moderate–Severe Traumatic Brain Injury. *Brain* **2020**, *143*, 3685–3698. [CrossRef]
45. Santurro, A.; Simone, D.; Choucha, A.; Morena, D.; Consalvo, F.; Romano, D.; Terrasi, P.; Corrivetti, F.; Scrofani, R.; Narciso, N.; et al. Integrative Diagnostic and Prognostic Paradigms in Diffuse Axonal Injury: Insights from Clinical, Histopathological, Biomolecular, Radiological, and AI-Based Perspectives. *Int. J. Mol. Sci.* **2025**, *26*, 7808. [CrossRef] [PubMed]
46. Lin, T.-H.; Chiang, C.-W.; Perez-Torres, C.J.; Sun, P.; Wallendorf, M.; Schmidt, R.E.; Cross, A.H.; Song, S.-K. Diffusion MRI Quantifies Early Axonal Loss in the Presence of Nerve Swelling. *J. Neuroinflamm.* **2017**, *14*, 78. [CrossRef] [PubMed]
47. Schallerer, C.; Neuschmid, S.; Ehrlich, B.E.; McGuone, D. Calpain in Traumatic Brain Injury: From Cinderella to Central Player. *Cells* **2025**, *14*, 1253. [CrossRef] [PubMed]
48. Liu, Y.; Zhang, Y.; Wang, X.; Zhang, H.; You, X.; Liao, F.-F.; Xu, H. Intracellular Trafficking of Presenilin 1 Is Regulated by β-Amyloid Precursor Protein and Phospholipase D1. *J. Biol. Chem.* **2009**, *284*, 12145–12152. [CrossRef] [PubMed]
49. Rodrigues, E.M.; Weissmiller, A.M.; Goldstein, L.S.B. Enhanced β-Secretase Processing Alters APP Axonal Transport and Leads to Axonal Defects. *Hum. Mol. Genet.* **2012**, *21*, 4587–4601. [CrossRef]
50. Albayram, O.; Angeli, P.; Bernstein, E.; Baxley, S.; Gao, Z.; Lu, K.P.; Zhou, X.Z. Targeting Prion-like Cis Phosphorylated Tau Pathology in Neurodegenerative Diseases. *J. Alzheimer's Dis. Park.* **2018**, *8*, 443. [CrossRef]
51. McKenzie, K.J.; McLellan, D.R.; Gentleman, S.M.; Maxwell, W.L.; Gennarelli, T.A.; Graham, D.I. Is β-APP a Marker of Axonal Damage in Short-Surviving Head Injury? *Acta Neuropathol.* **1996**, *92*, 608–613. [CrossRef]
52. Bruggeman, G.F.; Haitsma, I.K.; Dirven, C.M.F.; Volovici, V. Traumatic Axonal Injury (TAI): Definitions, Pathophysiology and Imaging—A Narrative Review. *Acta Neurochir.* **2020**, *163*, 31–44. [CrossRef]
53. Johnson, V.E.; Stewart, J.E.; Begbie, F.D.; Trojanowski, J.Q.; Smith, D.H.; Stewart, W. Inflammation and White Matter Degeneration Persist for Years after a Single Traumatic Brain Injury. *Brain* **2013**, *136*, 28–42. [CrossRef]
54. Graham, N.S.N.; Cole, J.H.; Bourke, N.J.; Schott, J.M.; Sharp, D.J. Distinct Patterns of Neurodegeneration after TBI and in Alzheimer's Disease. *Alzheimer's Dement.* **2023**, *19*, 3065–3077. [CrossRef]
55. Rutgers, D.R.; Toulgoat, F.; Cazejust, J.; Fillard, P.; Lasjaunias, P.; Ducreux, D. White Matter Abnormalities in Mild Traumatic Brain Injury: A Diffusion Tensor Imaging Study. *Am. J. Neuroradiol.* **2007**, *29*, 514–519. [CrossRef] [PubMed]
56. Hayes, J.P.; Bigler, E.D.; Verfaellie, M. Traumatic Brain Injury as a Disorder of Brain Connectivity. *J. Int. Neuropsychol. Soc.* **2016**, *22*, 120–137. [CrossRef] [PubMed]
57. Cole, J.H.; Jolly, A.; de Simoni, S.; Bourke, N.; Patel, M.C.; Scott, G.; Sharp, D.J. Spatial Patterns of Progressive Brain Volume Loss after Moderate-Severe Traumatic Brain Injury. *Brain* **2018**, *141*, 822–836. [CrossRef] [PubMed]
58. Poudel, G.R.; Dominguez, J.F.; Verhelst, H.; Linden, C.V.; Deblaere, K.; Jones, D.K.; Cerin, E.; Vingerhoets, G.; Caeyenberghs, K. Network Diffusion Modeling Predicts Neurodegeneration in Traumatic Brain Injury. *Ann. Clin. Transl. Neurol.* **2020**, *7*, 270–279. [CrossRef]
59. Mckee, A.C.; Daneshvar, D.H. Chapter 4—The Neuropathology of Traumatic Brain Injury. In *Handbook of Clinical Neurology*; Elsevier: Amsterdam, The Netherlands, 2026. Available online: https://www.sciencedirect.com/science/article/abs/pii/B9780444528926000040 (accessed on 8 July 2025).
60. Permenter, C.M.; Sherman, A.L. Postconcussive Syndrome. Available online: https://www.ncbi.nlm.nih.gov/books/NBK534786/ (accessed on 8 July 2025).
61. Pellot, J.E.; De Jesus, O. Cerebral Contusion. Available online: https://www.ncbi.nlm.nih.gov/books/NBK562147/ (accessed on 5 August 2025).
62. Ginsburg, J.; Huff, J.S. Traumatic Brain Injury. Available online: https://www.ncbi.nlm.nih.gov/books/NBK557861/ (accessed on 8 July 2025).
63. Koerte, I.K.; Hufschmidt, J.; Muehlmann, M.; Lin, A.P.; Shenton, M.E. Advanced Neuroimaging of Mild Traumatic Brain Injury. Available online: https://www.ncbi.nlm.nih.gov/books/NBK326714/ (accessed on 27 July 2025).
64. Dong, X.; Wang, Y.; Qin, Z. Molecular Mechanisms of Excitotoxicity and Their Relevance to Pathogenesis of Neurodegenerative Diseases. *Acta Pharmacol. Sin.* **2009**, *30*, 379–387. [CrossRef]
65. Hardingham, G.E.; Fukunaga, Y.; Bading, H. Extrasynaptic NMDARs Oppose Synaptic NMDARs by Triggering CREB Shut-off and Cell Death Pathways. *Nat. Neurosci.* **2002**, *5*, 405–414. [CrossRef]
66. Liang, D.; Bhatta, S.; Gerzanich, V.; Simard, J.M. Cytotoxic Edema: Mechanisms of Pathological Cell Swelling. *Neurosurg. Focus* **2007**, *22*, 379–387. [CrossRef]

67. Archie, S.R.; Al Shoyaib, A.; Cucullo, L. Blood-Brain Barrier Dysfunction in CNS Disorders and Putative Therapeutic Targets: An Overview. *Pharmaceutics* **2021**, *13*, 1779. [CrossRef]
68. Yang, C.; Hawkins, K.E.; Doré, S.; Candelario-Jalil, E. Neuroinflammatory Mechanisms of Blood-Brain Barrier Damage in Ischemic Stroke. *Am. J. Physiol. Cell Physiol.* **2019**, *316*, C135–C153. [CrossRef]
69. Donat, C.K.; Scott, G.; Gentleman, S.M.; Sastre, M. Microglial Activation in Traumatic Brain Injury. *Front. Aging Neurosci.* **2017**, *9*, 209. [CrossRef]
70. Seehaus, A.; Roebroeck, A.; Bastiani, M.; Fonseca, L.; Bratzke, H.; Lori, N.; Vilanova, A.; Goebel, R.; Galuske, R. Histological Validation of High-Resolution DTI in Human Post Mortem Tissue. *Front. Neuroanat.* **2015**, *9*, 98. [CrossRef] [PubMed]
71. Tsitsopoulos, P.P.; Marklund, N. Amyloid-β Peptides and Tau Protein as Biomarkers in Cerebrospinal and Interstitial Fluid Following Traumatic Brain Injury: A Review of Experimental and Clinical Studies. *Front. Neurol.* **2013**, *4*, 49. [CrossRef] [PubMed]
72. Tagge, C.A.; Fisher, A.M.; Minaeva, O.V.; Gaudreau-Balderrama, A.; Moncaster, J.A.; Zhang, X.-L.; Wojnarowicz, M.W.; Casey, N.; Lu, H.; Kokiko-Cochran, O.N.; et al. Concussion, Microvascular Injury, and Early Tauopathy in Young Athletes after Impact Head Injury and an Impact Concussion Mouse Model. *Brain* **2018**, *141*, 422–458. [CrossRef] [PubMed]
73. Kilinc, D.; Gallo, G.; Barbee, K.A. Mechanical Membrane Injury Induces Axonal Beading through Localized Activation of Calpain. *Exp. Neurol.* **2009**, *219*, 553–561. [CrossRef]
74. McKee, A.C.; Alosco, M.L.; Huber, B.R. Repetitive Head Impacts and Chronic Traumatic Encephalopathy. *Neurosurg. Clin. N. Am.* **2016**, *27*, 529–535. [CrossRef]
75. Martland, H.S. Punch Drunk. *J. Am. Med. Assoc.* **1928**, *91*, 1103. [CrossRef]
76. Baugh, C.M.; Robbins, C.A.; Stern, R.A.; McKee, A.C. Current Understanding of Chronic Traumatic Encephalopathy. *Curr. Treat. Options Neurol.* **2014**, *16*, 306. [CrossRef]
77. Yi, J.; Padalino, D.J.; Chin, L.S.; Montenegro, P.; Cantu, R.C. Chronic Traumatic Encephalopathy. *Curr. Sports Med. Rep.* **2013**, *12*, 28–32. [CrossRef]
78. Castellani, R.J.; Perry, G. Dementia Pugilistica Revisited. *J. Alzheimer's Dis.* **2017**, *60*, 1209–1221. [CrossRef]
79. Corsellis, J.A.N.; Bruton, C.J.; Freeman-Browne, D. The Aftermath of Boxing. *Psychol. Med.* **1973**, *3*, 270–303. [CrossRef]
80. McKee, A.C.; Cantu, R.C.; Nowinski, C.J.; Hedley-Whyte, E.T.; Gavett, B.E.; Budson, A.E.; Santini, V.E.; Lee, H.-S.; Kubilus, C.A.; Stern, R.A. Chronic Traumatic Encephalopathy in Athletes: Progressive Tauopathy after Repetitive Head Injury. *J. Neuropathol. Exp. Neurol.* **2009**, *68*, 709–735. [CrossRef] [PubMed]
81. McKee, A.C. The Spectrum of Disease in Chronic Traumatic Encephalopathy. *Brain* **2012**, *136*, 43–64. [CrossRef] [PubMed]
82. McKee, A.C.; Stein, T.D.; Kiernan, P.T.; Alvarez, V.E. The Neuropathology of Chronic Traumatic Encephalopathy. *Brain Pathol.* **2015**, *25*, 350–364. [CrossRef] [PubMed]
83. Osterman, C.; Hamlin, D.; Suter, C.M.; Affleck, A.J.; Gloss, B.S.; Turner, C.P.; Faull, R.L.M.; Stein, T.D.; McKee, A.; Buckland, M.E.; et al. Perivascular Glial Reactivity Is a Feature of Phosphorylated Tau Lesions in Chronic Traumatic Encephalopathy. *Acta Neuropathol.* **2025**, *149*, 16. [CrossRef]
84. Falcon, B.; Zivanov, J.; Zhang, W.; Murzin, A.G.; Garringer, H.J.; Vidal, R.; Crowther, R.A.; Newell, K.L.; Ghetti, B.; Goedert, M.; et al. Novel Tau Filament Fold in Chronic Traumatic Encephalopathy Encloses Hydrophobic Molecules. *Nature* **2019**, *568*, 420–423. [CrossRef]
85. Hay, J.; Johnson, V.E.; Smith, D.H.; Stewart, W. Chronic Traumatic Encephalopathy: The Neuropathological Legacy of Traumatic Brain Injury. *Annu. Rev. Pathol. Mech. Dis.* **2016**, *11*, 21–45. [CrossRef]
86. Murray, H.C.; Osterman, C.; Bell, P.; Vinnell, L.; Curtis, M.A. Neuropathology in Chronic Traumatic Encephalopathy: A Systematic Review of Comparative Post-Mortem Histology Literature. *Acta Neuropathol. Commun.* **2022**, *10*, 108. [CrossRef]
87. Nicks, R.; Shah, A.; Stathas, S.A.; Kirsch, D.; Horowitz, S.M.; Saltiel, N.; Calderazzo, S.M.; Butler, D.; Cormier, K.A.; Aytan, N.; et al. Neurodegeneration in the Cortical Sulcus Is a Feature of Chronic Traumatic Encephalopathy and Associated with Repetitive Head Impacts. *Acta Neuropathol.* **2024**, *148*, 79. [CrossRef]
88. Fukuda, A.M.; Badaut, J. Aquaporin 4: A Player in Cerebral Edema and Neuroinflammation. *J. Neuroinflamm.* **2012**, *9*, 279. [CrossRef]
89. Gardner, A.; Iverson, G.L.; McCrory, P. Chronic Traumatic Encephalopathy in Sport: A Systematic Review. *Br. J. Sports Med.* **2013**, *48*, 84–90. [CrossRef]
90. Stern, R.A.; Daneshvar, D.H.; Baugh, C.M.; Seichepine, D.R.; Montenigro, P.H.; Riley, D.O.; Fritts, N.G.; Stamm, J.M.; Robbins, C.A.; McHale, L.; et al. Clinical Presentation of Chronic Traumatic Encephalopathy. *Neurology* **2013**, *81*, 1122–1129. [CrossRef]
91. Fesharaki-Zadeh, A. Chronic Traumatic Encephalopathy: A Brief Overview. *Front. Neurol.* **2019**, *10*, 713. [CrossRef]
92. Gavett, B.E.; Stern, R.A.; McKee, A.C. Chronic Traumatic Encephalopathy: A Potential Late Effect of Sport-Related Concussive and Subconcussive Head Trauma. *Clin. Sports Med.* **2011**, *30*, 179–188. [CrossRef]
93. Tharmaratnam, T.; Iskandar, M.A.; Tabobondung, T.C.; Tobbia, I.; Gopee-Ramanan, P.; Tabobondung, T.A. Chronic Traumatic Encephalopathy in Professional American Football Players: Where Are We Now? *Front. Neurol.* **2018**, *9*, 445. [CrossRef] [PubMed]

94. Montenigro, P.H.; Baugh, C.M.; Daneshvar, D.H.; Mez, J.; Budson, A.E.; Au, R.; Katz, D.I.; Cantu, R.C.; Stern, R.A. Clinical Subtypes of Chronic Traumatic Encephalopathy: Literature Review and Proposed Research Diagnostic Criteria for Traumatic Encephalopathy Syndrome. *Alzheimer's Res. Ther.* **2014**, *6*, 68. [CrossRef] [PubMed]
95. Mez, J.; Daneshvar, D.H.; Kiernan, P.T.; Abdolmohammadi, B.; Alvarez, V.E.; Huber, B.R.; Alosco, M.L.; Solomon, T.M.; Nowinski, C.J.; McHale, L.; et al. Clinicopathological Evaluation of Chronic Traumatic Encephalopathy in Players of American Football. *JAMA* **2017**, *318*, 360–370. [CrossRef] [PubMed]
96. Alosco, M.L.; Mez, J.; Tripodis, Y.; Kiernan, P.T.; Abdolmohammadi, B.; Murphy, L.; Kowall, N.W.; Stein, T.D.; Huber, B.R.; Goldstein, L.E.; et al. Age of First Exposure to Tackle Football and Chronic Traumatic Encephalopathy. *Ann. Neurol.* **2018**, *83*, 886–901. [CrossRef]
97. Cherry, J.D.; Mez, J.; Crary, J.F.; Tripodis, Y.; Alvarez, V.E.; Mahar, I.; Huber, B.R.; Alosco, M.L.; Nicks, R.; Abdolmohammadi, B.; et al. Variation in TMEM106B in Chronic Traumatic Encephalopathy. *Acta Neuropathol. Commun.* **2018**, *6*, 115. [CrossRef]
98. Al-Rubaie, A. Traumatic Brain Injury and Dementia: Mechanisms, Risk Stratification, and Clinical Management. *J. Clin. Neurol.* **2025**, *21*, 265. [CrossRef]
99. van Amerongen, S.; Pulukuri, S.V.; Tuz-Zahra, F.; Tripodis, Y.; Cherry, J.D.; Bernick, C.; Geda, Y.E.; Wethe, J.V.; Katz, D.I.; Alosco, M.L.; et al. Inflammatory Biomarkers for Neurobehavioral Dysregulation in Former American Football Players: Findings from the DIAGNOSE CTE Research Project. *J. Neuroinflamm.* **2024**, *21*, 46. [CrossRef]
100. Alosco, M.L.; Tripodis, Y.; Fritts, N.G.; Heslegrave, A.; Baugh, C.M.; Conneely, S.; Mariani, M.; Martin, B.M.; Frank, S.; Mez, J.; et al. Cerebrospinal Fluid Tau, Aβ, and STREM2 in Former National Football League Players: Modeling the Relationship between Repetitive Head Impacts, Microglial Activation, and Neurodegeneration. *Alzheimer's Dement.* **2018**, *14*, 1159–1170. [CrossRef]
101. Wolburg, H.; Noell, S.; Mack, A.; Wolburg-Buchholz, K.; Fallier-Becker, P. Brain Endothelial Cells and the Glio-Vascular Complex. *Cell Tissue Res.* **2008**, *335*, 75–96. [CrossRef]
102. Li, Z.; Li, M.; Fang, Z.; Wang, H. Immunological Mechanisms and Therapeutic Strategies in Cerebral Ischemia—Reperfusion Injury: From Inflammatory Response to Neurorepair. *Int. J. Mol. Sci.* **2025**, *26*, 8336. [CrossRef] [PubMed]
103. Zhao, Z.-A.; Yan, L.; Wen, J.; Satyanarayanan, S.K.; Yu, F.; Lu, J.; Liu, Y.U.; Su, H. Cellular and Molecular Mechanisms in Vascular Repair after Traumatic Brain Injury: A Narrative Review. *Burn. Trauma* **2023**, *11*, tkad033. [CrossRef] [PubMed]
104. Argaw, A.T.; Gurfein, B.T.; Zhang, Y.; Zameer, A.; John, G.R. VEGF-Mediated Disruption of Endothelial CLN-5 Promotes Blood-Brain Barrier Breakdown. *Proc. Natl. Acad. Sci. USA* **2009**, *106*, 1977–1982. [CrossRef] [PubMed]
105. Winkler, E.A.; Sagare, A.P.; Zlokovic, B.V. The Pericyte: A Forgotten Cell Type with Important Implications for Alzheimer's Disease? *Brain Pathol.* **2014**, *24*, 371–386. [CrossRef]
106. Sun, B.; Li, L.; Harris, O.A.; Luo, J. Blood-Brain Barrier Disruption: A Pervasive Driver and Mechanistic Link between Traumatic Brain Injury and Alzheimer's Disease. *Transl. Neurodegener.* **2025**, *14*, 16. [CrossRef]
107. Sweeney, M.D.; Sagare, A.P.; Zlokovic, B.V. Blood–Brain Barrier Breakdown in Alzheimer Disease and Other Neurodegenerative Disorders. *Nat. Rev. Neurol.* **2018**, *14*, 133–150. [CrossRef]
108. Ikonomovic, M.D.; Mi, Z.; Abrahamson, E.E. Disordered APP Metabolism and Neurovasculature in Trauma and Aging: Combined Risks for Chronic Neurodegenerative Disorders. *Ageing Res. Rev.* **2017**, *34*, 51–63. [CrossRef]
109. Nelson, A.R.; Sweeney, M.D.; Sagare, A.P.; Zlokovic, B.V. Neurovascular Dysfunction and Neurodegeneration in Dementia and Alzheimer's Disease. *Biochim. Biophys. Acta (BBA)—Mol. Basis Dis.* **2016**, *1862*, 887–900. [CrossRef]
110. Vázquez-Rosa, E.; Shin, M.-K.; Dhar, M.; Chaubey, K.; Cintrón-Pérez, C.J.; Tang, X.; Liao, X.; Miller, E.; Koh, Y.; Barker, S.; et al. P7C3-A20 Treatment One Year after TBI in Mice Repairs the Blood-Brain Barrier, Arrests Chronic Neurodegeneration, and Restores Cognition. *Proc. Natl. Acad. Sci. USA* **2020**, *117*, 27667–27675. [CrossRef] [PubMed]
111. Zhao, T.; Pan, P.; Jia, Y.; Zhou, Y.; Zhang, X.; Guan, H.; Li, Q.; Zhou, Y. Crossing Pathological Boundaries: Multi-Target Restoration of the Neurovascular Unit in Alzheimer's and Vascular Dementia—From Modern Therapeutics to Traditional Chinese Medicine. *Aging Dis.* 2025, *online ahead of print*. [CrossRef]
112. Manukjan, N.; Chau, S.; Caiment, F.; van Herwijnen, M.; Smeets, H.J.; Fulton, D.; Ahmed, Z.; Blankesteijn, W.M.; Foulquier, S. Wnt7a Decreases Brain Endothelial Barrier Function via β-Catenin Activation. *Mol. Neurobiol.* **2023**, *61*, 4854–4867. [CrossRef] [PubMed]
113. Özen, I.; Hamdeh, S.A.; Ruscher, K.; Marklund, N. Traumatic Brain Injury Causes Early Aggregation of Beta-Amyloid Peptides and NOTCH3 Reduction in Vascular Smooth Muscle Cells of Leptomeningeal Arteries. *Acta Neuropathol.* **2025**, *149*, 10. [CrossRef]
114. Hayes, G.; Pinto, J.; Sparks, S.N.; Wang, C.; Suri, S.; Bulte, D.P. Vascular Smooth Muscle Cell Dysfunction in Neurodegeneration. *Front. Neurosci.* **2022**, *16*, 1010164. [CrossRef]
115. Baracaldo-Santamaría, D.; Ariza-Salamanca, D.F.; Corrales-Hernández, M.G.; Pachón-Londoño, M.J.; Hernandez-Duarte, I.; Calderon-Ospina, C.-A. Revisiting Excitotoxicity in Traumatic Brain Injury: From Bench to Bedside. *Pharmaceutics* **2022**, *14*, 152. [CrossRef]
116. Guerriero, R.M.; Giza, C.C.; Rotenberg, A. Glutamate and GABA Imbalance Following Traumatic Brain Injury. *Curr. Neurol. Neurosci. Rep.* **2015**, *15*, 27. [CrossRef]

 https://doi.org/10.3390/ijms27031570

117. Kirdajova, D.B.; Kriska, J.; Tureckova, J.; Anderova, M. Ischemia-Triggered Glutamate Excitotoxicity from the Perspective of Glial Cells. *Front. Cell. Neurosci.* **2020**, *14*, 51. [CrossRef]
118. Angeloni, C.; Prata, C.; Sega, F.V.D.; Piperno, R.; Hrelia, S. Traumatic Brain Injury and NADPH Oxidase: A Deep Relationship. *Oxidative Med. Cell. Longev.* **2015**, *2015*, 370312. [CrossRef]
119. Toledano-Díaz, A.; Álvarez, M.I.; Toledano, A. The Relationships between Neuroglial Alterations and Neuronal Changes in Alzheimer's Disease, and the Related Controversies I: Gliopathogenesis and Glioprotection. *J. Cent. Nerv. Syst. Dis.* **2022**, *14*, 11795735221123896. [CrossRef]
120. Feissner, R.F.; Skalska, J.; E Gaum, W.; Sheu, S.-S. Crosstalk Signaling between Mitochondrial Ca^{2+} and ROS. *Front. Biosci.* **2009**, *14*, 1197–1218. [CrossRef] [PubMed]
121. Wei, Z.; Yu, H.; Zhao, H.; Wei, M.; Xing, H.; Pei, J.; Yang, Y.; Ren, K. Broadening Horizons: Ferroptosis as a New Target for Traumatic Brain Injury. *Burn. Trauma* **2024**, *12*, tkad051. [CrossRef] [PubMed]
122. Li, J.; Jia, B.; Xu, Y.; Zhao, Y.; Wang, S.; Yang, R.; Su, L.; Zeng, X.; Li, Q.; Luo, C. Inhibition of MitoNEET Ameliorates Traumatic Brain Injury-Induced Ferroptosis and Cognitive Dysfunction by Stabilizing Dihydroorotate Dehydrogenase. *Exp. Neurol.* **2025**, *389*, 115235. [CrossRef] [PubMed]
123. Mantash, S.; Aboulouard, S.; Dakik, H.; Zirem, Y.; Ziane-Chaouche, L.; Nehme, A.; Mallah, K.; El-Kurdi, M.; Ramadan, N.; Fournier, I.; et al. Uncovering Injury-Specific Proteomic Signatures and Neurodegenerative Risks in Single and Repetitive Traumatic Brain Injury. *Signal Transduct. Target. Ther.* **2025**, *10*, 195. [CrossRef]
124. Zou, Z.; Li, L.; Schäfer, N.; Huang, Q.; Maegele, M.; Gu, Z. Endothelial Glycocalyx in Traumatic Brain Injury Associated Coagulopathy: Potential Mechanisms and Impact. *J. Neuroinflamm.* **2021**, *18*, 134. [CrossRef]
125. Zhang, J.; Zhang, F.; Dong, J. Coagulopathy Induced by Traumatic Brain Injury: Systemic Manifestation of a Localized Injury. *Blood* **2018**, *131*, 2001–2006. [CrossRef]
126. Carminita, E.; Tourn, J.; Crescence, L.; Brouilly, N.; Merrill-Skoloff, G.; Mazharian, A.; Dubois, C.; Panicot-Dubois, L. A Thrombus Is Formed by a Gradient of Platelet Activation and Procoagulant Endothelium. *Res. Pract. Thromb. Haemost.* **2023**, *7*, 102209. [CrossRef]
127. Barry, M.; Pati, S. Targeting Repair of the Vascular Endothelium and Glycocalyx after Traumatic Injury with Plasma and Platelet Resuscitation. *Matrix Biol. Plus* **2022**, *14*, 100107. [CrossRef]
128. Xu, L.; Ware, J.B.; Kim, J.J.; Shahim, P.; Silverman, E.; Magdamo, B.; Dabrowski, C.; Wesley, L.; Le, M.D.; Morrison, J.; et al. Arterial Spin Labeling Reveals Elevated Cerebral Blood Flow with Distinct Clusters of Hypo- and Hyperperfusion after Traumatic Brain Injury. *J. Neurotrauma* **2021**, *38*, 2538–2548. [CrossRef]
129. Toth, P.; Szarka, N.; Farkas, E.; Ezer, E.; Czeiter, E.; Amrein, K.; Ungvari, Z.; Hartings, J.A.; Buki, A.; Koller, A. Traumatic Brain Injury-Induced Autoregulatory Dysfunction and Spreading Depression-Related Neurovascular Uncoupling: Pathomechanisms, Perspectives, and Therapeutic Implications. *Am. J. Physiol.-Heart Circ. Physiol.* **2016**, *311*, H1118–H1131. [CrossRef]
130. Hartmann, D.A.; Coelho-Santos, V.; Shih, A.Y. Pericyte Control of Blood Flow across Microvascular Zones in the Central Nervous System. *Annu. Rev. Physiol.* **2021**, *84*, 331–354. [CrossRef]
131. Grubb, S.; Lauritzen, M.; Aalkjær, C. Brain Capillary Pericytes and Neurovascular Coupling. *Comp. Biochem. Physiol. Part A Mol. Integr. Physiol.* **2021**, *254*, 110893. [CrossRef] [PubMed]
132. Hamilton, N.B. Pericyte-Mediated Regulation of Capillary Diameter: A Component of Neurovascular Coupling in Health and Disease. *Front. Neuroenergetics* **2010**, *2*, 5. [CrossRef] [PubMed]
133. Ding, Y.; Klomparens, K. Updates on the Association of Brain Injury and Alzheimer's Disease. *Brain Circ.* **2020**, *6*, 65. [CrossRef] [PubMed]
134. Bamel, K.; Panwar, A.; Kumar, M.; Kumar, S.; Sharma, V.; Sharma, A. Traumatic Brain Injury as a Precursor to Neurodegenerative Diseases: Mechanisms Linking TBI to Alzheimer's Disease. *Brain Disord.* **2025**, *18*, 100232. [CrossRef]
135. Dong, G.; Ma, C.C.; Mao, S.; Brown, K.S.-M.; Naik, S.M.; McDonough, G.A.; Wijethunga, S.P.; Kim, J.; Kirkham, S.L.; Shao, D.D.; et al. Diverse Somatic Genomic Alterations in Single Neurons in Chronic Traumatic Encephalopathy. *Science* **2025**, *390*, eadu1351. [CrossRef]
136. Liu, C.-C.; Kanekiyo, T.; Xu, H.; Bu, G. Apolipoprotein E and Alzheimer Disease: Risk, Mechanisms and Therapy. *Nat. Rev. Neurol.* **2013**, *9*, 106–118. Correction in *Nat. Rev. Neurol.* **2013**, *9*, 184. [CrossRef]
137. Mayeux, R.; Ottman, R.; Maestre, G.; Ngai, C.; Tang, M.X.; Ginsberg, H.; Chun, M.; Tycko, B.; Shelanski, M. Synergistic Effects of Traumatic Head Injury and Apolipoprotein-Epsilon 4 in Patients with Alzheimer's Disease. *Neurology* **1995**, *45*, 555–557. [CrossRef]
138. McFadyen, C.A.; Zeiler, F.A.; Newcombe, V.; Synnot, A.; Steyerberg, E.; Gruen, R.L.; Rosand, J.; Palotie, A.; Maas, A.I.R.; Menon, D.K. Apolipoprotein E4 Polymorphism and Outcomes from Traumatic Brain Injury: A Living Systematic Review and Meta-Analysis. *J. Neurotrauma* **2019**, *38*, 1124–1136. [CrossRef]
139. Blumenfeld, J.; Yip, O.; Kim, M.J.; Huang, Y. Cell Type-Specific Roles of APOE4 in Alzheimer Disease. *Nat. Rev. Neurosci.* **2024**, *25*, 91–110. [CrossRef]

 https://doi.org/10.3390/ijms27031570

140. LoBue, C.; Woon, F.L.; Rossetti, H.C.; Hynan, L.S.; Hart, J.; Cullum, C.M. Traumatic Brain Injury History and Progression from Mild Cognitive Impairment to Alzheimer Disease. *Neuropsychology* **2018**, *32*, 401–409. [CrossRef] [PubMed]
141. Xu, J.; Duan, J.; Cai, Z.; Arai, C.; Di, C.; Venters, C.C.; Xu, J.; Jones, M.; So, B.-R.; Dreyfuss, G. TOMM40-APOE Chimera Linking Alzheimer's Highest Risk Genes: A New Pathway for Mitochondria Regulation and APOE4 Pathogenesis. *bioRxiv* **2024**. [CrossRef]
142. Sweeney, N.; Kim, T.Y.; Morrison, C.T.; Li, L.; Acosta, D.; Liang, J.; Datla, N.V.; Fitzgerald, J.A.; Huang, H.; Liu, X.; et al. Neuronal BAG3 Attenuates Tau Hyperphosphorylation, Synaptic Dysfunction, and Cognitive Deficits Induced by Traumatic Brain Injury via the Regulation of Autophagy-Lysosome Pathway. *Acta Neuropathol.* **2024**, *148*, 52. [CrossRef] [PubMed]
143. Grimm, M.O.W.; Mett, J.; Stahlmann, C.P.; Haupenthal, V.J.; Zimmer, V.C.; Hartmann, T. Neprilysin and Aβ Clearance: Impact of the APP Intracellular Domain in NEP Regulation and Implications in Alzheimer's Disease. *Front. Aging Neurosci.* **2013**, *5*, 98. [CrossRef]
144. Saxena, S.K.; Ansari, S.; Maurya, V.K.; Kumar, S.; Sharma, D.; Malhotra, H.S.; Tiwari, S.; Srivastava, C.; Paweska, J.T.; Abdel-Moneim, A.S.; et al. Neprilysin-Mediated Amyloid Beta Clearance and Its Therapeutic Implications in Neurodegenerative Disorders. *ACS Pharmacol. Transl. Sci.* **2024**, *7*, 3645–3657. [CrossRef]
145. Guerreiro, R.; Wojtas, A.; Bras, J.; Carrasquillo, M.; Rogaeva, E.; Majounie, E.; Cruchaga, C.; Sassi, C.; Kauwe, J.S.K.; Younkin, S.; et al. TREM2 Variants in Alzheimer's Disease. *N. Engl. J. Med.* **2013**, *368*, 117–127. [CrossRef]
146. Luo, J. TGF-β as a Key Modulator of Astrocyte Reactivity: Disease Relevance and Therapeutic Implications. *Biomedicines* **2022**, *10*, 1206. [CrossRef]
147. Macura, I.J.; Zivanovic, A.; Perovic, M.; Ciric, J.; Major, T.; Kanazir, S.; Ivkovic, S. The Expression of Major Facilitator Superfamily Domain-Containing Protein2a (Mfsd2a) and Aquaporin 4 Is Altered in the Retinas of a 5xFAD Mouse Model of Alzheimer's Disease. *Int. J. Mol. Sci.* **2023**, *24*, 14092. [CrossRef]
148. Ben-Zvi, A.; Lacoste, B.; Kur, E.; Andreone, B.J.; Mayshar, Y.; Yan, H.; Gu, C. Mfsd2a Is Critical for the Formation and Function of the Blood–Brain Barrier. *Nature* **2014**, *509*, 507–511. [CrossRef]
149. Salehi, A.; Jullienne, A.; Baghchechi, M.; Hamer, M.; Walsworth, M.; Donovan, V.; Tang, J.; Zhang, J.H.; Pearce, W.J.; Obenaus, A. Up-Regulation of Wnt/β-Catenin Expression Is Accompanied with Vascular Repair after Traumatic Brain Injury. *J. Cereb. Blood Flow Metab.* **2017**, *38*, 274–289. [CrossRef]
150. Zou, D.; Hu, W.; Qin, J.; Wei, Z.; Wang, D.; Li, L. Rapid Orderly Migration of Neutrophils after Traumatic Brain Injury Depends on MMP9/13. *Biochem. Biophys. Res. Commun.* **2021**, *579*, 161–167. [CrossRef] [PubMed]
151. Wu, J.; He, J.; Tian, X.; Zhong, J.; Li, H.; Sun, X. Activation of the Hedgehog Pathway Promotes Recovery of Neurological Function after Traumatic Brain Injury by Protecting the Neurovascular Unit. *Transl. Stroke Res.* **2020**, *11*, 720–733. [CrossRef] [PubMed]
152. Razak, L.H.A.; Denis, T.; Murugiah, Y.; Yoong, W.K.; Idris, Z.; Senik, M.H. The Effect of Traumatic Brain Injury on Memory. *Malays. J. Med. Sci.* **2024**, *31*, 52–74. [CrossRef] [PubMed]
153. Paterno, R.; Folweiler, K.A.; Cohen, A.S. Pathophysiology and Treatment of Memory Dysfunction after Traumatic Brain Injury. *Curr. Neurol. Neurosci. Rep.* **2017**, *17*, 52. [CrossRef]
154. Ozga, J.E.; Povroznik, J.M.; Engler-Chiurazzi, E.B.; Haar, C.V. Executive (Dys) Function after Traumatic Brain Injury: Special Considerations for Behavioral Pharmacology. *Behav. Pharmacol.* **2018**, *29*, 617–637. [CrossRef]
155. Cardoso, M.G.D.F.; Faleiro, R.M.; de Paula, J.J.; Kummer, A.; Caramelli, P.; Teixeira, A.L.; de Souza, L.C.; Miranda, A.S. Cognitive Impairment Following Acute Mild Traumatic Brain Injury. *Front. Neurol.* **2019**, *10*, 198. [CrossRef]
156. Howlett, J.R.; Nelson, L.D.; Stein, M.B. Mental Health Consequences of Traumatic Brain Injury. *Biol. Psychiatry* **2021**, *91*, 412–420. [CrossRef]
157. Chan, A.; Ouyang, J.; Nguyen, K.; Jones, A.; Basso, S.; Karasik, R. Traumatic Brain Injuries: A Neuropsychological Review. *Front. Behav. Neurosci.* **2024**, *18*, 1326115. [CrossRef]
158. Sundström, A.; Nilsson, L.-G.; Cruts, M.; Adolfsson, R.; Van Broeckhoven, C.; Nyberg, L. Increased Risk of Dementia Following Mild Head Injury for Carriers but Not for Non-Carriers of the APOE ε4 Allele. *Int. Psychogeriatr.* **2006**, *19*, 159. [CrossRef]
159. Li, Y.; Li, Y.; Li, X.; Zhang, S.; Zhao, J.; Zhu, X.; Tian, G. Head Injury as a Risk Factor for Dementia and Alzheimer's Disease: A Systematic Review and Meta-Analysis of 32 Observational Studies. *PLoS ONE* **2017**, *12*, e0169650. [CrossRef]
160. Gardner, R.C.; Burke, J.F.; Nettiksimmons, J.; Kaup, A.; Barnes, D.E.; Yaffe, K. Dementia Risk after Traumatic Brain Injury vs Nonbrain Trauma. *JAMA Neurol.* **2014**, *71*, 1490. [CrossRef] [PubMed]
161. Winter, L.; Moriarty, H.; Robinson, K.M.; Leiby, B.E.; Schmidt, K.; Whitehouse, C.R.; Swanson, R.L. Age Suppresses the Association between Traumatic Brain Injury Severity and Functional Outcomes: A Study Using the NIDILRR TBIMS Dataset. *J. Head Trauma Rehabil.* **2024**, *39*, E582–E590. [CrossRef] [PubMed]
162. Zhang, X.; Pan, Y.; Wu, T.; Zhao, W.; Zhang, H.; Ding, J.; Ji, Q.; Jia, X.; Li, X.; Lee, Z.; et al. Brain Age Prediction Using Interpretable Multi-Feature-Based Convolutional Neural Network in Mild Traumatic Brain Injury. *NeuroImage* **2024**, *297*, 120751. [CrossRef] [PubMed]

163. Gupte, R.; Brooks, W.; Vukas, R.; Pierce, J.; Harris, J. Sex Differences in Traumatic Brain Injury: What We Know and What We Should Know. *J. Neurotrauma* **2019**, *36*, 3063–3091. [CrossRef]
164. Farace, E.; Alves, W.M. Do Women Fare Worse? A Metaanalysis of Gender Differences in Outcome after Traumatic Brain Injury. *Neurosurg. Focus* **2000**, *8*, 1–8. [CrossRef]
165. Broshek, D.K.; Kaushik, T.; Freeman, J.R.; Erlanger, D.; Webbe, F.; Barth, J.T. Sex Differences in Outcome Following Sports-Related Concussion. *J. Neurosurg.* **2005**, *102*, 856–863. [CrossRef]
166. Schneider, A.L.C.; Barber, J.; Temkin, N.; Gardner, R.C.; Manley, G.; Diaz-Arrastia, R.; Sandsmark, D. Associations of Preexisting Vascular Risk Factors with Outcomes after Traumatic Brain Injury: A TRACK-TBI Study. *J. Head Trauma Rehabil.* **2022**, *38*, E88–E98. [CrossRef]
167. Csiszar, A.; Tarantini, S.; Fulop, G.A.; Kiss, T.; Valcarcel-Ares, M.N.; Galvan, V.; Ungvari, Z.; Yabluchanskiy, A. Hypertension Impairs Neurovascular Coupling and Promotes Microvascular Injury: Role in Exacerbation of Alzheimer's Disease. *GeroScience* **2017**, *39*, 359–372. [CrossRef]
168. Tatara, Y.; Shimada, R.; Kibayashi, K. Effects of Preexisting Diabetes Mellitus on the Severity of Traumatic Brain Injury. *J. Neurotrauma* **2021**, *38*, 886–902. [CrossRef]
169. Wu, A.; Molteni, R.; Ying, Z.; Gomez-Pinilla, F. A Saturated-Fat Diet Aggravates the Outcome of Traumatic Brain Injury on Hippocampal Plasticity and Cognitive Function by Reducing Brain-Derived Neurotrophic Factor. *Neuroscience* **2003**, *119*, 365–375. [CrossRef]
170. Sivandzade, F.; Alqahtani, F.; Sifat, A.; Cucullo, L. The Cerebrovascular and Neurological Impact of Chronic Smoking on Post-Traumatic Brain Injury Outcome and Recovery: An in Vivo Study. *J. Neuroinflamm.* **2020**, *17*, 133. [CrossRef] [PubMed]
171. Hu, H.; Fu, J.-T.; Bi, Y.; Ma, Y.; Huang, Y.; Xin, W.; Tan, L.; Yu, J. Tau Pathologies Mediate the Association of Cigarette Smoking with Cognitive Impairment in Older Adults without Dementia: The CABLE Study. *J. Alzheimer's Dis.* **2022**, *86*, 1849–1859. [CrossRef] [PubMed]
172. Brzecka, A.; Leszek, J.; Ashraf, G.M.; Ejma, M.; Ávila-Rodriguez, M.F.; Yarla, N.S.; Tarasov, V.V.; Chubarev, V.N.; Samsonova, A.N.; Barreto, G.E.; et al. Sleep Disorders Associated with Alzheimer's Disease: A Perspective. *Front. Neurosci.* **2018**, *12*, 330. [CrossRef] [PubMed]
173. Houle, S.; Tapp, Z.; Dobres, S.; Ahsan, S.; Reyes, Y.; Cotter, C.; Mitsch, J.; Zimomra, Z.; Peng, J.; Rowe, R.K.; et al. Sleep Fragmentation after Traumatic Brain Injury Impairs Behavior and Conveys Long-Lasting Impacts on Neuroinflammation. *Brain Behav. Immun.—Health* **2024**, *38*, 100797. [CrossRef]
174. Green, T.R.F.; Ortiz, J.B.; Wonnacott, S.; Williams, R.J.; Rowe, R.K. The Bidirectional Relationship between Sleep and Inflammation Links Traumatic Brain Injury and Alzheimer's Disease. *Front. Neurosci.* **2020**, *14*, 894. [CrossRef]
175. Daigle, B.; Bevilacqua, L.M.; Noël, V.P.; Lebel, M.; Côté, D.; Lévesque, M.; Menard, C. Neurogliovascular Alterations in Brain Pathologies across Lifespan. *Glial Health Res.* **2024**, *1*, 100004. [CrossRef]
176. Zetterberg, H.; Blennow, K. Fluid Biomarkers for Mild Traumatic Brain Injury and Related Conditions. *Nat. Rev. Neurol.* **2016**, *12*, 563–574. [CrossRef]
177. Singh, A.K.; Asif, S.; Pandey, D.K.; Chaudhary, A.; Kapoor, V.; Verma, P.K. Biomarkers in Acute Traumatic Brain Injury: A Systematic Review and Meta-Analysis. *Cureus* **2024**, *16*, e63020. [CrossRef]
178. Huebschmann, N.A.; Luoto, T.M.; Karr, J.E.; Berghem, K.; Blennow, K.; Zetterberg, H.; Ashton, N.J.; Simrén, J.; Posti, J.P.; Gill, J.M.; et al. Comparing Glial Fibrillary Acidic Protein (GFAP) in Serum and Plasma Following Mild Traumatic Brain Injury in Older Adults. *Front. Neurol.* **2020**, *11*, 1054. [CrossRef]
179. Seidenfaden, S.-C.; Kjerulff, J.L.; Juul, N.; Kirkegaard, H.; Møller, M.F.; Münster, A.-M.B.; Bøtker, M.T. Temporal Changes in Serum S100B Levels from Prehospital to Early In-Hospital Sampling in Patients Suffering Traumatic Brain Injury. *Front. Neurol.* **2022**, *13*, 80015. [CrossRef]
180. Spaziani, G.; Rozzi, G.; Baroni, S.; Simeoni, B.; Racco, S.; Barone, F.; Fuorlo, M.; Franceschi, F.; Covino, M. From the Emergency Department to Follow-Up: Clinical Utility of Biomarkers in Mild Traumatic Brain Injury. *Emerg. Care Med.* **2025**, *2*, 45. [CrossRef]
181. Whitehouse, D.P.; Monteiro, M.; Czeiter, E.; Vyvere, T.V.; Valerio, F.; Ye, Z.; Amrein, K.; Kamnitsas, K.; Xu, H.; Yang, Z.; et al. Relationship of Admission Blood Proteomic Biomarkers Levels to Lesion Type and Lesion Burden in Traumatic Brain Injury: A CENTER-TBI Study. *eBioMedicine* **2022**, *75*, 103777. [CrossRef] [PubMed]
182. Shahim, P.; Tegner, Y.; Marklund, N.; Blennow, K.; Zetterberg, H. Neurofilament Light and Tau as Blood Biomarkers for Sports-Related Concussion. *Neurology* **2018**, *90*, e1780–e1788. [CrossRef] [PubMed]
183. Li, L.M.; Kodosaki, E.; Heslegrave, A.; Zetterberg, H.; Graham, N.; Zimmerman, K.; Soreq, E.; Parker, T.; Garbero, E.; Moro, F.; et al. High-Dimensional Proteomic Analysis for Pathophysiological Classification of Traumatic Brain Injury. *Brain* **2024**, *148*, 1015–1030. [CrossRef]
184. Graham, N.S.N.; Zimmerman, K.A.; Hain, J.; Rooney, E.; Lee, Y.; Giovane, M.D.; Parker, T.; Wilson, M.G.; Malhotra, P.; David, M.C.B.; et al. Biomarker Evidence of Neurodegeneration in Mid-Life Former Rugby Players. *Brain* **2025**, *148*, 2684–2697. [CrossRef]

 https://doi.org/10.3390/ijms27031570

185. Fantoni, E.; Collij, L.; Alves, I.L.; Buckley, C.; Farrar, G. The Spatial-Temporal Ordering of Amyloid Pathology and Opportunities for PET Imaging. *J. Nucl. Med.* **2019**, *61*, 166–171. [CrossRef]
186. Jack, C.R.; Knopman, D.S.; Jagust, W.J.; Petersen, R.C.; Weiner, M.W.; Aisen, P.S.; Shaw, L.M.; Vemuri, P.; Wiste, H.J.; Weigand, S.D.; et al. Tracking Pathophysiological Processes in Alzheimer's Disease: An Updated Hypothetical Model of Dynamic Biomarkers. *Lancet Neurol.* **2013**, *12*, 207–216. [CrossRef]
187. Thal, D.R.; Beach, T.G.; Zanette, M.; Heurling, K.; Chakrabarty, A.; Ismail, A.; Smith, A.P.L.; Buckley, C. [^{18}F] Flutemetamol Amyloid Positron Emission Tomography in Preclinical and Symptomatic Alzheimer's Disease: Specific Detection of Advanced Phases of Amyloid-β Pathology. *Alzheimer's Dement.* **2015**, *11*, 975–985. [CrossRef]
188. Eagle, S.R.; Puccio, A.; Svirsky, S.; Mountz, J.; Laymon, C.; Borasso, A.; Henry, L.; Okonkwo, D.O. Identifying a Biological Signature of Trauma-Related Neurodegeneration Following Repeated Traumatic Brain Injuries Compared with Healthy Controls. *Neurotrauma Rep.* **2025**, *6*, 560–568. [CrossRef]
189. Macdonald, C.; Dikranian, K.; Song, S.; Bayly, P.; Holtzman, D.; Brody, D. Detection of Traumatic Axonal Injury with Diffusion Tensor Imaging in a Mouse Model of Traumatic Brain Injury. *Exp. Neurol.* **2007**, *205*, 116–131. [CrossRef]
190. Zimmerman, K.A.; Laverse, E.; Samra, R.; Lopez, M.Y.; Jolly, A.E.; Bourke, N.J.; Graham, N.S.N.; Patel, M.C.; Hardy, J.; Kemp, S.; et al. White Matter Abnormalities in Active Elite Adult Rugby Players. *Brain Commun.* **2021**, *3*, fcab133. [CrossRef]
191. Wojtowicz, M.; Gardner, A.J.; Stanwell, P.; Zafonte, R.; Dickerson, B.C.; Iverson, G.L. Cortical Thickness and Subcortical Brain Volumes in Professional Rugby League Players. *NeuroImage Clin.* **2018**, *18*, 377–381. [CrossRef] [PubMed]
192. Zimmerman, K.A.; Hain, J.A.; Graham, N.S.N.; Rooney, E.J.; Lee, Y.; Del-Giovane, M.; Parker, T.D.; Friedland, D.; Cross, M.J.; Kemp, S.; et al. Prospective Cohort Study of Long-Term Neurological Outcomes in Retired Elite Athletes: The Advanced BiomaRker, Advanced Imaging and Neurocognitive (BRAIN) Health Study Protocol. *BMJ Open* **2024**, *14*, e082902. [CrossRef] [PubMed]
193. Raji, C.A.; Meysami, S.; Porter, V.R.; Merrill, D.A.; Mendez, M.F. Diagnostic Utility of Brain MRI Volumetry in Comparing Traumatic Brain Injury, Alzheimer Disease and Behavioral Variant Frontotemporal Dementia. *BMC Neurol.* **2024**, *24*, e082902. [CrossRef] [PubMed]
194. Shetty, T.; Raince, A.; Manning, E.; Tsiouris, A.J. Imaging in Chronic Traumatic Encephalopathy and Traumatic Brain Injury. *Sports Health A Multidiscip. Approach* **2015**, *8*, 26–36. [CrossRef]
195. Legramante, J.M.; Minieri, M.; Belli, M.; Giovannelli, A.; Agnoli, A.; Bajo, D.; Bellincampi, L.; Maria, A.; Terrinoni, A.; Pieri, M.; et al. Evaluation of GFAP/UCH-L1 Biomarkers for Computed Tomography Exclusion in Mild Traumatic Brain Injury (MTBI). *Int. J. Emerg. Med.* **2024**, *17*, 164. [CrossRef]
196. Park, Y.; Kc, N.; Paneque, A.; Cole, P.D. Tau, Glial Fibrillary Acidic Protein, and Neurofilament Light Chain as Brain Protein Biomarkers in Cerebrospinal Fluid and Blood for Diagnosis of Neurobiological Diseases. *Int. J. Mol. Sci.* **2024**, *25*, 6295. [CrossRef]
197. Blennow, K.; Zetterberg, H. Biomarkers for Alzheimer's Disease: Current Status and Prospects for the Future. *J. Intern. Med.* **2018**, *284*, 643–663. [CrossRef]
198. Palmqvist, S.; Janelidze, S.; Stomrud, E.; Zetterberg, H.; Karl, J.; Zink, K.; Bittner, T.; Mattsson, N.; Eichenlaub, U.; Blennow, K.; et al. Performance of Fully Automated Plasma Assays as Screening Tests for Alzheimer Disease–Related β-Amyloid Status. *JAMA Neurol.* **2019**, *76*, 1060–1069. [CrossRef]
199. Ashton, N.J.; Leuzy, A.; Karikari, T.K.; Mattsson-Carlgren, N.; Dodich, A.; Boccardi, M.; Corre, J.; Drzezga, A.; Nordberg, A.; Ossenkoppele, R.; et al. The Validation Status of Blood Biomarkers of Amyloid and Phospho-Tau Assessed with the 5-Phase Development Framework for AD Biomarkers. *Eur. J. Nucl. Med. Mol. Imaging* **2021**, *48*, 2140–2156. [CrossRef]
200. Yadikar, H.; Kobeissy, F.H.; Robertson, C.; Tsetsou, S.; Williamson, J.B.; Lamb, D.G.; Wagner, A.K.; Kibaugh, T.; Kao, S.-H.; Kou, Z.; et al. Early Cerebrospinal Fluid Elevations of PTau-217 in Severe Traumatic Brain Injury Subjects. *Front. Neurol.* **2025**, *16*, 1632679. [CrossRef]
201. Bogoslovsky, T.; Wilson, D.; Chen, Y.; Hanlon, D.; Gill, J.; Jeromin, A.; Song, L.; Moore, C.; Gong, Y.; Kenney, K.; et al. Increases of Plasma Levels of Glial Fibrillary Acidic Protein, Tau, and Amyloid β up to 90 Days after Traumatic Brain Injury. *J. Neurotrauma* **2017**, *34*, 66–73. [CrossRef] [PubMed]
202. Ramos-Cejudo, J.; Wisniewski, T.; Marmar, C.; Zetterberg, H.; Blennow, K.; de Leon, M.J.; Fossati, S. Traumatic Brain Injury and Alzheimer's Disease: The Cerebrovascular Link. *EBioMedicine* **2018**, *28*, 21–30. [CrossRef] [PubMed]
203. Johnson, V.E.; Stewart, W.; Smith, D.H. Traumatic Brain Injury and Amyloid-β Pathology: A Link to Alzheimer's Disease? *Nat. Rev. Neurosci.* **2010**, *11*, 361–370. [CrossRef] [PubMed]
204. Friberg, S.; Lindblad, C.; Zeiler, F.A.; Zetterberg, H.; Granberg, T.; Svenningsson, P.; Piehl, F.; Thelin, E.P. Fluid Biomarkers of Chronic Traumatic Brain Injury. *Nat. Rev. Neurol.* **2024**, *20*, 671–684. [CrossRef]
205. Graham, N.S.; Blissitt, G.; Zimmerman, K.; Orton, L.; Friedland, D.; Coady, E.; Laban, R.; Veleva, E.; Heslegrave, A.J.; Zetterberg, H.; et al. Poor Long-Term Outcomes and Abnormal Neurodegeneration Biomarkers after Military Traumatic Brain Injury: The ADVANCE Study. *J. Neurol. Neurosurg. Psychiatry* **2024**, *96*, 105–113. [CrossRef]

 https://doi.org/10.3390/ijms27031570

206. Asken, B.M.; Tanner, J.A.; VandeVrede, L.; Mantyh, W.G.; Casaletto, K.B.; Staffaroni, A.M.; La Joie, R.; Iaccarino, L.; Soleimani-Meigooni, D.; Rojas, J.C.; et al. Plasma P-Tau181 and P-Tau217 in Patients with Traumatic Encephalopathy Syndrome with and without Evidence of Alzheimer Disease Pathology. *Neurology* **2022**, *99*, e594–e604. [CrossRef]
207. Andersson, E.; Öst, M.; Dalla, K.; Zetterberg, H.; Blennow, K.; Nellgård, B. Acute-Phase Neurofilament Light and Glial Fibrillary Acidic Proteins in Cerebrospinal Fluid Predict Long-Term Outcome after Severe Traumatic Brain Injury. *Neurocrit. Care* **2024**, *41*, 813–827. [CrossRef]
208. Heuvel, F.O.; Li, Z.; Riedel, D.; Halbgebauer, S.; Oeckl, P.; Mayer, B.; Gotzman, N.; Shultz, S.; Semple, B.; Tumani, H.; et al. Dynamics of Synaptic Damage in Severe Traumatic Brain Injury Revealed by Cerebrospinal Fluid SNAP-25 and VILIP-1. *J. Neurol. Neurosurg. Psychiatry* **2024**, *95*, 1158–1167. [CrossRef]
209. Schindler, S.E.; Li, Y.; Todd, K.W.; Herries, E.M.; Henson, R.L.; Gray, J.D.; Wang, G.; Graham, D.L.; Shaw, L.M.; Trojanowski, J.Q.; et al. Emerging Cerebrospinal Fluid Biomarkers in Autosomal Dominant Alzheimer's Disease. *Alzheimer's Dement.* **2019**, *15*, 655–665. [CrossRef]
210. Harting, M.T.; Jimenez, F.; Adams, S.D.; Mercer, D.W.; Cox, C.S. Acute, Regional Inflammatory Response after Traumatic Brain Injury: Implications for Cellular Therapy. *Surgery* **2008**, *144*, 803–813. [CrossRef] [PubMed]
211. Needham, E.J.; Stoevesandt, O.; Thelin, E.P.; Zetterberg, H.; Zanier, E.R.; Al Nimer, F.; Ashton, N.J.; Outtrim, J.G.; Newcombe, V.F.J.; Mousa, H.S.; et al. Complex Autoantibody Responses Occur Following Moderate to Severe Traumatic Brain Injury. *J. Immunol.* **2021**, *207*, 90–100. [CrossRef]
212. Walker, A.; Chapin, B.; Abisambra, J.; DeKosky, S.T. Association between Single Moderate to Severe Traumatic Brain Injury and Long-Term Tauopathy in Humans and Preclinical Animal Models: A Systematic Narrative Review of the Literature. *Acta Neuropathol. Commun.* **2022**, *10*, 13. [CrossRef] [PubMed]
213. Zanier, E.R.; Bertani, I.; Sammali, E.; Pischiutta, F.; Chiaravalloti, M.A.; Vegliante, G.; Masone, A.; Corbelli, A.; Smith, D.H.; Menon, D.K.; et al. Induction of a Transmissible Tau Pathology by Traumatic Brain Injury. *Brain* **2018**, *141*, 2685–2699. [CrossRef] [PubMed]
214. Johnson, V.E.; Stewart, W.; Smith, D.H. Widespread Tau and Amyloid-Beta Pathology Many Years after a Single Traumatic Brain Injury in Humans. *Brain Pathol.* **2011**, *22*, 142–149. [CrossRef] [PubMed]
215. Risling, M.; Smith, D.; Stein, T.D.; Thelin, E.P.; Zanier, E.R.; Ankarcrona, M.; Nilsson, P. Modelling Human Pathology of Traumatic Brain Injury in Animal Models. *J. Intern. Med.* **2019**, *285*, 594–607. [CrossRef]
216. Moloney, C.M.; Lowe, V.J.; Murray, M.E. Visualization of Neurofibrillary Tangle Maturity in Alzheimer's Disease: A Clinicopathologic Perspective for Biomarker Research. *Alzheimer's Dement.* **2021**, *17*, 1554–1574. [CrossRef]
217. Collins-Praino, L.E.; Corrigan, F. Does Neuroinflammation Drive the Relationship between Tau Hyperphosphorylation and Dementia Development Following Traumatic Brain Injury? *Brain Behav. Immun.* **2017**, *60*, 369–382. [CrossRef]
218. Griffin, W.T.; Sheng, J.G.; Gentleman, S.M.; Graham, D.I.; Mrak, R.E.; Roberts, G.W. Microglial Interleukin-Lα Expression in Human Head Injury: Correlations with Neuronal and Neuritic β-Amyloid Precursor Protein Expression. *Neurosci. Lett.* **1994**, *176*, 133–136. [CrossRef]
219. Ising, C.; Venegas, C.; Zhang, S.; Scheiblich, H.; Schmidt, S.V.; Vieira-Saecker, A.; Schwartz, S.; Albasset, S.; McManus, R.M.; Tejera, D.; et al. NLRP3 Inflammasome Activation Drives Tau Pathology. *Nature* **2019**, *575*, 669–673. [CrossRef]
220. Cai, L.; Gong, Q.; Qi, L.; Xu, T.; Suo, Q.; Li, X.; Wang, W.; Jing, Y.; Yang, D.; Xu, Z.; et al. ACT001 Attenuates Microglia-Mediated Neuroinflammation after Traumatic Brain Injury via Inhibiting AKT/NFκB/NLRP3 Pathway. *Cell Commun. Signal.* **2022**, *20*, 56. [CrossRef]
221. Li, J.; Du, M.; Dong, G.; Hu, D.; Zhang, H.; Yu, Z.; Lu, Y.; Yang, Y. Early Intervention with Pericyte Fli-1 Post-TBI Attenuates Hippocampal BBB Disruption and Subsequent Neuroinflammation with Neurological Deficits. *Biochim. Biophys. Acta-Mol. Basis Dis.* **2025**, *1871*, 167671. [CrossRef] [PubMed]
222. Mouzon, B.C.; Bachmeier, C.; Ferro, A.; Ojo, J.-O.; Crynen, G.; Acker, C.M.; Davies, P.; Mullan, M.; Stewart, W.; Crawford, F. Chronic Neuropathological and Neurobehavioral Changes in a Repetitive Mild Traumatic Brain Injury Model. *Ann. Neurol.* **2014**, *75*, 241–254. [CrossRef] [PubMed]
223. Ayerra, L.; Shumilov, K.; Ni, A.; Aymerich, M.S.; Friess, S.H.; Celorrio, M. Chronic Traumatic Brain Injury Induces Neurodegeneration, Neuroinflammation, and Cognitive Deficits in a T Cell-Dependent Manner. *Brain Res.* **2025**, *1850*, 149446. [CrossRef] [PubMed]
224. Mei, Y.; She, F.; Zhang, L.; Kim, G.; Li, R.; Zheng, X.; Wang, Z.; Chen, R.; Wang, L.; Chen, D.; et al. Zipper-Interacting Protein Kinase Mediates Neuronal Cell Death and Cognitive Dysfunction in Traumatic Brain Injury via Regulating DEDD. *Cell Death Dis.* **2025**, *16*, 151. [CrossRef]
225. Lisi, I.; Moro, F.; Mazzone, E.; Marklund, N.; Pischiutta, F.; Kobeissy, F.; Mao, X.; Corrigan, F.; Helmy, A.; Nasrallah, F.; et al. Exploiting Blood-Based Biomarkers to Align Preclinical Models with Human Traumatic Brain Injury. *Brain* **2024**, *148*, 1062–1080. [CrossRef]

 https://doi.org/10.3390/ijms27031570

226. Smith, D.J.; Chen, X.R.; Nonaka, M.; Trojanowski, J.Q.; Lee, V.-Y.; Saatman, K.E.; Leoni, M.J.; Xu, B.-N.; Wolf, J.A.; Meaney, D.F. Accumulation of Amyloid β and Tau and the Formation of Neurofilament Inclusions Following Diffuse Brain Injury in the Pig. *J. Neuropathol. Exp. Neurol.* **1999**, *58*, 982–992. [CrossRef]
227. Liu, C.-C.; Zhao, N.; Fu, Y.; Wang, N.; Linares, C.; Tsai, C.-W.; Bu, G. ApoE4 Accelerates Early Seeding of Amyloid Pathology. *Neuron* **2017**, *96*, 1024–1032.e3. [CrossRef]
228. Uryu, K.; Laurer, H.; McIntosh, T.; Praticò, D.; Martinez, D.; Leight, S.; Lee, V.M.-Y.; Trojanowski, J.Q. Repetitive Mild Brain Trauma Accelerates Aβ Deposition, Lipid Peroxidation, and Cognitive Impairment in a Transgenic Mouse Model of Alzheimer Amyloidosis. *J. Neurosci.* **2002**, *22*, 446–454. [CrossRef]
229. Mielke, M.M.; Ransom, J.E.; Mandrekar, J.; Turcano, P.; Savica, R.; Brown, A.W. Traumatic Brain Injury and Risk of Alzheimer's Disease and Related Dementias in the Population. *J. Alzheimer's Dis.* **2022**, *88*, 1049–1059. [CrossRef]
230. Russell, E.R.; Mackay, D.F.; Lyall, D.; Stewart, K.; MacLean, J.A.; Robson, J.; Pell, J.P.; Stewart, W. Neurodegenerative Disease Risk among Former International Rugby Union Players. *J. Neurol. Neurosurg. Psychiatry* **2022**, *93*, 1262–1268. [CrossRef] [PubMed]
231. Hume, P.A.; Theadom, A.; Lewis, G.N.; Quarrie, K.L.; Brown, S.R.; Hill, R.; Marshall, S.W. A Comparison of Cognitive Function in Former Rugby Union Players Compared with Former Non-Contact-Sport Players and the Impact of Concussion History. *Sports Med.* **2016**, *47*, 1209–1220. [CrossRef] [PubMed]
232. Snowden, T.M.; Hinde, A.K.; Reid, H.M.O.; Christie, B.R. Does Mild Traumatic Brain Injury Increase the Risk for Dementia? A Systematic Review and Meta-Analysis. *J. Alzheimer's Dis.* **2020**, *78*, 757–775. [CrossRef] [PubMed]
233. Mavroudis, I.; Kazis, D.; Petridis, F.E.; Balmus, I.-M.; Papaliagkas, V.; Ciobica, A. The Association between Traumatic Brain Injury and the Risk of Cognitive Decline: An Umbrella Systematic Review and Meta-Analysis. *Brain Sci.* **2024**, *14*, 1188. [CrossRef]
234. Barnes, D.E.; Kaup, A.; Kirby, K.A.; Byers, A.L.; Diaz-Arrastia, R.; Yaffe, K. Traumatic Brain Injury and Risk of Dementia in Older Veterans. *Neurology* **2014**, *83*, 312–319. [CrossRef]
235. Saltiel, N.; Tripodis, Y.; Menzin, T.; Olaniyan, A.; Baucom, Z.; Yhang, E.; Palmisano, J.N.; Martin, B.; Uretsky, M.; Nair, E.; et al. Relative Contributions of Mixed Pathologies to Cognitive and Functional Symptoms in Brain Donors Exposed to Repetitive Head Impacts. *Ann. Neurol.* **2024**, *95*, 314–324. [CrossRef]
236. Brett, B.L.; Temkin, N.; Barber, J.; Okonkwo, D.O.; Stein, M.; Bodien, Y.; Corrigan, J.; Diaz-Arrastia, R.; Giacino, J.T.; McCrea, M.; et al. Long-Term Multi-Domain Patterns of Change Following Traumatic Brain Injury: A Track-Tbi Long Study. *Neurology* **2023**, *101*, e740–e753. [CrossRef]
237. Hammond, F.M.; Hart, T.; Bushnik, T.; Corrigan, J.D.; Sasser, H. Change and Predictors of Change in Communication, Cognition, and Social Function between 1 and 5 Years after Traumatic Brain Injury. *J. Head Trauma Rehabil.* **2004**, *19*, 314–328. [CrossRef]
238. Hanrahan, J.G.; Burford, C.; Nagappan, P.; Adegboyega, G.; Rajkumar, S.; Kolias, A.; Helmy, A.; Hutchinson, P.J. Is Dementia More Likely Following Traumatic Brain Injury? A Systematic Review. *J. Neurol.* **2023**, *270*, 3022–3051. [CrossRef]
239. Roberts, G.W.; Gentleman, S.M.; Lynch, A.; Graham, D.I. BA4 Amyloid Protein Deposition in Brain after Head Trauma. *Lancet* **1991**, *338*, 1422–1423. [CrossRef]
240. Dams-O'Connor, K.; Gibbons, L.E.; Bowen, J.D.; McCurry, S.M.; Larson, E.B.; Crane, P.K. Risk for Late-Life Re-Injury, Dementia and Death among Individuals with Traumatic Brain Injury: A Population-Based Study. *J. Neurol. Neurosurg. Psychiatry* **2012**, *84*, 177–182. [CrossRef]
241. Daneshvar, D.H.; Goldstein, L.E.; Kiernan, P.T.; Stein, T.D.; McKee, A.C. Post-Traumatic Neurodegeneration and Chronic Traumatic Encephalopathy. *Mol. Cell. Neurosci.* **2015**, *66*, 81–90. [CrossRef] [PubMed]
242. Guo, Z.; Cupples, L.A.; Kurz, A.; Auerbach, S.H.; Volicer, L.; Chui, H.; Green, R.C.; Sadovnick, A.D.; Duara, R.; DeCarli, C.; et al. Head Injury and the Risk of AD in the MIRAGE Study. *Neurology* **2000**, *54*, 1316–1323. [CrossRef] [PubMed]
243. Arciniega, H.; Baucom, Z.H.; Tuz-Zahra, F.; Tripodis, Y.; John, O.; Carrington, H.; Kim, N.; Knyazhanskaya, E.E.; Jung, L.B.; Breedlove, K.; et al. Brain Morphometry in Former American Football Players: Findings from the DIAGNOSE CTE Research Project. *Brain* **2024**, *147*, 3596–3610. [CrossRef] [PubMed]
244. Vaibhav, K.; Gulhane, M.; Ahluwalia, P.; Kumar, M.; Ahluwalia, M.; Rafiq, A.M.; Amble, V.; Zabala, M.G.; Miller, J.B.; Goldman, L.; et al. Single Episode of Moderate to Severe Traumatic Brain Injury Leads to Chronic Neurological Deficits and Alzheimer's-like Pathological Dementia. *GeroScience* **2024**, *46*, 5439–5457. [CrossRef]
245. Hicks, A.J.; Ponsford, J.L.; Spitz, G.; Dore, V.; Krishnadas, N.; Roberts, C.; Rowe, C.C. β-Amyloid and Tau Imaging in Chronic Traumatic Brain Injury: A Cross-Sectional Study. *Neurology* **2022**, *99*, e1131–e1141. [CrossRef]
246. Fedak, K.M.; Bernal, A.; Capshaw, Z.A.; Gross, S. Applying the Bradford Hill Criteria in the 21st Century: How Data Integration Has Changed Causal Inference in Molecular Epidemiology. *Emerg. Themes Epidemiol.* **2015**, *12*, 14. [CrossRef]
247. Washington, P.M.; Villapol, S.; Burns, M.P. Polypathology and Dementia after Brain Trauma: Does Brain Injury Trigger Distinct Neurodegenerative Diseases, or Should They Be Classified Together as Traumatic Encephalopathy? *Exp. Neurol.* **2016**, *275*, 381–388. [CrossRef]

https://doi.org/10.3390/ijms27031570

248. Emrani, S.; Koutures, A.; Tripodis, Y.; Uretsky, M.; Abdolmohammadi, B.; Nowinski, C.; Daneshvar, D.H.; Dwyer, B.; Katz, D.I.; Goldstein, L.E.; et al. Characterizing White Matter and Vascular Pathologies in Brain Donors Exposed to Repetitive Head Impacts. *Acta Neuropathol.* **2025**, *149*, 24. [CrossRef]
249. Chong, F.P.; Ng, K.Y.; Koh, R.Y.; Chye, S.M. Tau Proteins and Tauopathies in Alzheimer's Disease. *Cell. Mol. Neurobiol.* **2018**, *38*, 965–980. [CrossRef]
250. McGroarty, J.; Salinas, S.; Evans, H.; Jimenez, B.; Tran, V.; Kadavakollu, S.; Vashist, A.; Atluri, V. Inflammasome-Mediated Neuroinflammation: A Key Driver in Alzheimer's Disease Pathogenesis. *Biomolecules* **2025**, *15*, 676. [CrossRef]
251. Seplovich, G.; Bouchi, Y.; Pablo, J.; Munoz, J.C.; Reisner, A.; Blackwell, L.; Mechref, Y.; Wang, K.K.; Tyndall, J.A.; Tharakan, B.; et al. Inflammasome Links Traumatic Brain Injury, Chronic Traumatic Encephalopathy, and Alzheimer's Disease. *Neural Regen. Res.* **2024**, *20*, 1644–1664. [CrossRef] [PubMed]
252. Powell, J.R.; Zong, X.; Weinstein, J.M.; DeLellis, S.M.; Kane, S.F.; Means, G.E.; Mihalik, J.P. Mild Traumatic Brain Injury and Career Stage Associate with Visible Perivascular Spaces in Special Operations Forces Soldiers. *Ann. Biomed. Eng.* **2024**, *52*, 2812–2817. [CrossRef] [PubMed]
253. Tinney, E.M.; España-Irla, G.; Warren, A.E.L.; Whitehurst, L.N.; Stillman, A.M.; Hillman, C.H.; Morris, T.P. Axonal Injury Sleep Disturbances, and Memory Following Traumatic Brain Injury. *Ann. Clin. Transl. Neurol.* **2024**, *11*, 2314–2326. [CrossRef] [PubMed]
254. Kriukova, K.; Boswell, M.; Asifriyaz, T.; Gong, J.; Duncan, D.; Vespa, P. Changes in Hippocampal Volume after Traumatic Brain Injury (TBI). *Res. Sq.* 2024, *preprint*. [CrossRef]
255. Chia, Y.C.; Anjum, C.E.; Yee, H.R.; Kenisi, Y.; Chan, M.K.S.; Wong, M.B.F.; Pan, S.Y. Stem Cell Therapy for Neurodegenerative Diseases: How Do Stem Cells Bypass the Blood-Brain Barrier and Home to the Brain? *Stem Cells Int.* **2020**, *2020*, e8889061. [CrossRef]
256. Xia, H.; Zhou, W.; Li, D.; Peng, F.; Wang, C.; Yu, L.; Du, J.; Zheng, Y.; Sang, Y.; Zhang, Y.; et al. Engineering Neural Stem Cells with Micropatches for Improved Therapy of Traumatic Brain Injury. *Angew. Chem. (Int. Ed. Engl.)* **2025**, *64*, e202512804 [CrossRef]
257. Lu, M.-H.; Ji, W.-L.; Chen, H.; Sun, Y.-Y.; Zhao, X.-Y.; Wang, F.; Shi, Y.; Hu, Y.-N.; Liu, B.-X.; Wu, J.; et al. Intranasal Transplantation of Human Neural Stem Cells Ameliorates Alzheimer's Disease-like Pathology in a Mouse Model. *Front. Aging Neurosci.* **2021**, *13*, 350103. [CrossRef]
258. Roncalli, J.G.; Tongers, J.; Renault, M.-A.; Losordo, D.W. Endothelial Progenitor Cells in Regenerative Medicine and Cancer: A Decade of Research. *Trends Biotechnol.* **2008**, *26*, 276–283. [CrossRef]
259. Huang, X.-T.; Zhang, Y.-Q.; Li, S.-J.; Li, S.-H.; Tang, Q.; Wang, Z.-T.; Dong, J.-F.; Zhang, J.-N. Intracerebroventricular Transplantation of *Ex Vivo* Expanded Endothelial Colony-Forming Cells Restores Blood–Brain Barrier Integrity and Promotes Angiogenesis of Mice with Traumatic Brain Injury. *J. Neurotrauma* **2013**, *30*, 2080–2088. [CrossRef]
260. Zhang, S.; Zhi, Y.; Li, F.; Huang, S.; Gao, H.; Han, Z.; Ge, X.; Li, D.; Chen, F.; Kong, X.; et al. Transplantation of in Vitro Cultured Endothelial Progenitor Cells Repairs the Blood-Brain Barrier and Improves Cognitive Function of APP/PS1 Transgenic AD Mice. *J. Neurol. Sci.* **2018**, *387*, 6–15. [CrossRef]
261. de Vasconcelos, P.; Lacerda, J.F. Hematopoietic Stem Cell Transplantation for Neurological Disorders: A Focus on Inborn Errors of Metabolism. *Front. Cell. Neurosci.* **2022**, *16*, 895511. [CrossRef]
262. Mishra, P.; Silva, A.; Sharma, J.; Nguyen, J.; Pizzo, D.; Hinz, D.; Sahoo, D.; Cherqui, S. Rescue of Alzheimer's Disease Phenotype in a Mouse Model by Transplantation of Wild-Type Hematopoietic Stem and Progenitor Cells. *Cell Rep.* **2023**, *42*, 112956. [CrossRef] [PubMed]
263. Via, A.G.; Frizziero, A.; Oliva, F. Biological Properties of Mesenchymal Stem Cells from Different Sources. *Muscles Ligaments Tendons J.* **2012**, *2*, 154–162. [PubMed]
264. Menge, T.; Zhao, Y.; Zhao, J.; Wataha, K.; Gerber, M.; Zhang, J.; Letourneau, P.; Redell, J.; Shen, L.; Wang, J.; et al. Mesenchymal Stem Cells Regulate Blood-Brain Barrier Integrity through TIMP3 Release after Traumatic Brain Injury. *Sci. Transl. Med.* **2012**, *4*, 161ra150. [CrossRef] [PubMed]
265. Eyvari-Brooshghalan, S.; Haddadi, R.; Shahidi, S.; Ghaderi, S.; Rashno, M.; Kalantari, A.; Salehi, I.; Komaki, A.; Sarihi, A. Acute Treatment with Fucoidan Ameliorates Traumatic Brain Injury-Induced Neurological Damages and Memory Deficits in Rats: Role of BBB Integrity, Microglial Activity, Neuroinflammation, and Oxidative Stress. *Mol. Neurobiol.* **2024**, *62*, 5990–6013. [CrossRef]
266. Lin, C.; Chao, H.; Li, Z.; Xu, X.; Liu, Y.; Hou, L.; Liu, N.; Ji, J. Melatonin Attenuates Traumatic Brain Injury-Induced Inflammation: A Possible Role for Mitophagy. *J. Pineal Res.* **2016**, *61*, 177–186. [CrossRef]
267. Ding, K.; Xu, J.; Wang, H.; Zhang, L.; Wu, Y.; Li, T. Melatonin Protects the Brain from Apoptosis by Enhancement of Autophagy after Traumatic Brain Injury in Mice. *Neurochem. Int.* **2015**, *91*, 46–54. [CrossRef]
268. Rehman, S.U.; Ikram, M.; Ullah, N.; Alam, S.I.; Park, H.Y.; Badshah, H.; Choe, K.; Kim, M.O. Neurological Enhancement Effects of Melatonin against Brain Injury-Induced Oxidative Stress, Neuroinflammation, and Neurodegeneration via AMPK/CREB Signaling. *Cells* **2019**, *8*, 760. [CrossRef]
269. Gunara, S.O.; Akgun, M.Y.; Seker, U.; Ucar, E.A.; Ates, O.; Basar, I. Neuroprotection Unveiled: Melatonin Mitigates Apoptotic Pathways in Traumatic Brain Injury. *Front. Neurol.* **2025**, *16*, 1551449. [CrossRef]

270. Carteri, R.B. Mitochondria as a Therapeutic Target: Focusing on Traumatic Brain Injury. *J. Integr. Neurosci.* **2025**, *24*, 25292. [CrossRef]
271. Riou, A.; Broeglin, A.; Grimm, A. Mitochondrial Transplantation in Brain Disorders: Achievements, Methods, and Challenges. *Neurosci. Biobehav. Rev.* **2024**, *169*, 105971. [CrossRef]
272. Alshial, E.E.; Abdulghaney, M.I.; Wadan, A.-H.S.; Abdellatif, M.; Ramadan, N.; Suleiman, A.; Waheed, N.; Abdellatif, M.; Mohammed, H.S. Mitochondrial Dysfunction and Neurological Disorders: A Narrative Review and Treatment Overview. *Life Sci.* **2023**, *334*, 122257. [CrossRef] [PubMed]
273. Bamshad, C.; Najafi-Ghalehlou, N.; Pourmohammadi-Bejarpasi, Z.; Tomita, K.; Kuwahara, Y.; Sato, T.; Feizkhah, A.; Roushnadeh, A.M.; Roudkenar, M.H. Mitochondria: How Eminent in Ageing and Neurodegenerative Disorders? *Hum. Cell* **2022**, *36*, 41–61. [CrossRef] [PubMed]
274. Alkaslasi, M.R.; Lloyd, E.Y.H.; Gable, A.S.; Silberberg, H.; Yarur, H.E.; Tsai, V.S.; Sohn, M.; Margolin, G.; Tejeda, H.A.; Le, C.E. The Transcriptional Response of Cortical Neurons to Concussion Reveals Divergent Fates after Injury. *Nat. Commun.* **2025**, *16*, 1097. [CrossRef] [PubMed]
275. Zhang, G.; Diamante, G.; Ahn, I.S.; Palafox-Sanchez, V.; Cheng, J.; Cheng, M.; Ying, Z.; Wang, S.S.-M.; Abuhanna, K.D.; Phi, N.; et al. Thyroid Hormone T4 Mitigates Traumatic Brain Injury in Mice by Dynamically Remodeling Cell Type Specific Genes, Pathways, and Networks in Hippocampus and Frontal Cortex. *Biochim. Biophys. Acta Mol. Basis Dis.* **2024**, *1870*, 167344. [CrossRef]
276. Bernal, J. Thyroid Hormones and Brain Development. *Vitam. Horm.* **2005**, *71*, 95–122.
277. Mele, C.; Marchi, L.D.; Pitino, R.; Costantini, L.; Cavigiolo, B.; Caputo, M.; Marzullo, P.; Aimaretti, G. The Interplay between Thyrotropic Axis, Neurological Complications, and Rehabilitation Outcomes in Patients with Traumatic Brain Injury. *Best Pract. Res. Clin. Endocrinol. Metab.* **2025**, *39*, 102001. [CrossRef]
278. Crupi, R.; Paterniti, I.; Campolo, M.; Di Paola, R.; Cuzzocrea, S.; Esposito, E. Exogenous T3 Administration Provides Neuroprotection in a Murine Model of Traumatic Brain Injury. *Pharmacol. Res.* **2013**, *70*, 80–89. [CrossRef]
279. Corps, K.N.; Roth, T.L.; McGavern, D.B. Inflammation and Neuroprotection in Traumatic Brain Injury. *JAMA Neurol.* **2015**, *72*, 355. [CrossRef]
280. Needham, E.J.; Helmy, A.; Zanier, E.R.; Jones, J.L.; Coles, A.J.; Menon, D.K. The Immunological Response to Traumatic Brain Injury. *J. Neuroimmunol.* **2019**, *332*, 112–125. [CrossRef] [PubMed]
281. Czerpaniak, K.; do Nascimento, L.F.; Guo, T.; Zhang, J.; Liu, X.; Sarzaeim, M.; Fine, Z.D.; Cao, Y.-Q. Low-Dose Interleukin-2 Reverses Traumatic Brain Injury-Induced Cognitive Deficit and Pain in a Murine Model. *Ann. Neurol.* **2024**, *96*, 508–525. [CrossRef] [PubMed]
282. Wang, P.; Okada-Rising, S.; Scultetus, A.H.; Bailey, Z.S. The Relevance and Implications of Monoclonal Antibody Therapies on Traumatic Brain Injury Pathologies. *Biomedicines* **2024**, *12*, 2698. [CrossRef] [PubMed]
283. Kondo, A.; Shahpasand, K.; Mannix, R.; Qiu, J.; Moncaster, J.; Chen, C.-H.; Yao, Y.; Lin, Y.-M.; Driver, J.A.; Sun, Y.; et al. Antibody against Early Driver of Neurodegeneration Cis P-Tau Blocks Brain Injury and Tauopathy. *Nature* **2015**, *523*, 431–436. [CrossRef] [PubMed]
284. Cognacq, G.; Attwood, J.E.; DeLuca, G.C. Traumatic Brain Injury and Alzheimer's Disease: A Shared Neurovascular Hypothesis. *Neurosci. Insights* **2025**, *20*, 26331055251323292. [CrossRef] [PubMed]
285. Vella, M.A.; Crandall, M.L.; Patel, M.B. Acute Management of Traumatic Brain Injury. *Surg. Clin. N. Am.* **2017**, *97*, 1015–1030. [CrossRef]
286. The CRASH-3 Trial Collaborators. Effects of Tranexamic Acid on Death, Disability, Vascular Occlusive Events and Other Morbidities in Patients with Acute Traumatic Brain Injury (CRASH-3): A Randomised, Placebo-Controlled Trial. *Lancet* **2019**, *394*, 1713–1723. [CrossRef]
287. Alnemari, A.M.; Krafcik, B.M.; Mansour, T.R.; Gaudin, D. A Comparison of Pharmacologic Therapeutic Agents Used for the Reduction of Intracranial Pressure after Traumatic Brain Injury. *World Neurosurg.* **2017**, *106*, 509–528. [CrossRef]
288. Roberts, I. Barbiturates for Acute Traumatic Brain Injury. *Cochrane Database Syst. Rev.* **2000**, *2000*, CD000033. [CrossRef]
289. Thompson, K.; Pohlmann-Eden, B.; Campbell, L.A.; Abel, H. Pharmacological Treatments for Preventing Epilepsy Following Traumatic Head Injury. *Cochrane Database Syst. Rev.* **2015**, *2015*, CD009900. [CrossRef]
290. Tani, J.; Wen, Y.-T.; Hu, C.-J.; Sung, J.-Y. Current and Potential Pharmacologic Therapies for Traumatic Brain Injury. *Pharmaceuticals* **2022**, *15*, 838. [CrossRef]
291. Edwards, P.; Arango, M.; Balica, L.; Cottingham, R.; El-Sayed, H.; Farrell, B.; Fernandes, J.; Gogichaisvili, T.; Golden, N.; Hartzenberg, B.; et al. Final Results of MRC CRASH, a Randomised Placebo-Controlled Trial of Intravenous Corticosteroid in Adults with Head Injury—Outcomes at 6 Months. *Lancet* **2005**, *365*, 1957–1959. [CrossRef]
292. Zafonte, R.; Friedewald, W.T.; Lee, S.M.; Levin, B.; Diaz-Arrastia, R.; Ansel, B.; Eisenberg, H.; Timmons, S.D.; Temkin, N.; Novack, T.; et al. The Citicoline Brain Injury Treatment (COBRIT) Trial: Design and Methods. *J. Neurotrauma* **2009**, *26*, 2207–2216. [CrossRef]

293. Espinoza, T.R.; Wright, D.W. The Role of Progesterone in Traumatic Brain Injury. *J. Head Trauma Rehabil.* **2011**, *26*, 497–499. [CrossRef]
294. Yue, J.; Burke, J.; Upadhyayula, P.; Winkler, E.; Deng, H.; Robinson, C.; Pirracchio, R.; Suen, C.; Sharma, S.; Ferguson, A.; et al. Selective Serotonin Reuptake Inhibitors for Treating Neurocognitive and Neuropsychiatric Disorders Following Traumatic Brain Injury: An Evaluation of Current Evidence. *Brain Sci.* **2017**, *7*, 93. [CrossRef]
295. Duarte, D.C.; Duarte, J.C.; Ocampo González, Á.A.; García, J.F.C. Psychiatric Disorders in Post-Traumatic Brain Injury Patients: A Scoping Review. *Heliyon* **2023**, *9*, e12905. [CrossRef]

https://doi.org/10.3390/ijms27031570

International Journal of
Molecular Sciences

MDPI

Article

Early Blood–Brain Barrier Impairment as a Pathological Hallmark in a Novel Model of Closed-Head Concussive Brain Injury (CBI) in Mice

Stefan J. Blaschke [1,2,*], Nora Rautenberg [1,2], Heike Endepols [3,4,5], Aileen Jendro [1], Jens Konrad [6], Susan Vlachakis [1], Dirk Wiedermann [7], Michael Schroeter [1,2], Bernd Hoffmann [6], Rudolf Merkel [6], Niklas Marklund [8], Gereon R. Fink [1,2] and Maria A. Rueger [1,2]

1 Department of Neurology, Faculty of Medicine and University Hospital, University of Cologne, 50923 Cologne, Germany; nora.rautenberg@uk-koeln.de (N.R.); ajendro@smail.uni-koeln.de (A.J.); michael.schroeter@uk-koeln.de (M.S.); gereon.fink@uk-koeln.de (G.R.F.); maria.rueger@uk-koeln.de (M.A.R.)
2 Cognitive Neuroscience Section, Institute of Neuroscience and Medicine (INM-3), Research Centre Juelich, 52428 Juelich, Germany
3 Institute of Radiochemistry and Experimental Molecular Imaging, Faculty of Medicine and University Hospital Cologne, University of Cologne, 50937 Cologne, Germany; hendepo1@uni-koeln.de
4 Department of Nuclear Medicine, Faculty of Medicine and University Hospital Cologne, University of Cologne, 50931 Cologne, Germany
5 Nuclear Chemistry, Institute of Neuroscience and Medicine (INM-5), Research Centre Juelich, 52428 Juelich, Germany
6 Mechanobiology, Institute of Biological Information Processing (IBI-2), Research Centre Juelich, 52425 Juelich, Germany; j.konrad@fz-juelich.de (J.K.); b.hoffmann@fz-juelich.de (B.H.); r.merkel@fz-juelich.de (R.M.)
7 Multimodal Imaging Group, Max Planck Institute for Metabolism Research, 50931 Cologne, Germany; dirk.wiedermann@sf.mpg.de
8 Department of Clinical Sciences Lund, Neurosurgery, Lund University, 221 85 Lund, Sweden; niklas.marklund@med.lu.se
* Correspondence: stefan.blaschke@uk-koeln.de

Abstract: Concussion, caused by a rotational acceleration/deceleration injury mild enough to avoid structural brain damage, is insufficiently captured in recent preclinical models, hampering the relation of pathophysiological findings on the cellular level to functional and behavioral deficits. We here describe a novel model of unrestrained, single vs. repetitive concussive brain injury (CBI) in male C56Bl/6j mice. Longitudinal behavioral assessments were conducted for up to seven days afterward, alongside the evaluation of structural cerebral integrity by in vivo magnetic resonance imaging (MRI, 9.4 T), and validated ex vivo by histology. Blood–brain barrier (BBB) integrity was analyzed by means of fluorescent dextran- as well as immunoglobulin G (IgG) extravasation, and neuroinflammatory processes were characterized both in vivo by positron emission tomography (PET) using [^{18}F]DPA-714 and ex vivo using immunohistochemistry. While a single CBI resulted in a defined, subacute neuropsychiatric phenotype, longitudinal cognitive testing revealed a marked decrease in spatial cognition, most pronounced in mice subjected to CBI at high frequency (every 48 h). Functional deficits were correlated to a parallel disruption of the BBB, (R^2 = 0.29, $p < 0.01$), even detectable by a significant increase in hippocampal uptake of [^{18}F]DPA-714, which was not due to activation of microglia, as confirmed immunohistochemically. Featuring a mild but widespread disruption of the BBB without evidence of macroscopic damage, this model induces a characteristic neuro-psychiatric phenotype that correlates to the degree of BBB disruption. Based on these findings, the BBB may function as both a biomarker of CBI severity and as a potential treatment target to improve recovery from concussion.

Keywords: mild traumatic brain injury; mice; repetitive concussion; blood–brain barrier; neuroinflammation; PET

1. Introduction

Traumatic brain injury (TBI) remains a common and potentially devastating disease, especially impacting younger populations, with a high degree of mortality and morbidity [1]. The past decades have highlighted the multifaceted pathophysiological spectrum of TBI [2]. The mildest form of TBI—the mere concussion—caused by a rotational acceleration/deceleration injury, occurs most frequently, and by definition, it does not cause any macroscopic structural brain damage [3]. However, in recent years, it has become increasingly obvious that concussions—especially when they occur repeatedly as commonly observed in contact sports—may induce concussive brain injury (CBI) with a substantial impact on life and well-being, particularly in the juvenile and young adult group [4]. Accumulating evidence suggests that post-concussive symptoms (i.e., chronic headaches, dizziness, or impairment of memory and concentration) not only persist in a subgroup of patients over weeks to months (i.e., the "miserable minority" [5,6]) but may also lead to secondary deterioration due to accelerated neurodegeneration, especially after repetitive trauma [7,8]. Since conventional neuroimaging does not detect structural abnormalities after a concussion, CBI and its related persistent post-concussion syndrome are not easily diagnosed, and most healthcare providers are unsure about its prognosis, leading to uncertainty in both patients and healthcare providers [9]. Consequently, persistent symptoms after CBI might be trivialized or suspected as malingering [10], leading to inadequate medical care and chronification. Thus, while a diagnostic arsenal is urgently needed to diagnose CBI and estimate an individual prognosis for CBI patients, the underlying pathophysiological hallmarks and the (variable) temporal course of CBI remain poorly understood.

To uncover underlying disease mechanisms, preclinical models of various degrees of TBI have been assessed. Nonetheless, there is an astonishing lack of animal models for the characteristic CBI injury mechanism. Most often, a mild form of TBI is induced by a focal blow to the fixated head of the experimental animal, with scaled mechanical intensity [11]. However, this does not exactly reflect the situation of, e.g., the impact on the head when boxing or playing soccer (header), since a blow to the unfixated head causes its rotation [12]—therefore inducing a more diffuse and multifocal brain trauma while avoiding skull fractures [11].

Among the few preclinical models of rotational closed-head injury described to date, a heterogeneous pathophysiology is highlighted, mainly featuring neuroinflammation [13,14], blood–brain barrier (BBB) disruption, and microvascular injury [15,16], as well as electrophysiological alterations [16,17]. To date, pathophysiological findings on the focal/cellular level have not been comprehensively correlated to functional/behavioral deficits in the respective animal models.

We here aimed to characterize a novel model of CBI featuring a mild rotational brain injury in the mouse. To alleviate the need for specific biomarkers of disease severity and biological recovery, we specifically assessed the effects of CBI on BBB disruption and neuroinflammation. In this way, we specifically evaluated a second-generation translocator protein (TSPO) tracer, $[^{18}F]$DPA-714, as a surrogate marker of secondary brain injury, corroborating in vivo data by ex vivo histology and correlating those parameters to the disease severity of the individual behavioral phenotype.

2. Results

2.1. Novel Mouse Model of Rotational Closed-Head Concussive Brain Injury (CBI)

We aimed to establish a highly reproducible mouse model of CBI that resembles key hallmarks of concussion observed in humans. Thus, we applied (i) an external force on the (ii) freely moving (iii) closed skull in (iv) a controlled fashion that (v) induced a rotational acceleration–deceleration force on the brain, (vi) without causing structural lesions such as brain contusion or bone fractures.

Additionally, to prevent the presumably (neuro-)protective effect of anesthesia [18] and preserve neuronal activity at the timepoint of injury, we induced CBI in conscious sedation with medetomidine, a drug widely used to preserve functional network activity in neuroimaging [19].

Inducing the concussion, we observed a reproducible head rotation (Figure 1C), with only minor variations depending on the exact position of the mouse head (mean rotation 22°, SD = 6.7°). The impact parameters were chosen based on previous finite element analyses in mice [20] to reproduce similar biomechanical forces as in human CBI. To mitigate the focal strain on the skull and the brain, we used a rounded impactor tip that was additionally softened by a custom-made silicon cap. Hitting the skull with this tip at 5 mm/s did not cause a skull fracture in any of the mice, while a higher impactor speed of 6 mm/s led to occasional skull fractures. Most noteworthy, the assessment of the structural integrity of the brain one day after the last concussion, using in vivo T2-weighted MRI (Figure 1D) as well as ex vivo HE-staining (Figure 1E), indicated no signs of macroscopic tissue disruption, contusions, or microbleeds in any of the mice.

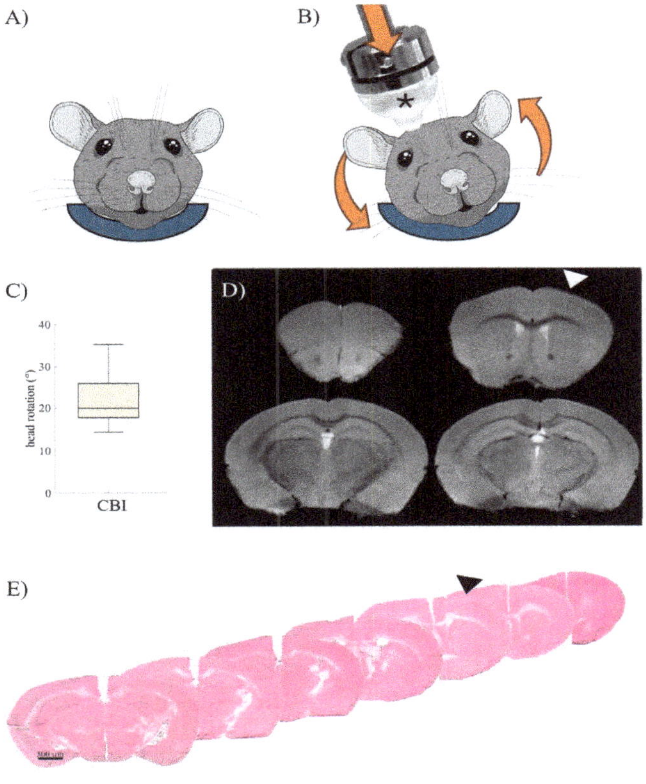

Figure 1. To induce the closed-head acceleration–deceleration concussive brain injury (CBI), mice were placed in a foam bed ((**A**) blue). The head was struck by an electromagnetically driven impactor, softened with a custom-made 5 mm silicon tip ((**B**), *). Thus, the head, limiting linear translation by the foam bed, was rotated counterclockwise in the coronal plane by about 20° (**C**). As assessed by early structural MRI ((**D**) T2 weighted TurboRARE), even repetitive concussions at high frequency (repCBI-hf) did not cause cerebral contusions or microbleeds, as additionally verified by hematoxylin and eosin staining ((**E**) impact location indicated by the arrowhead).

2.2. CBI Induces a Neuro-Psychiatric Phenotype

We next characterized the clinical phenotype of our novel CBI model. Using a surrogate parameter for CBI-induced loss of consciousness (LoC), animals receiving a concussion required a significantly longer time to regain full alertness after this procedure and rise from a supine position (Figure 2A, 234 s, [171, 297]) compared with the sham group

(127 s, [89, 165]; U = 4.6, p = 0.03). Over the next five days, mice suffering from CBI displayed a higher overall activity level, as indicated by their extensive use of nestlet material (Figure 2B, CBI: mean AUC = 63.2, [56.9, 69.5], sham: mean AUC = 48.8, [36.2, 61.4]; U = 6.6, p = 0.01). In the Open Field, the control (sham) group of mice evenly explored the entire arena (Figure 2C; exploration ratio = −0.82, [−0.87, −0.78]), while mice suffering from CBI kept moving along its walls and avoiding the center zone, suggesting increased anxiety (Figure 2D,E; exploration ratio = −0.89, [−0.93, −0.85], U = 5.4, p = 0.02). Consistent with the result on overall motor activity, the overall distance traveled by CBI mice was not significantly decreased (CBI: mean distance = 4.6 m, [3.6, 5.6]; sham: 4.1, [2.8, 5.4]), despite their avoidance of the center zone (Figure 2F).

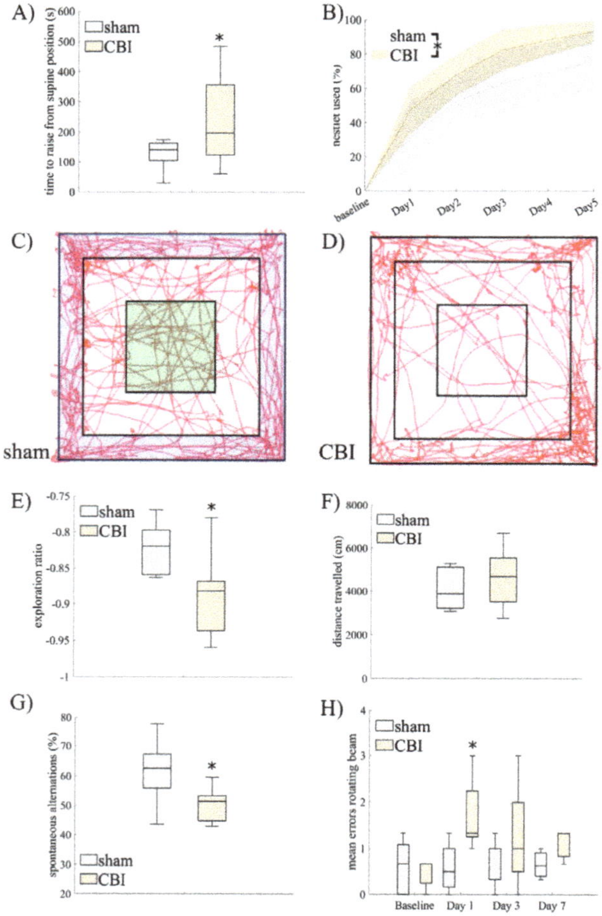

Figure 2. A single CBI induced a significantly longer latency to rise from a supine position after the trauma (**A**). Over the first days after CBI, animals exhibited significantly higher overall activity levels (**B**), as measured by the amount of nesting materials used over time. In the Open Field, relative exploration of the center zone ((**C**), green) was compared to exploration of the outer border zone (blue). While healthy (sham) animals (**C**) explored all regions evenly, CBI mice (**D**) showed a significant decrease in exploration of the center zone and instead kept to the walls of the arena, suggesting anxiety (**E**), while their overall locomotor activity was not reduced (**F**). Mice suffering from CBI displayed a significant decrease in spatial cognition in the spontaneous alternation test (Y-maze, (**G**)) and a transient lack of motor control (rotating beam, (**H**)). (* $p < 0.05$ compared with sham, n (sham) = 6, n (CBI) = 9).

On the level of spatial cognition, mice underperformed in the Y-maze test after CBI (48.7, [41.7, 55.7]) compared with healthy (sham) animals (Figure 2G; 61.5, [46.4, 76.6]; U = 4.7, $p = 0.04$). Cognitive impairment assessed as spontaneous alternations in the Y-maze was corroborated by similar results in the Novel Object Recognition (NOR) Test, albeit with a non-significant decrease in the recognition rate after CBI (0.26, [0.12, 0.39]; U = 2.9, $p = 0.08$; Figure S1B) compared with the sham group (0.51, [0.06, 0.97]). However, there was a significant correlation between individual scores in the Y-maze and NOR Test ($R^2 = 0.65$, $p < 0.01$, Figure S1C). Motor control as assessed by the rotating beam test revealed a mild motor impairment one day after CBI (Figure 2H). That impairment spontaneously ameliorated over the first week after a concussion, indicated by a significant main effect of the trauma group assignment ($F(1,40) = 6.0$, $p = 0.02$) and a significant difference between groups one day after injury in the simple effects analysis.

2.3. CBI Leads to Sustained Impairment in Spatial Cognition

In order to characterize the effects of repetitive CBI, groups of mice were subjected to several concussions in different intervals, leading to different periods of recovery between impacts (Figure 3A). Five consecutive concussions, either (i) every 48 h (repCBI-hf, orange) or (ii) every week (repCBI-lf, blue), were compared to (iii) a one-time CBI (CBI, light grey), and to (iv) sham-injured healthy mice (sham, white; Figure 3A).

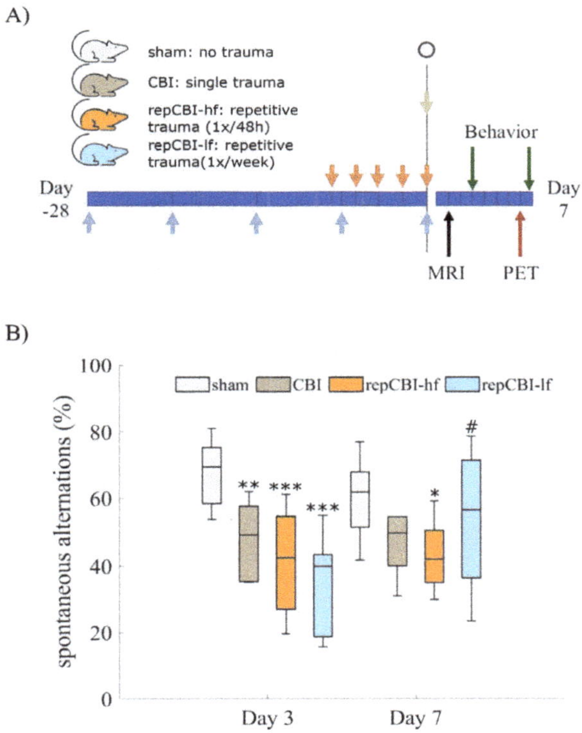

Figure 3. (**A**) To characterize the effects of repetitive CBI, mice received five consecutive concussions either every 48 h (repCBI-hf) or once a week (repCBI-lf) and were compared to a group of mice receiving a single CBI or sham surgery only. (**B**) CBI induced a significant decrease in spatial cognition in all groups compared with sham three days after the last trauma. However, seven days after (the last) CBI, cognitive impairment was still visible only after a single CBI (light grey), as well as after concussions delivered at a high frequency (orange), while the low-frequency CBI group (blue) improved most rapidly. (* $p < 0.05$; ** $p < 0.01$, *** $p < 0.001$ compared with sham, # $p < 0.05$ compared to day 3, n = 10).

Based on spontaneous alternations in the Y-maze test, a mixed ANOVA with repeated measures over time and trauma as a between-group factor, revealed a significant effect of trauma on spatial cognition (F(3,24) = 7.9, $p < 0.001$), with a significant interaction effect of time and trauma (F(3,24) = 5.8, $p = 0.004$). Mice receiving a single CBI exhibited a subacute impairment in spatial cognition after three days (CBI: 48.1, [36.2, 59.9], sham: 67.7, [59.3, 76.0]), which slightly improved after seven days (Figure 3B, CBI: 46.6, [32.0, 75.7], sham: 60.2, [50.5, 69.8]). A particularly long-lasting decrease in spatial cognition was observed in the group of mice suffering from repetitive CBI at short (48 h) intervals (repCBI-hf, Figure 3B; Day 3: 41.2, [28.2, 54.1], Day 7: 43.1, [34.5, 51.6]). Interestingly, mice subjected to repetitive CBI at weekly intervals—allowing longer recovery phases between concussions—recovered most rapidly from their last concussion, reaching mostly their baseline cognitive level 7 days after their last concussion (repCBI-lf, Figure 3B, Day 3: 35.4, [19.3, 51.5], Day 7: 53.9, [32.0, 75.7]).

2.4. Repetitive CBI Causes Lasting Disruption of the Blood–Brain Barrier (BBB)

The dynamic effect of repetitive CBI on the integrity of the BBB over time was evaluated by infusion of TxRed-labeled dextran during the terminal perfusion of mice. One day after a set of five concussions every 48 h, dye leakage—causing a blurry fluorescent signal—appeared most prominent in the hippocampus (Figure 4A) compared with sham mice, where the dextran signal was located intravascularly for the most part (Figure 4B). Quantification of this observation based on the fluorescence intensity of labeled dextran (Figure 4C) demonstrated a significant effect of trauma (H(3,15) = 9.3, $p = 0.026$) with the largest effect of CBI on BBB integrity one day after the (last) trauma (mean intensity Day 1: 38.1, [32.0, 44.2], sham: 51.0, [40.0, 61.9]; $p = 0.012$), with incomplete recovery over the course of seven days (Figure 4D; mean intensity Day 7: 42.4, [33.7, 51.1]).

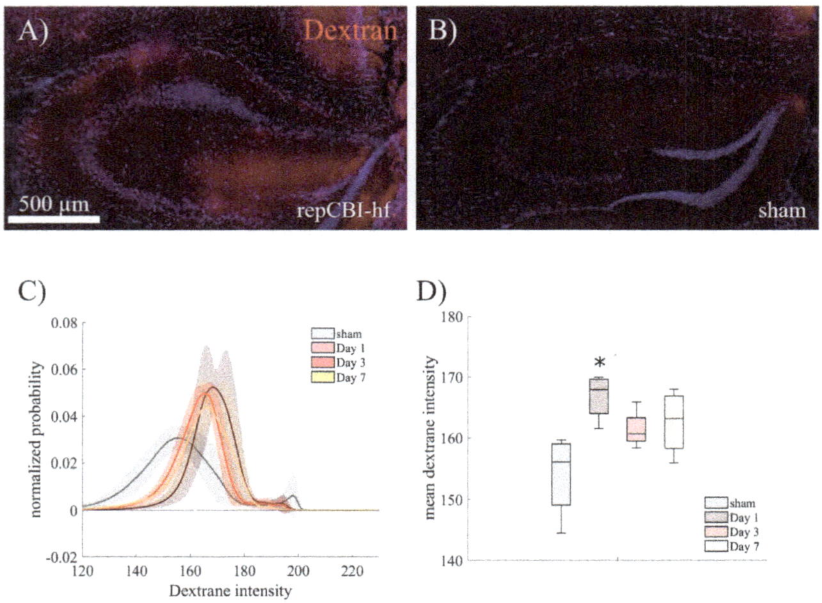

Figure 4. TxRed-labeled dextran (40 kDa) was infused one, three, or seven days after repetitive CBI at high frequency to evaluate the effects of CBI on the temporal dynamics of blood–brain barrier impairment. CBI animals showed a significant dye leakage one day (**A**) after the last concussion compared with sham animals (**B**). This effect was detectable over the first week after CBI (**C**,**D**). (* $p < 0.05$ compared with sham, n = 4).

2.5. CBI-Induced Impairment of the BBB Is Linked to the Behavioral Phenotype

We next characterized the region-specific extent of BBB disruption to evaluate an interrelationship with brain function. Therefore, we used the extravasation of (endogenous) immunoglobulins (IgG) as a surrogate parameter for BBB dysfunction one week after the (last) concussion.

Corroborating the results obtained with dextran (cmp. Figure 4), mice subjected to repetitive CBI (Figure 5A) showed a profound and multilocular impairment of the BBB compared with sham animals (Figure 5B). This effect was most pronounced after repetitive CBI in high frequency, i.e., five consecutive concussions every 48 h (Figure 5C–E). Most interestingly, brain regions implied in the behavioral impairments detailed above (cmp. Figures 2 and 3) were most prominently affected. In particular, we saw a significant BBB disruption in the hippocampus (Figure 5C), in the primary motor cortex (Figure 5D), and in the thalamus (Figure 5E). In particular, levels of IgG extravasation in the hippocampus correlated well with the impairment in spatial cognition as assessed by the Y-maze test ($R^2 = 0.18$, $p = 0.01$, Spearman correlation, Figure 5F).

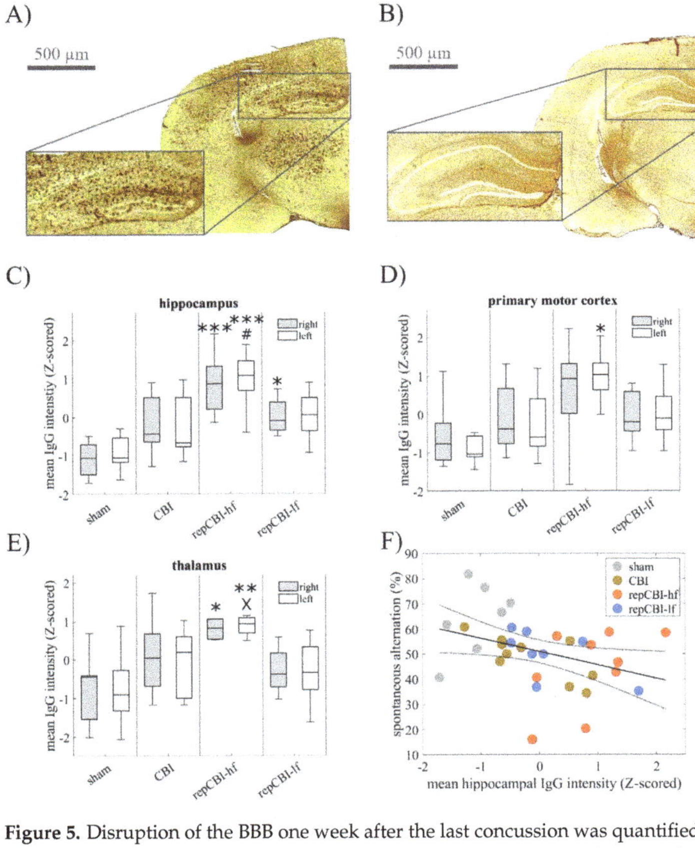

Figure 5. Disruption of the BBB one week after the last concussion was quantified via extravasation of (endogenous) immunoglobulins (IgG) as a surrogate parameter. Compared with sham animals (**A**), mice subjected to repetitive CBI showed a profound and multilocular impairment in BBB function (**B**), especially within the lateral cortex, hippocampus, and thalamus of both hemispheres. This effect was most pronounced after repetitive CBI in high frequency, prominently affecting the brain regions responsible for the behavioral impairment detailed above: hippocampus (**C**), thalamus (**D**), and primary motor cortex (**E**). Consequently, levels of IgG extravasation correlated with impairment in spatial cognition (**F**). (* $p < 0.05$; ** $p < 0.01$; *** $p < 0.001$ compared with sham, # $p < 0.05$ compared with mTBI, X $p < 0.05$ compared with rmTBI-low; n [sham, CBI, rmCBI-hf, rmCBI-lf] = 8, 10, 9, 8).

2.6. Detection of Neuroinflammation by Positron Emission Tomography (PET) Is Distorted by BBB Disruption

In order to evaluate neuroinflammatory processes following CBI, we used positron emission tomography (PET) and the second-generation TSPO tracer [^{18}F]DPA714 to detect and quantify immune cell activation. We indeed observed an increased uptake of [^{18}F]DPA714 (Figure 6A), with significantly elevated standard uptake values (SUVs), especially in the hippocampus after repetitive CBI, one week after the last concussion (Figure 6B).

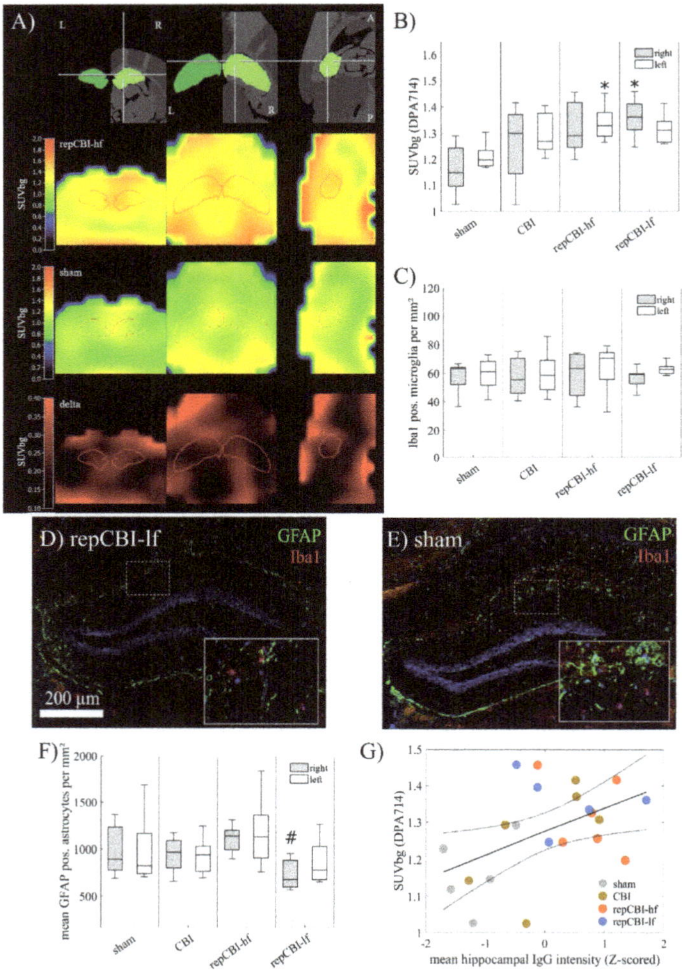

Figure 6. (**A**) Second-generation TSPO tracer [^{18}F]DPA714 showed an increased uptake, especially after repetitive CBI (intensity normalization to background signal, R = right, L = left, A = anterior, P = posterior), with a significant bilateral increase in hippocampal uptake (**B**) one week after (the last) concussion (n = 6). However, at this time, radiotracer accumulation was not explained by elevated signs of activated Iba-1+ microglia (**C**), evaluated after single or repetitive CBI (**D**) compared to the sham group (**E**). Likewise, the amount of GFAP+ reactive astrocytes (**F**) in the hippocampus was not significantly altered compared to the sham group, with a significant difference between the groups of repetitive trauma (n [sham, CBI, rmCBI-hf, rmCBI-lf] = 8, 10, 9, 8). Thus, [^{18}F]DPA714 accumulation was potentially explained by the disturbance of the BBB one week after concussion, with a good correlation between tracer uptake and IgG extravasation in the hippocampus (**G**). (* $p < 0.05$ compared with sham, # $p < 0.05$ compared with repCBI-hf).

However, immunohistological analyses revealed that tracer uptake was not due to an increase in activated Iba-1+ microglia (Figure 6C), as quantified in the hippocampus in either group of mice (Figure 6D,E). While we found a significant group effect of trauma on astrocyte density (H(3,33) = 9.5, $p = 0.02$, Figure 6F), the effect was not due to a difference compared to sham injured animals (99.2 cells/mm^2, [75.2, 123.3]) but a significant decrease in astrocyte density after low-frequency repetitive CBI (repCBI-lf: 73.3 cells/mm^2, [60.2, 86.4]) compared with the high-frequency group (repCBI-hf: 104.5 cells/mm^2, [81.6, 127.5]; $p = 0.01$). On the other hand, [^{18}F]DPA714 radiotracer accumulation correlated well with the extent of BBB disruption, as assessed by IgG extravasation ($R^2 = 0.21$, $p = 0.03$, Spearman correlation, Figure 6G), suggesting unspecific tracer accumulation as a result of increased BBB permeability. However, further analyses of the time–activity curves did not show a difference in the temporal tracer dynamics (Figure S1I), warranting further investigation.

3. Discussion

We here characterize a novel mouse model of CBI, featuring a rotational brain injury without macroscopic structural damage. Our model reveals a characteristic neuropsychiatric phenotype resembling CBI in patients or athletes that correlates to the amount of BBB disruption, thus suggesting a biomarker for disease severity and, potentially, for disease prognosis in the future.

Mice subjected to our novel CBI model display a hypervigilant phenotype, with increased anxiety, a sustained impairment in spatial cognition, and a mild and short-lived disturbance of motor control. This phenotype resembles part of the clinical spectrum in patients with CBI [5,6]. Interestingly, we saw a faster recovery of spatial cognition in animals subjected to weekly concussions, both in comparison with mice suffering from concussions at a higher frequency (every 48 h), even compared to mice subjected to only a single (one-time) concussion. This suggests a pre-conditioning effect of preceding traumata [21], but only when sufficient time for recovery was granted between concussions. This finding has high relevance for sports-associated concussions, as it may impact current return-to-play guidelines. Similar to our findings, in an alternative mouse model of non-rotational TBI induced by a weight drop, faster recovery of cognitive function occurred as assessed by Morris Water Maze, when mice were exposed to TBI at monthly intervals as compared with higher TBI frequencies [22].

As a key finding on the structural level, we here report a robust and multifocal BBB breakdown in the subacute phase after CBI, which was most pronounced after repetitive concussions. BBB integrity plays an important role in maintaining a healthy neural state, while disruption of the BBB constitutes a known driver of disease progression, e.g., in ischemic stroke, multiple sclerosis, or neurodegenerative disease [23]. While widely observed after all forms of TBI [24], the role of BBB impairment in the development of functional/behavioral deficits as well as in the recovery of function remains to be established. Various preclinical models of mild TBI have reported disturbance of the BBB, e.g., in swine [25], rats [26], and mice [27]. Moreover, indirect evidence by means of gadolinium contrast extravasation on MR-imaging was observed in repetitive sports-associated concussions in humans [28].

By means of both the infusion of labeled dextran as well as staining for IgG extravasation, we here describe an ongoing disruption of the BBB even seven days after CBI. On the same note, ongoing impairment of the BBB, detectable hours after trauma [29], was described for at least days [30–32]—in one study, for up to two weeks [33]—after repetitive trauma in various preclinical models. Using laser speckle imaging, Lynch et al. demonstrated impairment in cerebrovascular reactivity several months after repetitive mild TBI [34]. Corresponding to our result on the effect of CBI frequency on the extent of BBB damage (cmp. Figure 5), intraindividual longitudinal non-invasive monitoring of BBB function by contrast-enhanced MRI over the course of three mild head impacts in rats revealed an increasing degree of BBB permeability [35]. While most murine closed-head injury models highlight some kind of TBI-induced BBB alterations [36], Wu et al. described a model of rotational acceleration–deceleration in the sagittal plane, explicitly lacking Evans blue extravasation or IgG leakage from vessels [37]. Although considerable differences in the mechanical parameters used for trauma induction render a comparative interpretation difficult, these divergences suggest a relevance of the direction of head rotation for the type of brain injury induced.

In contrast to the robust BBB breakdown, we did not observe a significant alteration in astrocyte density within the hippocampus compared to sham-injured animals. Of note, we noted a slight difference in astrocyte density across CBI groups. While astrocyte density was reduced in the group of animals receiving CBI at low (weekly) frequency, it was considerably higher in mice subjected to CBI every 48 h. We argue that on top of a potential mechanical disruption of the neurovascular interface, CBI might induce relevant reactive astrocytosis only at high but not at low frequency. A recent study reported an atypical astrocytic response in the weight-drop model of mild TBI in areas of sustained BBB impairment that was detectable for months after the trauma [27]. Otherwise, glial dysfunction with reduced hippocampal GFAP expression levels was previously reported in preclinical depression models of chronic psychosocial stress [38]. Overall, further analyses of the temporal and molecular underpinnings of the effects of CBI on astrocyte function are certainly warranted.

The tightly regulated dynamic BBB serves as a core component of the neurovascular unit that is intimately involved in neuroinflammatory processes, by leaking neurotoxic products as well as hematogeneous pro-inflammatory immune cells into the brain, while, at the same time, being disrupted by neuroinflammatory mediators in a reciprocal fashion [39]. Thus, after TBI, BBB impairment may originate (primarily) from mechanical disruption of capillaries, and then be maintained (secondarily) by persistent neuroinflammation. However, recent evidence from preclinical models of mild TBI, e.g., in rats [26] or swine [25], highlights BBB impairment as a key pathological hallmark even in the absence of neuroinflammation, indicating the existence of complementary mechanisms. Astrocytic activation may be induced by the trauma itself as well as a secondary phenomenon to protein extravasation, change in neurological activity [40], rise of extracellular potassium, and spreading depression [41,42].

Chronic and aberrant neuroinflammation is a hallmark of many neurological disorders without a primary immune-mediated etiology, such as stroke, Parkinson's disease, Alzheimer's disease, and epilepsy, rendering it a promising future therapeutic target [43,44]. In moderate to severe TBI leading to structural brain damage, neuroinflammatory processes have been well characterized, even in the (sterile) setting without injury to the dura mater [45]. However, the presence and time course of neuroinflammation after pure concussion—without macroscopic structural lesions to the brain—remains a matter of constant debate [46]. We and others have previously shown that brain-endogenous immune cells, i.e., microglia, respond to even subtle mechanical cues, e.g., stretch [47] or mere tissue elasticity [48], and thus hypothesized that microglia activation would be seen in CBI. A few studies observing microglia activation after mild TBI/CBI described simultaneous neuronal injury/axonal degeneration, even if detectable only ex vivo [15,49], or did not perform histological evaluation of structural damage [50]. Our novel CBI model did not cause any structural lesion detectable in vivo (MRI) or ex vivo (histologically) and also found no evidence of activated microglia, albeit observing the accumulation of the second-generation TSPO radiotracer [^{18}F]DPA-714 one week after CBI. Ligands to the 18 kDa translocator protein TSPO, an outer membrane molecule highly upregulated in microglia, have been widely used to monitor neuroinflammation in, e.g., stroke [51] or neurodegeneration [52], but other sources of TSPO accumulation, e.g., increased neuronal activity [53], reactive astrocytes [54], or endothelial cells [55], have been debated to contribute to the PET signal. We here observed radiotracer accumulation to coincide with BBB disruption, but a longitudinal study with different time points of PET-imaging (including kinetic modeling) would be required to verify a causal connection. TSPO accumulation due to BBB disruption may overcast more subtle and putatively transient microglia activation. Recently, TSPO has been linked to microglia proliferation and increased density in CNS pathology but not an activated state of microglia state per se [56,57]. Histological analyses of higher temporal granularity will be needed to deeper elucidate the astro- and microglial response to our novel model of CBI.

Altogether, the present findings highlight the potential of blood–brain barrier-derived biomarkers to also assess and monitor the clinical course of traumatic brain injuries in humans. Specifically, we here addressed the diffuse secondary disease progression of acceleration/deceleration injuries that is particularly difficult to capture [58]. Currently, blood–based biomarkers reflecting neuronal (i.e., ubiquitin carboxy-terminal hydrolase L1, UCH-L1) or astrocyte (i.e., glial fibrillary acidic protein, GFAP) have emerged, mostly differentiating more severe forms of TBI, but not (mild) concussion [59]. To this end, further analyses of the detailed microscopic and molecular damage of the neurovascular interface in the presented model might help derive biomarkers specific to concussion/CBI.

Notably, we only examined male mice. Several studies have described a sex-dependent difference in the severity and incidence of concussion in athletes [60]. Likewise, behavioral impairment after repetitive CBI, i.e., cued fear memory, was different in female compared with male mice [61]. The differential effect of sex and the potential influence of hormonal fluctuations need to be addressed in further studies.

Overall, we here introduce a preclinical model of CBI that constitutes a promising tool for translational investigation of concussion and its secondary complications, i.e., prolonged post-concussion syndrome and chronic traumatic encephalopathy. Different from other preclinical studies claiming to induce "mild TBI", our model explicitly evokes a diffuse and multifocal BBB impairment corresponding to a characteristic neuro-psychiatric phenotype but without any evidence of focal cortical damage. Furthermore, our study comprises evidence that neuroimaging, as exemplified here by TSPO-PET, can detect early and disease-specific features that might be a valuable tool in longitudinal studies addressing individual vulnerability to secondary CBI-induced neurodegeneration.

4. Materials and Methods

4.1. Experimental Procedures

All animal procedures followed the German Laws for Animal Protection and were approved by the animal care committee and governmental authorities (Landesamt für Natur, Umwelt und Verbraucherschutz North Rhine-Westphalia, LANUV; AZ 81-02.04.2020.A058). Experiments were performed and reported in accordance with the ARRIVE guidelines. Animals were socially housed under a fixed 12 h light/dark cycle with ad libitum access to food and water. A total of 52 male C56BL/6J mice (either 7 or 11 weeks old, 21–26 g), were used for all experiments. All experimental procedures were performed either under anesthesia with isoflurane (4% for induction, 1.5–2% for maintenance in a 65%/35% nitrous oxide/oxygen atmosphere) supplied via a face mask, or in conscious sedation with medetomidine (0.1 mg/kg), as stated below. Mice received analgesia by tramadol (100 mg/mL) via drinking water, starting two days before and continuing for three days after CBI. Body temperature was measured via a fiber optic rectal probe (SA Instruments) and kept constant at 37 ± 1.0 °C by a water-circulating system (MedRes, Cologne, Germany).

4.2. Concussive Brain Injury (CBI)

For the establishment of the concussion model, a group of 16 mice received either a single concussion or sham surgery. CBI was induced by an adaptation of the controlled cortical injury model [11], but omitting prior opening of the skull. Instead, the closed skull was hit using a 5 mm impactor tip, softened by a custom-made silpuran cover (Figure 1A; ratio 1:1; stiffness 830 kPa). In detail, the anesthetized mice were positioned prone in a custom-made head holder made of polyethylene, to allow free rotational movement but at the same time limiting linear movements of the head (Figure 1B). Using a stereotactic frame, the impactor tip was lowered onto the exposed skull over the right primary motor cortex (AP +0.5 mm, ML 1 mm) in the protracted position, lightly touching the skull. After retraction, the tip was lowered by 3 mm (=penetration depth, dwell time = 0.1 s).

To minimize the impact of anesthesia on cognition, anesthesia was next switched to conscious sedation with the alpha agonist medetomidine at 0.1 mg/kg (Domitor®, Elanco, Greenfield, Indiana, USA, suspended in 250 µL of NaCl), and Isofluran was reduced to 0.5%. After 10 min, concussion was induced (impact velocity 5 mm/s, dwell time 0.1 s), and anesthesia immediately antagonized (Antisedan®, 0.5 mg/kg, Elanco, Greenfield, Indiana, USA, suspended in 250 µL of NaCl). The animals were returned to their home cages in a supine position, and the time until the first effort to rise from this position was measured as a surrogate marker to quantify loss of consciousness (LoC). Sham animals received the same treatment but instead of lowering, the impactor was raised by 3 mm, thus missing the animal on striking. Two animals died during surgery (one sham, one CBI) because of anesthesia-related complications.

To approximate the CBI-induced head rotation, experiments were filmed (GoProHero5, GoPro, San Mateo, CA, USA) at 240 frames per second. Head rotation was quantified by tracking of three well-visible fixpoints (eyes and nose) and calculated using Kinovea software (https://www.kinovea.org (accessed on 1 November 2023); vers. 0.9.4).

4.3. Behavioral Test

As a measure of the overall activity rate, nest-building activity was assessed. For this, animals received nestlet material of pre-defined weight, made of compacted paper (compact crinklets natural, SAFE® Enrichment, Rosenberg, Germany). The unused (still compacted) material was weighed daily, and its reduction (i.e., the material used for nest-building activities), was normalized (%) to the original pellet.

Spontaneous locomotor activity and anxiety behavior were assessed five days after trauma using an open field cage (40 cm × 40 cm × 30 cm) equally lit from above with dimmed light (70–80 lux). Mice were allowed to freely explore the arena for ten minutes. Animal movement was automatically tracked (EthoVisionXt, Noldus, Weganingen, The Netherlands). For analyses, the arena was segmented into an outer "border" zone (Figure 2C, blue) and an inner "center" zone (Figure 2C, green). Analyses included the time spent in the border—resp. the center zone, as well as the overall distance covered. Additionally, we calculated an exploration ratio as time spent in the center zone relative to the border zone, using the formula (timecenter − timeborder)/(timecenter + timeborder).

To evaluate cognitive deficits, spontaneous alternation was measured using a Y-maze five days after CBI. Animals entered the Y-maze in its center, and their free exploration of the three arms was monitored for five minutes. The ratio of correct alternations, consisting of a consecutive visit to all three arms, was computed as a measure of cognition. Moreover, as an analysis of sensitivity, the novel object recognition test was additionally performed in a subgroup of mice. Four days after CBI, animals were familiarized with a first object (plastic bottle filled with blue liquid, Figure S1A right) for ten minutes within an open field arena described above. On the next day, animals were again placed in the arena equipped with the known object as well as a novel object (magic cube, Figure S1A left). The time spent with the novel relative to the known object (recognition ratio) was computed as (timenovel − timeknown)/(timenovel + timeknown).

In order to assess motor control, the rotating beam test was performed 1, 3, and 7 days after CBI. Mice were encouraged to run over a fiberglass beam of 120 cm in length and 13 mm in diameter, located 60 cm above a table covered with bubble cushions and rotating at 6 rpm. Mice were pre-trained three times before CBI, and the last trial was used as the baseline. The total time of traversal and the amount of foot slips were measured.

4.4. Repetitive CBI

To assess the impact of repetitive CBI, two different groups receiving five consecutive concussions were compared (Figure 3A). One group (blue arrows in Figure 3A, "repCBI-lf") received one concussion every week, starting from the age of 7 weeks; the other group (orange arrows in Figure 3A, "repCBI-hf") received one concussion every 48 h, starting at the age of 11 weeks. Both concussion groups were compared to animals receiving four sham surgeries and finally a single concussion (light grey arrow in Figure 3A, "CBI"), as well as to a sham group receiving 5 sham surgeries without any concussion (dark grey arrow in Figure 3A, "sham").

Sample sizes were calculated a priori, based on the results of the prior exploratory trials, to yield 80% power to detect a significant cognitive impairment (Kruskal–Wallis Test) with a p-value < 0.05.

4.5. Structural Magnetic Resonance Imaging

To exclude gross structural damage, structural magnetic resonance imaging (MRI) was performed one day after the last concussion (black arrow in Figure 3A), using a small-animal 9.4 T horizontal MRI system (BioSpec; Bruker BioSpin, Ettlingen, Germany) with a 20 cm bore diameter and actively shielded gradient coils (BGA12S2, 600 mT/m; Bruker BioSpin). Radio frequency (RF) excitation and signal reception were performed with a 1H quadrature cryo-genic surface coil (CryoProbe, Bruker BioSpin). Image acquisition was accomplished using ParaVision 6.0.1 software (Bruker BioSpin). A T2-weighted spin echo (turbo-RARE, 24 slices, field of view (FOV) of 17.5 $\times$ 17.5 mm^2, RARE factor = 8, repetition time (TR) = 5500 ms, echo time (TE) = 32.5 ms) sequence was performed and visually inspected.

4.6. PET Acquisition and Image Analysis

Positron emission tomography (PET) using [^{18}F]DPA-714, a second-generation TSPO ligand, was conducted in six animals per group, six days after the last concussion (red arrow in Figure 3A). The tracer was synthesized as described previously [62]. Prior to tracer application, animals were anesthetized with isoflurane, and a catheter for tracer injection was inserted into the lateral tail vein. Mice were placed in a custom-made animal holder (medres® GmbH, Cologne, Germany) with the head fixed in a tooth bar. Scans were conducted in list mode using a Focus 220 micro-PET scanner (CTI-Siemens, Erlangen, Germany; resolution at the center of field of view: 1.4 mm). Data acquisition over 30 min started immediately after intravenous tracer injection (activity: 8–12 MBq in 125 μL). A 10-min transmission scan using a 57Co point source was conducted for attenuation correction. After imaging, the catheters were removed, and the mice were returned to their home cages.

After full 3D rebinning, summed images were reconstructed (voxel size: 0.47 $\times$ 0.47 $\times$ 0.80 mm). VINCI 4.72 for Windows (Max Planck Institute for Metabolism Research, Cologne, Germany) was used for further analyses. After co-registration, images were intensity-normalized to a background signal. To this end, an elliptical volume of interest (VOI) of 4 mm^3 was placed inside a background region within the midbrain of each individual image. Each image was divided by the mean value of the background VOI, resulting in the "background-standardized uptake value ratio" (SUVbg). Mean intensity values within a hippocampus mask, separated for each hemisphere, were used for further analyses. Additionally, time–activity curves were compared between groups (Figure S1D).

4.7. Histology and Immunofluorescence

Seven days after the last concussion, animals were deeply anesthetized and systemically perfused with 20 mL ice-cold phosphate-buffered saline (PBS, Gibco™ PBS, Thermo Fisher Scientific, Darmstadt, Germany) and subsequent endovascular fixation with 20 mL paraformaldehyde (PFA, ROTI®Histofix 4%, Carl Roth GmbH + Co. KG, Karlsruhe, Germany) infused constantly at 9 mL/min. Brains were immediately shock-frozen in methylbutane at −60 °C. All staining techniques were performed on coronal brain slices of 10 μm thickness.

For assessment of structural damage, hematoxylin and eosin (H&E) staining was performed. Slides were stained for 20 min with Mayer's hemalaum solution (Mayers

Hämalaunlösung 109249, Merck®, Rahway, NJ, USA), differentiated in ethanol (70%) before staining with eosin (Eosin G-Lösung 0.5%, Carl Roth®, Karlsruhe, Germany) for 3 min, and subsequently washed with distilled water. Afterward, the slices were dipped for 2 min each in an ascending alcohol series and coverslipped with Entellan.

Immunoglobulin G (IgG) extravasation was examined by staining for mouse IgG. After antigen retrieval using 0.3% H_2O_2, slices were incubated with 1% normal horse serum (NHS) for one hour to prevent unspecific antibody binding. Biotinylated anti-mouse IgG (BA-2000, Vector Laboratories, Newark, NJ, USA) antibody was applied (1:500 in PBS) for 1.5 h. Antibody binding was visualized by an avidin/biotin-based peroxidase system (Vectastain ABC kit, Vector Laboratories) and the chromogen diaminobenzidine (DAB). One experimental group of eight mice was stained at the same time by an investigator blinded to group allocation. From the region of interest, mean color intensities were assessed using MATLAB (2020b, MathWorks, Natick, MA, USA). Because of the different coloration across trials, Z-transformation was performed.

To visualize microglia and astrocytes, staining for Iba-1 and GFAP were performed, respectively. Additional post-fixation in PFA for 30 min was performed before antigen retrieval by incubation in citrate buffer (pH = 6) at 80 °C for 20 min. To prevent unspecific antibody binding, slices were incubated with 5% donkey serum and 0.25% Triton X (Tx) in PBS for one hour. Primary antibodies (Iba1, rabbit monoclonal antibody, 1:200, ab178847, Abcam, Cambridge, U.K.; GFAP, mouse monoclonal antibody, 1:200, MAB360, Sigma) were incubated in 1% donkey serum, 0.25% Tx, and PBS overnight before incubation with a secondary antibody (Alexa Fluor donkey anti-rabbit IgG or Alexa Fluor donkey anti-mouse, 1:500, Invitrogen, Karlsruhe, Germany) diluted in 1% NDS and 0.25% Tx in PBS. Slices were counterstained with Hoechst (1:500) for 10 min. Images of the hippocampus were acquired using an inverted fluorescence microscope (BZ 9000, Keyence, Osaka, Japan), with a 40× magnification, and Iba-1 and GFAP-positive microglia were counted manually by a blinded investigator.

4.8. Dextran Extravasation

To detect blood–brain barrier disruption at various timepoints, mice receiving five repetitive CBIs ("repCBI-hf") were sacrificed either one, three, or seven days after the last trauma compared to a group of mice receiving sham surgeries. Under low-light conditions, mice were perfused with Texas-red labeled dextran (40 kDa, D1829, Invitrogen, Waltham, MA, USA) solved in saline containing heparin [63]. Brains were extracted and shock-frozen in methylbuthane. Using a cryostat, 30 µm thick coronal slices across the hippocampus were cut. In addition to counterstaining with Hoechst as described above, capillaries were stained using Lectin (Lycopersicon Esculentum (Tomato) Lectin, DyLight 488, 1:400, Invitrogen L32470). Sections were imaged using a confocal microscope (TCS SP8, Leica, Wetzlar, Germany).

A region of interest (ROI) outlining the hippocampus was manually drawn. Using a custom script written using MATLAB (2020b, MathWorks, Massachusetts), a mask was generated based on the Lectin-positive capillaries (Figure S1G) to mask out intravascular dextran signal (Figure S1H). The remaining images were used to compute histograms of intensity values across the red color channel (RGB brightness intensities between 0 and 255) and its mean intensity values were compared across groups.

4.9. Statistical Analyses

Data normality was assessed by visualization of histograms and analysis of the Kolmogorov–Smirnov test. For analyses of differences across groups, a One-way ANOVA or Kruskal–Wallis test was used in the case of normally or not normally distributed data, respectively. For analysis of repetitively measured behavioral data, a mixed ANOVA with the kind of trauma applied as a between-group factor and time as a within-group factor was modeled. Spearman correlation was used to analyze associations between features. A p-value of less than 0.05 was considered significant.

Supplementary Materials: The supporting information can be downloaded at: https://www.mdpi.com/article/10.3390/ijms25094837/s1.

Author Contributions: Conceptualization, S.J.B., N.M. and M.A.R.; data curation, S.J.B., D.W. and M.A.R.; formal analysis, S.J.B., H.E. and A.J.; funding acquisition, M.S. and G.R.F.; investigation, S.J.B., N.R., H.E. and S.V.; methodology, S.J.B., N.R., H.E., J.K. and D.W.; project administration, S.J.B. and M.A.R.; resources, H.E., D.W., M.S. and R.M.; software, S.J.B.; supervision, M.S., B.H., R.M., N.M., G.R.F. and M.A.R.; validation, H.E., S.V. and D.W.; visualization, S.J.B., N.R., H.E. and A.J.; writing—original draft, S.J.B.; writing—review and editing, H.E., M.S., B.H., R.M., N.M., G.R.F. and M.A.R. All authors discussed the results and contributed to this study. All authors have read and agreed to the published version of the manuscript.

Funding: This research was funded by the Deutsche Forschungsgemeinschaft (DFG, German Research Foundation): project ID 431549029–SFB 1451 X01, and the Marga-und-Walter-Boll-Stiftung (#210-10-15). SJB was supported by the "Cologne Clinician Scientist Program" (CCSP), funded by the DFG (FI 773/15-1).

Institutional Review Board Statement: The animal study protocol followed the German Laws for Animal Protection and was approved by the animal care committee and governmental authorities (Landesamt für Natur, Umwelt und Verbraucherschutz North Rhine-Westphalia, LANUV; AZ 81-02.04.2020.A058).

Informed Consent Statement: Not applicable.

Data Availability Statement: As part of an ongoing investigation, all data that support the findings of this study and all custom-written MATLAB codes are available from the corresponding author upon reasonable request.

Acknowledgments: We thank Simon Schlömer, Martin Lieb, and Christina Balla for their valuable technical support.

Conflicts of Interest: The authors declare no conflicts of interest.

Abbreviations

BBB	blood-brain barrier
CBI	concussive brain injury
GFAP	glial fibrillary acidic protein
IgG	immunoglobulin G
LoC	loss of consciousness
MRI	magnetic resonance imaging
PET	positron emission tomography
repCBI-hf	repetitive CBI—high frequency
repCBI-lf	repetitive CBI—low frequency
SUV	standard uptake values
TBI	traumatic brain injury
TSPO	translocator protein
VOI	volume of interest

References

1. Dewan, M.C.; Rattani, A.; Gupta, S.; Baticulon, R.E.; Hung, Y.C.; Punchak, M.; Agrawal, A.; Adeleye, A.O.; Shrime, M.G.; Rubiano, A.M.; et al. Estimating the global incidence of traumatic brain injury. *J. Neurosurg.* **2018**, *130*, 1080–1097. [CrossRef]
2. Maas, A.I.R.; Menon, D.K.; Manley, G.T.; Abrams, M.; Åkerlund, C.; Andelic, N.; Aries, M.; Bashford, T.; Bell, M.J.; Bodien, Y.G.; et al. Traumatic brain injury: Progress and challenges in prevention, clinical care, and research. *Lancet Neurol.* **2022**, *21*, 1004–1060. [CrossRef]
3. McCrory, P.; Meeuwisse, W.; Dvořák, J.; Aubry, M.; Bailes, J.; Broglio, S.; Cantu, R.C.; Cassidy, D.; Echemendia, R.J.; Castellani, R.J.; et al. Consensus statement on concussion in sport-the 5(th) international conference on concussion in sport held in Berlin, October 2016. *Br. J. Sports Med.* **2017**, *51*, 838–847.
4. Langer, L.; Levy, C.; Bayley, M. Increasing Incidence of Concussion: True Epidemic or Better Recognition? *J. Head Trauma Rehabil.* **2020**, *35*, E60–E66. [CrossRef] [PubMed]
5. McInnes, K.; Friesen, C.L.; MacKenzie, D.E.; Westwood, D.A.; Boe, S.G. Mild Traumatic Brain Injury (mTBI) and chronic cognitive impairment: A scoping review. *PLoS ONE* **2017**, *12*, e0174847. [CrossRef] [PubMed]

6. Nelson, L.D.; Temkin, N.R.; Dikmen, S.; Barber, J.; Giacino, J.T.; Yuh, E.; Levin, H.S.; McCrea, M.A.; Stein, M.B.; Mukherjee, P.; et al. Recovery After Mild Traumatic Brain Injury in Patients Presenting to US Level I Trauma Centers: A Transforming Research and Clinical Knowledge in Traumatic Brain Injury (TRACK-TBI) Study. *JAMA Neurol.* **2019**, *76*, 1049–1059. [CrossRef] [PubMed]
7. Barnes, D.E.; Byers, A.L.; Gardner, R.C.; Seal, K.H.; Boscardin, W.J.; Yaffe, K. Association of Mild Traumatic Brain Injury With and Without Loss of Consciousness With Dementia in US Military Veterans. *JAMA Neurol.* **2018**, *75*, 1055–1061. [CrossRef]
8. Mackay, D.F.; Russell, E.R.; Stewart, K.; MacLean, J.A.; Pell, J.P.; Stewart, W. Neurodegenerative Disease Mortality among Former Professional Soccer Players. *N. Engl. J. Med.* **2019**, *381*, 1801–1808. [CrossRef] [PubMed]
9. Bigler, E.D. Neuropsychology and clinical neuroscience of persistent post-concussive syndrome. *J. Int. Neuropsychol. Soc.* **2008**, *14*, 1–22. [CrossRef]
10. Silver, J.M. Effort, exaggeration and malingering after concussion. *J. Neurol. Neurosurg. Psychiatry* **2012**, *83*, 836–841. [CrossRef]
11. Bodnar, C.N.; Roberts, K.N.; Higgins, E.K.; Bachstetter, A.D. A Systematic Review of Closed Head Injury Models of Mild Traumatic Brain Injury in Mice and Rats. *J. Neurotrauma* **2019**, *36*, 1683–1706. [CrossRef] [PubMed]
12. McKee, A.C.; Daneshvar, D.H. The neuropathology of traumatic brain injury. *Handb. Clin. Neurol.* **2015**, *127*, 45–66. [PubMed]
13. Petraglia, A.L.; Plog, B.A.; Dayawansa, S.; Dashnaw, M.L.; Czerniecka, K.; Walker, C.T.; Chen, M.; Hyrien, O.; Iliff, J.J.; Deane, R.; et al. The pathophysiology underlying repetitive mild traumatic brain injury in a novel mouse model of chronic traumatic encephalopathy. *Surg. Neurol. Int.* **2014**, *5*, 184. [CrossRef] [PubMed]
14. Khuman, J.; Meehan, W.P., 3rd; Zhu, X.; Qiu, J.; Hoffmann, U.; Zhang, J.; Giovannone, E.; Lo, E.H.; Whalen, M.J. Tumor necrosis factor alpha and Fas receptor contribute to cognitive deficits independent of cell death after concussive traumatic brain injury in mice. *J. Cereb. Blood Flow Metab. Off. J. Int. Soc. Cereb. Blood Flow Metab.* **2011**, *31*, 778–789. [CrossRef] [PubMed]
15. Tagge, C.A.; Fisher, A.M.; Minaeva, O.V.; Gaudreau-Balderrama, A.; Moncaster, J.A.; Zhang, X.L.; Wojnarowicz, M.W.; Casey, N.; Lu, H.; Kokiko-Cochran, O.N.; et al. Concussion, microvascular injury, and early tauopathy in young athletes after impact head injury and an impact concussion mouse model. *Brain A J. Neurol.* **2018**, *141*, 422–458. [CrossRef] [PubMed]
16. Parker, E.; Aboghazleh, R.; Mumby, G.; Veksler, R.; Ofer, J.; Newton, J.; Smith, R.; Kamintsky, L.; Jones, C.M.A.; O'Keeffe, E.; et al. Concussion susceptibility is mediated by spreading depolarization-induced neurovascular dysfunction. *Brain A J. Neurol.* **2022**, *145*, 2049–2063. [CrossRef] [PubMed]
17. Petraglia, A.L.; Plog, B.A.; Dayawansa, S.; Chen, M.; Dashnaw, M.L.; Czerniecka, K.; Walker, C.T.; Viterise, T.; Hyrien, O.; Iliff, J.J.; et al. The spectrum of neurobehavioral sequelae after repetitive mild traumatic brain injury: A novel mouse model of chronic traumatic encephalopathy. *J. Neurotrauma* **2014**, *31*, 1211–1224. [CrossRef] [PubMed]
18. Statler, K.D.; Alexander, H.; Vagni, V.; Holubkov, R.; Dixon, C.E.; Clark, R.S.; Jenkins, L.; Kochanek, P.M. Isoflurane exerts neuroprotective actions at or near the time of severe traumatic brain injury. *Brain Res.* **2006**, *1076*, 216–224. [CrossRef] [PubMed]
19. Bukhari, Q.; Schroeter, A.; Cole, D.M.; Rudin, M. Resting State fMRI in Mice Reveals Anesthesia Specific Signatures of Brain Functional Networks and Their Interactions. *Front. Neural Circuits* **2017**, *11*, 5. [CrossRef]
20. Chen, Y.; Mao, H.; Yang, K.H.; Abel, T.; Meaney, D.F. A modified controlled cortical impact technique to model mild traumatic brain injury mechanics in mice. *Front. Neurol.* **2014**, *5*, 100. [CrossRef]
21. Yokobori, S.; Mazzeo, A.T.; Hosein, K.; Gajavelli, S.; Dietrich, W.D.; Bullock, M.R. Preconditioning for traumatic brain injury. *Transl. Stroke Res.* **2013**, *4*, 25–39. [CrossRef] [PubMed]
22. Meehan, W.P., 3rd; Zhang, J.; Mannix, R.; Whalen, M.J. Increasing recovery time between injuries improves cognitive outcome after repetitive mild concussive brain injuries in mice. *Neurosurgery* **2012**, *71*, 885–891. [CrossRef] [PubMed]
23. Xiao, M.; Xiao, Z.J.; Yang, B.; Lan, Z.; Fang, F. Blood-Brain Barrier: More Contributor to Disruption of Central Nervous System Homeostasis Than Victim in Neurological Disorders. *Front. Neurosci.* **2020**, *14*, 764. [CrossRef]
24. Alluri, H.; Wiggins-Dohlvik, K.; Davis, M.L.; Huang, J.H.; Tharakan, B. Blood-brain barrier dysfunction following traumatic brain injury. *Metab. Brain Dis.* **2015**, *30*, 1093–1104. [CrossRef] [PubMed]
25. Johnson, V.E.; Weber, M.T.; Xiao, R.; Cullen, D.K.; Meaney, D.F.; Stewart, W.; Smith, D.H. Mechanical disruption of the blood-brain barrier following experimental concussion. *Acta Neuropathol.* **2018**, *135*, 711–726. [CrossRef]
26. Hiles-Murison, B.; Lavender, A.P.; Hackett, M.J.; Armstrong, J.J.; Nesbit, M.; Rawlings, S.; McGonigle, T.; Warnock, A.; Lam, V.; Mamo, J.C.L.; et al. Blood-brain barrier disruption and ventricular enlargement are the earliest neuropathological changes in rats with repeated sub-concussive impacts over 2 weeks. *Sci. Rep.* **2021**, *11*, 9261. [CrossRef] [PubMed]
27. George, K.K.; Heithoff, B.P.; Shandra, O.; Robel, S. Mild Traumatic Brain Injury/Concussion Initiates an Atypical Astrocyte Response Caused by Blood-Brain Barrier Dysfunction. *J. Neurotrauma* **2022**, *39*, 211–226. [CrossRef]
28. O'Keeffe, E.; Kelly, E.; Liu, Y.; Giordano, C.; Wallace, E.; Hynes, M.; Tiernan, S.; Meagher, A.; Greene, C.; Hughes, S.; et al. Dynamic Blood-Brain Barrier Regulation in Mild Traumatic Brain Injury. *J. Neurotrauma* **2020**, *37*, 347–356. [CrossRef] [PubMed]
29. Yang, S.H.; Gustafson, J.; Gangidine, M.; Stepien, D.; Schuster, R.; Pritts, T.A.; Goodman, M.D.; Remick, D.G.; Lentsch, A.B. A murine model of mild traumatic brain injury exhibiting cognitive and motor deficits. *J. Surg. Res.* **2013**, *184*. 981–988. [CrossRef]
30. Logsdon, A.F.; Meabon, J.S.; Cline, M.M.; Bullock, K.M.; Raskind, M.A.; Peskind, E.R.; Banks, W.A.; Cook, D.G. Blast exposure elicits blood-brain barrier disruption and repair mediated by tight junction integrity and nitric oxide dependent processes. *Sci. Rep.* **2018**, *8*, 11344. [CrossRef]
31. Ichkova, A.; Rodriguez-Grande, B.; Zub, E.; Saudi, A.; Fournier, M.L.; Aussudre, J.; Sicard, P.; Obenaus, A.; Marchi, N.; Badaut, J. Early cerebrovascular and long-term neurological modifications ensue following juvenile mild traumatic brain injury in male mice. *Neurobiol. Dis.* **2020**, *141*, 104952. [CrossRef] [PubMed]

32. Yates, N.J.; Lydiard, S.; Fehily, B.; Weir, G.; Chin, A.; Bartlett, C.A.; Alderson, J.; Fitzgerald, M. Repeated mild traumatic brain injury in female rats increases lipid peroxidation in neurons. *Exp. Brain Res.* **2017**, *235*, 2133–2149. [CrossRef] [PubMed]
33. Adams, C.; Bazzigaluppi, P.; Beckett, T.L.; Bishay, J.; Weisspapir, I.; Dorr, A.; Mester, J.R.; Steinman, J.; Hirschler, L.; Warnking, J.M.; et al. Neurogliovascular dysfunction in a model of repeated traumatic brain injury. *Theranostics* **2018**, *8*, 4824–4836. [CrossRef] [PubMed]
34. Lynch, C.E.; Eisenbaum, M.; Algamal, M.; Balbi, M.; Ferguson, S.; Mouzon, B.; Saltiel, N.; Ojo, J.; Diaz-Arrastia, R.; Mullan, M.; et al. Impairment of cerebrovascular reactivity in response to hypercapnic challenge in a mouse model of repetitive mild traumatic brain injury. *J. Cereb. Blood Flow Metab. Off. J. Int. Soc. Cereb. Blood Flow Metab.* **2021**, *41*, 1362–1378. [CrossRef] [PubMed]
35. Leaston, J.; Qiao, J.; Harding, I.C.; Kulkarni, P.; Gharagouzloo, C.; Ebong, E.; Ferris, C.F. Quantitative Imaging of Blood-Brain Barrier Permeability Following Repetitive Mild Head Impacts. *Front. Neurol.* **2021**, *12*, 729464. [CrossRef] [PubMed]
36. Wu, Y.; Wu, H.; Guo, X.; Pluimer, B.; Zhao, Z. Blood-Brain Barrier Dysfunction in Mild Traumatic Brain Injury: Evidence From Preclinical Murine Models. *Front. Physiol.* **2020**, *11*, 1030. [CrossRef] [PubMed]
37. Wu, L.; Chung, J.Y.; Saith, S.; Tozzi, L.; Buckley, E.M.; Sanders, B.; Franceschini, M.A.; Lule, S.; Izzy, S.; Lok, J.; et al. Repetitive head injury in adolescent mice: A role for vascular inflammation. *J. Cereb. Blood Flow Metab. Off. J. Int. Soc. Cereb. Blood Flow Metab.* **2019**, *39*, 2196–2209. [CrossRef] [PubMed]
38. Araya-Callís, C.; Hiemke, C.; Abumaria, N.; Flugge, G. Chronic psychosocial stress and citalopram modulate the expression of the glial proteins GFAP and NDRG2 in the hippocampus. *Psychopharmacology* **2012**, *224*, 209–222. [CrossRef]
39. Takata, F.; Nakagawa, S.; Matsumoto, J.; Dohgu, S. Blood-Brain Barrier Dysfunction Amplifies the Development of Neuroinflammation: Understanding of Cellular Events in Brain Microvascular Endothelial Cells for Prevention and Treatment of BBB Dysfunction. *Front. Cell. Neurosci.* **2021**, *15*, 661838. [CrossRef]
40. Burda, J.E.; Bernstein, A.M.; Sofroniew, M.V. Astrocyte roles in traumatic brain injury. *Exp. Neurol.* **2016**, *275 Pt 3*, 305–315. [CrossRef]
41. Soldozy, S.; Sharifi, K.A.; Desai, B.; Giraldo, D.; Yeghyayan, M.; Liu, L.; Norat, P.; Sokolowski, J.D.; Yağmurlu, K.; Park, M.S.; et al. Cortical Spreading Depression in the Setting of Traumatic Brain Injury. *World Neurosurg.* **2020**, *134*, 50–57. [CrossRef] [PubMed]
42. Schroeter, M.; Schiene, K.; Kraemer, M.; Hagemann, G.; Weigel, H.; Eysel, U.T.; Witte, O.W.; Stoll, G. Astroglial responses in photochemically induced focal ischemia of the rat cortex. *Exp. Brain Res.* **1995**, *106*, 1–6. [CrossRef] [PubMed]
43. Shichita, T.; Ooboshi, H.; Yoshimura, A. Neuroimmune mechanisms and therapies mediating post-ischaemic brain injury and repair. *Nat. Rev. Neurosci.* **2023**, *24*, 299–312. [CrossRef] [PubMed]
44. Mishra, A.; Bandopadhyay, R.; Singh, P.K.; Mishra, P.S.; Sharma, N.; Khurana, N. Neuroinflammation in neurological disorders: Pharmacotherapeutic targets from bench to bedside. *Metab. Brain Dis.* **2021**, *36*, 1591–1626. [CrossRef] [PubMed]
45. Wofford, K.L.; Loane, D.J.; Cullen, D.K. Acute drivers of neuroinflammation in traumatic brain injury. *Neural Regen. Res.* **2019**, *14*, 1481–1489. [PubMed]
46. Gard, A.; Vedung, F.; Piehl, F.; Khademi, M.; Wernersson, M.P.; Rorsman, I.; Tegner, Y.; Pessah-Rasmussen, H.; Ruscher, K.; Marklund, N. Cerebrospinal fluid levels of neuroinflammatory biomarkers are increased in athletes with persistent post-concussive symptoms following sports-related concussion. *J. Neuroinflammation* **2023**, *20*, 189. [CrossRef] [PubMed]
47. Shaughness, M.; Byrnes, K. Assessment of the Effects of Stretch-Injury on Primary Rat Microglia. *Mol. Neurobiol.* **2021**, *58*, 3545–3560. [CrossRef] [PubMed]
48. Blaschke, S.J.; Demir, S.; König, A.; Abraham, J.A.; Vay, S.U.; Rabenstein, M.; Olschewski, D.N.; Hoffmann, C.; Hoffmann, M.; Hersch, N.; et al. Substrate Elasticity Exerts Functional Effects on Primary Microglia. *Front. Cell. Neurosci.* **2020**, *14*, 590500. [CrossRef] [PubMed]
49. Israel, I.; Ohsiek, A.; Al-Momani, E.; Albert-Weissenberger, C.; Stetter, C.; Mencl, S.; Buck, A.K.; Kleinschnitz, C.; Samnick, S.; Sirén, A.-L. Combined [^{18}F]DPA-714 micro-positron emission tomography and autoradiography imaging of microglia activation after closed head injury in mice. *J. Neuroinflammation* **2016**, *13*, 140. [CrossRef]
50. Drieu, A.; Lanquetin, A.; Prunotto, P.; Gulhan, Z.; Pédron, S.; Vegliante, G.; Tolomeo, D.; Serrière, S.; Vercouillie, J.; Galineau, L.; et al. Persistent neuroinflammation and behavioural deficits after single mild traumatic brain injury. *J. Cereb. Blood Flow Metab.* **2022**, *42*, 2216–2229. [CrossRef]
51. Orhii, P.D.; Haque, M.E.; Fujita, M.; Selvaraj, S. Advances in PET imaging of ischemic stroke. *Front. Stroke* **2023**, *1*, 1093386. [CrossRef]
52. Tournier, B.B.; Tsartsalis, S.; Ceyzériat, K.; Garibotto, V.; Millet, P. In Vivo TSPO Signal and Neuroinflammation in Alzheimer's Disease. *Cells* **2020**, *9*, 1941. [CrossRef] [PubMed]
53. Notter, T.; Schalbetter, S.M.; Clifton, N.E.; Mattei, D.; Richetto, J.; Thomas, K.; Meyer, U.; Hall, J. Neuronal activity increases translocator protein (TSPO) levels. *Mol. Psychiatry* **2020**, *26*, 2025–2037. [CrossRef] [PubMed]
54. Sonia, L.; Martine, G.; Anne-Sophie, H.; Fanny, P.; Marion, D.; Nadja Van, C.; Lucile Ben, H.; Vincent, L.; Philippe, R.; Frédéric, D.; et al. Reactive Astrocytes Overexpress TSPO and Are Detected by TSPO Positron Emission Tomography Imaging. *J. Neurosci.* **2012**, *32*, 10809.
55. Guilarte, T.R.; Rodichkin, A.N.; McGlothan, J.L.; Acanda De La Rocha, A.M.; Azzam, D.J. Imaging neuroinflammation with TSPO: A new perspective on the cellular sources and subcellular localization. *Pharmacol. Ther.* **2022**, *234*, 108048. [CrossRef] [PubMed]

56. Nutma, E.; Gebro, E.; Marzin, M.C.; van der Valk, P.; Matthews, P.M.; Owen, D.R.; Amor, S. Activated microglia do not increase 18 kDa translocator protein (TSPO) expression in the multiple sclerosis brain. *Glia* **2021**, *69*, 2447–2458. [CrossRef] [PubMed]
57. Wood, H. TSPO levels in multiple sclerosis lesions reflect microglial density rather than activation state. *Nat. Reviews. Neurol.* **2021**, *17*, 462. [CrossRef]
58. de Macedo Filho, L.; Figueredo, L.F.; Villegas-Gomez, G.A.; Arthur, M.; Pedraza-Ciro, M.C.; Martins, H.; Kanawati Neto, J.; Hawryluk, G.J.; Amorim, R.L.O. Pathophysiology-Based Management of Secondary Injuries and Insults in TBI. *Biomedicines* **2024**, *12*, 520. [CrossRef]
59. Bertozzi, G.; Maglietta, F.; Sessa, F.; Scoto, E.; Cipolloni, L.; Di Mizio, G.; Salerno, M.; Pomara, C. Traumatic Brain Injury: A Forensic Approach: A Literature Review. *Curr. Neuropharmacol.* **2020**, *18*, 538–550. [CrossRef]
60. Hannah, T.C.; Li, A.Y.; Spiera, Z.; Kuohn, L.; Dai, J.; McAuley, F.; Ali, M.; Durbin, J.R.; Dreher, N.; Marayati, N.F.; et al. Sex-Related Differences in the Incidence, Severity, and Recovery of Concussion in Adolescent Student-Athletes Between 2009 and 2019. *Am. J. Sports Med.* **2021**, *49*, 1929–1937. [CrossRef]
61. Tucker, L.B.; Winston, B.S.; Liu, J.; Velosky, A.G.; Fu, A.H.; Grillakis, A.A.; McCabe, J.T. Sex differences in cued fear responses and parvalbumin cell density in the hippocampus following repetitive concussive brain injuries in C57BL/6J mice. *PLoS ONE* **2019**, *14*, e0222153. [CrossRef] [PubMed]
62. Endepols, H.; Anglada-Huguet, M.; Mandelkow, E.; Schmidt, Y.; Krapf, P.; Zlatopolskiy, B.D.; Neumaier, B.; Mandelkow, E.-M.; Drzezga, A. Assessment of the In Vivo Relationship Between Cerebral Hypometabolism, Tau Deposition, TSPO Expression, and Synaptic Density in a Tauopathy Mouse Model: A Multi-tracer PET Study. *Mol. Neurobiol.* **2022**, *59*, 3402–3413. [CrossRef] [PubMed]
63. Natarajan, R.; Northrop, N.; Yamamoto, B. Fluorescein Isothiocyanate (FITC)-Dextran Extravasation as a Measure of Blood-Brain Barrier Permeability. *Curr. Protoc. Neurosci.* **2017**, *79*, 9.58.1–9.58.15. [CrossRef] [PubMed]

International Journal of
Molecular Sciences

MDPI

Review

Assessing Mild Traumatic Brain Injury-Associated Blood–Brain Barrier (BBB) Damage and Restoration Using Late-Phase Perfusion Analysis by 3D ASL MRI: Implications for Predicting Progressive Brain Injury in a Focused Review

Charles R. Joseph

Department of Neurology and Internal Medicine, College of Osteopathic Medicine, Liberty University, Lynchburg, VA 24502, USA; crjoseph@liberty.edu

Abstract: Mild traumatic brain injury (mTBI) is a common occurrence around the world, associated with a variety of blunt force and torsion injuries affecting all age groups. Most never reach medical attention, and the identification of acute injury and later clearance to return to usual activities is relegated to clinical evaluation—particularly in sports injuries. Advanced structural imaging is rarely performed due to the usual absence of associated acute anatomic/hemorrhagic changes. This review targets physiologic imaging techniques available to identify subtle blood–brain barrier dysfunction and white matter tract shear injury and their association with chronic traumatic encephalopathy. These techniques provide needed objective measures to assure recovery from injury in those patients with persistent cognitive/emotional symptoms and in the face of repetitive mTBI.

Keywords: mild traumatic brain injury; chronic traumatic encephalopathy; blood–brain barrier dysfunction; shear white matter injury; arterial spin labeling MRI; diffusion tensor imaging MRI; 18-fluorodeoxyglucose positron emission tomography; dynamic contrast-enhanced (DCE) MR perfusion

1. Introduction

Mild traumatic brain injury (mTBI) is a pervasive problem worldwide and typically managed without hospitalization thus unreported. The estimated prevalence rate of 369 per 100,000 individuals underestimates the scope of the problem as a result [1–3]. Most head injuries are mild (90%) defined by clinical functionality scales such as the Glasgow coma Scale where a score of 13–15/15 is defined as mild [4]. The scale includes basic tests of the level of consciousness as defined by verbal, eye opening, and motor responses to commands or noxious stimuli. It is an invaluable tool for assessing a TBI and for follow-up of moderate or severe TBI. More subtle brain injury, however, is not captured by this scale, as it is meant as a simple stratification tool for the rapid assessment and prognosis of more serious head trauma [4].

The assessment of mTBI is currently the sole purview of clinical assessment to categorize acute cognitive or other subtle neurologic deficit, along with behavioral/emotional effects [5]. Likewise, return to normal learning or sport activities are based on these measures [6]. Since anatomic imaging studies in mTBI are almost invariably normal, they are often not employed [7–10]. However, developing a non-invasive physiologic measure of acute injury and subsequent recovery would provide a critical objective assessment of a safe return to sport/usual activities.

1.1. Background

1.1.1. Minor Closed Head Injury

Mild traumatic brain injury (mTBI) is defined as a traumatic brain injury resulting in brief if any loss of consciousness with associated transient cognitive (thinking, memory, general confusion), neurologic (vertigo, delayed reaction times), and/or behavioral (depression, aggression) dysfunctions, and often headache [11]. The most frequent causes include contact sport, blast, and acceleration–deceleration injuries. Sports injuries are acutely assessed on the field acutely or within 72 h using validated clinical testing instruments such as the Sport Concussion Assessment Tool (SKAT 6), Glasgow coma scale, and Vestibular Ocular Motor Screening (VOMS) [5,12,13], along with inclusion of the testing measures of cognitive function (orientation, immediate and delayed memory recall, concentration, emotionality, depression, anxiety, speech, and task specific processing, as well as coordination, balance, eye movements, and headache) [14]. A more comprehensive version for office use, SCOAT6, has also been validated for use within 72 h of injury. In addition to testing the SKAT 6 items, it catalogues prior head injuries, cognitive and emotional history, current medications, and a detailed neurologic examination. Further, it includes recommended strategies for a return to learning and return to sports activities [15,16]. Currently, corroborating physiologic testing is not commonly employed.

To that end, the physiologic consequences of an mTBI will be reviewed, and methods to identify those changes using imaging modalities will be discussed. The potential combined use of both clinical and physiologic testing may improve the accuracy of diagnosis and readiness to return to usual activities and predict persistent physiologic dysfunction. The latter when associated with prior mTBIs may potentially lead to chronic progressive disease—chronic traumatic encephalopathy (CTE)—years later, as seen in professional contact sports athletes [17].

1.1.2. Pathophysiology

Given the brain is a soft delicate organ contained within a closed box, sudden applied blunt trauma and inertial forces of acceleration, deceleration, or rotational trauma dictate the severity and location of cortical, white matter tract, and vasculature injury. Unlike moderate or severe TBI, mTBI injury is associated with minimal or no disruption of these structures in most cases, but not all [18,19]. That said, the physiologic consequences of disrupted electrolyte pumps and blood flow acutely post head injury have been well delineated with an acute impairment of electrolyte (sodium potassium, calcium, and magnesium) distribution, notably mitochondrial calcium sequestration with altered metabolism in neurons and astroglia and reduced local blood flow [20–22]. This results in an acute glycolysis hypermetabolism to restore normal electrolyte balance [22]. Disruption of the blood–brain barrier (BBB) even in mild traumatic brain injury (mTBI) is reflective of this acute phase of injury [22–26]. The duration of this varies, dependent upon location and the severity of injury, leading to hypometabolism in the affected regions which can persist for weeks or longer [23,25,26]. The chronic phase of injury is largely thought to be related to persistent physiologic alterations from persistent activated microglia and their effect on the neurovascular unit [24,25,27–29]. The focus of this article principally addresses the BBB dysfunction and identifying imaging techniques. Additionally, techniques demonstrating shear force white matter tract injury will be discussed.

1.1.3. Shear Injury in TBI

The most troublesome and unfortunately permanent sequela of TBI is the result of shear injury to white matter tracts [18,29,30]. The extent and severity of damage escalates proportional to the force and location of impact with compiled additional effects from the edema mass effect and inflammatory upregulation. In mTBI, the general assumption is an absence of substantive anatomic injury [17,18]. This, however, belies subtle changes in personality and emotionality, particularly in adolescence, from damage to developing limbic connections [31–33].

The underpinnings of shear injury are quite understandable with sudden mechanical deformation of the brain tearing apart connections [26–29,33,34]. The longer-term damage though is caused by the subsequent mass effect from edema, hemorrhage, or later from inflammation [22,26–29,34–40]. To understand the latter influences, animal models have been employed using a variety of species, most commonly rodents and, rarely, large animals [37–39]. Translatability from rodent to human physiology is problematic given the vast differences in brain structure and physiologic responses to injury [39,40].

The current medical state of the art possesses no means of restoring shear tract injury. The prevention of delayed injury by reducing the mass effect and swelling preserve life and may reduce secondary shear injury [28,30–33]. But, knowledge of how best to limit the secondary tract damage will require additional mechanistic study in higher animals (swine or primates) with a closer anatomic affiliation to the human brain [39,40].

Too often, the die is cast for potential clinical recovery based on the initial extent of shear force injury [37–40]. That said, identifying the severity of tract injury has potential importance in prognostic discussions with the patient and family.

1.1.4. Blood–Brain Barrier Dysfunction in Acute and Chronic TBI

In mTBI, the microcirculation (capillary system) BBB is disrupted acutely, causing reduced mean capillary transit time (cMTT) and reduced glymphatic flow (GF) in the region(s) of impact both coup and contrecoup with limited or no white matter tract injury [36–38]. The effect of this disruption is leakage and the trapping of normally restricted substances into the brain and thus reduced clearance out of the brain. This results in a measurable reduction in the clearance rate of intraparenchymal fluid which can be identified by imaging techniques (see below). In animal studies, BBB disruption and leakage of Immunoglobulin G IGG locally persisted for up to a month following moderate TBI with associated delayed microhemorrhages at those sites [19,34,37–39].

A brief summary of the complexity of the BBB structure and function is presented here. The details of the cellular interactions, signaling mechanisms, and what is known of the basic biology are beyond the scope of this review but are referenced [24,41–51].

The blood–brain barrier is dependent on a complex interplay of cells within the neurovascular unit (NVU), including the endothelial cells, pericytes, and astrocyte end feet [34,43–45,50]. Signaling within this triad in homeostasis allows for the transport of needed electrolyte, glucose, and specific lipids and proteins while excluding common blood constituents that are toxic to the complex interstitial environment [41–50]. The interaction and communication among elements of the NVU, extracellular matrix (ECM), and microglia dictate the presence, expression, and integrity of the various tight junction (TJ) proteins [52,53]. The presence of specific inflammatory cytokines may result in either a proinflammatory milieux or, counterintuitively, a restorative one [25,54–58]. For example, endothelial cell-derived IL1 β acutely downregulates the TJ protein ZO-1/occludens, thus increasing the BBB permeability, but by promoting the expression of Pentraxin 3 it later enhances its restoration [57]. The relative balance of local microglial proinflammatory versus homeostatic regulatory influences also determines the BBB integrity [59–63]. That balance is dictated by neuronal and astrocyte signaling which is influenced by the presence of cell injury and local inflammation [43,60].

Fundamental questions remain regarding defining the restorative pathways post injury and the long-term consequences of persistent leakage [59–63]. In TBI, BBB leakage is caused by proinflammatory microglial–endothelial signaling and to shear forces disrupting the endothelial glycocalyx, the latter allowing the leakage of IGG and the upregulation of inflammatory cytokines causing the local conversion of microglia to a proinflammatory state [51,64–66]. The activated microglia express the complement (C3a) fragment which in turn upregulates C3aR in capillary endothelial cells, altering their phenotype to a proinflammatory/immune cell attractant state with associated disruption of intercellular tight junctions, thus triggering additional BBB leakage [67]. In the chronic phase of injury, the presence of activated microglia induces BBB dysfunction, local chronic inflammatory changes, and neuronal dysfunction [23,35,65]. The persistence of this especially with multiple mTBI results in the perivascular accumulation of pTau within astroglia and neurons in the sulcal depths, the hallmark of CTE [65–68].

Restoration of normal BBB integrity following mTBI generally occurs quickly in youth; however, in some, the persistence of activated microglia can persist for months or longer resulting in a reduction in the threshold for additional mTBI-related pathology with clinical accompaniment [68–70]. With advancing age, recovery is less complete and the progression of both inflammation with BBB leakage and cognitive decline following TBI is more likely [71–74]. Further, the lower threshold and persistence of mTBI in young women is well recognized and postulated to be influenced by sex hormones and reduced muscle mass compared to their male counterparts [75,76]. The mechanisms of the BBB repair process in humans remain incompletely delineated, in particular the specifics of the pathways active in youth but lost in the circumstances of repetitive TBI or in the normal aging process [66,70–73].

The cumulative effects of multiple mTBI events cause ongoing inflammatory upregulation resulting in both permanent microvascular changes and a persistent upregulation of inflammation within the neuropil [61,65,74]. This induces altered synthesis, degradation, accumulation, and transsynaptic spread of toxic misfolded proteins, predominantly hyperphosphorylated Tau (pTau), but also Lewy body proteins, TDP-43, and to a delayed and minor degree β amyloid [61,74]. The consequence is the well-described high incidence of chronic traumatic encephalopathy (CTE) in professional contact sports athletes [77–81]. The clinical correlates include aggression, depression, suicide ideation, impulsivity, cognitive decline, and parkinsonian features [69,77–81]. To reduce the likelihood of progressive dementia from cumulative brain injury, the addition of objective physiologic testing in mTBI should prove invaluable in conjunction with clinical assessment to identify the early chronic inflammatory conversion.

Given the importance of maintaining BBB integrity, there is likely more than one pathway regulating restoration post injury [66,71]. That said, the duration and speed of clinical recovery are correlated with the patient's age, location and magnitude of force, and prior head injuries [30–32,72,73]. The inverse correlation of the severity of the head injury and the restoration of BBB integrity is understandably related to anatomic and vascular disruption [73,77]. The caveat to this general rule in adolescence has been shown via clinical evidence of emotional dysfunction even in mTBI which can be persistent, suggesting limbic system tract injury [68,73,76].

Of particular interest is why these mechanisms rapidly repair the BBB leakages in young patients with a single mTBI but flag in the aging brain even without injury. The answer may lie in the long-term presence of proinflammatory microglia and ECM constituent protein alterations which increase local BBB permeability [70–73]. Alternatively, the upregulation of inflammatory cytokines related to other unrelated health issues may contribute [79–82]. Conversely, is it a failure of the normal restorative signaling pathways lost in the aging process that are the root cause? Or perhaps a combination all three mechanisms? This question has profound implications for treatment of the underpinnings of other diseases such as neurogenerative diseases where the loss of BBB integrity is the nexus of their development [63,79–84]. Further, of the two pathologic features of traumatic brain injury, the restoration of BBB integrity as opposed to shear injury appears to be the most amenable to intervention.

To assess the presence of BBB leakage, physiologic imaging methods are required. Likewise, the assessment of future treatment modalities to monitor outcomes of clinical treatment trials addressing BBB repair requires the same [71,78,85,86]. In the next sections, the current most usable and clinically available methods of anatomic and physiologic imaging will be discussed along with their roles in identifying specific pathologic features of head injury.

1.2. Structural Imaging in mTBI

Although anatomic imaging in singular mTBI is infrequently conducted and if performed is almost invariably normal, it clearly has a place in the evaluation of patients with multiple mTBIs and of course more severe injury. In moderate or severe acute brain injury, CT head is the study of choice for the evaluation of bleeding or mass effect [18,86,87]. In evaluating subacute or chronic injury, using the benefit of MRI capabilities for defining atrophy with T1 sequences, white matter lesions with fluid attenuated inversion recovery (FLAIR), and prior micro or macro hemorrhages with susceptibility weighted images (SWIs) are the gold standards [85–87].

1.3. Physiologic Imaging

The assessment of TBI potentially takes three forms: anatomic, clinical cognitive/neurologic examination, and physiologic testing. In mTBI, structural evaluation with CT or MRI is nearly always normal, whereas the clinical cognitive assessments are far more revealing. However, subtle changes in processing, memory, personality, and emotional stability may miss subclinical physiologic injuries. Current clinically available methodologies for assessing the physiologic change in BBB function include dynamic contrast-enhanced MRI (DCE-MRI) and arterial spin labeling (ASL) MRI [85]. The evaluation of WM tract shear injury can be assessed directly with MRI tractography–diffusion tensor imaging (DTI) [88]. Positron emission tomography (PET) assesses the reduction in glucose utilization reflective of a regional loss of normal metabolic function [22,85]. Indirect means include functional MRI imaging (not discussed further as it is not commonly used clinically). The pros and cons of each method will be discussed below.

1.3.1. Diffusion Tensor Imaging

Diffusion tensor imaging is a non-invasive method of investigating white matter tract integrity, which leverages the imaging of water movement within the brain [88,89]. Water in the absence of obstruction will move randomly in all directions (isotropy). In the normal brain, however, water is forced to move parallel to white matter tract obstructions (anisotropy). With shearing force injury, these neat pathways are interrupted and hence the water present in those locations moves more isotopically. Diffusion imaging techniques can identify that disruption (Figure 1). Technical improvement to this methodology is under development. The current methods available lack sensitivity and are hampered by long scan times [89,90]. Nonetheless, as methods ameliorating persistent BBB leakage develop, knowledge of the extent of untreatable shear injury is of value in prognosticating ultimate neurologic recovery.

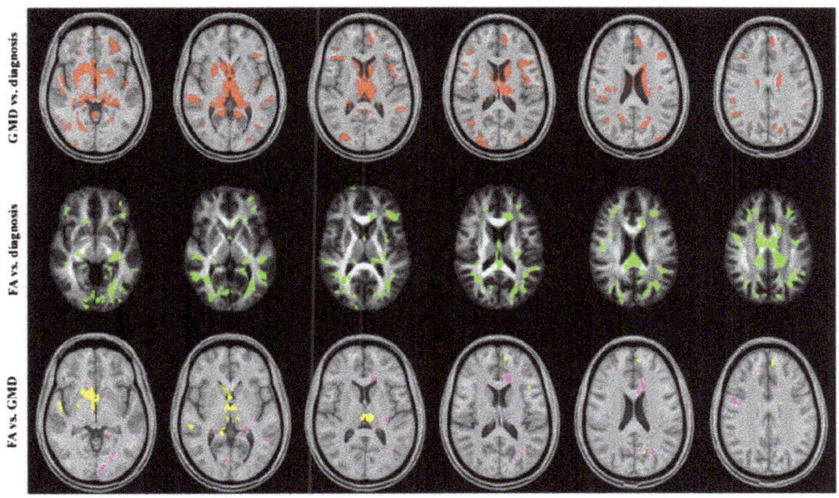

Figure 1. Sparse canonical correlation analyses (SCCA) of T1-weighted MP-RAGE and 30-direction diffusion tensor images (DTI) datasets are used to quantify the traumatically induced disruption of WM and cortical networks. The cohort includes 17 controls and 16 patients with TBI (age and gender matched). Each patient had a history of non-penetrating TBI of at least moderate severity. White matter integrity is assessed by DTI, and fractional anisotropy (FA) maps are generated. Separately,

probabilistic segmentation of the T1-weighted imaging is performed to assess gray matter integrity. Variation in brain shape across subjects is normalized by diffeomorphically mapping these data into a population-specific template space. Image processing steps rely on the Camino and ANTs (Advanced Normalization Tools) neuroimage analysis open-source toolkits. SCCA demonstrates significant differences between the control and patient groups in both the FA ($p < 0.002$) and gray matter ($p < 0.01$) that are widespread but largely focus on thalamocortical networks related to the limbic system. Using SCCA identified regions, a strong correlation is identified between degree of injury in WM and GM within the patient group. Figure courtesy of James R. Stone. Reference [91]. Gandy et al., Reference [91]. © 2014 Gandy et al.; licensee BioMed Central Ltd. This is an open access article distributed under the terms of the Creative Commons Attribution License (http://creativecommons.org/licenses/by/2.0, accessed on 29 July 2024) which permits unrestricted use, distribution, and reproduction in any medium, provided the original work is properly credited. The Creative Commons Public Domain Dedication waiver (http://creativecommons.org/publicdomain/zero/1.0/ accessed on 29 July 2024) applies to the data made available in this article, unless otherwise stated.

1.3.2. Dynamic Contrast-Enhanced (DCE) MRI

Dynamic contrast-enhanced (DCE) MRI directly measures the leakage of contrast agent into the neuropil in excess of its natural infusibility. Given gadolinium contrast agents are mostly excluded from passage through the BBB normally, their passage increases with its disruption. The technique assesses the plasma volume and interstitial space and can calculate a volumetric transfer constant from vascular to interstitial space [92–94]. The excess accumulation of contrast in the interstitial space beyond the calculated diffusion amount is indicative of BBB leakage. This method allows the direct visualization of BBB leakage but is invasive (requires gadolinium contrast injection) and thus is not suitable for serial measurements over time, given the concern of retained gadolinium contrast within the CNS [92–96] (Figure 2). Specific software to sort out normal contrast diffusion through the BBB from that related to excess leakage is also required [93,94]. Depending on patient circumstances though, its use as a diagnostic one-off study when coupled with DTI may prove helpful in prognosticating recovery in those slow to do so. The combination would reveal the extent of BBB leakage and white matter tract injury.

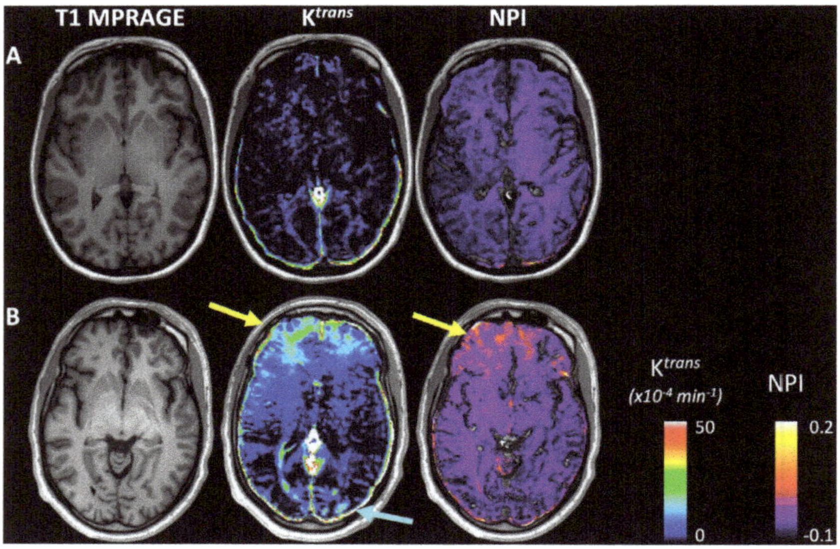

Figure 2. Individual subject examples from a representative control (**A**) and subacute TBI subject (**B**). The TBI non-lesional patient shows similar mild elevations in K^{trans} and NPI in the anterior frontal

white matter (yellow arrows) and additional small areas of K^{trans} elevation in the occipital lobes (blue arrow). Ware et al., Ref [97]. The Authors. Published by Elsevier Inc. Amsterdam, The Netherlands. This is an open access article under the CCBY-NC-ND license (https://creativecommons.org/licenses/by-nc-nd/4.0/) accessed on 29 July 2024.

1.3.3. 18-Fluorodeoxyglucose Positron Emission Tomography (FDG-PET)

Utilizing radioactively labeled fluorodeoxyglucose (^{18}FDG) and measuring the cellular uptake thereof via positron emission tomography (PET) indirectly measures metabolic activity [22,98,99]. With less metabolic activity, there is a reduced uptake of labeled glucose as occurs in acute mTBI [98,99]. In the acute phase of severe injury, there is an increased uptake of ^{18}FDG corresponding to the increased glycolysis required to reestablish electrolyte balance [100,101]. In the chronic phase of injury, there is a reduced uptake of glucose corresponding to reduced local activity [101–108]. When overlayed on CT or MRI brain images, the localization of abnormalities is enhanced (Figure 3). Sensitivity, specificity, cost, and scan times limit its utility.

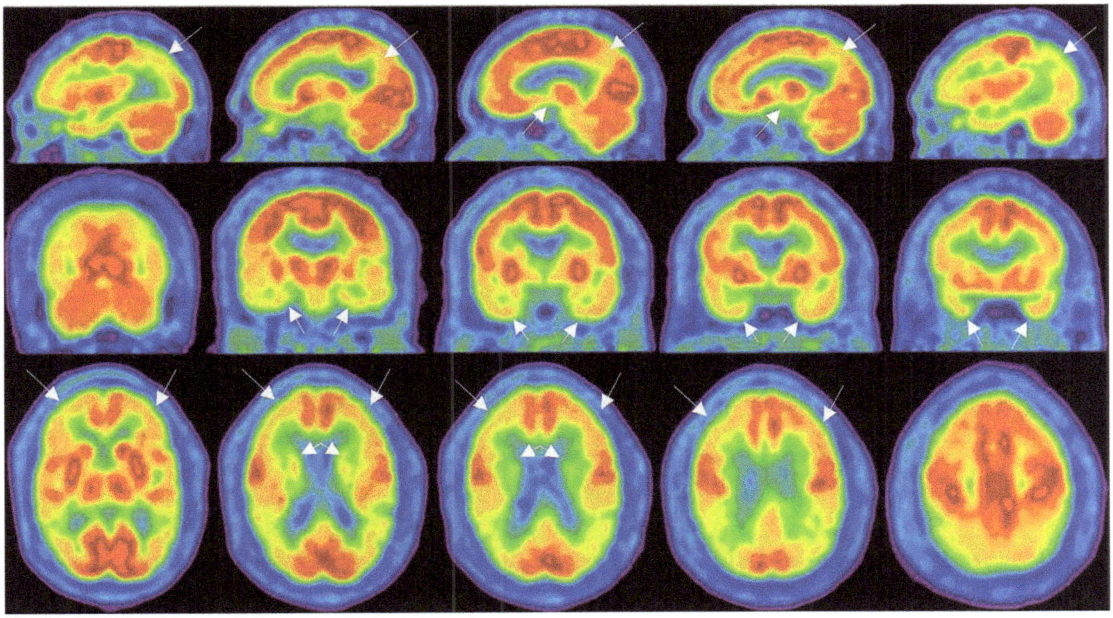

Figure 3. FDG-PET scan of the brain. (**Top**) Sagittal views (left to right) illustrate hypometabolism in the bilateral posterior parietotemporal regions, posterior cingulate, and precuneus. (**Middle**) Coronal views (posterior to anterior) show decreased uptake in the medial temporal lobes. (**Bottom**) Axial views (inferior to superior) demonstrate hypometabolism in the frontal lobe and anterior temporal poles. Morgan, et al. Ref [109]. This is an open access article under the CC BY-NC-ND license accessed on 29 July 2024.

The application of tau PET scans will not be discussed here as they would not be employed in mTBI but in cases of repetitive TBI when the question of CTE arises. The reader is referred to excellent reviews of these techniques [95,106–108,110–122].

1.3.4. Arterial Spin Labeling MRI

Another method to indirectly determine BBB leakage utilizes the clearance rate of magnetically labelled protons using 3D ASL MRI technology. This technique leverages the physiologic effects of altered perfusion with delayed capillary mean transit time (cMTT) and reduced glymphatic outflow in the late phase of the perfusion cycle, correlating with the location of BBB leakage [123] (Figure 4). It is a non-invasive method using blood protons magnetically spin labeled in the neck and, after a specified post-labeling delay (PLD), residual signal averages are sampled in the desired cortical regions.

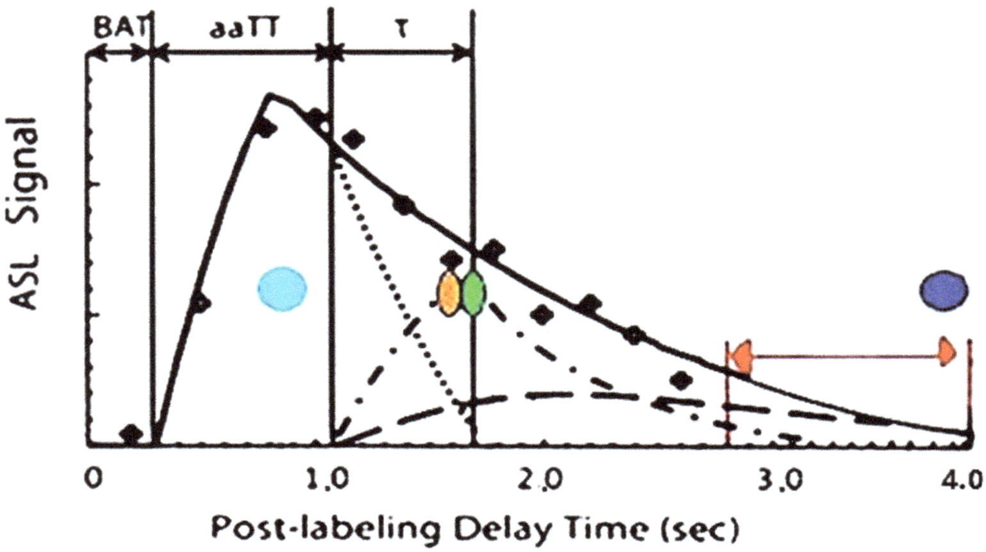

Figure 4. Composition of the ASL signal/time from its various components: arterial (dotted line), capillary (dot and dash line), and extra capillary (dashed line) spaces. The TI (acquisition) times utilized in this study are between the red arrow limits. Note that the signal composition is nearly completely extravascular (interstitial) water, which is normally cleared by the glymphatic system. The T1 times (63% signal decay) of major signal contributors are indicated by the colored dots. Magenta dot = 800–850 ms T1 of white matter. Orange dot = 1650 ms T1 of blood. Green dot = 1700 ms T1 of gray matter (all values are for 3T). Dark blue dot = 3800 ms T1 of water (CSF fluid); (all values are for 3T). BAT = bolus arrival time; aaTT = artery–artery arrival time; τ = peak capillary arrival time. Adapted with the publisher's permission (Wiley) from reference [124].

By using multiple sequential PLD's in the late phase of the perfusion cycle, linear analysis can be employed to determine the rate of clearance (slope of the line) [123,125]. In regions of BBB leakage, there are localized changes in blood flow that correlate with diminished clearance rates as can be seen in preclinical dementia [126]. Since the signal measured is largely from trapped labeled water protons in the late phase of perfusion, it is much more sensitive in determining minor BBB leakages than gadolinium (large molecule)-based DCE [127].

Since this is a non-invasive technique, it is suitable for sequential studies without risk of exogenous contrast. Currently, when combined with FLAIR and susceptibility weighted images (SWIs), the scan time is about $18^{1/2}$ min [123]. ASL MRI requires a 3 Tesla magnet and is limited to a reduced S/N. The latter can be circumvented by measuring a large field of view and obtaining multiple PLDs. Using 3D data acquisition allows for multiple regions of the brain to be investigated in one study. For example, our usage in mTBI acquired clearance rates from six brain regions, homologous temporal, frontal, and parietal regions [123,125,126]. Newer ASL MRI sequences allow for acquiring multiple PLDs in one sequence, reducing scan times to under 5 min [127]. This technique has significant potential in following the progress of BBB repair or alternatively persistent leakage. In our pilot study of mTBI in college-age athletes, both reduced glymphatic clearance post-acute injury and the restoration of BBB integrity as measured by this technique correlated with high sensitivity with clinical recovery [125]. Future studies evaluating more serious head injury sequentially may shed light on those patients with persistent dysfunction which sets the stage for progressive dementia (CTE) [126,127]. In addition, this objective measure could help to determine when and if an injured athlete or soldier could safely return to their sport or active duty. The limitations of this technique include the indirect method of determination (the results are graphic in nature post analysis as opposed to specific images of leakage) and a low S/N ratio. The latter is compensated for by analyzing multiple data points and acquiring signal from a large field of view FOV in the region of interest (ROI) chosen for analysis. Given the test is non-invasive as well as time- and cost-efficient, this method is suitable for serial studies post TBI in the assessment of BBB dysfunction [63,125].

1.3.5. Summary of Techniques

Table 1 below summarizes the imaging target, relative advantages, and disadvantages of the four physiologic imaging techniques presented in evaluating mTBI. They provide objective information regarding the state of brain and vascular function at the time they are employed post head injury. The techniques will continue to improve and become more widely available providing both prognostic information. They will also be essential as outcome measures in future treatment trials.

Table 1. Methods of BBB and shear injury imaging modalities and limitations.

Imaging Procedure	Target	Sensitivity	Limitations
Diffusion Tensor Imaging (DTI) MRI	Diffusion of water-white matter tracts	Not clear, older techniques problematic; New techniques in development	Technique specific sensitivity, scan time
Dynamic Contrast-Enhanced (DCE) MRI	BBB leakage of contrast	Sensitive for BBB leakage	Requires gadolinium contrast, long scan times, not suitable for serial studies due to cerebral accumulation of GD^{+}
^{18}FDG-PET	Glucose utilization metabolic activity	Can identify the brief hypermetabolism acutely (animal studies) and chronic hypometabolism (human) chronically across severity spectrum	Cost, long scan times, availability of technology
3D ASL MRI	Perfusion/Diffusion of labeled protons	Highly sensitive in a small clinical trial in mTBI determining altered cMTT/BBB leakage with delayed BBB clearance/recovery	Requires a 3T or greater field strength, low signal, hand-drawn ROI, requires a larger series for confirmation of sensitivity

Table 1 summarizes the DTI in identifying shear tract injury, which if present correlates with the persistence of neurologic deficit and recovery depending on its extent. DCE and ASL MRI identify the loss of BBB integrity: the former directly and the latter indirectly. ^{18}FDG-PET measures metabolic activity and thus could be negatively influenced by either shear tract injury or BBB leakage.

2. Conclusions

With the mounting evidence of severe late life repercussions (CTE) related to repetitive mTBI among professional athletes in contact sports, there is a need to evaluate the physiology leveraging the loss of BBB integrity post injury that, if persistent, correlates with progressive changes in CTE [39,60,79,85,128–132]. By adopting physiologic testing measures, a safe return to the sport could be more accurately predicted in mTBI. Since the potential of both subtle white matter tact shear injury and persistent BBB leakage occur simultaneously, the assessment of damage requires attention to both, which translates into white matter tract imaging as well as an assessment of BBB integrity [24]. Further, as we uncover the mechanisms of BBB repair going forward and can leverage that knowledge therapeutically, identifying and assessing the outcomes of future treatments will be possible. The benefits of identifying BBB dysfunction spill over from TBI to preclinical stages of neurodegenerative disease (NG), where the loss of its integrity is at the nexus of development and a prime target for early intervention.

Funding: This research was funded by Liberty University Funding MR2508-MR2508-B3C98-711099-20.

Acknowledgments: The author wants to thank student and faculty colleagues who were instrumental in the ASL MRI clinical research projects referred to in this article.

Conflicts of Interest: The author has no competing issues or conflicts of interest in the writing of this manuscript. The author is solely responsible for the content of this article.

Abbreviations

TBI	mild traumatic brain injury
TJ protein ZO-1	tight junction protein Zona Occludens 1
C3a	complement C3a protein
C3aR	complement C3a receptor protein
pTau	hyperphosphorylated Tau
TDP-43	transactive response DNA binding protein-43
FLAIR	fluid attenuated inversion recovery
SWI	susceptibility weighted images
ASL MRI	arterial spin labeling MRI
PLD	post-labeling delay
DTI MRI	diffusion tensor imaging
DCE MRI	dynamic contrast-enhanced MRI
PET	positron emission tomography
S/N	signal to noise
SKAT 6	Sport Concussion Assessment Tool
VOMS	Vestibular Ocular Motor Screening
SCOAT 6	Sports Concussion Office Assessment Tool
CTE	chronic traumatic encephalopathy
cMTT	mean capillary transit time
GF	glymphatic flow
BBB	blood–brain barrier
IGG	immunoglobulin G
NVU	the neurovascular unit
ECM	extracellular matrix
IL1 β	interleukin 1 beta

References

1. Maas, A.I.; Menon, D.K.; Manley, G.T.; Abrams, M.; Åkerlund, C.; Andelic, N.; Aries, M.; Bashford, T.; Bell, M.J.; Bodien, Y.G.; et al. Traumatic brain injury: Progress and challenges in prevention, clinical care, and research. *Lancet Neurol.* **2022**, *21*, 1004–1060. [CrossRef] [PubMed]
2. Forrest, R.H.J.; Henry, J.D.; McGarry, P.J.; Marshall, R.N. Mild traumatic brain injury in New Zealand: Factors influencing post-concussion symptom recovery time in a specialised concussion service. *J. Prim. Health Care* **2018**, *10*, 159. Available online: https://www.publish.csiro.au/hc/pdf/HC17071 (accessed on 21 June 2024). [CrossRef]
3. Fatuki, T.A.; Zvonarev, V.; Rodas, A.W. Prevention of traumatic brain injury in the United States: Significance, new findings, and practical applications. *Cureus* **2020**, *12*, e11225. [CrossRef]
4. Reith, F.C.; Lingsma, H.F.; Gabbe, B.J.; Lecky, F.E.; Roberts, I.; Maas, A.I. Differential effects of the Glasgow Coma Scale Score and its Components: An analysis of 54,069 patients with traumatic brain injury. *Injury* **2017**, *48*, 1932–1943. [CrossRef]
5. Dessy, A.M.; Yuk, F.J.; Maniya, A.Y.; Gometz, A.; Rasouli, J.J.; Lovell, M.R.; Choudhri, T.F. Review of Assessment Scales for Diagnosing and Monitoring Sports-related Concussion. *Cureus* **2017**, *9*, e1922. Available online: https://pmc.ncbi.nlm.nih.gov/articles/PMC5802754/ (accessed on 21 June 2024). [CrossRef]
6. Polinder, S.; Cnossen, M.C.; Real, R.G.L.; Covic, A.; Gorbunova, A.; Voormolen, D.C.; Master, C.L.; Haagsma, J.A.; Diaz-Arrastia, R.; Von Steinbuechel, N. A Multidimensional Approach to Post-concussion Symptoms in Mild Traumatic Brain Injury. *Front. Neurol.* **2018**, *9*, 1113. Available online: https://www.frontiersin.org/articles/10.3389/fneur.2018.01113/full (accessed on 21 June 2024). [CrossRef] [PubMed]

7. Williams, R.M.; Puetz, T.W.; Giza, C.C.; Broglio, S.P. Concussion Recovery Time Among High School and Collegiate Athletes: A Systematic Review and Meta-Analysis. *Sports Med.* **2015**, *45*, 893–903. Available online: https://www.ncbi.nlm.nih.gov/pmc/articles/PMC4441834/ (accessed on 21 June 2024). [CrossRef]
8. McCrea, M.; Guskiewicz, K.; Randolph, C.; Barr, W.B.; Hammeke, T.A.; Marshall, S.W.; Powell, M.R.; Ahn, K.W.; Wang, Y.; Kelly, J.P. Incidence, clinical course, and predictors of prolonged recovery time following sport-related concussion in high school and college athletes. *J. Int. Neuropsychol. Soc.* **2013**, *19*, 22–33. [CrossRef] [PubMed]
9. Wintermark, M.; Sanelli, P.C.; Anzai, Y.; Tsiouris, A.J.; Whitlow, C.T.; Druzgal, T.J.; Gean, A.D.; Lui, Y.W.; Norbash, A.M.; Raji, C.; et al. Imaging Evidence and Recommendations for Traumatic Brain Injury: Conventional Neuroimaging Techniques. *J. Am. Coll. Radiol.* **2015**, *12*, e1–e14. [CrossRef]
10. Eierud, C.; Craddock, R.C.; Fletcher, S.; Aulakh, M.; King-Casas, B.; Kuehl, D.; LaConte, S.M. Neuroimaging after mild traumatic brain injury: Review and meta-analysis. *NeuroImage Clin.* **2014**, *4*, 283–294. [CrossRef]
11. Roberts, M.A.; Manshadi, F.; Bushnell, D.; Hines, M. Neurobehavioural dysfunction following mild traumatic brain injury in childhood: A case report with positive findings on positron emission tomography (PET). *Brain Inj.* **1995**, *9*, 427–436. [CrossRef] [PubMed]
12. Echemendia, R.J.; Brett, B.L.; Broglio, S.; Davis, G.A.; Giza, C.C.; Guskiewicz, K.M.; Harmon, K.G.; Herring, S.; Howell, D.R.; Master, C.L.; et al. Introducing the sport concussion assessment tool 6 (SCAT6). *Br. J. Sports Med.* **2023**, *57*, 619–621. [CrossRef] [PubMed]
13. Patricios, J.; Schneider, G.M.; van Ierssel, J.; Purcell, L.K.; Davis, G.A.; Echemendia, R.J.; Frémont, P.; Fuller, G.W.; Herring, S.; Harmon, K.G.; et al. Sport concussion office assessment tool 6. *Br. J. Sports Med.* **2023**, *57*, 651–667. [CrossRef] [PubMed]
14. Voormolen, D.C.; Cnossen, M.C.; Polinder, S.; Von Steinbuechel, N.; Vos, P.E.; Haagsma, J.A. Divergent classification methods of post-concussion syndrome after mild traumatic brain injury: Prevalence rates, risk factors, and functional outcome. *J. Neurotrauma* **2018**, *35*, 1233–1241. [CrossRef] [PubMed]
15. Meier, T.B.; Bellgowan, P.S.; Singh, R.; Kuplicki, R.; Polanski, D.W.; Mayer, A.R. Recovery of cerebral blood flow following sports related concussion. *JAMA Neurol.* **2015**, *72*, 530–538. [CrossRef]
16. Lee, B.G.; Leavitt, M.J.; Bernick, C.B.; Leger, G.C.; Rabinovici, G.; Banks, S.J. A systematic review of positron emission tomography of tau, amyloid beta, and neuroinflammation in chronic traumatic encephalopathy: The evidence to date. *J. Neurotrauma* **2018**, *35*, 2015–2024. [CrossRef]
17. Mez, J.; Daneshvar, D.H.; Kiernan, P.T.; Abdolmohammadi, B.; Alvarez, V.E.; Huber, B.R.; Alosco, M.L.; Solomon, T.M.; Nowinski, C.J.; McHale, L.; et al. Clinicopathological evaluation of chronic traumatic encephalopathy in players of American football. *JAMA* **2017**, *318*, 360–370. [CrossRef]
18. Ilvesmäki, T.; Luoto, T.M.; Hakulinen, U.; Brander, A.; Ryymin, P.; Eskola, H.; Iverson, G.L.; Öhman, J. Acute mild traumatic brain injury is not associated with white matter change on diffusion tensor imaging. *Brain* **2014**, *137*, 1876–1882. [CrossRef]
19. Lee, H.; Wintermark, M.; Gean, A.D.; Ghajar, J.; Manley, G.T.; Mukherjee, P. Focal lesions in acute mild traumatic brain injury and neurocognitive outcome: CT versus 3T MRI. *J. Neurotrauma* **2008**, *25*, 1049–1056. [CrossRef]
20. Hiebert, J.B.; Shen, Q.; Thimmesch, A.R.; Pierce, J.D. Traumatic brain injury and mitochondrial dysfunction. *Am. J. Med. Sci.* **2015**, *350*, 132–138. [CrossRef]
21. Vagnozzi, R.; Tavazzi, B.; Signoretti, S.; Amorini, A.M.; Belli, A.; Cimatti, M.; Delfini, R.; Di Pietro, V.; Finocchiaro, A.; Lazzarino, G.; et al. Temporal window of metabolic brain vulnerability to concussions: Mitochondrial-related impairment part I. *Neurosurgery* **2007**, *61*, 379–389. [CrossRef] [PubMed]
22. Bergsneider, M.; Hovda, D.A.; Shalmon, E.; Kelly, D.F.; Vespa, P.M.; Martin, N.A.; Phelps, M.E.; McArthur, D.L.; Caron, M.J.; Kraus, J.F.; et al. Cerebral hyperglycolysis following severe traumatic brain injury in humans: A positron emission tomography study. *J. Neurosurg.* **1997**, *86*, 241–251. [CrossRef] [PubMed]
23. Hay, J.R.; Johnson, V.E.; Young, A.M.; Smith, D.H.; Stewart, W. Blood-brain barrier disruption is an early event that may persist for many years after traumatic brain injury in humans. *J. Neuropathol. Exp. Neurol.* **2015**, *74*, 1147–1157. [PubMed]
24. Johnson, V.E.; Stewart, J.E.; Begbie, F.D.; Trojanowski, J.Q.; Smith, D.H.; Stewart, W. Inflammation and white matter degeneration persist for years after a single traumatic brain injury. *Brain* **2013**, *136*, 28–42. [CrossRef]
25. Abrahamson, E.E.; Ikonomovic, M.D. Brain injury-induced dysfunction of the blood brain barrier as a risk for dementia. *Exp. Neurol.* **2020**, *328*, 113257. [CrossRef]
26. O'Keeffe, E.; Kelly, E.; Liu, Y.; Giordano, C.; Wallace, E.; Hynes, M.; Tiernan, S.; Meagher, A.; Greene, C.; Hughes, S.; et al. Dynamic blood–brain barrier regulation in mild traumatic brain injury. *J. Neurotrauma* **2020**, *37*, 347–356. [CrossRef]
27. Elder, G.A.; Gama Sosa, M.A.; De Gasperi, R.; Stone, J.R.; Dickstein, D.L.; Haghighi, F.; Hof, P.R.; Ahlers, S.T. Vascular and inflammatory factors in the pathophysiology of blast-induced brain injury. *Front. Neurol.* **2015**, *6*, 48. [CrossRef]
28. Thurgur, H.; Pinteaux, E. Microglia in the neurovascular unit: Blood–brain barrier–microglia interactions after central nervous system disorders. *Neuroscience* **2019**, *405*, 55–67. [CrossRef]
29. Christman, C.W.; Grady, M.S.; Walker, S.A.; Holloway, K.L.; Povlishock, J.T. Ultrastructural studies of diffuse axonal injury in humans. *J. Neurotrauma* **1994**, *11*, 173–186. [CrossRef]
30. El Sayed, T.; Mota, A.; Fraternali, F.; Ortiz, M. Biomechanics of traumatic brain injury. *Comput. Methods Appl. Mech. Eng.* **2008**, *197*, 4692–4701. [CrossRef]

31. Wang, Y.; Nencka, A.S.; Meier, T.B.; Guskiewicz, K.; Mihalik, J.P.; Alison Brooks, M.; Saykin, A.J.; Koch, K.M.; Wu, Y.-C.; Nelson, L.D.; et al. Cerebral blood flow in acute concussion: Preliminary ASL findings from the NCAA-DoD CARE consortium. *Brain Imaging Behav.* **2019**, *13*, 1375–1385. [CrossRef]
32. Pavlovic, D.; Pekic, S.; Stojanovic, M.; Popovic, V. Traumatic brain injury: Neuropathological, neurocognitive and neurobehavioral sequelae. *Pituitary* **2019**, *22*, 270–282. [CrossRef] [PubMed]
33. Meaney, D.F.; Smith, D.H. Biomechanics of concussion. *Clin. Sports Med.* **2011**, *30*, 19–31. [CrossRef] [PubMed]
34. Donat, C.K.; Yanez Lopez, M.; Sastre, M.; Baxan, N.; Goldfinger, M.; Seeamber, R.; Müller, F.; Davies, P.; Hellyer, P.; Siegkas, P.; et al. From biomechanics to pathology: Predicting axonal injury from patterns of strain after traumatic brain injury. *Brain* **2021**, *144*, 70–91. [CrossRef] [PubMed]
35. Glushakova, O.Y.; Johnson, D.; Hayes, R.L. Hayes. Delayed increases in microvascular pathology after experimental traumatic brain injury are associated with prolonged inflammation, blood–brain barrier disruption, and progressive white matter damage. *J. Neurotrauma* **2014**, *31*, 1180–1193. [CrossRef] [PubMed]
36. Gullotta, G.S.; Costantino, G.; Sortino, M.A.; Spampinato, S.F. Microglia and the blood–brain barrier: An external player in acute and chronic neuroinflammatory conditions. *Int. J. Mol. Sci.* **2023**, *24*, 9144. [CrossRef]
37. Haider, M.N.; Leddy, J.J.; Hinds, A.L.; Aronoff, N.; Rein, D.; Poulsen, D.; Willer, B.S. Intracranial pressure changes after mild traumatic brain injury: A systematic review. *Brain Inj.* **2018**, *32*, 809–815. [CrossRef]
38. Ojo, J.O.; Mouzon, B.; Algamal, M.; Leary, P.; Lynch, C.; Abdullah, L.; Evans, J.; Mullan, M.; Bachmeier, C.; Stewart, W.; et al. Chronic repetitive mild traumatic brain injury results in reduced cerebral blood flow, axonal injury, gliosis, and increased T-tau and tau oligomers. *J. Neuropathol. Exp. Neurol.* **2016**, *75*, 636–655. [CrossRef]
39. Barzó, P.; Marmarou, A.; Fatouros, P.; Corwin, F.; Dunbar, J. Magnetic resonance imaging—Monitored acute blood-brain barrier changes in experimental traumatic brain injury. *J. Neurosurg.* **1996**, *85*, 1113–1121. [CrossRef]
40. Morganti-Kossmann, M.C.; Yan, E.; Bye, N. Animal models of traumatic brain injury: Is there an optimal model to reproduce human brain injury in the laboratory? *Injury* **2010**, *41*, S10–S13. [CrossRef] [PubMed]
41. Johnson, V.E.; Meaney, D.F.; Cullen, D.K.; Smith, D.H. Smith. Animal models of traumatic brain injury. *Handb. Clin. Neurol.* **2015**, *127*, 115–128. [PubMed]
42. Engelhardt, B.; Liebner, S. Novel insights into the development and maintenance of the blood–brain barrier. *Cell Tissue Res.* **2014**, *355*, 687–699. [CrossRef] [PubMed]
43. Iadecola, C. The neurovascular unit coming of age: A journey through neurovascular coupling in health and disease. *Neuron* **2017**, *96*, 17–42. [CrossRef] [PubMed]
44. Bhowmick, S.; D'Mello, V.; Caruso, D.; Wallerstein, A.; Abdul-Muneer, P. Impairment of pericyte-endothelium crosstalk leads to blood-brain barrier dysfunction following traumatic brain injury. *Exp. Neurol.* **2019**, *317*, 260–270. [CrossRef]
45. Sweeney, M.D.; Kisler, K.; Montagne, A.; Toga, A.W.; Zlokovic, B.V. The role of brain vasculature in neurodegenerative disorders. *Nat. Neurosci.* **2018**, *21*, 1318–1331. [CrossRef]
46. Sweeney, M.D.; Zhao, Z.; Montagne, A.; Nelson, A.R.; Zlokovic, B.V. Bloodbrain barrier: From physiology to disease and back. *Physiol. Rev.* **2019**, *99*, 21–78. [CrossRef]
47. Andjelkovic, A.V.; Situ, M.; Citalan-Madrid, A.F.; Stamatovic, S.M.; Xiang, J.; Keep, R.F. Blood-brain barrier dysfunction in normal aging and neurodegeneration: Mechanisms, impact, and treatments. *Stroke* **2023**, *54*, 661–672. [CrossRef] [PubMed]
48. Kadry, H.; Noorani, B.; Cucullo, L. A blood–brain barrier overview on structure, function, impairment, and biomarkers of integrity. *Fluids Barriers CNS* **2020**, *17*, 69. [CrossRef]
49. Zhao, L.; Tannenbaum, A.; Bakker, E.N.; Benveniste, H. Physiology of glymphatic solute transport and waste clearance from the brain. *Physiology* **2022**, *37*, 349–362. [CrossRef]
50. Preston, J.E.; Abbott, N.J.; Begley, D.J. Transcytosis of macromolecules at the blood–brain barrier. *Adv. Pharmacol.* **2014**, *71*, 147–163.
51. van Leeuwen, E.; Hampton, M.B.; Smyth LC, D. Redox signaling and regulation of the blood-brain barrier. *Int. J. Biochem. Cell Biol.* **2020**, *125*, 105794. [CrossRef] [PubMed]
52. Baeten, K.M.; Akassoglou, K. Extracellular matrix and matrix receptors in blood–brain barrier formation and stroke. *Dev. Neurobiol.* **2011**, *71*, 1018–1039. [CrossRef]
53. Lochhead, J.J.; Yang, J.; Ronaldson, P.T.; Davis, T.P. Structure, function, and regulation of the blood-brain barrier tight junction in central nervous system disorders. *Front. Physiol.* **2020**, *11*, 914. [CrossRef]
54. Hudson, N.; Campbell, M. Tight junctions of the neurovascular unit. *Front. Mol. Neurosci.* **2021**, *14*, 752781. [CrossRef] [PubMed]
55. Woodcock, T.; Morganti-Kossmann, C. The role of markers of inflammation in traumatic brain injury. *Front. Neurol.* **2013**, *4*, 18. [CrossRef]
56. Ziebell, J.M.; Morganti-Kossmann, M.C. Involvement of pro-and anti-inflammatory cytokines and chemokines in the pathophysiology of traumatic brain injury. *Neurotherapeutics* **2010**, *7*, 22–30. [CrossRef]
57. Zheng, R.Z.; Lee, K.Y.; Qi, Z.X.; Wang, Z.; Xu, Z.Y.; Wu, X.H.; Mao, Y. Neuroinflammation following traumatic brain injury: Take it seriously or not. *Front. Immunol.* **2022**, *13*, 855701. [CrossRef] [PubMed]
58. Zhou, C.; Chen, H.; Zheng, J.F.; Guo, Z.D.; Huang, Z.J.; Wu, Y.; Zhong, J.J.; Sun, X.C.; Cheng, C.J. Pentraxin 3 contributes to neurogenesis after traumatic brain injury in mice. *Neural Regen. Res.* **2020**, *15*, 2318–2326.

59. Shabab, T.; Khanabdali, R.; Moghadamtousi, S.Z.; Kadir, H.A.; Mohan, G. Neuroinflammation pathways: A general review. *Int. J. Neurosci.* **2017**, *127*, 624–633. [CrossRef]
60. Morganti-Kossmann, M.C.; Rancan, M.; Otto, V.I.; Stahel, P.F.; Kossmann, T. Role of cerebral inflammation after traumatic brain injury: A revisited concept. *Shock* **2001**, *16*, 165–177. [CrossRef]
61. Doherty, C.P.; O'Keefe, E.; Wallace, E.; Loftus, T.; Keaney, J.; Kealy, J.; Humphries, M.M.; Molloy, M.G.; Meaney, J.F.; Farrell, M.; et al. Blood–brain barrier dysfunction as a hallmark pathology in chronic traumatic encephalopathy. *J. Neuropathol. Exp. Neurol.* **2016**, *75*, 656–662. [CrossRef] [PubMed]
62. Abdul-Muneer, P.M.; Chandra, N.; Haorah, J. Interactions of oxidative stress and neurovascular inflammation in the pathogenesis of traumatic brain injury. *Mol. Neurobiol.* **2015**, *51*, 966–979. [CrossRef] [PubMed]
63. Hattori, N.; Huang, S.-C.; Wu, H.-M.; Liao, W.; Glenn, T.C.; Vespa, P.M.; E Phelps, M.; A Hovda, D.; Bergsneider, M. Acute changes in regional cerebral 18F-FDG kinetics in patients with traumatic brain injury. *J. Nucl. Med.* **2004**, *45*, 775–783.
64. Kempuraj, D.; Ahmed, M.E.; Selvakumar, G.P.; Thangavel, R.; Dhaliwal, A.S.; Dubova, I.; Mentor, S.; Premkumar, K.; Saeed, D.; Zahoor, H.; et al. Brain injury–mediated neuroinflammatory response and Alzheimer's disease. *Neuroscientist* **2020**, *26*, 134–155. [CrossRef]
65. Zhang, J.; Puvenna, V.; Janigro, D. Biomarkers of traumatic brain injury and their relationship to pathology. In *Translational Research in Traumatic Brain Injury*; CRC Press/Taylor and Francis Group: Boca Raton, FL, USA, 2016.
66. Ramlackhansingh, A.F.; Brooks, D.J.; Greenwood, R.J.; Bose, S.K.; Turkheimer, F.E.; Kinnunen, K.M.; Gentleman, S.; Heckemann, R.A.; Gunanayagam, K.; Gelosa, G.; et al. Inflammation after trauma: Microglial activation and traumatic brain injury. *Ann. Neurol.* **2011**, *70*, 374–383. [CrossRef]
67. Profaci, C.P.; Munji, R.N.; Pulido, R.S.; Daneman, R. The blood–brain barrier in health and disease: Important unanswered questions. *J. Exp. Med.* **2020**, *217*, e20190062. [CrossRef]
68. Propson, N.E.; Roy, E.R.; Litvinchuk, A.; Köhl, J.; Zheng, H. Endothelial C3a receptor mediates vascular inflammation and blood-brain barrier permeability during aging. *J. Clin. Investig.* **2021**, *131*, e144348. [CrossRef] [PubMed]
69. Nakajima, Y.; Horiuchi, Y.; Kamata, H.; Yukawa, M.; Kuwabara, M.; Tsubokawa, T. Distinct time courses of secondary brain damage in the hippocampus following brain concussion and contusion in rats. *Tohoku J. Exp. Med.* **2010**, *221*, 229–235. [CrossRef]
70. Ryan, L.M.; Warden, D.L. Post concussion syndrome. *Int. Rev. Psychiatry* **2003**, *15*, 310–316. [CrossRef] [PubMed]
71. Prins, M.L.; Alexander, D.; Giza, C.C.; Hovda, D.A. Repeated mild traumatic brain injury: Mechanisms of cerebral vulnerability. *J. Neurotrauma* **2013**, *30*, 30–38. [CrossRef]
72. Joseph, C.R. Progressive Age-Associated Blood–Brain Barrier Leak/Dysfunction-Nexus of Neurodegenerative Disease Using MRI Markers to Identify Preclinical Disease and Potential New Targets for Future Treatments. *Diagnostics* **2024**, *14*, 726. [CrossRef] [PubMed]
73. Laurer, H.L.; Bareyre, F.M.; Lee, V.M.; Trojanowski, J.Q.; Longhi, L.; Hoover, R.; Saatman, K.E.; Raghupathi, R.; Hoshino, S.; Grady, M.S.; et al. Mild head injury increasing the brain's vulnerability to a second concussive impact. *J. Neurosurg.* **2001**, *95*, 859–870. [CrossRef] [PubMed]
74. Yuan, Y.; Sun, J.; Dong, Q.; Cui, M. Blood–brain barrier endothelial cells in neurodegenerative diseases: Signals from the "barrier". *Front. Neurosci.* **2023**, *17*, 1047778. [CrossRef] [PubMed]
75. Sorby-Adams, A.; Marcoionni, A.; Dempsey, E.; Woenig, J.; Turner, R. The role of neurogenic inflammation in blood-brain barrier disruption and development of cerebral oedema following acute central nervous system (CNS) injury. *Int. J. Mol. Sci.* **2017**, *18*, 1788. [CrossRef]
76. Farace, E.; Alves, W.M. Do women fare worse? A metanalysis of gender differences in outcome after traumatic brain injury. *Neurosurg. Focus.* **2000**, *8*, e6. [CrossRef]
77. Bennett, E.R.; Reuter-Rice, K.; Laskowitz, D.T. Genetic influences in traumatic brain injury. In *Translational Research in Traumatic Brain Injury*; CRC Press/Taylor and Francis Group: Boca Raton, FL, USA, 2016.
78. McKee, A.C.; Stein, T.D.; Nowinski, C.J.; Stern, R.A.; Daneshvar, D.H.; Alvarez, V.E.; Lee, H.-S.; Hall, G.; Wojtowicz, S.M.; Baugh, C.M.; et al. The spectrum of disease in chronic traumatic encephalopathy. *Brain* **2013**, *136*, 43–64. [CrossRef] [PubMed]
79. Ayubcha, C.; Revheim, M.E.; Newberg, A.; Moghbel, M.; Rojulpote, C.; Werner, T.J.; Alavi, A. A critical review of radiotracers in the positron emission tomography imaging of traumatic brain injury: FDG, tau, and amyloid imaging in mild traumatic brain injury and chronic traumatic encephalopathy. *Eur. J. Nucl. Med. Mol. Imaging* **2021**, *48*, 623–641. [CrossRef]
80. Lenihan, M.W.; Jordan, B.D. The clinical presentation of chronic traumatic encephalopathy. *Curr. Neurol. Neurosci. Rep.* **2015**, *15*, 23. [CrossRef]
81. Karton, C.; Hoshizaki, T.B. Concussive and subconcussive brain trauma: The complexity of impact biomechanics and injury risk in contact sport. *Handb. Clin. Neurol.* **2018**, *158*, 39–49.
82. Bang, S.A.; Song, Y.S.; Moon, B.S.; Lee, B.C.; Lee, H.-Y.; Kim, J.-M.; Kim, S.-E. Neuropsychological, metabolic, and GABAA receptor studies in subjects with repetitive traumatic brain injury. *J. Neurotrauma* **2016**, *33*, 1005–1014. [CrossRef]
83. Wooten, D.W.; Ortiz-Teran, L.; Zubcevik, N.; Zhang, X.; Huang, C.; Sepulcre, J.; Atassi, N.; Johnson, K.A.; Zafonte, R.D.; El Fakhri, G. Multi-modal signatures of tau pathology, neuronal fiber integrity, and functional connectivity in traumatic brain injury. *J. Neurotrauma* **2019**, *36*, 3233–3243. [CrossRef]

84. Knox, E.G.; Aburto, M.R.; Clarke, G.; Cryan, J.F.; O'Driscoll, C.M. The blood-brain barrier in aging and neurodegeneration. *Mol. Psychiatry* **2022**, *27*, 2659–2673. [CrossRef]
85. Lendahl, U.; Nilsson, P.; Betsholtz, C. Emerging links between cerebrovascular and neurodegenerative diseases—A special role for pericytes. *EMBO Rep.* **2019**, *20*, e48070. [CrossRef] [PubMed]
86. Li, L.; Liang, J.; Fu, H. An update on the association between traumatic brain injury and Alzheimer's disease: Focus on Tau pathology and synaptic dysfunction. *Neurosci. Biobehav. Rev.* **2021**, *120*, 372–386. [CrossRef] [PubMed]
87. Zhao, N.; Chung, T.D.; Guo, Z.; Jamieson, J.J.; Liang, L.; Linville, R.M.; Pessell, A.F.; Wang, L.; Searson, P.C. The influence of physiological and pathological perturbations on blood-brain barrier function. *Front. Neurosci.* **2023**, *17*, 1289894. [CrossRef] [PubMed]
88. Elschot, E.P.; Backes, W.H.; Postma, A.A.; van Oostenbrugge, R.J.; Staals, J.; Rouhl, R.P.; Jansen, J.F. A comprehensive view on MRI techniques for imaging blood-brain barrier integrity. *Investig. Radiol.* **2021**, *56*, 10–19. [CrossRef] [PubMed]
89. Griffin, A.D.; Turtzo, L.C.; Parikh, G.Y.; Tolpygo, A.; Lodato, Z.; Moses, A.D.; Nair, G.; Perl, D.P.; Edwards, N.A.; Dardzinski, B.J.; et al. Traumatic microbleeds suggest vascular injury and predict disability in traumatic brain injury. *Brain* **2019**, *142*, 3550–3564. [CrossRef]
90. Sparks, P.; Lawrence, T.; Hinze, S. Neuroimaging in the diagnosis of chronic traumatic encephalopathy: A systematic review. *Clin. J. Sport. Med.* **2020**, *30*, S1–S10. [CrossRef]
91. Gandy, S.; Ikonomovic, M.D.; Mitsis, E.; Elder, G.; Ahlers, S.T.; Barth, J.; Stone, J.R.; DeKosky, S.T. Chronic traumatic encephalopathy: Clinical-biomarker correlations and current concepts in pathogenesis. *Mol. Neurodegener.* **2014**, *9*, 37. [CrossRef]
92. Ranzenberger, L.R.; Das, J.M.; Snyder, T. Diffusion Tensor Imaging. In *StatPearls*; StatPearls Publishing: Treasure Island, FL, USA, 2024. [PubMed]
93. Kim, E.; Yoo, R.E.; Seong, M.Y.; Oh, B.M. A systematic review and data synthesis of longitudinal changes in white matter integrity after mild traumatic brain injury assessed by diffusion tensor imaging in adults. *Eur. J. Radiol.* **2022**, *147*, 110117. [CrossRef]
94. Yoen, H.; Yoo, R.E.; Choi, S.H.; Kim, E.; Oh, B.M.; Yang, D.; Hwang, I.; Kang, K.M.; Yun, T.J.; Kim, J.H.; et al. Blood-brain barrier disruption in mild traumatic brain injury patients with post-concussion syndrome: Evaluation with region-based quantification of dynamic contrast-enhanced MR imaging parameters using automatic whole-brain segmentation. *Korean J. Radiol.* **2021**, *22*, 118. [CrossRef]
95. Ware, J.B.; Sinha, S.; Morrison, J.; Walter, A.E.; Gugger, J.J.; Schneider, A.L.; Dabrowski, C.; Zamore, H.; Wesley, L.; Magdamo, B.; et al. Dynamic contrast enhanced MRI for characterization of blood-brain-barrier dysfunction after traumatic brain injury. *NeuroImage Clin.* **2022**, *36*, 103236. [CrossRef] [PubMed]
96. Oh, S.S.; Lee, E.H.; Kim, J.H.; Seo, Y.B.; Choo, Y.J.; Park, J.; Chang, M.C. The use of dynamic contrast-enhanced magnetic resonance imaging for the evaluation of blood-brain barrier disruption in traumatic brain injury: What is the evidence? *Brain Sci.* **2021**, *11*, 775. [CrossRef] [PubMed]
97. Zaltsman, S.L.; Sharabi, S.; Guez, D.; Daniels, D.; Cooper, I.; Shemesh, C.; Atrakchi, D.; Ravid, O.; Omesi, L.; Rand, D.; et al. Application of Delayed Contrast Extravasation Magnetic Resonance Imaging for Depicting Subtle Blood–Brain Barrier Disruption in a Traumatic Brain Injury Model. *J. Neurotrauma* **2024**, *41*, 430–446. [CrossRef] [PubMed]
98. Iyad, N.; Ahmad, M.S.; Alkhatib, S.G.; Hjouj, M. Gadolinium contrast agents-challenges and opportunities of a multidisciplinary approach: Literature review. *Eur. J. Radiol. Open* **2023**, *11*, 100503. [CrossRef]
99. Rasschaert, M.; Weller, R.O.; Schroeder, J.A.; Brochhausen, C.; Idée, J.M. Retention of gadolinium in brain parenchyma: Pathways for speciation, access, and distribution. A critical review. *J. Magn. Reson. Imaging* **2020**, *52*, 1293–1305. [CrossRef]
100. Stanescu, A.L.; Shaw, D.W.; Murata, N.; Murata, K.; Rutledge, J.C.; Maloney, E.; Maravilla, K.R. Brain tissue gadolinium retention in pediatric patients after contrast-enhanced magnetic resonance exams: Pathological confirmation. *Pediatr. Radiol.* **2020**, *50*, 388–396. [CrossRef] [PubMed]
101. Byrnes, K.R.; Wilson, C.M.; Brabazon, F.; Von Leden, R.; Jurgens, J.S.; Oakes, T.R.; Selwyn, R.G. FDG-PET imaging in mild traumatic brain injury: A critical review. *Front. Neuroenergetics* **2014**, *5*, 13. [CrossRef]
102. Gandy, S.; DeKosky, S.T. [18F]-T807 tauopathy PET imaging in chronic traumatic encephalopathy. *F1000Research* **2014**, *3*, 229. [CrossRef]
103. Selwyn, R.; Hockenbury, N.; Jaiswal, S.; Mathur, S.; Armstrong, R.C.; Byrnes, K.R. Mild traumatic brain injury results in depressed cerebral glucose uptake: An 18FDG PET Study. *J. Neurotrauma* **2013**, *30*, 1943–1953. [CrossRef]
104. Wu, H.M.; Huang, S.C.; Hattori, N.; Glenn, T.C.; Vespa, P.M.; Yu, C.L.; Hovda, D.A.; Phelps, M.E.; Bergsneider, M. Selective metabolic reduction in gray matter acutely following human traumatic brain injury. *J. Neurotrauma* **2004**, *21*, 149–161. [CrossRef] [PubMed]
105. Hermanides, J.; Hong, Y.T.; Trivedi, M.; Outtrim, J.; Aigbirhio, F.; Nestor, P.J.; Guilfoyle, M.; Winzeck, S.; Newcombe, V.F.; Das, T.; et al. Metabolic derangements are associated with impaired glucose delivery following traumatic brain injury. *Brain* **2021**, *144*, 3492–3504. [CrossRef] [PubMed]
106. Jalloh, I.; Carpenter, K.L.H.; Helmy, A.; Carpenter, T.A.; Menon, D.K.; Hutchinson, P.J. Glucose metabolism following human traumatic brain injury: Methods of assessment and pathophysiological findings. *Metab. Brain Dis.* **2015**, *30*, 615–632. [CrossRef] [PubMed]

107. Worley, G.; Hoffman, J.M.; Paine, S.S.; Kalman, S.L.; Claerhout, S.J.; Boyko, O.B.; Kandt, R.S.; Santos, C.C.; Hanson, M.W.; Oakes, W.J.; et al. 18-Fluorodeoxyglucose positron emission tomography in children and adolescents with traumatic brain injury. *Dev. Med. Child Neurol.* **1995**, *37*, 213–220. [CrossRef]
108. Bergsneider, M.; Hovda, D.A.; Lee, S.M.; Kelly, D.F.; McArthur, D.L.; Vespa, P.M.; Lee, J.H.; Huang, S.C.; Martin, N.A.; Phelps, M.E.; et al. Dissociation of cerebral glucose metabolism and level of consciousness during the period of metabolic depression following human traumatic brain injury. *J. Neurotrauma* **2000**, *17*, 389–401. [CrossRef]
109. Morgan, R.; Prosapio, J.; Kara, S.; Sonty, S.; Youssef, P.; Nedd, K. Preliminary clinical diagnostic criteria for chronic traumatic encephalopathy: A case report and literature review. *Interdiscip. Neurosurg.* **2021**, *26*, 101290. [CrossRef]
110. Buchsbaum, M.S.; Simmons, A.N.; DeCastro, A.; Farid, N.; Matthews, S.C. Clusters of low 18F-fluorodeoxyglucose uptake voxels in combat veterans with traumatic brain injury and post-traumatic stress disorder. *J. Neurotrauma* **2015**, *32*, 1736–1750. [CrossRef]
111. Komura, A.; Kawasaki, T.; Yamada, Y.; Uzuyama, S.; Asano, Y.; Shinoda, J. Cerebral glucose metabolism in patients with chronic mental and cognitive sequelae after a single blunt mild traumatic brain injury without visible brain lesions. *J. Neurotrauma* **2019**, *36*, 641–649. [CrossRef]
112. Leuzy, A.; Chiotis, K.; Lemoine, L.; Gillberg, P.-G.; Almkvist, O.; Rodriguez-Vieitez, E.; Nordberg, A. Tau PET imaging in neurodegenerative tauopathies—Still a challenge. *Mol. Psychiatry* **2019**, *24*, 1112–1134. [CrossRef]
113. Shivamurthy, V.K.; Tahari, A.K.; Marcus, C.; Subramaniam, R.M. Brain FDG PET and the diagnosis of dementia. *Am. J. Roentgenol.* **2015**, *204*, W76–W85. [CrossRef]
114. Moghbel, M.C.; Saboury, B.; Basu, S.; Metzler, S.D.; Torigian, D.A.; Långström, B.; Alavi, A. Amyloid-β imaging with PET in Alzheimer's Disease: Is It Feasible with Current Radiotracers and Technologies? Springer: Berlin/Heidelberg, Germany, 2012.
115. Saint-Aubert, L.; Lemoine, L.; Chiotis, K.; Leuzy, A.; Rodriguez-Vieitez, E.; Nordberg, A. Tau PET imaging: Present and future directions. *Mol. Neurodegener.* **2017**, *12*, 19. [CrossRef] [PubMed]
116. Barrio, J.R.; Small, G.W.; Wong, K.P.; Huang, S.C.; Liu, J.; Merrill, D.A.; Giza, C.C.; Fitzsimmons, R.P.; Omalu, B.; Bailes, J.; et al. In vivo characterization of chronic traumatic encephalopathy using [F-18] FDDNP PET brain imaging. *Proc. Natl. Acad. Sci. USA* **2015**, *112*, E2039–E2047. [CrossRef] [PubMed]
117. Harada, R.; Okamura, N.; Furumoto, S.; Tago, T.; Yanai, K.; Arai, H.; Kudo, Y. Characteristics of tau and its ligands in PET imaging. *Biomolecules* **2016**, *6*, 7. [CrossRef] [PubMed]
118. Choi, S.R.; Schneider, J.A.; Bennett, D.A.; Beach, T.G.; Bedell, B.J.; Zehntner, S.P.; Krautkramer, M.J.; Kung, H.F.; Skovronsky, D.M.; Hefti, F.; et al. Correlation of amyloid PET ligand florbetapir F 18 binding with Aβ aggregation and neuritic plaque deposition in postmortem brain tissue. *Alzheimer Dis. Assoc. Disord.* **2012**, *26*, 8–16. [CrossRef]
119. Lister-James, J.; Pontecorvo, M.J.; Clark, C.; Joshi, A.D.; Mintun, M.A.; Zhang, W.; Lim, N.; Zhuang, Z.; Golding, G.; Choi, S.R.; et al. Florbetapir f-18: A histopathologically validated Beta-amyloid positron emission tomography imaging agent. In *Seminars in Nuclear Medicine*; Elsevier: Amsterdam, The Netherlands, 2011; pp. 300–304.
120. Baker, S.L.; Harrison, T.M.; Maass, A.; La Joie, R.; Jagust, W.J. Effect of off-target binding on 18F-flortaucipir variability in healthy controls across the life span. *J. Nucl. Med.* **2019**, *60*, 1444–1451. [CrossRef]
121. Dani, M.; Brooks, D.; Edison, P. Tau imaging in neurodegenerative diseases. *Eur. J. Nucl. Med. Mol. Imaging* **2016**, *43*, 1139–1150. [CrossRef]
122. Kantarci, K.; Lowe, V.J.; Boeve, B.F.; Senjem, M.L.; Tosakulwong, N.; Lesnick, T.G.; Spychalla, A.J.; Gunter, J.L.; Fields, J.A.; Graff-Radford, J.; et al. AV-1451 tau and β-amyloid positron emission tomography imaging in dementia with Lewy bodies. *Ann. Neurol.* **2017**, *81*, 58–67. [CrossRef]
123. Joseph, C.R.; Benhatzel, C.M.; Stern, L.J.; Hopper, O.M.; Lockwood, M.D. Pilot study utilizing MRI 3D TGSE PASL (arterial spin labeling) differentiating clearance rates of labeled protons in the CNS of patients with early Alzheimer disease from normal subjects. *Magn. Reson. Mater. Phys. Biol. Med.* **2020**, *33*, 559–568. [CrossRef] [PubMed]
124. Li, K.L.; Zhu, X.; Hylton, N.; Jahng, G.H.; Weiner, M.W.; Schuff, N. Four-phase single-capillary stepwise model for kinetics in arterial spin labeling MRI. *Magn. Reson. Med.* **2005**, *53*, 511–518. [CrossRef]
125. Joseph, C.R.; Lim, J.K.; Grohol, B.N.; Zivcevska, M.; Lencke, J.; Rich, E.D.; Arrasmith, C.J.; Dorman, I.S.; Clark, B.W.; Love, K.; et al. Identifying delay in glymphatic clearance of labeled protons post-acute head trauma utilizing 3D ASL MRI (arterial spin labeling): A pilot study. *Sci. Rep.* **2024**, *14*, 6188. [CrossRef]
126. Joseph, C.R.; Kreilach, A.; Reyna, V.A.; Kepler, T.A.; Taylor, B.V.; Kang, J.; McCorkle, D.; Rider, N.L. Utilizing Reduced Labeled Proton Clearance to Identify Preclinical Alzheimer Disease with 3D ASL MRI. *Case Rep. Neurol.* **2023**, *15*, 177. [CrossRef] [PubMed]
127. Zhao, M.Y.; Fan, A.P.; Chen, D.Y.T.; Sokolska, M.J.; Guo, J.; Ishii, Y.; Shin, D.D.; Khalighi, M.M.; Holley, D.; Halbert, K.; et al. Cerebrovascular reactivity measurements using simultaneous 15O-water PET and ASL MRI: Impacts of arterial transit time, labeling efficiency, and hematocrit. *Neuroimage* **2021**, *233*, 117955. [CrossRef] [PubMed]
128. Voorter, P.H.M.; van Dinther, M.; Jansen, W.J.; Postma, A.A.; Staals, J.; Jansen, J.F.A.; van Oostenbrugge, R.J.; van der Thiel, M.M.; Backes, W.H. Blood–brain barrier disruption and perivascular spaces in small vessel disease and neurodegenerative diseases: A review on MRI methods and insights. *J. Magn. Reson. Imaging* **2024**, *59*, 397–411. [CrossRef] [PubMed]
129. Haarbauer-Krupa, J.; Pugh, M.J.; Prager, E.M.; Harmon, N.; Wolfe, J.; Yaffe, K. Epidemiology of chronic effects of traumatic brain injury. *J. Neurotrauma* **2021**, *38*, 3235–3247. [CrossRef]

130. Snowden, T.M.; Hinde, A.K.; Reid, H.M.; Christie, B.R. Does mild traumatic brain injury increase the risk for dementia? A systematic review and meta-analysis. *J. Alzheimer's Dis.* **2020**, *78*, 757–775. [CrossRef]
131. Takahata, K.; Kimura, Y.; Sahara, N.; Koga, S.; Shimada, H.; Ichise, M.; Saito, F.; Moriguchi, S.; Kitamura, S.; Kubota, M.; et al. PET-detectable tau pathology correlates with long-term neuropsychiatric outcomes in patients with traumatic brain injury. *Brain* **2019**, *142*, 3265–3279. [CrossRef]
132. Gorgoraptis, N.; Li, L.M.; Whittington, A.; Zimmerman, K.A.; Maclean, L.M.; McLeod, C.; Ross, E.; Heslegrave, A.; Zetterberg, H.; Passchier, J.; et al. In vivo detection of cerebral tau pathology in long-term survivors of traumatic brain injury. *Sci. Transl. Med.* **2019**, *11*, eaaw1993. [CrossRef]

International Journal of
Molecular Sciences

MDPI

Review

Pathological Calcium Signaling in Traumatic Brain Injury and Alzheimer's Disease: From Acute Neuronal Injury to Chronic Neurodegeneration

Stephan Neuschmid [1,2], Carla Schallerer [1,2], Barbara E. Ehrlich [2,*] and Declan McGuone [3,*]

1 School of Medicine and Health, Technical University of Munich, 81675 Munich, Germany; stephan.neuschmid@yale.edu (S.N.); carla.schallerer@yale.edu (C.S.)
2 Department of Pharmacology, Yale School of Medicine, New Haven, CT 06510, USA
3 Department of Pathology, Yale School of Medicine, New Haven, CT 06510, USA
* Correspondence: barbara.ehrlich@yale.edu (B.E.E.); declan.mcguone@yale.edu (D.M.)

Abstract

Loss of calcium homeostasis, a shared feature of Alzheimer's Disease (AD) and Traumatic Brain Injury (TBI), activates enzyme-dependent cascades that promote protein misfolding, degrade synaptic architecture, impair axonal transport, and lead to neuronal death. Epidemiological studies identify TBI as a major risk factor for AD, yet the mechanistic basis for this association remains incompletely understood. Evidence from human and experimental studies implicate calcium dysregulation as a central link, triggering interconnected kinase, phosphatase, and protease networks that drive AD hallmark pathology, including amyloid-β (Aβ) accumulation and tau hyperphosphorylation. The calcium-dependent protease calpain is a key node in this network, regulating downstream enzyme activity, and cleaving essential scaffolding and signaling proteins. Selective vulnerability of the hippocampus and white matter to calcium-mediated damage may underlie cognitive deficits common to both conditions. In preclinical TBI and AD models, pharmacological inhibition of calcium-dependent enzymes confers neuroprotection. Recognizing disrupted calcium signaling as an upstream driver of post-traumatic neurodegeneration may enable early interventions to reduce AD risk among TBI survivors.

Keywords: axonal degeneration; beta-amyloid; calcium dysregulation; calpain; cognitive impairment; post-traumatic dementia; protein misfolding; tau; synaptic dysfunction; white matter injury

1. Introduction

Alzheimer's Disease (AD), the leading cause of dementia, is projected to affect more than 150 million people globally by 2050 [1]. Age and genetic factors such as the ε4 allele of apolipoprotein E (*APOE4*) are established risks, but decades of epidemiological evidence has also identified Traumatic Brain Injury (TBI) as an important, potentially modifiable environmental factor for AD and all-cause dementia (ACD) [2]. Repetitive TBI accelerates cognitive decline, and a dose–response relationship exists between injury severity and risk of dementia [3]. Risk magnitude is influenced by age at injury, baseline cognitive reserve, *APOE4* genotype, and trauma biomechanics [4]. Epidemiological data indicate a significant association between TBI and dementia risk, whereas findings regarding AD are less robust, inconsistent, and show substantial statistical and clinical heterogeneity (Table 1). These discrepancies are partly attributable to differences in diagnostic criteria for

TBI and AD, variations in study design, and incomplete reporting or characterization of trauma biomechanics, which may also introduce recall bias [5]. Differences in covariate adjustment and heterogeneity in the length of clinical follow-up (months to decades) limits conclusiveness and raises the possibility of reverse causation, for example, when dementia-related cognitive impairment increases the risk of falls and subsequent TBI. Despite inconsistent evidence for a direct causal link between TBI and AD, the high prevalence and frequent identification of TBI as a dementia risk factor warrants careful scrutiny and structured follow-up in clinical settings, particularly for repetitive or severe injuries. Because most TBI's are mild and often self-limiting, and because TBI is highly heterogenous, the lifetime incidence of TBI-related neurodegeneration remains uncertain [6]. Current data are largely heterogenous and retrospective, limiting causal inference and underscoring the need for larger, well-characterized prospective cohorts and biomarker-based studies.

Table 1. Epidemiological evidence supporting TBI as a risk factor for AD and Dementia.

Study Design	Sample Size	Outcome	Pooled Effect Size	95% CI	TBI Characteristics	Included Articles
Systematic Review & Meta-Analysis [7]	7,634,844	Dementia	RR: 1.66	1.42–1.93	No Restrictions (Mild to Severe)	32
Systematic Review & Meta-Analysis [8]	3,149,740	AD	RR: 1.18	1.11–1.25	Mild TBI	5
Systematic Review & Meta-Analysis [9]	2,820,181 8,684,485	AD Dementia	OR: 1.02 OR: 1.81	0.91–1.15 1.53–2.14	No Restrictions (Mild to Severe)	7 21
Meta-Analysis [10]	3,263,207	AD Dementia	OR: 1.03 OR: 1.93	0.06–16.33 1.47–2.55	No Restrictions (Mild to Severe)	18
Systematic Review & Meta-Analysis [11]	7,100,000	AD Dementia	HR: 1.30 HR: 1.95	0.88–1.91 1.55–2.45	No Restrictions (Severity-Dependent Risk)	13
Systematic Review & Meta-Analysis [12]	2,013,197	AD Dementia	RR: 1.51 RR: 1.63	1.26–1.80 1.34–1.99	No Restrictions (Any Head Injury or TBI)	32
Umbrella Systematic Review & Meta-Analysis [13]	20,684,373	AD Dementia	OR: 1.18 OR: 1.81	1.11–1.25 1.53–2.14	No Restrictions (Severity-Dependent Risk)	6
Meta-Analysis [5]	29,181 10,331	AD Dementia	OR: 1.40 OR: 1.36	1.02–1.90 0.84–2.19	Mild TBI	57
Systematic Review & Meta-Analysis [14]	2,351,334	Dementia	OR: 1.96	1.70–2.26	Mild TBI	21
Systematic Review & Meta-Analysis [2]	1,936,593 1,773,342	AD Dementia	OR: 1.60 OR: 1.79	1.44–1.77 1.66–1.92	No Restrictions (Any Head Injury or TBI)	107
Meta-Analysis [15]	4,289,548	AD AD	RR: 1.17 RR: 1.30	1.05–1.29 1.01–1.59	Any TBI (Moderate to Severe)	17

Summary of epidemiological studies assessing TBI as a risk factor for Alzheimer's Disease and all-cause dementia. The table lists systematic reviews and meta-analyses published over the past 10 years that examine TBI-related neurodegeneration. Reported data include total sample size, outcome (AD or Dementia), effect sizes with corresponding 95% confidence intervals, TBI exposure characteristics (e.g., severity), and number of included studies. Many reviews draw on overlapping primary cohorts or case–control studies; therefore, effect estimates are not additive. Abbreviations: CI: Confidence Interval; HR: Hazard Ratio; OR: Odds Ratio; RR: Risk Ratio.

The strong epidemiologic and mechanistic associations between TBI and AD have focused attention on shared biological pathways linking acute neuronal injury to chronic neurodegeneration. Disruption of neuronal calcium homeostasis has emerged as a leading candidate mechanism [16,17]. Under physiological conditions, neurons maintain a steep calcium gradient with extracellular concentrations (~1 mM) approximately 10,000 times higher than cytosolic levels (~100 nM), which allows tightly regulated signaling for synaptic transmission, plasticity, gene transcription, and cell survival [18]. In TBI, this gradient collapses within seconds of injury due to excitotoxicity, mitochondrial and endoplasmic reticulum (ER) dysfunction, and membrane disruption [17]. In AD, calcium overload arises through mechanisms that include Aβ-induced calcium influx, tau-related organelle stress, and altered pump or channel function [19] (Figure 1). Loss of calcium regulation across neuronal subcellular compartments (e.g., the axon, synapse, and ER) activates overlapping kinase, phosphatase, and protease cascades which in turn promote Aβ and tau pathology [20]. Sustained intracellular calcium elevation is neurotoxic and consistently associated with neurodegeneration in human postmortem studies [21].

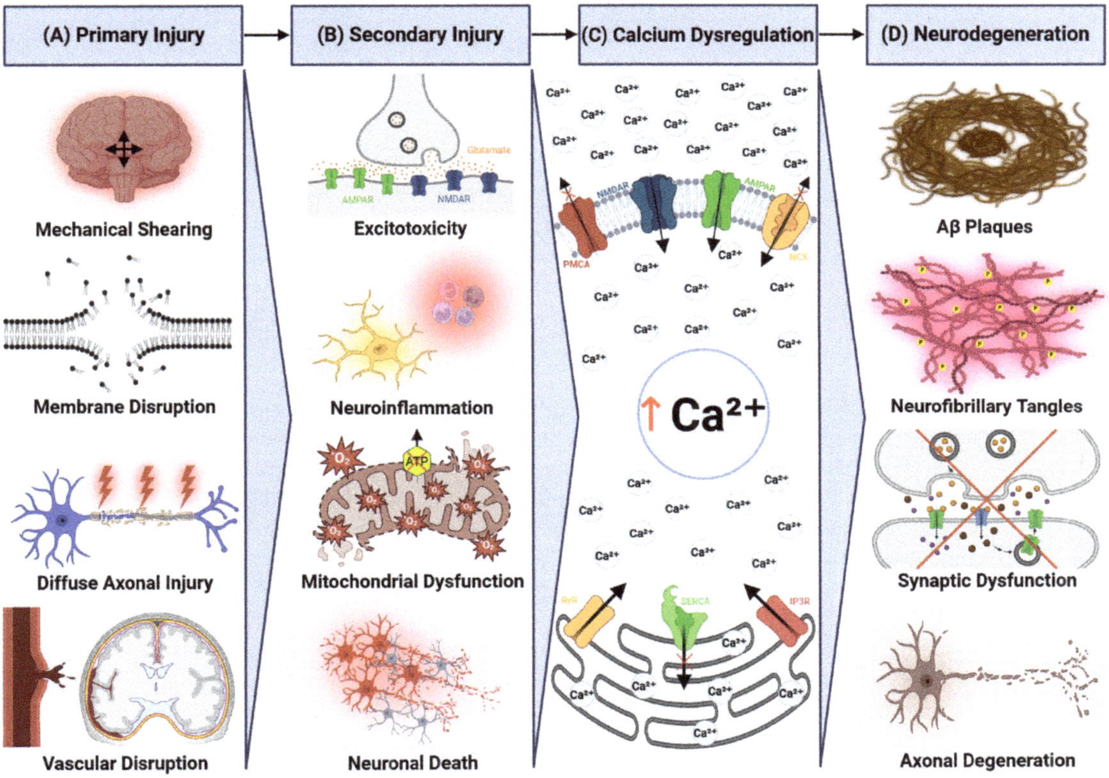

Figure 1. Schematic overview of TBI-induced neurodegeneration, highlighting calcium dysregulation as a central driver. (**A**) Primary injury mechanisms, including mechanical shearing, membrane disruption, diffuse axonal injury, and vascular disruption, cause initial damage that subsequently triggers (**B**) secondary injury cascades. These involve excitotoxicity, neuroinflammation, and mitochondrial dysfunction, promoting reactive oxygen species (ROS) production (labeled as O_2^{-}), ATP depletion, neuronal death, and (**C**) disruption of calcium homeostasis. Elevated cytosolic calcium concentration (marked with red arrow) results from increased influx, reduced extrusion, and enhanced release from intracellular stores. (**D**) Sustained calcium imbalance activates chronic neurodegenerative pathways, resulting in Aβ and tau accumulation, synaptic dysfunction, and progressive axonal degeneration. Black arrows indicate physiological direction of transport, while red crosses denote dysfunction. Created with BioRender.com (https://www.biorender.com, accessed on 20 August 2025).

This review examines evidence linking calcium dysregulation after TBI to downstream neurodegenerative processes relevant to AD, with an emphasis on calcium-activated kinases (CDK5, GSK3β, PKC, CaMKII, ERK, and DYRK1A), phosphatases (PP2A and CaN), and proteases (calpains). Although these enzymes have been studied individually, their integrated network dynamics and contributions to vulnerability or resilience after TBI remain poorly defined [22]. Pharmacological inhibition of calcium-dependent enzymes reduces AD pathology in multiple preclinical models [23–25]. Collectively, these observations identify impaired calcium signaling as a central mechanistic link between TBI and chronic neurodegeneration.

2. Physiological Calcium Signaling and Enzymatic Regulation

Calcium, a ubiquitous second messenger, regulates synaptic transmission, plasticity, survival, and enzymatic activity. Its homeostasis depends on a highly coordinated network spanning distinct cellular compartments to ensure precise spatial and temporal control [26].

Following neurotransmitter binding or membrane depolarization, calcium enters the neuron through N-methyl-D-aspartate receptors (NMDARs), α-amino-3-hydroxy 5-methyl-4-isoxazolepropionate receptors (AMPARs), store-operated calcium (SOC) channels, transient receptor potential (TRP) channels, and voltage-gated calcium channels (VGCCs) [18]. In contrast, signal termination is mediated by sodium/calcium exchangers (NCXs) and the plasma membrane calcium-ATPase (PMCA), which extrude calcium to the extracellular space and prevent toxic accumulation [27].

In the ER, calcium modulates the inositol 1,4,5-trisphosphate receptor (IP3R) and ryanodine receptor (RyR) to promote further calcium release, a process known as calcium-induced calcium release (CICR) [28]. Restoration requires sarcoplasmic–endoplasmic reticulum calcium-ATPase (SERCA) pumps, which maintain the ER as the major calcium store [29].

Mitochondria buffer calcium using the mitochondrial calcium uniporter (MCU) and release it through mitochondrial NCX (mNCX), preventing overload and permeability transition [30]. Crosstalk between organelles occurs at mitochondria-associated ER membranes (MAMs), which coordinate calcium transfer from the ER to mitochondria [31]. Cytosolic buffering proteins (e.g., calmodulin, calbindin, parvalbumin, and S100) fine-tune signal amplitude and duration [32]. Together, this architecture generates spatially restricted calcium microdomains at presynaptic terminals, postsynaptic densities, and MAMs, enabling localized calcium transients without global calcium elevation [31,33].

Disruption of this network after TBI or in AD results in sustained calcium elevation, loss of microdomain compartmentalization, and aberrant activation of calcium-dependent enzymes [19] (Table 2). These enzymes include kinases, phosphatases, and proteases regulated by calcium or calcium-sensitive co-factors. Once activated, kinases phosphorylate targets, phosphatases counteract these modifications to maintain dynamic regulation, and proteases such as calpains cleave structural and signaling proteins [34,35]. Their combined activity shapes downstream signaling to determine neuronal fate.

Table 2. Network of calcium-regulated enzymes implicated in TBI-induced neurodegeneration.

Enzyme	Calcium-Dependence	Physiological Function	Major Targets	Pathological Consequence	Enzyme Interactions	Refs.
CaMKII	Direct (Calcium/CaM)	Memory Consolidation, Neurotransmitter Release, Synaptic Plasticity	AMPAR, APP, CREB, CRMP2, NF-κB, NMDAR, Tau	AD Hallmark Pathology, Axonal Damage, Synaptic Dysfunction	Calpain CaN CDK5 GSK3β	[36,37]
CDK5	Indirect (Calpain)	Cytoskeletal Dynamics, Neuronal Development, Synaptic Plasticity	APP, BACE1, CRMP2, DRP1, Dynamin-1, Neurofilaments, PSD-95, STAT, Tau	AD Hallmark Pathology, Axonal Damage, Synaptic Dysfunction	Calpain CaMKII ERK GSK3β PP2A	[38–40]
DYRK1A	Indirect (Calpain)	Neuronal Development, Synaptic Plasticity, Transcriptional Control	APP, Dynamin-1, NFAT, Presenilin, STAT, Tau	AD Hallmark Pathology, Neuroinflammation, Synaptic Dysfunction	Calpain, CaN, GSK3β.	[41,42]
ERK	Indirect (Upstream via MAPK cascade, Calpain)	Inflammatory Response, Neuronal Survival, Synaptic Plasticity	α-Secretase, DRP1, NF-κB, STAT, Tau	AD Hallmark Pathology, Neuroinflammation, Oxidative Stress	Calpain CDK5 GSK3β	[43,44]

Table 2. *Cont.*

Enzyme	Calcium-Dependence	Physiological Function	Major Targets	Pathological Consequence	Enzyme Interactions	Refs.
GSK3β	Indirect (Calpain)	Cell Cycle Regulation, Neuronal Development, Synaptic Plasticity	APP, BACE1, CREB, CRMP2, DRP1, Dynamin-1, Kinesin, NFAT, NF-κB, Presenilin, PSD-95, STAT, Tau	AD Hallmark Pathology, Axonal Damage, Synaptic Dysfunction.	Calpain CaMKII CaN CDK5 DYRK1A ERK PKC PP2A	[45–47]
PKC	Direct/Indirect (Isoform dependent, Calpain)	Cell Signaling, Memory Formation, Neuronal Survival	AMPAR, APP, CREB, GAP43, NF-κB, NMDAR, STAT, Tau	AD Hallmark Pathology, Protective (e.g., PKCε) vs. Toxic (e.g., PKCδ), Structural Damage, Synaptic Dysfunction	Calpain GSK3β PP2A	[48–50]
CaN	Direct (Calcium/CaM, Calpain)	Inflammatory Response, Neuronal Survival, Synaptic Plasticity	AMPAR, BAD, DRP1, Dynamin 1, MAP, NFAT, NF-κB, NMDAR, Tau	AD Hallmark Pathology, Axonal Damage, Synaptic Dysfunction	Calpain CaMKII DYRK1A GSK3β	[51,52]
PP2A	Indirect (Calcium-binding subunits, Calpain)	Major Tau Phosphatase, Neuronal Survival, Signal Transduction	AMPAR, APP, CREB, DRP1, Neurofilaments, NF-κB, NMDAR, Tau	Decreased Activity: AD Hallmark Pathology, Synaptic Dysfunction	Calpain CDK5 GSK3β PKC	[53–55]
Calpain	Direct (Calcium)	Apoptosis Regulation, Cytoskeletal Remodeling, Signal Transduction	AMPAR, APP, CRMP2, DRP1, Dynamin-1, GAP43, IP3R, MAP, NCX, Neurofilaments, NMDAR, p35, PMCA, Presenilin, PSD-95, Spectrin, Tau, Tubulin, VGCC	Aβ Pathology, Axonal Damage, Enzyme Modulation, Synaptic Dysfunction, Tau Fragmentation	CaMKII CaN CDK5 DYRK1A ERK GSK3β PKC PP2A	[56,57]

Comprehensive overview of calcium-regulated enzymes implicated in trauma-related neurodegeneration and their pathological roles. The table summarizes calcium-dependence (direct or indirect), physiological functions, major molecular targets, and pathological consequences of calcium-modulated enzymes. It also highlights key enzymatic interactions, emphasizing the interconnected signaling network that collectively drives hallmark neuropathological features. Abbreviations: APP: Amyloid Precursor Protein; CaM: Calmodulin; CaMKII: Calcium/Calmodulin-Dependent Protein Kinase II; CaN: Calcineurin; CDK5: Cyclin-Dependent Kinase 5; CREB: cAMP Response Element-Binding Protein; CRMP2: Collapsin Response Mediator Protein 2; DRP1: Dynamin-Related Protein 1; DYRK1A: Dual-Specificity Tyrosine Phosphorylation-Regulated Kinase 1A; ERK: Extracellular Signal-Regulated Kinase; GAP43: Growth Associated Protein 43; GSK3β: Glycogen Synthase Kinase 3β; MAP: Microtubule Associated Protein; NFAT: Nuclear Factor of Activated T Cells; NF-κB: Nuclear Factor Kappa-B; PKC: Protein Kinase C; PP2A: Protein Phosphatase 2A; PSD-95: Postsynaptic Density Protein 95; Refs: References; STAT: Signal Transducer and Activator of Transcription.

3. Calcium Hypothesis

The calcium hypothesis of AD proposes that intracellular calcium dysregulation serves as an initiating driver of neurodegeneration, preceding and amplifying proteinopathies that define AD, including Aβ plaques and tau-containing neurofibrillary tangles (NFTs) [58]. Although age and genetic susceptibility (e.g., *APOE4*, presenilin mutations) remain major determinants of risk, converging experimental and human evidence demonstrates that TBI disrupts calcium homeostasis in ways that recapitulate early AD-like changes, providing a mechanistic link between acute injury and chronic degeneration [17,19]. Three core criteria support the calcium hypothesis: (1) calcium imbalance occurs prior to AD neuropathology and clinical symptoms; (2) disrupted calcium signaling acts as a convergent pathway for diverse AD risk factors, and (3) bidirectional interactions between calcium dysregulation and Aβ and tau pathology create a self-perpetuating cycle of injury and degeneration [59,60]. Genetic data support an upstream role for calcium dysregulation. Familial presenilin mutations increase ER calcium leak, leading to aberrant calcium release and neuronal hyperexcitability that precedes Aβ and tau pathology [61]. Despite this, the underlying molecular mechanisms remain incompletely defined [62], and no effective disease-modifying therapies have emerged [63]. This gap has renewed interest in calcium dysregulation as a central link between TBI and neurodegeneration.

In TBI, mechanical membrane disruption permits uncontrolled calcium influx [64], compounded by NCX reversal [65]. The resulting depolarization opens VGCCs and initiates excitotoxicity, where excess glutamate release drives further calcium entry through

NMDAR and calcium-permeable AMPAR (CP-AMPAR), lacking the GluA2 subunit [66]. Metabotropic glutamate receptors (mGluRs) stimulate calcium release through IP3 signaling and calcium triggers CICR, progressively diminishing ER stores [67,68]. This depletion disrupts lysosomal calcium signaling by altering its pH and protein clearance, thereby aggravating AD-related proteinopathies [69]. Mitochondrial dysfunction and ATP shortage further compromise PMCA and SERCA activity, exacerbating calcium overload [70].

In AD, presenilin mutations and aging impair calcium buffering, receptor density, and organelle function [26]. These mutations sensitize IP3Rs [71], increase RyR expression [72], impair SERCA function [73], and alter ER leak channels [74], producing sustained calcium release and neuronal hyperexcitability. Presenilin mutant models develop neurodegeneration even in the absence of Aβ, supporting an upstream role for calcium dysregulation [75]. *APOE4* genotype further exacerbates calcium-driven synaptic loss, cognitive decline, and both tau and amyloid pathology following TBI and in AD [76,77].

Human genetic evidence further supports a calcium-centered mechanism by linking TBI and AD risk through shared polygenic architecture. A genome-wide association study (GWAS) from the VA Million Veteran Program (111,494 TBI cases) identified top signals including *NCAM1*, *APOE*, *FTO*, and *FOXP2*, with a single-nucleotide polymorphism (SNP)-based heritability of 0.060. Using bivariate mixer model (MiXeR) analysis, the authors demonstrated that AD harbored fewer risk variants (lower polygenicity), but each had stronger detectable effects (higher discoverability) compared to TBI, and that approximately 60% of influential AD variants are shared with TBI, despite near-zero genome-wide genetic correlation, implying mixed effect directions and shared architecture not fully captured by correlation analysis alone [78]. A separate GWAS, evaluating post-injury outcomes in the CENTER-TBI/TRACK-TBI cohort (5268 TBI cases), estimated a liability-scale heritability of 0.26 but found no genome-wide significant associations [79]. The authors attributed the null genome-wide findings to limited power and emphasized a need for larger, ancestrally diverse cohorts, harmonization of TBI phenotypes, and future replication studies. Although no canonical calcium-channel coding genes reached genome-wide significance in either GWAS, sub-threshold calcium-relevant signals in TBI and pathway-level findings in AD support calcium-related biology as a shared feature, supporting polygenic approaches that aggregate sub-threshold effects across calcium-regulatory genes, for example, using PGS-Depot or PGSFusion, to stratify TBI survivors by predicted risk of calcium-stress-related neurodegeneration [80].

Calcium imbalance drives ER stress, mitochondrial dysfunction, oxidative stress, impaired autophagy, lysosomal dysfunction, and neuroinflammation, all of which contribute to Aβ pathology [81–84]. Once Aβ accumulates, it further disrupts calcium homeostasis through several mechanisms, including membrane insertion to form calcium-permeable pores [85], NMDAR modulation (particularly NR2B subtypes) [86], and dysregulation of RyR and IP3R signaling [87,88]. NR2B upregulation correlates with hippocampal degeneration and cognitive decline [89], and pharmacological inhibition mitigates many of these effects [90].

Aβ oligomers induce oxidative stress [91] and mitochondrial injury [92] through NMDAR activation. L-type VGCCs mediate Aβ-induced calcium influx [93], and their inhibition is neuroprotective [94]. Aβ also promotes AMPAR internalization, impairing synaptic function [95], although selective AMPAR subtypes may exert neuroprotective effects by enhancing non-amyloidogenic (α-secretase-mediated) APP processing [96].

In both TBI and AD, calcium overload activates calpains, which cleave calcium-regulatory proteins and worsen dysregulation. Calpain degrades NCX and PMCA, impairing extrusion [97,98], and cleaves IP3R to enhance ER calcium release [99]. Additional substrates include VGCCs [100], AMPARs [101], and NMDARs [102], further perpetuating excitotoxicity and synaptic dysfunction. The centrality of these proteins is underscored by the fact that nearly all have been pharmacologically targeted in preclinical models, with several advancing to clinical trials and some incorporated into current FDA-approved therapeutic strategies [103].

4. From APP to Aβ: Calcium-Mediated Aβ Pathology

Following TBI, calcium dysregulation activates enzymatic cascades that promote amyloidogenic APP processing, resulting in early Aβ accumulation through mechanisms overlapping with core molecular events implicated AD [104]. Calcium-dependent processes triggered by TBI therefore contribute directly to AD-related neuropathology, establishing calcium imbalance as a plausible mechanistic link between acute neuronal injury and chronic neurodegeneration.

APP, a type I transmembrane glycoprotein enriched in neurons, can undergo two mutually exclusive proteolytic pathways (Figure 2). In the non-amyloidogenic pathway, α-secretase (primarily ADAM10) cleaves APP within the Aβ domain, precluding Aβ formation [105]. In contrast, amyloidogenic processing begins with β-secretase (BACE1) cleavage, generating the C-terminal fragment C99, which is subsequently cleaved by γ-secretase (with presenilin as a catalytic subunit), predominantly producing Aβ40 and Aβ42 [106]. Aβ42 is particularly prone to aggregation and neurotoxicity [107].

TBI induces rapid APP upregulation and Aβ accumulation, particularly Aβ42, in the brain and cerebrospinal fluid (CSF) within hours of injury [108–110]. Animal models demonstrate that TBI-induced Aβ oligomers share conformational features with those in AD [111], and are associated with hippocampal atrophy and cognitive impairment [112]. Experimental TBI accelerates intra-axonal Aβ accumulation, lipid peroxidation, tau pathology, and cognitive impairment in transgenic models [113,114]. Both BACE1 expression and γ-secretase components, including presenilin-1, are upregulated in the hippocampus and cortex within 24–72 h of injury [115–117], and their inhibition reduces Aβ accumulation and cognitive decline [118].

Elevated intracellular calcium influences multiple steps of the amyloidogenic cascade. Calcium enhances acidification of endosomal and lysosomal compartments, optimizing BACE1 catalytic environment [119–121], and calcium/CaM binding further increases BACE1 enzymatic activity by approximately 2.5-fold in vitro [122].

The calcium-dependent protease calpain acts as a central mediator of post-traumatic neurodegenerative signaling. Normally restrained and tightly regulated by the endogenous inhibitor calpastatin [123], calpain becomes activated within minutes after TBI, persisting for days to weeks [56]. In AD, calpain is similarly elevated [124], where it promotes Aβ pathology by upregulating BACE1 expression [125]. Inhibition of calpain reduces BACE1 activity, Aβ accumulation, neuroinflammation, and memory deficits [25,125].

Following both controlled cortical impact [126] and blast exposure [127], calpain cleaves the CDK5 activator p35 to p25, resulting in sustained CDK5 activation and altered substrate specificity [128]. CDK5/p25 phosphorylates APP at crucial site Thr668, thereby enhancing its trafficking and susceptibility to BACE1 cleavage [129,130]. It also phosphorylates BACE1 directly, and the transcription factor STAT3, which in turn further upregulates BACE1 and presenilin-1 expression [131–133]. These effects are reversible with CDK5 inhibition and have been observed in transgenic models and human AD brains [23,131].

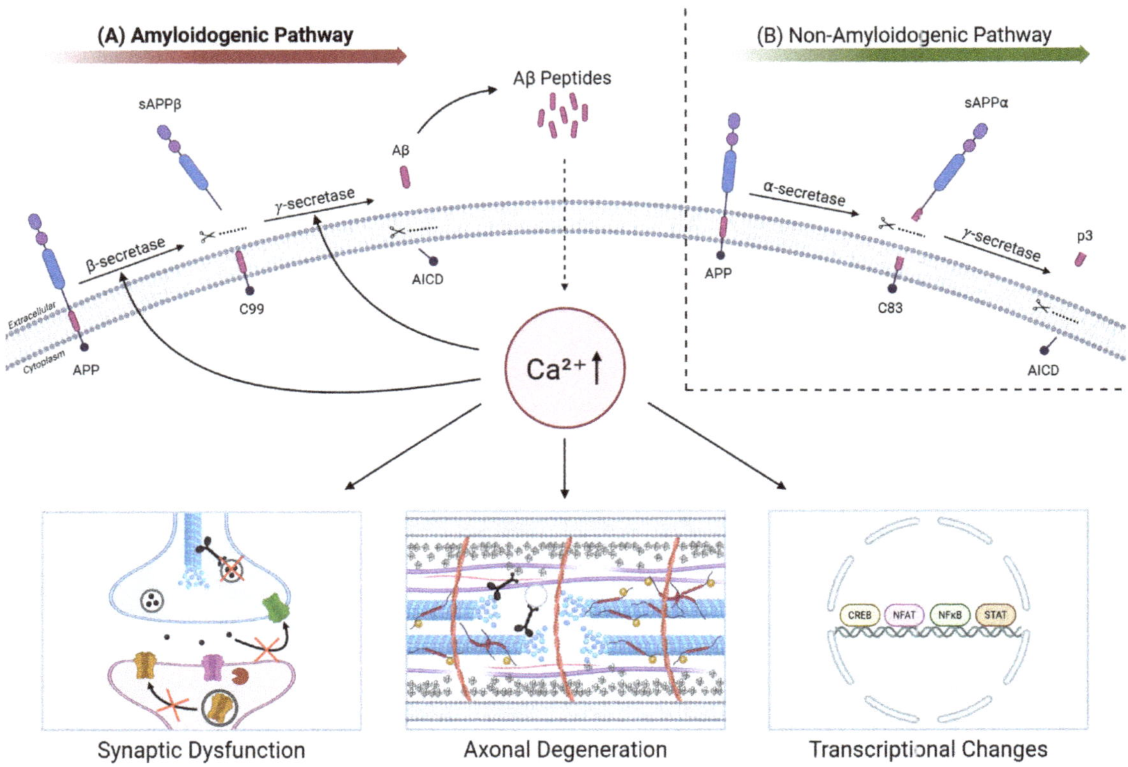

Figure 2. Integrative model of APP processing and Aβ-mediated calcium dysregulation driving downstream pathology. (**A**) In the amyloidogenic pathway, β-secretase generates a soluble N-terminal (sAPPβ) and C-terminal fragment (C99), which γ-secretase processes into pathological Aβ peptides and APP intracellular domain (AICD). Aβ accumulation in turn triggers calcium dysregulation, activating enzymatic cascades that cleave or aberrantly phosphorylate structural and signaling proteins. These changes impair vesicle trafficking and recycling, reduce postsynaptic receptor density, destabilize axons and synapses, and alter transcriptional regulation—ultimately promoting synaptic dysfunction, axonal degeneration, and persistent transcriptional changes. Additionally, Aβ-induced calcium dysregulation further enhances amyloidogenic processing, establishing a self-perpetuating cycle. Black arrows depict a simplified representation of pathological calcium signaling and downstream pathologies triggered by Aβ. (**B**) In the non-amyloidogenic pathway, α-secretase cleaves APP within the Aβ domain to generate sAPPα and C83. Subsequent γ-secretase cleavage yields p3 peptide and AICD. Created with BioRender.com (https://www.biorender.com, accessed on 20 August 2025).

GSK3β activity, enhanced by p25 binding [134] and calpain-mediated cleavage [135], further drives amyloidogenesis. GSK3β phosphorylates APP at Thr668, and BACE1 at Thr252, promoting Aβ accumulation [136,137], whereas inhibition of GSK3β reduces Aβ hallmark pathology [138]. GSK3β activity is increased in AD brains [139] and strengthened by familial AD presenilin-1 mutations [140,141]. After TBI, GSK3β is transiently inhibited by protein kinase B (PKB)-mediated phosphorylation [142], but becomes persistently activated in the subacute phase, contributing to secondary injury [143].

PKC activity is dynamically elevated post-TBI [144], and has been identified as an early and persistent kinase targeting multiple AD-associated core proteins [145]. Calcium directly activates conventional PKC isoforms [48], whereas others undergo calpain-mediated cleavage to alter activity [146]. PKCε promotes non-amyloidogenic α-secretase cleavage, but is reduced in AD, possibly due to inhibition by GSK3β or Aβ peptides [147,148]. In contrast, PKCδ, activated in ischemia and AD [149], correlates with BACE1 expression, and enhances Aβ production. Inhibition or knockout of PKCδ reduces Aβ pathology [150]. Other PKC isoforms such as PKC-λ/ι promote BACE1 transcription by phosphorylation of NF-κB [151]. CaMKII may also contribute by phosphorylating APP at Thr654/Ser655 and activating NF-κB, although in vivo evidence remains limited [152,153].

Phosphatases oppose amyloidogenic processing. PP2A dephosphorylates APP at Thr668, suppressing Aβ production [154], but its activity is diminished in TBI and AD, leading to sustained APP phosphorylation and BACE1-mediated cleavage [155,156]. Aβ inhibits PP2A, amplifying cell injury [157]. PP2A reactivation attenuates amyloidogenic processing by modulating BACE1 and presenilin [158], while also reducing both astrogliosis and expression of the senescence marker p21 [159]. CaN, activated by calcium/CaM binding or calpain-mediated cleavage [160], enhances BACE1 expression through NFAT signaling promoting excitotoxicity and neuroinflammation, whereas inhibition reduces Aβ pathology and cognitive decline [161–163].

MAPK/ERK, activated downstream of calcium and calpain [164], increases BACE1 activity through DRP1 and STAT1 phosphorylation [165,166]. DRP1 is also phosphorylated by CDK5 and GSK3β, and dephosphorylated by PP2A and CaN, thereby linking aberrant mitochondrial fission with increased neuronal vulnerability to Aβ-induced toxicity [167]. Aβ oligomers further activate ERK, establishing a self-amplifying cycle [168]. DYRK1A phosphorylates APP and presenilin, leading to increased enzymatic activity and Aβ accumulation [169,170]. Calcium overload also promotes ER calcium leak, mitochondrial uptake, and unfolded protein response (UPR) activation, which upregulates BACE1 through eukaryotic initiation factor 2α (eIF2α) [171,172].

Together, these pathways define a calcium-APP-Aβ axis activated by TBI and sustained in AD. Central mediators include calpain, CDK5, GSK3β, PKC isoforms, DYRK1A, MAPK/ERK, PP2A, and CaN, all of which influence BACE1 or γ-secretase activity through transcriptional or posttranslational modifications. Their activation by calcium dysregulation supports the view that calcium imbalance is not a consequence, but rather a proximal driver of Aβ pathology in both acute injury and chronic neurodegeneration.

5. Tau Pathology: Calcium-Dependent Disruption of Microtubule Homeostasis

Hyperphosphorylated tau accumulation is a defining feature of AD and other tauopathies including chronic traumatic encephalopathy (CTE), which is associated with repetitive mild TBI [173]. Tau, encoded by the *MAPT* gene on chromosome 17, exists as six isoforms, generated by alternative splicing [174]. Under physiological conditions, tau stabilizes microtubules and supports axonal transport through interactions with dynein and kinesin motor proteins. These functions are regulated by posttranslational modifications, particularly phosphorylation at multiple serine and threonine residues [175]. Under pathological conditions, tau becomes hyperphosphorylated, detaches from microtubules, and aggregates into β-sheet-rich paired helical filaments (PHFs) that form intracellular NFTs [176]. This process is driven by an imbalance between kinases and phosphatases, many of which are calcium-sensitive. Disruption of calcium homeostasis after TBI promotes tau phosphorylation and aggregation, linking acute injury to chronic tau pathology [177] (Figure 3).

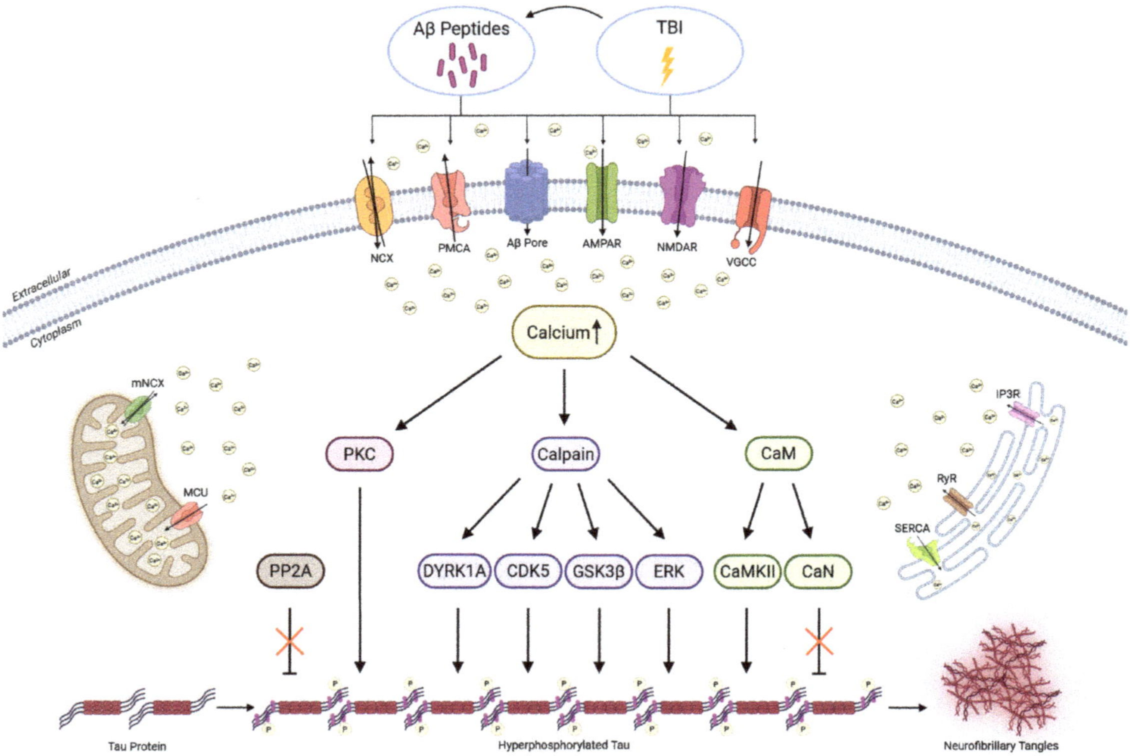

Figure 3. TBI- and Aβ-induced calcium dysregulation activates calcium-dependent signaling cascades that promote tau hyperphosphorylation and NFT formation. Excess calcium influx from the extracellular space and increased release from intracellular stores (ER and mitochondria) cause sustained cytosolic calcium elevation. This overload activates kinases (PKC, DYRK1A, CDK5, GSK3β, ERK, CaMKII), proteases (calpains), and co-factors (CaM), while impairing phosphatases (PP2A, CaN), thereby shifting the balance towards phosphorylation. This enzymatic imbalance promotes tau hyperphosphorylation, detachment, misfolding, aggregation, and ultimately NFT formation. Black arrows indicate a potential activation pattern of the interconnected enzyme network triggered by calcium dysregulation, whereas red crosses denote loss of physiological dephosphorylation capacity. Labeled calcium-signaling proteins mediate cytosolic calcium overload, with black arrows indicating the physiological direction of calcium transport. Created with BioRender.com (https://www.biorender.com, accessed on 20 August 2025).

Calcium dysregulation and tau pathology reinforce each other in a self-perpetuating cycle. Hyperphosphorylated tau reduces nuclear calcium levels and perturbs ER and mitochondrial calcium handling and communication, while suppressing CREB-mediated transcription, increasing neuronal vulnerability [178,179]. Age-related loss of the calcium-buffering protein calbindin correlates with increased NFT formations [180]. RyR-mediated calcium leak promotes tau hyperphosphorylation [181], whereas, in vitro, tau impairs mitochondrial calcium extrusion and promotes overload by mNCX inhibition, causing caspase-mediated death [182]. Pathological tau can also induce aberrant calcium influx, generating spontaneous calcium oscillations [182].

Experimental TBI models demonstrate early tau pathology, with abnormal phosphorylation detectable within 24 h and persisting for days [113]. Postmortem human studies confirm elevated levels of hyperphosphorylated tau after TBI [117]. Although NFTs are typically absent in acute injury [110], they can develop years later in survivors [183,184]. Even a single moderate TBI increases long-term risk of chronic tauopathy [185]. Beyond phosphorylation, tau acetylation promotes mislocalization and aggregation, with overlapping modifications observed in TBI and AD [186].

5.1. Kinase Hyperactivity

Multiple calcium-sensitive kinases implicated in AD are activated after TBI, where they conspire to drive tau hyperphosphorylation and NFT formation through reciprocal priming and feed-forward amplification.

GSK3β functions as a central effector: GSK3β phosphorylates more than 40 serine/threonine residues on tau, destabilizing microtubules and promoting detachment [20]. Pathological tau in turn alters GSK3β acetylation to prevent degradation, creating a feed-forward loop [187]. Calpain cleavage removes GSK3β's autoinhibitory N-terminal domain, generating truncated forms with increased activity [188]. These fragments correlate with calpain activation and tau phosphorylation in AD and injury models [189], and also interact with PP2A, promoting GSK3β dephosphorylation and further enhancing its activity [190]. Presenilin-1 binds both tau and GSK3β, facilitating spatial proximity and linking tau phosphorylation with amyloidogenic APP processing [140]. Pharmacological inhibition (e.g., lithium) reduces tau phosphorylation and improves cognition in TBI models [191].

CDK5 acts as a priming kinase: CDK5 phosphorylates tau at residues that enhance subsequent GSK3β-mediated phosphorylation, accelerating and amplifying hyperphosphorylation [192]. Apart from direct tau phosphorylation [193], CDK5/p25 modifies hundreds of other substrates involved in calcium signaling and neurodegeneration, thereby exacerbating NFT formation and calcium dysregulation [194]. Elevated p25 levels are associated with early-onset AD pathology [195], and CDK5 inhibition confers neuroprotection in rotational TBI models [194].

CaMKII propagates tau pathology: Direct calcium/CaM binding induces CaMKII autophosphorylation at Thr286, increasing activity and driving redistribution from the cytosol to membranes within 30 min of TBI [196]. CaMKII phosphorylates tau at multiple sites [197], and promotes additional calcium influx by phosphorylating CP-AMPARs [198]. Calpain-generated CaMKII fragments display constitutive activity [199], priming tau for further phosphorylation by GSK3β and CDK5 [200]. Persistent CaMKII activation impairs synaptic plasticity and memory consolidation in transgenic mice, compounding post-injury cognitive decline [201].

DYRK1A and ERK serve as auxiliary kinases: DYRK1A phosphorylates tau [202], potentially priming it for subsequent GSK3β phosphorylation [203]. This cascade is amplified by calpain-mediated cleavage, which generates hyperactive DYRK1A fragments that are elevated in human AD brains and CSF [204]. ERK similarly phosphorylates tau at multiple sites [205] and is upregulated early in AD, correlating with NFT progression [206]. Both kinases promote neuroinflammation by activating cytokine pathways. Pharmacological inhibition with calcium channel blockers (e.g., lomerizine) reduces tau pathology and dampens proinflammatory responses, likely by indirectly modulating GSK3β and DYRK1A activity, although lomerizine is not a selective DYRK1A inhibitor [207].

PKC modulates context-dependent effects: Because certain isoforms have been implicated in both neuroprotection and neurotoxicity, PKC's overall role remains context-dependent. Some isoforms phosphorylate tau [48], whereas others inhibit GSK3β, thereby promoting sensorimotor recovery and structural remodeling after TBI [208]. PKC activation reduces tau phosphorylation in experimental models, suggesting that targeted modulation may offer therapeutic benefit [209].

5.2. Phosphatase Failure

In AD, phosphatase activity is markedly reduced, particularly PP2A, which shows decreased expression, inhibition, and altered regulation, thereby failing to counterbalance kinase hyperactivity and shifting the balance towards pathological tau phosphorylation [210].

PP2A is responsible for approximately 70% of tau dephosphorylation in the human brain [211]. Both TBI and AD exhibit reduced PP2A activity, leading to sustained tau hyperphosphorylation [212]. Injury severity correlates with tau phosphorylation levels and with increased GSK3β and decreased PP2A expression, highlighting their inverse regulation [213]. Pharmacological activation of PP2A (e.g., sodium selenate) reduces tau phosphorylation and improves cognition in both TBI and AD models [212,214]. PP2A also suppresses ERK signaling; therefore, its loss amplifies tau pathology across multiple kinase pathways [215]. Calpain-mediated cleavage of PP2A's α4 regulatory subunit directly impairs its assembly and function [216]. Additionally, acidic pH, in injured or degenerating tissue, activates asparaginyl endopeptidase (AEP), which cleaves the endogenous PP2A inhibitor SET/I2^{PP2A}. The truncated inhibitor translocates to the cytosol, where it further suppresses PP2A activity, and promotes tau hyperphosphorylation and amyloidogenic processing [217,218].

CaN plays a paradoxical role in tau phosphorylation. Although it can directly dephosphorylate tau, calpain-mediated truncation removes its autoinhibitory domain, producing constitutively active fragments [219]. These fragments indirectly result in tau phosphorylation, potentially by modulating GSK3β activity [24]. Truncated CaN correlates with NFT density in AD brains [160], and remains persistently elevated for weeks after TBI, particularly in hippocampal and cortical regions [220]. The impact of CaN inhibition is context-dependent and in some models, it paradoxically exacerbates tau phosphorylation [221]. CaN also contributes to synaptic and neuronal loss by dephosphorylating the pro-apoptotic protein Bcl-2-associated death promoter (BAD), enabling its mitochondrial translocation and initiation of programmed cell death [222].

5.3. Proteolytic Cleavage

In addition to phosphorylation, tau undergoes proteolytic cleavage, generating fragments with increased aggregation potential and toxicity. Calpain and caspases, both activated downstream of calcium dysregulation, are central mediators of this process.

Calpain cleaves tau to generate neuron-specific N224 and 17-kDa fragments. N224, detectable in CSF, has been proposed as a candidate AD biomarker, because its levels correlate with cognitive decline [223,224]. The neurotoxic 17-kDa fragment damages the cytoskeleton, perturbs axonal transport, and contributes to synaptic dysfunction [225]. Calpain also disrupts mitochondrial dynamics by cleaving DRP1 and triggers the UPR, thereby amplifying neuronal stress [226,227]. In experimental TBI, calpain activity scales with cell death, and pharmacological inhibition blocks tau phosphorylation, delays NFT formation, reduces lesion volume, and preserves axonal integrity [228,229].

Caspases, activated by Aβ and potentially calpain, cleave tau at Asp421, producing truncated forms with high aggregation propensity that localize to early NFTs [230,231]. These fragments correlate with disease progression [232], and drive both tau-dependent [233] and tau-independent [234] cell death pathways.

Proteolytic cleavage acts at multiple points both upstream and downstream in the pathological cascade of tau phosphorylation. These interactions, driven by calcium overload, enzyme activation, fragment production, mitochondrial dysfunction, and cytoskeletal collapse, link acute injury to chronic propagation of tau pathology.

6. Synaptic Dysfunction

Synaptic loss is one of the strongest pathological correlates of cognitive decline in TBI and AD, often preceding overt hallmarks such as amyloid plaques or NFTs [235]. Because synaptic function, particularly long-term potentiation (LTP), depends on tightly regulated spatiotemporal calcium signaling, synapses are highly vulnerable to calcium imbalance [33]. Under physiological conditions, NMDAR-mediated calcium influx activates CaMKII, which promotes insertion of GluA1-containing AMPAR subunits into the postsynaptic membrane, strengthening synaptic efficacy [236].

In TBI and AD, sustained calcium dysregulation drives persistent calpain activation, which cleaves GluA1, NMDAR subunits, and scaffolding proteins such as PSD-95, disrupting LTP [237,238]. Calpain also degrades cytoskeletal components (spectrin, tubulin, MAPs, etc.), compromising synaptic integrity and vesicle trafficking [57]. Cleavage of presynaptic protein GAP43 impairs neuronal plasticity [239], and processing of dynamin-1 disrupts synaptic vesicle recycling and memory function [240]. These effects are compounded by CDK5, DYRK1A, and GSK3β-mediated phosphorylation of dynamin-1, which dysregulates endocytosis and alters dendritic spine morphology [41,241]. Spatial correlations between calpain activation, hippocampal neuronal loss, and cognitive impairment are seen in TBI [242], while inhibition of calpain restores LTP and CREB activity, preserving memory-associated transcription [243,244].

CaMKII dysregulation shows region-specific expression changes. Although its upregulation is spatially associated with Aβ plaques [245], transcriptomic studies reveal significant synaptic downregulation in both CTE and AD [246]. Reduced CaMKII activity limits GluA1 trafficking and activity, leading to maladaptive plasticity and cognitive decline [247]. Conversely, pharmacological activation enhances CREB activation, reduces synaptic damage, and improves memory performance [248].

GSK3β overactivity further impairs synaptic plasticity [249], by reducing presynaptic glutamate release, disrupting vesicle recycling, suppressing NMDAR expression, and promoting Aβ-dependent dendritic spine loss, possibly through downregulation of CREB target genes [250,251]. CRMP2, which normally regulates neuronal growth and cytoskeletal organization [252], is inactivated by GSK3β or CDK5, leading to dendritic spine simplification and cognitive deficits [253]. GSK3β-mediated phosphorylation of kinesin light chains also reduces motor motility and disrupts vesicle delivery required for synaptic function [254]. CDK5/p25 activity correlates with tau phosphorylation at Thr217, which is strongly linked to synaptic protein loss, disrupted synaptic and axonal integrity, and cognitive decline, whereas its inhibition reduces these deficits [255].

The PKC family is closely associated with memory formation [50]. PKC activators (e.g., bryostatin1) protect against cognitive deficits in TBI models, possibly by upregulating ADAM10 and downregulating BACE1, thereby reducing Aβ accumulation [256]. In contrast, PKCα gain-of-function mutations in AD enhance catalytic activity and are linked to reduced spine density and impaired cognition [257].

CaN is likewise dysregulated in TBI and AD. Pathological activation by NMDARs and Aβ oligomers promotes NFAT-dependent signaling and AMPAR internalization, weakening synaptic transmission and structural integrity [258,259]. Both tau and Aβ depress CREB activity through CaN-dependent and independent mechanisms, further disrupting memory-related transcription and LTP [260,261]. Inhibition of CaN (e.g., FK506) prevents dendritic spine loss, cortical injury, and AD-related pathology in rodent TBI and AD models [24,262]. Because CaN plays a central role in synaptic depression [69], its blockage also reverses LTP deficits and improves cognition in transgenic mice, rendering it a promising therapeutic or preventive target [263].

In both TBI and AD, calcium dysregulation activates kinases, phosphatases, and proteases that collectively disrupt synaptic architecture, receptor density, axonal transport, and activity-dependent gene expression, leading to LTP failure and cognitive decline.

7. Axonal Degeneration

Axonal degeneration represents an early and prominent finding in TBI and AD, initiated by calcium influx and propagated by downstream proteolytic and kinase cascades [264]. In TBI, mechanical membrane disruption permits rapid calcium entry, triggering a wave of enzymatic pathways that degrade the cytoskeleton, impair axonal transport, and promote progressive fragmentation [265]. In AD, similar calcium-dependent transport defects and axonal swellings occur in proximity to neuritic plaques, pointing to shared upstream mechanisms [266].

Calpain is central to this process by cleaving axonal and myelin-associated structural proteins, including spectrin, neurofilaments, and MAPs [57]. Calpain also generates the characteristic 145-kDa spectrin breakdown product (SBDP145) fragment, a candidate biomarker in both TBI and AD [267]. Blocking calpain or calcium influx using inhibitors or calcium channel blockers reduces axonal degeneration and lesion volume in preclinical TBI models [229,268], although incomplete recovery suggests additional calcium-dependent mechanisms are likely involved [269].

Kinases also contribute to axonal degeneration. Pathological CDK5/p25 drives white matter atrophy with focal tau aggregation [270], whereas selective inhibition partially reverses these effects [271]. Combined inhibition of CDK5 and GSK3β offers greater protection than either alone [272], indicating additive or synergistic effects. GSK3β-mediated phosphorylation of CRMP2 compromises cytoskeletal stabilization [252], whereas calpain-dependent CRMP2 cleavage prevents kinesin binding, further disrupting axonal transport [273]. Kinesin downregulation has been observed in early stages of axonal degeneration and is closely associated with local Aβ plaque formation [274]. CaMKII phosphorylation of CRMP2 also contributes to axonal swellings [275].

Calcium dysregulation activates CaN which exacerbates axonal degeneration, partly through mitochondrial impairment [276]. Mitochondrial dysfunction accelerates axonal loss by opening the mitochondrial permeability transition pore (mPTP), releasing calcium and ROS, and triggering irreversible axonal degeneration [277]. CaN inhibition (e.g., cyclosporin A) preserves mitochondrial function and prevents mPTP opening, thereby limiting axonal damage [277,278]. Mitochondrial energy failure disrupts axonal transport, leading to co-accumulation of APP, BACE1, and presenilin in injured axons, creating a permissive microenvironment for amyloidogenic processing [117]. Aβ peptides persist in damaged white matter adjacent to axonal damage months after TBI [279], correlating with increased presenilin expression [280]. Presenilin inhibition reduces white matter injury and improves cognition in animal models [118]. Hyperphosphorylated tau destabilizes the microtubule network, further disrupting APP axonal transport [281], increasing amyloidogenic processing, and reinforcing amyloid toxicity in a self-reinforcing loop [282,283]. Reducing tau improves trafficking and decreases Aβ toxicity [284].

Together, mitochondrial collapse, axonal transport blockade, protease and kinase activation, and amyloid feedback progressively damage axons, leading to chronic white matter degeneration that contributes to synaptic and cognitive deficits in TBI and AD [285].

8. Conclusions

Calcium dysregulation is a common and early upstream event in TBI and AD, initiating a network of calcium-dependent signaling cascades that drive neurodegeneration and promote the transition from acute injury to chronic decline. Experimental and human studies consistently show that calcium disturbances appear within minutes to hours after TBI and persistently impair neuronal and glial function. Although brief calcium-dependent enzymatic activation supports physiological function, sustained overload destabilizes homeostasis, sensitizing neurons to degeneration and providing a convergent pathway linking TBI and AD.

Disrupted calcium homeostasis shifts the equilibrium between kinases and phosphatases, promoting tau hyperphosphorylation and amyloidogenic processing, and setting the stage for amyloid plaque and NFT formation. The selective vulnerability of the hippocampus and cerebral white matter to calcium-driven injury likely underlies some of the long-term cognitive sequelae common to both conditions.

Among these mechanisms, calpain has emerged as a central regulator. Preclinical studies show that inhibiting calcium-dependent enzymes exerts neuroprotective effects, but translation into clinical therapies remains elusive. Clarifying the temporal and mechanistic links between calcium dysregulation in TBI and AD will be critical for developing targeted interventions capable of mitigating trauma-related neurodegeneration.

Future therapeutic strategies should aim for precise, combined inhibition of selected calcium-dependent enzymes to block or modulate pathological signaling, while still preserving normal cellular functions. One approach is to target downstream molecular players at critical branch points in the secondary injury cascade. Large, long-term prospective observational studies spanning decades are needed to more accurately define the epidemiological associations between TBI and dementia risk.

Revisiting the calcium hypothesis is timely, because many current AD therapies (e.g., memantine) act on calcium signaling pathways, whereas tau- or amyloid-directed treatments have yielded limited clinical benefit. Approaches to improve TBI classification and phenotyping, including explicit documentation of loss of consciousness, repetition, and injury severity, are essential for harmonizing clinical and experimental research. In parallel, polygenic-score approaches that aggregate small genetic effects across calcium-regulatory genes, using resources such as PGS-Depot and PGSFusion, could help to stratify TBI survivors at risk of calcium-related neurodegeneration. Together, these strategies may help clarify how calcium dysregulation links TBI to AD and ACD and provide a framework for precision-based interventions to prevent or slow TBI-related neurodegeneration.

Author Contributions: S.N. conducted the literature search and data collection, drafted the manuscript, and created the tables. C.S. designed the figures. D.M. and B.E.E. supervised, structured and edited the draft. All authors have read and agreed to the published version of the manuscript.

Funding: This research received no external funding.

Institutional Review Board Statement: Not applicable.

Informed Consent Statement: Not applicable.

Data Availability Statement: Not applicable, since no new data were created.

Acknowledgments: The authors would like to thank all those whose research contributed to the development of the papers cited in this review.

Conflicts of Interest: B.E.E is a cofounder of Osmol Therapeutics, a company that is targeting NCS1 for therapeutic purposes. All other authors declare no conflicts of interest.

Abbreviations

AD	Alzheimer's Disease
ADAM10	A Disintegrin and Metalloproteinase 10
ACD	All-Cause Dementia

AEP	Asparaginyl Endopeptidase
AICD	APP Intracellular Domain
AMPAR	α-Amino-3-Hydroxy 5-Methyl-4-Isoxazolepropionate Receptor
APOE4	Apolipoprotein ε4
APP	Amyloid Precursor Protein
ATP	Adenosine Triphosphate
Aβ	Amyloid-Beta
BACE1	Beta-site Amyloid Precursor Protein Cleaving Enzyme 1
BAD	Bcl-2-Associated Death Promoter
CaM	Calmodulin
CaMKII	Calcium/Calmodulin-Dependent Protein Kinase II
CaN	Calcineurin
CDK5	Cyclin-Dependent Kinase 5
CI	Confidence Interval
CICR	Calcium-Induced Calcium Release
CP-AMPAR	Calcium-Permeable AMPAR
CREB	cAMP Response Element-Binding Protein
CRMP2	Collapsin Response Mediator Protein 2
CSF	Cerebrospinal Fluid
CTE	Chronic Traumatic Encephalopathy
DRP1	Dynamin-Related Protein 1
DYRK1A	Dual-specificity tyrosine phosphorylation-regulated kinase 1A
eIF2α	Eukaryotic Initiation Factor 2α
ER	Endoplasmic Reticulum
ERK	Extracellular Signal-Regulated Kinase
FDA	Food and Drug Administration
FOXP2	Forkhead Box P2
FTO	Fat Mass and Obesity-Associated
GAP43	Growth-Associated Protein 43
GSK3β	Glycogen Synthase Kinase 3β
GWAS	Genome-Wide Association Study
HR	Hazard Ratio
IP3R	Inositol 1,4,5-Trisphosphate Receptor
LTP	Long-Term Potentiation
MAM	Mitochondria-Associated ER Membrane
MAP	Microtubule-Associated Protein
MAPT	Microtubule-Associated Protein Tau
MAPK	Mitogen-Activated Protein Kinase
MCU	Mitochondrial Calcium Uniporter
mGluR	Metabotropic Glutamate Receptor
mNCX	Mitochondrial Sodium/Calcium Exchangers
mPTP	Mitochondrial Permeability Transition Pore
NCAM1	Neural Cell Adhesion Molecule 1
NCX	Sodium/Calcium Exchangers
NFAT	Nuclear Factor of Activated T Cells
NFT	Neurofibrillary Tangles
NF-κB	Nuclear Factor Kappa-B
NMDAR	N-Methyl-D-Aspartate Receptor
OR	Odds Ratio
PHF	Paired Helical Filament
PKB	Protein Kinase B
PKC	Protein Kinase C
PMCA	Plasma Membrane Calcium-ATPase
PP2A	Protein Phosphatase 2A

PSD-95	Postsynaptic Density Protein 95
Refs.	References
ROS	Reactive Oxygen Species
RR	Risk Ratio
RyR	Ryanodine Receptor
SBDP145	145-kDa Spectrin Breakdown Product
SERCA	Sarcoplasmic–Endoplasmic Reticulum Calcium-ATPase
SNP	Single-Nucleotide Polymorphism
SOC	Store-Operated Calcium Channel
STAT	Signal Transducer and Activator of Transcription
TBI	Traumatic Brain Injury
TRP	Transient Receptor Potential Channel
UPR	Unfolded Protein Response
VGCC	Voltage-Gated Calcium Channels

References

1. GBD 2019 Dementia Forecasting Collaborators. Estimation of the global prevalence of dementia in 2019 and forecasted prevalence in 2050: An analysis for the Global Burden of Disease Study 2019. *Lancet Public Health* **2022**, *7*, e105–e125. [CrossRef]
2. Thompson, J.; Vasefi, M. Increased Risk of Dementia and Its Subtypes Following Various Forms of Acquired Brain Injury: A Meta-Analysis and Systematic Review. *Neuroscience* **2025**, *580*, 332–350. [CrossRef]
3. Nordstrom, A.; Nordstrom, P. Traumatic brain injury and the risk of dementia diagnosis: A nationwide cohort study. *PLoS Med.* **2018**, *15*, e1002496. [CrossRef] [PubMed]
4. Litke, R.; Garcharna, L.C.; Jiwani, S.; Neugroschl, J. Modifiable Risk Factors in Alzheimer Disease and Related Dementias: A Review. *Clin. Ther.* **2021**, *43*, 953–965. [CrossRef] [PubMed]
5. Perry, D.C.; Sturm, V.E.; Peterson, M.J.; Pieper, C.F.; Bullock, T.; Boeve, B.F.; Miller, B.L.; Guskiewicz, K.M.; Berger, M.S.; Kramer, J.H.; et al. Association of traumatic brain injury with subsequent neurological and psychiatric disease: A meta-analysis. *J. Neurosurg.* **2016**, *124*, 511–526. [CrossRef]
6. Whiteneck, G.G.; Cuthbert, J.P.; Corrigan, J.D.; Bogner, J.A. Prevalence of Self-Reported Lifetime History of Traumatic Brain Injury and Associated Disability: A Statewide Population-Based Survey. *J. Head Trauma Rehabil.* **2016**, *31*, E55–E62. [CrossRef]
7. Gardner, R.C.; Bahorik, A.; Kornblith, E.S.; Allen, I.E.; Plassman, B.L.; Yaffe, K. Systematic Review, Meta-Analysis, and Population Attributable Risk of Dementia Associated with Traumatic Brain Injury in Civilians and Veterans. *J. Neurotrauma* **2023**, *40*, 620–634. [CrossRef] [PubMed]
8. Graham, A.; Livingston, G.; Purnell, L.; Huntley, J. Mild Traumatic Brain Injuries and Future Risk of Developing Alzheimer's Disease: Systematic Review and Meta-Analysis. *J. Alzheimers Dis.* **2022**, *87*, 969–979. [CrossRef]
9. Gu, D.; Ou, S.; Liu, G. Traumatic Brain Injury and Risk of Dementia and Alzheimer's Disease: A Systematic Review and Meta-Analysis. *Neuroepidemiology* **2022**, *56*, 4–16. [CrossRef]
10. Huang, C.H.; Lin, C.W.; Lee, Y.C.; Huang, C.Y.; Huang, R.Y.; Tai, Y.C.; Wang, K.W.; Yang, S.N.; Sun, Y.T.; Wang, H.K. Is traumatic brain injury a risk factor for neurodegeneration? A meta-analysis of population-based studies. *BMC Neurol.* **2018**, *18*, 184. [CrossRef]
11. Leung, K.K.; Carr, F.M.; Russell, M.J.; Bremault-Phillips, S.; Triscott, J.A.C. Traumatic brain injuries among veterans and the risk of incident dementia: A systematic review & meta-analysis. *Age Ageing* **2022**, *51*, afab194. [CrossRef]
12. Li, Y.; Li, Y.; Li, X.; Zhang, S.; Zhao, J.; Zhu, X.; Tian, G. Head Injury as a Risk Factor for Dementia and Alzheimer's Disease: A Systematic Review and Meta-Analysis of 32 Observational Studies. *PLoS ONE* **2017**, *12*, e0169650. [CrossRef] [PubMed]
13. Mavroudis, I.; Kazis, D.; Petridis, F.E.; Balmus, I.M.; Papaliagkas, V.; Ciobica, A. The Association Between Traumatic Brain Injury and the Risk of Cognitive Decline: An Umbrella Systematic Review and Meta-Analysis. *Brain Sci.* **2024**, *14*, 1188. [CrossRef]
14. Snowden, T.M.; Hinde, A.K.; Reid, H.M.O.; Christie, B.R. Does Mild Traumatic Brain Injury Increase the Risk for Dementia? A Systematic Review and Meta-Analysis. *J. Alzheimers Dis.* **2020**, *78*, 757–775. [CrossRef] [PubMed]
15. Zhang, J.; Zhang, Y.; Zou, J.; Cao, F. A meta-analysis of cohort studies: Traumatic brain injury and risk of Alzheimer's Disease. *PLoS ONE* **2021**, *16*, e0253206. [CrossRef]
16. Baracaldo-Santamaria, D.; Avendano-Lopez, S.S.; Ariza-Salamanca, D.F.; Rodriguez-Giraldo, M.; Calderon-Ospina, C.A.; Gonzalez-Reyes, R.E.; Nava-Mesa, M.O. Role of Calcium Modulation in the Pathophysiology and Treatment of Alzheimer's Disease. *Int. J. Mol. Sci.* **2023**, *24*, 9067. [CrossRef]
17. Weber, J.T. Altered calcium signaling following traumatic brain injury. *Front. Pharmacol.* **2012**, *3*, 60. [CrossRef]

18. Brini, M.; Cali, T.; Ottolini, D.; Carafoli, E. Neuronal calcium signaling: Function and dysfunction. *Cell. Mol. Life Sci.* **2014**, *71*, 2787–2814. [CrossRef] [PubMed]
19. Wang, M.; Zhang, H.; Liang, J.; Huang, J.; Wu, T.; Chen, N. Calcium signaling hypothesis: A non-negligible pathogenesis in Alzheimer's disease. *J. Adv. Res.* **2025**, *in press*. [CrossRef]
20. Donison, N.; Palik, J.; Volkening, K.; Strong, M.J. Cellular and molecular mechanisms of pathological tau phosphorylation in traumatic brain injury: Implications for chronic traumatic encephalopathy. *Mol. Neurodegener.* **2025**, *20*, 56. [CrossRef]
21. Supnet, C.; Bezprozvanny, I. Neuronal calcium signaling, mitochondrial dysfunction, and Alzheimer's disease. *J. Alzheimers Dis.* **2010**, *20* (Suppl. S2), S487–S498. [CrossRef]
22. Naim, A.; Farooqui, A.M.; Badruddeen; Khan, M.I.; Akhtar, J.; Ahmad, A.; Islam, A. The Role of Kinases in Neurodegenerative Diseases: From Pathogenesis to Treatment. *Eur. J. Neurosci.* **2025**, *61*, e70156. [CrossRef] [PubMed]
23. Yang, W.; Xu, Q.Q.; Yuan, Q.; Xian, Y.F.; Lin, Z.X. Sulforaphene, a CDK5 Inhibitor, attenuates cognitive deficits in a transgenic mouse model of Alzheimer's disease via reducing Abeta Deposition, tau hyperphosphorylation and synaptic dysfunction. *Int. Immunopharmacol.* **2023**, *114*, 109504. [CrossRef]
24. Marcatti, M.; Tumurbaatar, B.; Borghi, M.; Guptarak, J.; Zhang, W.R.; Krishnan, B.; Kayed, R.; Fracassi, A.; Taglialatela, G. Inhibition of Calcineurin with FK506 Reduces Tau Levels and Attenuates Synaptic Impairment Driven by Tau Oligomers in the Hippocampus of Male Mouse Models. *Int. J. Mol. Sci.* **2024**, *25*, 9092. [CrossRef]
25. Medeiros, R.; Kitazawa, M.; Chabrier, M.A.; Cheng, D.; Baglietto-Vargas, D.; Kling, A.; Moeller, A.; Green, K.N.; LaFerla, F.M. Calpain inhibitor A-705253 mitigates Alzheimer's disease-like pathology and cognitive decline in aged 3xTgAD mice. *Am. J. Pathol.* **2012**, *181*, 616–625. [CrossRef]
26. Chandran, R.; Kumar, M.; Kesavan, L.; Jacob, R.S.; Gunasekaran, S.; Lakshmi, S.; Sadasivan, C.; Omkumar, R.V. Cellular calcium signaling in the aging brain. *J. Chem. Neuroanat.* **2019**, *95*, 95–114. [CrossRef]
27. Cali, T.; Brini, M.; Carafoli, E. The PMCA pumps in genetically determined neuronal pathologies. *Neurosci. Lett.* **2018**, *663*, 2–11. [CrossRef]
28. Ehrlich, B.E. Functional properties of intracellular calcium-release channels. *Curr. Opin. Neurobiol.* **1995**, *5*, 304–309. [CrossRef]
29. Britzolaki, A.; Saurine, J.; Klocke, B.; Pitychoutis, P.M. A Role for SERCA Pumps in the Neurobiology of Neuropsychiatric and Neurodegenerative Disorders. *Adv. Exp. Med. Biol.* **2020**, *1131*, 131–161. [CrossRef] [PubMed]
30. Islam, M.S. Calcium Signaling: From Basic to Bedside. *Adv. Exp. Med. Biol.* **2020**, *1131*, 1–6. [CrossRef] [PubMed]
31. Bononi, A.; Missiroli, S.; Poletti, F.; Suski, J.M.; Agnoletto, C.; Bonora, M.; De Marchi, E.; Giorgi, C.; Marchi, S.; Patergnani, S.; et al. Mitochondria-associated membranes (MAMs) as hotspot Ca(2+) signaling units. *Adv. Exp. Med. Biol.* **2012**, *740*, 411–437. [CrossRef] [PubMed]
32. Elies, J.; Yanez, M.; Pereira, T.M.C.; Gil-Longo, J.; MacDougall, D.A.; Campos-Toimil, M. An Update to Calcium Binding Proteins. *Adv. Exp. Med. Biol.* **2020**, *1131*, 183–213. [CrossRef]
33. Blackstone, C.; Sheng, M. Postsynaptic calcium signaling microdomains in neurons. *Front. Biosci.* **2002**, *7*, d872–d885. [CrossRef] [PubMed]
34. Martin, L.; Latypova, X.; Wilson, C.M.; Magnaudeix, A.; Perrin, M.L.; Yardin, C.; Terro, F. Tau protein kinases: Involvement in Alzheimer's disease. *Ageing Res. Rev.* **2013**, *12*, 289–309. [CrossRef]
35. Hassan, M.; Yasir, M.; Shahzadi, S.; Chun, W.; Kloczkowski, A. Molecular Role of Protein Phosphatases in Alzheimer's and Other Neurodegenerative Diseases. *Biomedicines* **2024**, *12*, 1097. [CrossRef]
36. Ghosh, A.; Giese, K.P. Calcium/calmodulin-dependent kinase II and Alzheimer's disease. *Mol. Brain* **2015**, *8*, 78. [CrossRef]
37. Kaushik, M.; Kaushik, P.; Parvez, S. Memory related molecular signatures: The pivots for memory consolidation and Alzheimer's related memory decline. *Ageing Res. Rev.* **2022**, *76*, 101577. [CrossRef]
38. Maitra, S.; Vincent, B. Cdk5-p25 as a key element linking amyloid and tau pathologies in Alzheimer's disease: Mechanisms and possible therapeutic interventions. *Life Sci.* **2022**, *308*, 120986. [CrossRef] [PubMed]
39. Lee, J.H.; Kim, K.T. Regulation of cyclin-dependent kinase 5 and p53 by ERK1/2 pathway in the DNA damage-induced neuronal death. *J. Cell Physiol.* **2007**, *210*, 784–797. [CrossRef]
40. Liu, S.L.; Wang, C.; Jiang, T.; Tan, L.; Xing, A.; Yu, J.T. The Role of Cdk5 in Alzheimer's Disease. *Mol. Neurobiol.* **2016**, *53*, 4328–4342. [CrossRef]
41. Smith, B.; Medda, F.; Gokhale, V.; Dunckley, T.; Hulme, C. Recent advances in the design, synthesis, and biological evaluation of selective DYRK1A inhibitors: A new avenue for a disease modifying treatment of Alzheimer's? *ACS Chem. Neurosci.* **2012**, *3*, 857–872. [CrossRef]
42. Abbassi, R.; Johns, T.G.; Kassiou, M.; Munoz, L. DYRK1A in neurodegeneration and cancer: Molecular basis and clinical implications. *Pharmacol. Ther.* **2015**, *151*, 87–98. [CrossRef] [PubMed]
43. Albert-Gasco, H.; Ros-Bernal, F.; Castillo-Gomez, E.; Olucha-Bordonau, F.E. MAP/ERK Signaling in Developing Cognitive and Emotional Function and Its Effect on Pathological and Neurodegenerative Processes. *Int. J. Mol. Sci.* **2020**, *21*, 4471. [CrossRef] [PubMed]

44. Khezri, M.R.; Yousefi, K.; Esmaeili, A.; Ghasemnejad-Berenji, M. The Role of ERK1/2 Pathway in the Pathophysiology of Alzheimer's Disease: An Overview and Update on New Developments. *Cell Mol. Neurobiol.* **2023**, *43*, 177–191. [CrossRef]
45. Zhao, J.; Wei, M.; Guo, M.; Wang, M.; Niu, H.; Xu, T.; Zhou, Y. GSK3: A potential target and pending issues for treatment of Alzheimer's disease. *CNS Neurosci. Ther.* **2024**, *30*, e14818. [CrossRef]
46. Lauretti, E.; Dincer, O.; Pratico, D. Glycogen synthase kinase-3 signaling in Alzheimer's disease. *Biochim. Biophys. Acta Mol. Cell Res.* **2020**, *1867*, 118664. [CrossRef] [PubMed]
47. Beurel, E.; Grieco, S.F.; Jope, R.S. Glycogen synthase kinase-3 (GSK3): Regulation, actions, and diseases. *Pharmacol. Ther.* **2015**, *148*, 114–131. [CrossRef]
48. Singh, N.; Nandy, S.K.; Jyoti, A.; Saxena, J.; Sharma, A.; Siddiqui, A.J.; Sharma, L. Protein Kinase C (PKC) in Neurological Health: Implications for Alzheimer's Disease and Chronic Alcohol Consumption. *Brain Sci.* **2024**, *14*, 554. [CrossRef]
49. Talman, V.; Pascale, A.; Jantti, M.; Amadio, M.; Tuominen, R.K. Protein Kinase C Activation as a Potential Therapeutic Strategy in Alzheimer's Disease: Is there a Role for Embryonic Lethal Abnormal Vision-like Proteins? *Basic Clin. Pharmacol. Toxicol.* **2016**, *119*, 149–160. [CrossRef]
50. Sun, M.K.; Alkon, D.L. The "memory kinases": Roles of PKC isoforms in signal processing and memory formation. *Prog. Mol. Biol. Transl. Sci.* **2014**, *122*, 31–59. [CrossRef]
51. Creamer, T.P. Calcineurin. *Cell Commun. Signal* **2020**, *18*, 137. [CrossRef] [PubMed]
52. Lim, D.; Tapella, L.; Dematteis, G.; Talmon, M.; Genazzani, A.A. Calcineurin Signalling in Astrocytes: From Pathology to Physiology and Control of Neuronal Functions. *Neurochem. Res.* **2023**, *48*, 1077–1090. [CrossRef]
53. Sontag, J.M.; Sontag, E. Protein phosphatase 2A dysfunction in Alzheimer's disease. *Front. Mol. Neurosci.* **2014**, *7*, 16. [CrossRef]
54. Dovega, R.; Tsutakawa, S.; Quistgaard, E.M.; Anandapadamanaban, M.; Low, C.; Nordlund, P. Structural and biochemical characterization of human PR70 in isolation and in complex with the scaffolding subunit of protein phosphatase 2A. *PLoS ONE* **2014**, *9*, e101846. [CrossRef] [PubMed]
55. Pan, J.; Zhou, L.; Zhang, C.; Xu, Q.; Sun, Y. Targeting protein phosphatases for the treatment of inflammation-related diseases: From signaling to therapy. *Signal Transduct. Target. Ther.* **2022**, *7*, 177. [CrossRef]
56. Schallerer, C.; Neuschmid, S.; Ehrlich, B.E.; McGuone, D. Calpain in Traumatic Brain Injury: From Cinderella to Central Player. *Cells* **2025**, *14*, 1253. [CrossRef]
57. Vosler, P.S.; Brennan, C.S.; Chen, J. Calpain-mediated signaling mechanisms in neuronal injury and neurodegeneration. *Mol. Neurobiol.* **2008**, *38*, 78–100. [CrossRef]
58. Khachaturian, Z.S. Calcium hypothesis of Alzheimer's disease and brain aging. *Ann. N. Y. Acad. Sci.* **1994**, *747*, 1–11. [CrossRef]
59. Stutzmann, G.E. The pathogenesis of Alzheimers disease is it a lifelong "calciumopathy"? *Neuroscientist* **2007**, *13*, 546–559. [CrossRef] [PubMed]
60. Demuro, A.; Parker, I.; Stutzmann, G.E. Calcium signaling and amyloid toxicity in Alzheimer disease. *J. Biol. Chem.* **2010**, *285*, 12463–12468. [CrossRef]
61. Supnet, C.; Bezprozvanny, I. Presenilins function in ER calcium leak and Alzheimer's disease pathogenesis. *Cell Calcium* **2011**, *50*, 303–309. [CrossRef]
62. Alzheimer's Association Calcium Hypothesis, W. Calcium Hypothesis of Alzheimer's disease and brain aging: A framework for integrating new evidence into a comprehensive theory of pathogenesis. *Alzheimers Dement* **2017**, *13*, 178–182.e17. [CrossRef] [PubMed]
63. Passeri, E.; Elkhoury, K.; Morsink, M.; Broersen, K.; Linder, M.; Tamayol, A.; Malaplate, C.; Yen, F.T.; Arab-Tehrany, E. Alzheimer's Disease: Treatment Strategies and Their Limitations. *Int. J. Mol. Sci.* **2022**, *23*, 13954. [CrossRef] [PubMed]
64. Wachtler, N.; O'Brien, R.; Ehrlich, B.E.; McGuone, D. Exploring Calcium Channels as Potential Therapeutic Targets in Blast Traumatic Brain Injury. *Pharmaceuticals* **2025**, *18*, 223. [CrossRef] [PubMed]
65. Floyd, C.L.; Gorin, F.A.; Lyeth, B.G. Mechanical strain injury increases intracellular sodium and reverses Na^+/Ca^{2+} exchange in cortical astrocytes. *Glia* **2005**, *51*, 35–46. [CrossRef]
66. Krishnamurthy, K.; Laskowitz, D.T. Cellular and Molecular Mechanisms of Secondary Neuronal Injury following Traumatic Brain Injury. In *Translational Research in Traumatic Brain Injury*; Laskowitz, D., Grant, G., Eds.; CRC Press: Boca Raton, FL, USA, 2016.
67. Chen, T.; Fei, F.; Jiang, X.F.; Zhang, L.; Qu, Y.; Huo, K.; Fei, Z. Down-regulation of Homer1b/c attenuates glutamate-mediated excitotoxicity through endoplasmic reticulum and mitochondria pathways in rat cortical neurons. *Free Radic. Biol. Med.* **2012**, *52*, 208–217. [CrossRef]
68. Weber, J.T.; Rzigalinski, B.A.; Ellis, E.F. Traumatic injury of cortical neurons causes changes in intracellular calcium stores and capacitative calcium influx. *J. Biol. Chem.* **2001**, *276*, 1800–1807. [CrossRef]
69. Webber, E.K.; Fivaz, M.; Stutzmann, G.E.; Griffioen, G. Cytosolic calcium: Judge, jury and executioner of neurodegeneration in Alzheimer's disease and beyond. *Alzheimers Dement.* **2023**, *19*, 3701–3717. [CrossRef]
70. Xiong, Y.; Gu, Q.; Peterson, P.L.; Muizelaar, J.P.; Lee, C.P. Mitochondrial dysfunction and calcium perturbation induced by traumatic brain injury. *J. Neurotrauma* **1997**, *14*, 23–34. [CrossRef]

71. Cheung, K.H.; Shineman, D.; Muller, M.; Cardenas, C.; Mei, L.; Yang, J.; Tomita, T.; Iwatsubo, T.; Lee, V.M.; Foskett, J.K. Mechanism of Ca^{2+} disruption in Alzheimer's disease by presenilin regulation of InsP3 receptor channel gating. *Neuron* **2008**, *58*, 871–883. [CrossRef]
72. Chan, S.L.; Mayne, M.; Holden, C.P.; Geiger, J.D.; Mattson, M.P. Presenilin-1 mutations increase levels of ryanodine receptors and calcium release in PC12 cells and cortical neurons. *J. Biol. Chem.* **2000**, *275*, 18195–18200. [CrossRef] [PubMed]
73. Green, K.N.; Demuro, A.; Akbari, Y.; Hitt, B.D.; Smith, I.F.; Parker, I.; LaFerla, F.M. SERCA pump activity is physiologically regulated by presenilin and regulates amyloid beta production. *J. Cell Biol.* **2008**, *181*, 1107–1116. [CrossRef] [PubMed]
74. Tu, H.; Nelson, O.; Bezprozvanny, A.; Wang, Z.; Lee, S.F.; Hao, Y.H.; Serneels, L.; De Strooper, B.; Yu, G.; Bezprozvanny, I. Presenilins form ER Ca^{2+} leak channels, a function disrupted by familial Alzheimer's disease-linked mutations. *Cell* **2006**, *126*, 981–993. [CrossRef]
75. Chui, D.H.; Tanahashi, H.; Ozawa, K.; Ikeda, S.; Checler, F.; Ueda, O.; Suzuki, H.; Araki, W.; Inoue, H.; Shirotani, K.; et al. Transgenic mice with Alzheimer presenilin 1 mutations show accelerated neurodegeneration without amyloid plaque formation. *Nat. Med.* **1999**, *5*, 560–564. [CrossRef]
76. Li, Y.; Macyczko, J.R.; Liu, C.C.; Bu, G. ApoE4 reduction: An emerging and promising therapeutic strategy for Alzheimer's disease. *Neurobiol. Aging* **2022**, *115*, 20–28. [CrossRef]
77. Zhou, W.; Xu, D.; Peng, X.; Zhang, Q.; Jia, J.; Crutcher, K.A. Meta-analysis of APOE4 allele and outcome after traumatic brain injury. *J. Neurotrauma* **2008**, *25*, 279–290. [CrossRef]
78. Merritt, V.C.; Maihofer, A.X.; Gasperi, M.; Chanfreau-Coffinier, C.; Stein, M.B.; Panizzon, M.S.; Hauger, R.L.; Logue, M.W.; Delano-Wood, L.; Nievergelt, C.M. Genome-wide association study of traumatic brain injury in U.S. military veterans enrolled in the VA million veteran program. *Mol. Psychiatry* **2024**, *29*, 97–111. [CrossRef]
79. Kals, M.; Kunzmann, K.; Parodi, L.; Radmanesh, F.; Wilson, L.; Izzy, S.; Anderson, C.D.; Puccio, A.M.; Okonkwo, D.O.; Temkin, N.; et al. A genome-wide association study of outcome from traumatic brain injury. *EBioMedicine* **2022**, *77*, 103933. [CrossRef]
80. Cao, C.; Zhang, S.; Wang, J.; Tian, M.; Ji, X.; Huang, D.; Yang, S.; Gu, N. PGS-Depot: A comprehensive resource for polygenic scores constructed by summary statistics based methods. *Nucleic Acids Res.* **2023**, *52*, D963–D971. [CrossRef] [PubMed]
81. Di Benedetto, G.; Burgaletto, C.; Bellanca, C.M.; Munafo, A.; Bernardini, R.; Cantarella, G. Role of Microglia and Astrocytes in Alzheimer's Disease: From Neuroinflammation to Ca(2+) Homeostasis Dysregulation. *Cells* **2022**, *11*, 2728. [CrossRef]
82. Ryan, K.C.; Ashkavand, Z.; Norman, K.R. The Role of Mitochondrial Calcium Homeostasis in Alzheimer's and Related Diseases. *Int. J. Mol. Sci.* **2020**, *21*, 9153. [CrossRef]
83. Lim, D.; Tapella, L.; Dematteis, G.; Genazzani, A.A.; Corazzari, M.; Verkhratsky, A. The endoplasmic reticulum stress and unfolded protein response in Alzheimer's disease: A calcium dyshomeostasis perspective. *Ageing Res. Rev.* **2023**, *87*, 101914. [CrossRef]
84. Zhang, H.; Bezprozvanny, I. "Dirty Dancing" of Calcium and Autophagy in Alzheimer's Disease. *Life* **2023**, *13*, 1187. [CrossRef] [PubMed]
85. Di Scala, C.; Yahi, N.; Boutemeur, S.; Flores, A.; Rodriguez, L.; Chahinian, H.; Fantini, J. Common molecular mechanism of amyloid pore formation by Alzheimer's beta-amyloid peptide and alpha-synuclein. *Sci. Rep.* **2016**, *6*, 28781. [CrossRef]
86. Raich, I.; Lillo, J.; Rebassa, J.B.; Capo, T.; Cordomi, A.; Reyes-Resina, I.; Pallas, M.; Navarro, G. Dual Role of NMDAR Containing NR2A and NR2B Subunits in Alzheimer's Disease. *Int. J. Mol. Sci.* **2024**, *25*, 4757. [CrossRef] [PubMed]
87. Demuro, A.; Parker, I. Cytotoxicity of intracellular abeta42 amyloid oligomers involves Ca^{2+} release from the endoplasmic reticulum by stimulated production of inositol trisphosphate. *J. Neurosci.* **2013**, *33*, 3824–3833. [CrossRef]
88. Goussakov, I.; Miller, M.B.; Stutzmann, G.E. NMDA-mediated Ca(2+) influx drives aberrant ryanodine receptor activation in dendrites of young Alzheimer's disease mice. *J. Neurosci.* **2010**, *30*, 12128–12137. [CrossRef] [PubMed]
89. Liu, Z.; Lv, C.; Zhao, W.; Song, Y.; Pei, D.; Xu, T. NR2B-containing NMDA receptors expression and their relationship to apoptosis in hippocampus of Alzheimer's disease-like rats. *Neurochem. Res.* **2012**, *37*, 1420–1427. [CrossRef]
90. Zhang, Z.; Guo, S.; Li, M.; Shao, K.; Xiao, B.; Jin, Q. Blocking the NR2B in the hippocampal dentate gyrus reduced the spatial memory deficits and apoptosis through the PERK-CHOP pathway in a rat model of sporadic Alzheimer's disease. *Behav. Brain Res.* **2025**, *493*, 115685. [CrossRef]
91. De Felice, F.G.; Velasco, P.T.; Lambert, M.P.; Viola, K.; Fernandez, S.J.; Ferreira, S.T.; Klein, W.L. Abeta oligomers induce neuronal oxidative stress through an N-methyl-D-aspartate receptor-dependent mechanism that is blocked by the Alzheimer drug memantine. *J. Biol. Chem.* **2007**, *282*, 11590–11601. [CrossRef]
92. Kelly, B.L.; Ferreira, A. beta-Amyloid-induced dynamin 1 degradation is mediated by N-methyl-D-aspartate receptors in hippocampal neurons. *J. Biol. Chem.* **2006**, *281*, 28079–28089. [CrossRef] [PubMed]
93. Ho, R.; Ortiz, D.; Shea, T.B. Amyloid-beta promotes calcium influx and neurodegeneration via stimulation of L voltage-sensitive calcium channels rather than NMDA channels in cultured neurons. *J. Alzheimers Dis.* **2001**, *3*, 479–483. [CrossRef]
94. Anekonda, T.S.; Quinn, J.F.; Harris, C.; Frahler, K.; Wadsworth, T.L.; Woltjer, R.L. L-type voltage-gated calcium channel blockade with isradipine as a therapeutic strategy for Alzheimer's disease. *Neurobiol. Dis.* **2011**, *41*, 62–70. [CrossRef]

95. Liu, S.J.; Gasperini, R.; Foa, L.; Small, D.H. Amyloid-beta decreases cell-surface AMPA receptors by increasing intracellular calcium and phosphorylation of GluR2. *J. Alzheimers Dis.* **2010**, *21*, 655–666. [CrossRef] [PubMed]
96. Hoey, S.E.; Buonocore, F.; Cox, C.J.; Hammond, V.J.; Perkinton, M.S.; Williams, R.J. AMPA receptor activation promotes non-amyloidogenic amyloid precursor protein processing and suppresses neuronal amyloid-beta production. *PLoS ONE* **2013**, *8*, e78155. [CrossRef]
97. Atherton, J.; Kurbatskaya, K.; Bondulich, M.; Croft, C.L.; Garwood, C.J.; Chhabra, R.; Wray, S.; Jeromin, A.; Hanger, D.P.; Noble, W. Calpain cleavage and inactivation of the sodium calcium exchanger-3 occur downstream of Abeta in Alzheimer's disease. *Aging Cell* **2014**, *13*, 49–59. [CrossRef]
98. Pottorf, W.J., 2nd; Johanns, T.M.; Derrington, S.M.; Strehler, E.E.; Enyedi, A.; Thayer, S.A. Glutamate-induced protease-mediated loss of plasma membrane Ca^{2+} pump activity in rat hippocampal neurons. *J. Neurochem.* **2006**, *98*, 1646–1656. [CrossRef]
99. Kopil, C.M.; Siebert, A.P.; Foskett, J.K.; Neumar, R.W. Calpain-cleaved type 1 inositol 1,4,5-trisphosphate receptor impairs ER Ca(2+) buffering and causes neurodegeneration in primary cortical neurons. *J. Neurochem.* **2012**, *123*, 147–158. [CrossRef]
100. Hell, J.W.; Westenbroek, R.E.; Breeze, L.J.; Wang, K.K.; Chavkin, C.; Catterall, W.A. N-methyl-D-aspartate receptor-induced proteolytic conversion of postsynaptic class C L-type calcium channels in hippocampal neurons. *Proc. Natl. Acad. Sci. USA* **1996**, *93*, 3362–3367. [CrossRef]
101. Yuen, E.Y.; Gu, Z.; Yan, Z. Calpain regulation of AMPA receptor channels in cortical pyramidal neurons. *J. Physiol.* **2007**, *580*, 241–254. [CrossRef] [PubMed]
102. Simpkins, K.L.; Guttmann, R.P.; Dong, Y.; Chen, Z.; Sokol, S.; Neumar, R.W.; Lynch, D.R. Selective activation induced cleavage of the NR2B subunit by calpain. *J. Neurosci.* **2003**, *23*, 11322–11331. [CrossRef]
103. Tong, B.C.; Wu, A.J.; Li, M.; Cheung, K.H. Calcium signaling in Alzheimer's disease & therapies. *Biochim. Biophys. Acta Mol. Cell Res.* **2018**, *1865*, 1745–1760. [CrossRef]
104. Chen, X.H.; Johnson, V.E.; Uryu, K.; Trojanowski, J.Q.; Smith, D.H. A lack of amyloid beta plaques despite persistent accumulation of amyloid beta in axons of long-term survivors of traumatic brain injury. *Brain Pathol.* **2009**, *19*, 214–223. [CrossRef]
105. O'Brien, R.J.; Wong, P.C. Amyloid precursor protein processing and Alzheimer's disease. *Annu. Rev. Neurosci.* **2011**, *34*, 185–204. [CrossRef]
106. Zhang, H.; Ma, Q.; Zhang, Y.W.; Xu, H. Proteolytic processing of Alzheimer's beta-amyloid precursor protein. *J. Neurochem.* **2012**, *120* (Suppl. S1), 9–21. [CrossRef] [PubMed]
107. Lin, Y.; Im, H.; Diem, L.T.; Ham, S. Characterizing the structural and thermodynamic properties of Abeta42 and Abeta40. *Biochem. Biophys. Res. Commun.* **2019**, *510*, 442–448. [CrossRef]
108. DeKosky, S.T.; Abrahamson, E.E.; Ciallella, J.R.; Paljug, W.R.; Wisniewski, S.R.; Clark, R.S.; Ikonomovic, M.D. Association of increased cortical soluble abeta42 levels with diffuse plaques after severe brain injury in humans. *Arch. Neurol.* **2007**, *64*, 541–544. [CrossRef]
109. Emmerling, M.R.; Morganti-Kossmann, M.C.; Kossmann, T.; Stahel, P.F.; Watson, M.D.; Evans, L.M.; Mehta, P.D.; Spiegel, K.; Kuo, Y.M.; Roher, A.E.; et al. Traumatic brain injury elevates the Alzheimer's amyloid peptide A beta 42 in human CSF. A possible role for nerve cell injury. *Ann. N. Y. Acad. Sci.* **2000**, *903*, 118–122. [CrossRef]
110. Ikonomovic, M.D.; Uryu, K.; Abrahamson, E.E.; Ciallella, J.R.; Trojanowski, J.Q.; Lee, V.M.; Clark, R.S.; Marion, D.W.; Wisniewski, S.R.; DeKosky, S.T. Alzheimer's pathology in human temporal cortex surgically excised after severe brain injury. *Exp. Neurol.* **2004**, *190*, 192–203. [CrossRef] [PubMed]
111. Johnson, V.E.; Stewart, W.; Smith, D.H. Traumatic brain injury and amyloid-beta pathology: A link to Alzheimer's disease? *Nat. Rev. Neurosci.* **2010**, *11*, 361–370. [CrossRef] [PubMed]
112. Smith, D.H.; Nakamura, M.; McIntosh, T.K.; Wang, J.; Rodriguez, A.; Chen, X.H.; Raghupathi, R.; Saatman, K.E.; Clemens, J.; Schmidt, M.L.; et al. Brain trauma induces massive hippocampal neuron death linked to a surge in beta-amyloid levels in mice overexpressing mutant amyloid precursor protein. *Am. J. Pathol.* **1998**, *153*, 1005–1010. [CrossRef]
113. Tran, H.T.; LaFerla, F.M.; Holtzman, D.M.; Brody, D.L. Controlled cortical impact traumatic brain injury in 3xTg-AD mice causes acute intra-axonal amyloid-beta accumulation and independently accelerates the development of tau abnormalities. *J. Neurosci.* **2011**, *31*, 9513–9525. [CrossRef]
114. Uryu, K.; Laurer, H.; McIntosh, T.; Pratico, D.; Martinez, D.; Leight, S.; Lee, V.M.; Trojanowski, J.Q. Repetitive mild brain trauma accelerates Abeta deposition, lipid peroxidation, and cognitive impairment in a transgenic mouse model of Alzheimer amyloidosis. *J. Neurosci.* **2002**, *22*, 446–454. [CrossRef] [PubMed]
115. Walker, K.R.; Kang, E.L.; Whalen, M.J.; Shen, Y.; Tesco, G. Depletion of GGA1 and GGA3 mediates postinjury elevation of BACE1. *J. Neurosci.* **2012**, *32*, 10423–10437. [CrossRef] [PubMed]
116. Blasko, I.; Beer, R.; Bigl, M.; Apelt, J.; Franz, G.; Rudzki, D.; Ransmayr, G.; Kampfl, A.; Schliebs, R. Experimental traumatic brain injury in rats stimulates the expression, production and activity of Alzheimer's disease beta-secretase (BACE-1). *J. Neural Transm (Vienna)* **2004**, *111*, 523–536. [CrossRef] [PubMed]

117. Uryu, K.; Chen, X.H.; Martinez, D.; Browne, K.D.; Johnson, V.E.; Graham, D.I.; Lee, V.M.; Trojanowski, J.Q.; Smith, D.H. Multiple proteins implicated in neurodegenerative diseases accumulate in axons after brain trauma in humans. *Exp. Neurol.* **2007**, *208*, 185–192. [CrossRef]
118. Loane, D.J.; Pocivavsek, A.; Moussa, C.E.; Thompson, R.; Matsuoka, Y.; Faden, A.I.; Rebeck, G.W.; Burns, M.P. Amyloid precursor protein secretases as therapeutic targets for traumatic brain injury. *Nat. Med.* **2009**, *15*, 377–379. [CrossRef]
119. Schutzmann, M.P.; Hasecke, F.; Bachmann, S.; Zielinski, M.; Hansch, S.; Schroder, G.F.; Zempel, H.; Hoyer, W. Endo-lysosomal Abeta concentration and pH trigger formation of Abeta oligomers that potently induce Tau missorting. *Nat. Commun.* **2021**, *12*, 4634. [CrossRef]
120. Lelouvier, B.; Puertollano, R. Mucolipin-3 regulates luminal calcium, acidification, and membrane fusion in the endosomal pathway. *J. Biol. Chem.* **2011**, *286*, 9826–9832. [CrossRef]
121. Cole, S.L.; Vassar, R. The Alzheimer's disease beta-secretase enzyme, BACE1. *Mol. Neurodegener.* **2007**, *2*, 22. [CrossRef]
122. Chavez, S.E.; O'Day, D.H. Calmodulin binds to and regulates the activity of beta-secretase (BACE1). *Curr. Res. Alzheimers Dis.* **2007**, *1*, 37–47.
123. Hanna, R.A.; Campbell, R.L.; Davies, P.L. Calcium-bound structure of calpain and its mechanism of inhibition by calpastatin. *Nature* **2008**, *456*, 409–412. [CrossRef] [PubMed]
124. Kurbatskaya, K.; Phillips, E.C.; Croft, C.L.; Dentoni, G.; Hughes, M.M.; Wade, M.A.; Al-Sarraj, S.; Troakes, C.; O'Neill, M.J.; Perez-Nievas, B.G. Upregulation of calpain activity precedes tau phosphorylation and loss of synaptic proteins in Alzheimer's disease brain. *Acta Neuropathol. Commun.* **2016**, *4*, 34. [CrossRef]
125. Liang, B.; Duan, B.Y.; Zhou, X.P.; Gong, J.X.; Luo, Z.G. Calpain activation promotes BACE1 expression, amyloid precursor protein processing, and amyloid plaque formation in a transgenic mouse model of Alzheimer disease. *J. Biol. Chem.* **2010**, *285*, 27737–27744. [CrossRef] [PubMed]
126. Yousuf, M.A.; Tan, C.; Torres-Altoro, M.I.; Lu, F.M.; Plautz, E.; Zhang, S.; Takahashi, M.; Hernandez, A.; Kernie, S.G.; Plattner, F.; et al. Involvement of aberrant cyclin-dependent kinase 5/p25 activity in experimental traumatic brain injury. *J. Neurochem.* **2016**, *138*, 317–327. [CrossRef]
127. Hernandez, A.; Tan, C.; Plattner, F.; Logsdon, A.F.; Pozo, K.; Yousuf, M.A.; Singh, T.; Turner, R.C.; Lucke-Wold, B.P.; Huber, J.D.; et al. Exposure to mild blast forces induces neuropathological effects, neurophysiological deficits and biochemical changes. *Mol. Brain* **2018**, *11*, 64. [CrossRef] [PubMed]
128. Kimura, T.; Ishiguro, K.; Hisanaga, S. Physiological and pathological phosphorylation of tau by Cdk5. *Front. Mol. Neurosci.* **2014**, *7*, 65. [CrossRef]
129. Lee, M.S.; Kao, S.C.; Lemere, C.A.; Xia, W.; Tseng, H.C.; Zhou, Y.; Neve, R.; Ahlijanian, M.K.; Tsai, L.H. APP processing is regulated by cytoplasmic phosphorylation. *J. Cell Biol.* **2003**, *163*, 83–95. [CrossRef]
130. Iijima, K.; Ando, K.; Takeda, S.; Satoh, Y.; Seki, T.; Itohara, S.; Greengard, P.; Kirino, Y.; Nairn, A.C.; Suzuki, T. Neuron-specific phosphorylation of Alzheimer's beta-amyloid precursor protein by cyclin-dependent kinase 5. *J. Neurochem.* **2000**, *75*, 1085–1091. [CrossRef]
131. Song, W.J.; Son, M.Y.; Lee, H.W.; Seo, H.; Kim, J.H.; Chung, S.H. Enhancement of BACE1 Activity by p25/Cdk5-Mediated Phosphorylation in Alzheimer's Disease. *PLoS ONE* **2015**, *10*, e0136950. [CrossRef]
132. Wen, Y.; Yu, W.H.; Maloney, B.; Bailey, J.; Ma, J.; Marie, I.; Maurin, T.; Wang, L.; Figueroa, H.; Herman, M.; et al. Transcriptional regulation of beta-secretase by p25/cdk5 leads to enhanced amyloidogenic processing. *Neuron* **2008**, *57*, 680–690. [CrossRef]
133. Liu, L.; Martin, R.; Kohler, G.; Chan, C. Palmitate induces transcriptional regulation of BACE1 and presenilin by STAT3 in neurons mediated by astrocytes. *Exp. Neurol.* **2013**, *248*, 482–490. [CrossRef]
134. Chow, H.M.; Guo, D.; Zhou, J.C.; Zhang, G.Y.; Li, H.F.; Herrup, K.; Zhang, J. CDK5 activator protein p25 preferentially binds and activates GSK3beta. *Proc. Natl. Acad. Sci. USA* **2014**, *111*, E4887–E4895. [CrossRef]
135. Ma, S.; Liu, S.; Huang, Q.; Xie, B.; Lai, B.; Wang, C.; Song, B.; Li, M. Site-specific phosphorylation protects glycogen synthase kinase-3beta from calpain-mediated truncation of its N and C termini. *J. Biol. Chem.* **2012**, *287*, 22521–22532. [CrossRef] [PubMed]
136. Pierrot, N.; Santos, S.F.; Feyt, C.; Morel, M.; Brion, J.P.; Octave, J.N. Calcium-mediated transient phosphorylation of tau and amyloid precursor protein followed by intraneuronal amyloid-beta accumulation. *J. Biol. Chem.* **2006**, *281*, 39907–39914. [CrossRef]
137. Velazquez Toledano, J.; Bello, M.; Correa Basurto, J.; Guerrero Gonzalez, I.; Pacheco-Yepez, J.; Rosales Hernandez, M.C. Determining Structural Changes for Ligand Recognition Between Human and Rat Phosphorylated BACE1 in Silico and Its Phosphorylation by GSK3beta at Thr252 by In Vitro Studies. *ACS Chem. Neurosci.* **2024**, *15*, 629–644. [CrossRef] [PubMed]
138. Ly, P.T.; Wu, Y.; Zou, H.; Wang, R.; Zhou, W.; Kinoshita, A.; Zhang, M.; Yang, Y.; Cai, F.; Woodgett, J.; et al. Inhibition of GSK3beta-mediated BACE1 expression reduces Alzheimer-associated phenotypes. *J. Clin. Investig.* **2013**, *123*, 224–235. [CrossRef]
139. Leroy, K.; Yilmaz, Z.; Brion, J.P. Increased level of active GSK-3beta in Alzheimer's disease and accumulation in argyrophilic grains and in neurones at different stages of neurofibrillary degeneration. *Neuropathol. Appl. Neurobiol.* **2007**, *33*, 43–55. [CrossRef]

140. Takashima, A.; Murayama, M.; Murayama, O.; Kohno, T.; Honda, T.; Yasutake, K.; Nihonmatsu, N.; Mercken, M.; Yamaguchi, H.; Sugihara, S. Presenilin 1 associates with glycogen synthase kinase-3β and its substrate tau. *Proc. Natl. Acad. Sci. USA* **1998**, *95*, 9637–9641. [CrossRef] [PubMed]
141. Kang, D.E.; Soriano, S.; Frosch, M.P.; Collins, T.; Naruse, S.; Sisodia, S.S.; Leibowitz, G.; Levine, F.; Koo, E.H. Presenilin 1 facilitates the constitutive turnover of β-catenin: Differential activity of Alzheimer's disease–linked PS1 mutants in the β-catenin–signaling pathway. *J. Neurosci.* **1999**, *19*, 4229–4237. [CrossRef] [PubMed]
142. Zhao, S.; Fu, J.; Liu, X.; Wang, T.; Zhang, J.; Zhao, Y. Activation of Akt/GSK-3beta/beta-catenin signaling pathway is involved in survival of neurons after traumatic brain injury in rats. *Neurol. Res.* **2012**, *34*, 400–407. [CrossRef]
143. Dash, P.K.; Johnson, D.; Clark, J.; Orsi, S.A.; Zhang, M.; Zhao, J.; Grill, R.J.; Moore, A.N.; Pati, S. Involvement of the glycogen synthase kinase-3 signaling pathway in TBI pathology and neurocognitive outcome. *PLoS ONE* **2011**, *6*, e24648. [CrossRef]
144. Celine, G.; Thomas, M. Temporal characterisation and electrophysiological implications of TBI-induced serine/threonine kinase activity in mouse cortex. *Cell Mol. Life Sci.* **2025**, *82*, 102. [CrossRef]
145. Tagawa, K.; Homma, H.; Saito, A.; Fujita, K.; Chen, X.; Imoto, S.; Oka, T.; Ito, H.; Motoki, K.; Yoshida, C.; et al. Comprehensive phosphoproteome analysis unravels the core signaling network that initiates the earliest synapse pathology in preclinical Alzheimer's disease brain. *Hum. Mol. Genet.* **2015**, *24*, 540–558. [CrossRef]
146. Kishimoto, A.; Mikawa, K.; Hashimoto, K.; Yasuda, I.; Tanaka, S.; Tominaga, M.; Kuroda, T.; Nishizuka, Y. Limited proteolysis of protein kinase C subspecies by calcium-dependent neutral protease (calpain). *J. Biol. Chem.* **1989**, *264*, 4088–4092. [CrossRef]
147. Lee, W.; Boo, J.H.; Jung, M.W.; Park, S.D.; Kim, Y.H.; Kim, S.U.; Mook-Jung, I. Amyloid beta peptide directly inhibits PKC activation. *Mol. Cell Neurosci.* **2004**, *26*, 222–231. [CrossRef] [PubMed]
148. Lucke-Wold, B.P.; Turner, R.C.; Logsdon, A.F.; Simpkins, J.W.; Alkon, D.L.; Smith, K.E.; Chen, Y.W.; Tan, Z.; Huber, J.D.; Rosen, C.L. Common mechanisms of Alzheimer's disease and ischemic stroke: The role of protein kinase C in the progression of age-related neurodegeneration. *J. Alzheimers Dis.* **2015**, *43*, 711–724. [CrossRef] [PubMed]
149. Lin, H.W.; Della-Morte, D.; Thompson, J.W.; Gresia, V.L.; Narayanan, S.V.; Defazio, R.A.; Raval, A.P.; Saul, I.; Dave, K.R.; Morris, K.C.; et al. Differential effects of delta and epsilon protein kinase C in modulation of postischemic cerebral blood flow. *Adv. Exp. Med. Biol.* **2012**, *737*, 63–69. [CrossRef] [PubMed]
150. Du, Y.; Zhao, Y.; Li, C.; Zheng, Q.; Tian, J.; Li, Z.; Huang, T.Y.; Zhang, W.; Xu, H. Inhibition of PKCdelta reduces amyloid-beta levels and reverses Alzheimer disease phenotypes. *J. Exp. Med.* **2018**, *215*, 1665–1677. [CrossRef]
151. Sajan, M.P.; Leitges, M.; Park, C.; Diamond, D.M.; Wu, J.; Hansen, B.C.; Duncan, M.A.; Apostolatos, C.A.; Apostolatos, A.H.; Kindy, M.; et al. Control of beta-Site Amyloid Precursor Protein-Cleaving Enzyme-1 Expression by Protein Kinase C-lambda/iota and Nuclear Factor kappa-B. *Curr. Alzheimer Res.* **2021**, *18*, 941–955. [CrossRef]
152. Snow, W.M.; Albensi, B.C. Neuronal Gene Targets of NF-kappaB and Their Dysregulation in Alzheimer's Disease. *Front. Mol. Neurosci.* **2016**, *9*, 118. [CrossRef]
153. Gandy, S.; Czernik, A.J.; Greengard, P. Phosphorylation of Alzheimer disease amyloid precursor peptide by protein kinase C and Ca^{2+}/calmodulin-dependent protein kinase II. *Proc. Natl. Acad. Sci. USA* **1988**, *85*, 6218–6221. [CrossRef] [PubMed]
154. Arribas, R.L.; Viejo, L.; Bravo, I.; Martinez, M.; Ramos, E.; Romero, A.; Garcia-Frutos, E.M.; Janssens, V.; Montiel, C.; de Los Rios, C. C-glycosides analogues of the okadaic acid central fragment exert neuroprotection via restoration of PP2A-phosphatase activity: A rational design of potential drugs for Alzheimer's disease targeting tauopathies. *Eur. J. Med. Chem.* **2023**, *251*, 115245. [CrossRef]
155. Grandizoli Saletti, P.; Casillas-Espinosa, P.M.; Panagiotis Lisgaras, C.; Bi Mowrey, W.; Li, Q.; Liu, W.; Brady, R.D.; Ali, I.; Silva, J.; Yamakawa, G.; et al. Tau Phosphorylation Patterns in the Rat Cerebral Cortex After Traumatic Brain Injury and Sodium Selenate Effects: An Epibios4rx Project 2 Study. *J. Neurotrauma* **2024**, *41*, 222–243. [CrossRef]
156. Voronkov, M.; Braithwaite, S.P.; Stock, J.B. Phosphoprotein phosphatase 2A: A novel druggable target for Alzheimer's disease. *Future Med. Chem.* **2011**, *3*, 821–833. [CrossRef] [PubMed]
157. Shentu, Y.P.; Huo, Y.; Feng, X.L.; Gilbert, J.; Zhang, Q.; Liuyang, Z.Y.; Wang, X.L.; Wang, G.; Zhou, H.; Wang, X.C.; et al. CIP2A Causes Tau/APP Phosphorylation, Synaptopathy, and Memory Deficits in Alzheimer's Disease. *Cell. Rep.* **2018**, *24*, 713–723. [CrossRef] [PubMed]
158. Wei, H.; Zhang, H.L.; Wang, X.C.; Xie, J.Z.; An, D.D.; Wan, L.; Wang, J.Z.; Zeng, Y.; Shu, X.J.; Westermarck, J.; et al. Direct Activation of Protein Phosphatase 2A (PP2A) by Tricyclic Sulfonamides Ameliorates Alzheimer's Disease Pathogenesis in Cell and Animal Models. *Neurotherapeutics* **2020**, *17*, 1087–1103. [CrossRef]
159. Uhari-Vaananen, J.; Etelainen, T.; Zuo, C.; De Lorenzo, F.; Kilpelainen, T.; Myohanen, T.T. A prolyl oligopeptidase ligand blocks memory deficit in a repeated mild traumatic brain injury model. *Exp. Neurol.* **2025**, *394*, 115454. [CrossRef]
160. Liu, F.; Grundke-Iqbal, I.; Iqbal, K.; Oda, Y.; Tomizawa, K.; Gong, C.X. Truncation and activation of calcineurin A by calpain I in Alzheimer disease brain. *J. Biol. Chem.* **2005**, *280*, 37755–37762. [CrossRef]
161. Abdul, H.M.; Furman, J.L.; Sama, M.A.; Mathis, D.M.; Norris, C.M. NFATs and Alzheimer's Disease. *Mol. Cell. Pharmacol.* **2010**, *2*, 7–14.

162. Furman, J.L.; Sama, D.M.; Gant, J.C.; Beckett, T.L.; Murphy, M.P.; Bachstetter, A.D.; Van Eldik, L.J.; Norris, C.M. Targeting astrocytes ameliorates neurologic changes in a mouse model of Alzheimer's disease. *J. Neurosci.* **2012**, *32*, 16129–16140. [CrossRef]
163. Sompol, P.; Furman, J.L.; Pleiss, M.M.; Kraner, S.D.; Artiushin, I.A.; Batten, S.R.; Quintero, J.E.; Simmerman, L.A.; Beckett, T.L.; Lovell, M.A.; et al. Calcineurin/NFAT Signaling in Activated Astrocytes Drives Network Hyperexcitability in Abeta-Bearing Mice. *J. Neurosci.* **2017**, *37*, 6132–6148. [CrossRef] [PubMed]
164. Veeranna; Kaji, T.; Boland, B.; Odrljin, T.; Mohan, P.; Basavarajappa, B.S.; Peterhoff, C.; Cataldo, A.; Rudnicki, A.; Amin, N.; et al. Calpain mediates calcium-induced activation of the erk1,2 MAPK pathway and cytoskeletal phosphorylation in neurons: Relevance to Alzheimer's disease. *Am. J. Pathol.* **2004**, *165*, 795–805. [CrossRef] [PubMed]
165. Yuan, Y.; Chen, J.; Ge, X.; Deng, J.; Xu, X.; Zhao, Y.; Wang, H. Activation of ERK-Drp1 signaling promotes hypoxia-induced Abeta accumulation by upregulating mitochondrial fission and BACE1 activity. *FEBS Open Bio* **2021**, *11*, 2740–2755. [CrossRef]
166. Cho, H.J.; Kim, S.K.; Jin, S.M.; Hwang, E.M.; Kim, Y.S.; Huh, K.; Mook-Jung, I. IFN-gamma-induced BACE1 expression is mediated by activation of JAK2 and ERK1/2 signaling pathways and direct binding of STAT1 to BACE1 promoter in astrocytes. *Glia* **2007**, *55*, 253–262. [CrossRef] [PubMed]
167. Wang, N.; Wang, X.; Lan, B.; Gao, Y.; Cai, Y. DRP1, fission and apoptosis. *Cell Death Discov.* **2025**, *11*, 150. [CrossRef]
168. Ghasemi, R.; Moosavi, M.; Zarifkar, A.; Rastegar, K.; Maghsoudi, N. The Interplay of Akt and ERK in Abeta Toxicity and Insulin-Mediated Protection in Primary Hippocampal Cell Culture. *J. Mol. Neurosci.* **2015**, *57*, 325–334. [CrossRef]
169. Ryu, Y.S.; Park, S.Y.; Jung, M.S.; Yoon, S.H.; Kwen, M.Y.; Lee, S.Y.; Choi, S.H.; Radnaabazar, C.; Kim, M.K.; Kim, H.; et al. Dyrk1A-mediated phosphorylation of Presenilin 1: A functional link between Down syndrome and Alzheimer's disease. *J. Neurochem.* **2010**, *115*, 574–584. [CrossRef]
170. Ryoo, S.R.; Cho, H.J.; Lee, H.W.; Jeong, H.K.; Radnaabazar, C.; Kim, Y.S.; Kim, M.J.; Son, M.Y.; Seo, H.; Chung, S.H. Dual-specificity tyrosine (Y)-phosphorylation regulated kinase 1A-mediated phosphorylation of amyloid precursor protein: Evidence for a functional link between Down syndrome and Alzheimer's disease. *J. Neurochem.* **2008**, *104*, 1333–1344. [CrossRef]
171. Uddin, M.S.; Tewari, D.; Sharma, G.; Kabir, M.T.; Barreto, G.E.; Bin-Jumah, M.N.; Perveen, A.; Abdel-Daim, M.M.; Ashraf, G.M. Molecular Mechanisms of ER Stress and UPR in the Pathogenesis of Alzheimer's Disease. *Mol. Neurobiol.* **2020**, *57*, 2902–2919. [CrossRef]
172. Moradi Majd, R.; Mayeli, M.; Rahmani, F. Pathogenesis and promising therapeutics of Alzheimer disease through eIF2alpha pathway and correspondent kinases. *Metab. Brain Dis.* **2020**, *35*, 1241–1250. [CrossRef]
173. Braun, N.J.; Yao, K.R.; Alford, P.W.; Liao, D. Mechanical injuries of neurons induce tau mislocalization to dendritic spines and tau-dependent synaptic dysfunction. *Proc. Natl. Acad. Sci. USA* **2020**, *117*, 29069–29079. [CrossRef] [PubMed]
174. Guo, T.; Noble, W.; Hanger, D.P. Roles of tau protein in health and disease. *Acta Neuropathol.* **2017**, *133*, 665–704. [CrossRef]
175. Kent, S.A.; Spires-Jones, T.L.; Durrant, C.S. The physiological roles of tau and Abeta: Implications for Alzheimer's disease pathology and therapeutics. *Acta Neuropathol.* **2020**, *140*, 417–447. [CrossRef]
176. Gonzalez, A.; Singh, S.K.; Churruca, M.; Maccioni, R.B. Alzheimer's Disease and Tau Self-Assembly: In the Search of the Missing Link. *Int. J. Mol. Sci.* **2022**, *23*, 4192. [CrossRef]
177. Iqbal, K.; Liu, F.; Gong, C.X.; Alonso Adel, C.; Grundke-Iqbal, I. Mechanisms of tau-induced neurodegeneration. *Acta Neuropathol.* **2009**, *118*, 53–69. [CrossRef]
178. Mahoney, R.; Ochoa Thomas, E.; Ramirez, P.; Miller, H.E.; Beckmann, A.; Zuniga, G.; Dobrowolski, R.; Frost, B. Pathogenic Tau Causes a Toxic Depletion of Nuclear Calcium. *Cell Rep.* **2020**, *32*, 107900. [CrossRef]
179. Cieri, D.; Vicario, M.; Vallese, F.; D'Orsi, B.; Berto, P.; Grinzato, A.; Catoni, C.; De Stefani, D.; Rizzuto, R.; Brini, M.; et al. Tau localises within mitochondrial sub-compartments and its caspase cleavage affects ER-mitochondria interactions and cellular Ca(2+) handling. *Biochim. Biophys. Acta Mol. Basis Dis.* **2018**, *1864*, 3247–3256. [CrossRef] [PubMed]
180. Ahmadian, S.S.; Rezvanian, A.; Peterson, M.; Weintraub, S.; Bigio, E.H.; Mesulam, M.M.; Geula, C. Loss of calbindin-D28K is associated with the full range of tangle pathology within basal forebrain cholinergic neurons in Alzheimer's disease. *Neurobiol. Aging* **2015**, *36*, 3163–3170. [CrossRef] [PubMed]
181. Datta, D.; Leslie, S.N.; Wang, M.; Morozov, Y.M.; Yang, S.; Mentone, S.; Zeiss, C.; Duque, A.; Rakic, P.; Horvath, T.L.; et al. Age-related calcium dysregulation linked with tau pathology and impaired cognition in non-human primates. *Alzheimers Dement.* **2021**, *17*, 920–932. [CrossRef]
182. Britti, E.; Ros, J.; Esteras, N.; Abramov, A.Y. Tau inhibits mitochondrial calcium efflux and makes neurons vulnerable to calcium-induced cell death. *Cell Calcium* **2020**, *86*, 102150. [CrossRef]
183. Johnson, V.E.; Stewart, W.; Smith, D.H. Widespread tau and amyloid-beta pathology many years after a single traumatic brain injury in humans. *Brain Pathol.* **2012**, *22*, 142–149. [CrossRef] [PubMed]
184. Zanier, E.R.; Bertani, I.; Sammali, E.; Pischiutta, F.; Chiaravalloti, M.A.; Vegliante, G.; Masone, A.; Corbelli, A.; Smith, D.H.; Menon, D.K.; et al. Induction of a transmissible tau pathology by traumatic brain injury. *Brain* **2018**, *141*, 2685–2699. [CrossRef] [PubMed]

185. Walker, A.; Chapin, B.; Abisambra, J.; DeKosky, S.T. Association between single moderate to severe traumatic brain injury and long-term tauopathy in humans and preclinical animal models: A systematic narrative review of the literature. *Acta Neuropathol. Commun.* **2022**, *10*, 13. [CrossRef]
186. Shin, M.K.; Vazquez-Rosa, E.; Koh, Y.; Dhar, M.; Chaubey, K.; Cintron-Perez, C.J.; Barker, S.; Miller, E.; Franke, K.; Noterman, M.F.; et al. Reducing acetylated tau is neuroprotective in brain injury. *Cell* **2021**, *184*, 2715–2732.e23. [CrossRef]
187. Zhou, Q.; Li, S.; Li, M.; Ke, D.; Wang, Q.; Yang, Y.; Liu, G.P.; Wang, X.C.; Liu, E.; Wang, J.Z. Human tau accumulation promotes glycogen synthase kinase-3beta acetylation and thus upregulates the kinase: A vicious cycle in Alzheimer neurodegeneration. *EBioMedicine* **2022**, *78*, 103970. [CrossRef]
188. Goni-Oliver, P.; Lucas, J.J.; Avila, J.; Hernandez, F. N-terminal cleavage of GSK-3 by calpain: A new form of GSK-3 regulation. *J. Biol. Chem.* **2007**, *282*, 22406–22413. [CrossRef]
189. Jin, N.; Yin, X.; Yu, D.; Cao, M.; Gong, C.X.; Iqbal, K.; Ding, F.; Gu, X.; Liu, F. Truncation and activation of GSK-3beta by calpain I: A molecular mechanism links to tau hyperphosphorylation in Alzheimer's disease. *Sci. Rep.* **2015**, *5*, 8187. [CrossRef]
190. Jin, N.; Wu, Y.; Xu, W.; Gong, C.X.; Iqbal, K.; Liu, F. C-terminal truncation of GSK-3beta enhances its dephosphorylation by PP2A. *FEBS Lett.* **2017**, *591*, 1053–1063. [CrossRef]
191. King, M.K.; Pardo, M.; Cheng, Y.; Downey, K.; Jope, R.S.; Beurel, E. Glycogen synthase kinase-3 inhibitors: Rescuers of cognitive impairments. *Pharmacol. Ther.* **2014**, *141*, 1–12. [CrossRef] [PubMed]
192. Li, T.; Hawkes, C.; Qureshi, H.Y.; Kar, S.; Paudel, H.K. Cyclin-dependent protein kinase 5 primes microtubule-associated protein tau site-specifically for glycogen synthase kinase 3beta. *Biochemistry* **2006**, *45*, 3134–3145. [CrossRef]
193. Noble, W.; Olm, V.; Takata, K.; Casey, E.; Mary, O.; Meyerson, J.; Gaynor, K.; LaFrancois, J.; Wang, L.; Kondo, T.; et al. Cdk5 is a key factor in tau aggregation and tangle formation in vivo. *Neuron* **2003**, *38*, 555–565. [CrossRef] [PubMed]
194. Umfress, A.; Chakraborti, A.; Priya Sudarsana Devi, S.; Adams, R.; Epstein, D.; Massicano, A.; Sorace, A.; Singh, S.; Iqbal Hossian, M.; Andrabi, S.A.; et al. Cdk5 mediates rotational force-induced brain injury. *Sci. Rep.* **2023**, *13*, 3394. [CrossRef] [PubMed]
195. Cruz, J.C.; Tseng, H.C.; Goldman, J.A.; Shih, H.; Tsai, L.H. Aberrant Cdk5 activation by p25 triggers pathological events leading to neurodegeneration and neurofibrillary tangles. *Neuron* **2003**, *40*, 471–483. [CrossRef]
196. Atkins, C.M.; Chen, S.; Alonso, O.F.; Dietrich, W.D.; Hu, B.R. Activation of calcium/calmodulin-dependent protein kinases after traumatic brain injury. *J. Cereb. Blood Flow Metab.* **2006**, *26*, 1507–1518. [CrossRef]
197. Yoshimura, Y.; Ichinose, T.; Yamauchi, T. Phosphorylation of tau protein to sites found in Alzheimer's disease brain is catalyzed by Ca^{2+}/calmodulin-dependent protein kinase II as demonstrated tandem mass spectrometry. *Neurosci. Lett.* **2003**, *353*, 185–188. [CrossRef] [PubMed]
198. Spaethling, J.M.; Klein, D.M.; Singh, P.; Meaney, D.F. Calcium-permeable AMPA receptors appear in cortical neurons after traumatic mechanical injury and contribute to neuronal fate. *J. Neurotrauma* **2008**, *25*, 1207–1216. [CrossRef]
199. Yoshimura, Y.; Nomura, T.; Yamauchi, T. Purification and characterization of active fragment of Ca^{2+}/calmodulin-dependent protein kinase II from the post-synaptic density in the rat forebrain. *J. Biochem.* **1996**, *119*, 268–273. [CrossRef]
200. Wang, J.Z.; Grundke-Iqbal, I.; Iqbal, K. Kinases and phosphatases and tau sites involved in Alzheimer neurofibrillary degeneration. *Eur. J. Neurosci.* **2007**, *25*, 59–68. [CrossRef]
201. Yasuda, M.; Mayford, M.R. CaMKII activation in the entorhinal cortex disrupts previously encoded spatial memory. *Neuron* **2006**, *50*, 309–318. [CrossRef]
202. Ryoo, S.-R.; Jeong, H.K.; Radnaabazar, C.; Yoo, J.-J.; Cho, H.-J.; Lee, H.-W.; Kim, I.-S.; Cheon, Y.-H.; Ahn, Y.S.; Chung, S.-H. DYRK1A-mediated Hyperphosphorylation of Tau. *J. Biol. Chem.* **2007**, *282*, 34850–34857. [CrossRef]
203. Woods, Y.L.; Cohen, P.; Becker, W.; Jakes, R.; Goedert, M.; Wang, X.; Proud, C.G. The kinase DYRK phosphorylates protein-synthesis initiation factor eIF2Bepsilon at Ser539 and the microtubule-associated protein tau at Thr212: Potential role for DYRK as a glycogen synthase kinase 3-priming kinase. *Biochem. J.* **2001**, *355*, 609–615. [CrossRef]
204. Moreau, M.; Carmona-Iragui, M.; Altuna, M.; Dalzon, L.; Barroeta, I.; Vilaire, M.; Durand, S.; Fortea, J.; Rebillat, A.S.; Janel, N. DYRK1A and Activity-Dependent Neuroprotective Protein Comparative Diagnosis Interest in Cerebrospinal Fluid and Plasma in the Context of Alzheimer-Related Cognitive Impairment in Down Syndrome Patients. *Biomedicines* **2022**, *10*, 1380. [CrossRef] [PubMed]
205. Reynolds, C.H.; Betts, J.C.; Blackstock, W.P.; Nebreda, A.R.; Anderton, B.H. Phosphorylation sites on tau identified by nanoelectrospray mass spectrometry: Differences in vitro between the mitogen-activated protein kinases ERK2, c-Jun N-terminal kinase and P38, and glycogen synthase kinase-3beta. *J. Neurochem.* **2000**, *74*, 1587–1595. [CrossRef] [PubMed]
206. Pei, J.-J.; Braak, H.; An, W.-L.; Winblad, B.; Cowburn, R.F.; Iqbal, K.; Grundke-Iqbal, I. Up-regulation of mitogen-activated protein kinases ERK1/2 and MEK1/2 is associated with the progression of neurofibrillary degeneration in Alzheimer's disease. *Mol. Brain Res.* **2002**, *109*, 45–55. [CrossRef] [PubMed]
207. Park, J.H.; Hwang, J.W.; Lee, H.J.; Jang, G.M.; Jeong, Y.J.; Cho, J.; Seo, J.; Hoe, H.S. Lomerizine inhibits LPS-mediated neuroinflammation and tau hyperphosphorylation by modulating NLRP3, DYRK1A, and GSK3alpha/beta. *Front. Immunol.* **2023**, *14*, 1150940. [CrossRef]

208. Zhang, B.; Li, Z.; Zhang, R.; Hu, Y.; Jiang, Y.; Cao, T.; Wang, J.; Gong, L.; Ji, L.; Mu, H.; et al. PKCgamma promotes axonal remodeling in the cortico-spinal tract via GSK3beta/beta-catenin signaling after traumatic brain injury. *Sci. Rep.* **2019**, *9*, 17078. [CrossRef]
209. Sun, M.K.; Alkon, D.L. Activation of protein kinase C isozymes for the treatment of dementias. *Adv. Pharmacol.* **2012**, *64*, 273–302. [CrossRef]
210. Braithwaite, S.P.; Stock, J.B.; Lombroso, P.J.; Nairn, A.C. Protein phosphatases and Alzheimer's disease. *Prog. Mol. Biol. Transl. Sci.* **2012**, *106*, 343–379. [CrossRef]
211. Liu, F.; Grundke-Iqbal, I.; Iqbal, K.; Gong, C.X. Contributions of protein phosphatases PP1, PP2A, PP2B and PP5 to the regulation of tau phosphorylation. *Eur. J. Neurosci.* **2005**, *22*, 1942–1950. [CrossRef]
212. Shultz, S.R.; Wright, D.K.; Zheng, P.; Stuchbery, R.; Liu, S.J.; Sashindranath, M.; Medcalf, R.L.; Johnston, L.A.; Hovens, C.M.; Jones, N.C.; et al. Sodium selenate reduces hyperphosphorylated tau and improves outcomes after traumatic brain injury. *Brain* **2015**, *138*, 1297–1313. [CrossRef]
213. Yang, W.J.; Chen, W.; Chen, L.; Guo, Y.J.; Zeng, J.S.; Li, G.Y.; Tong, W.S. Involvement of tau phosphorylation in traumatic brain injury patients. *Acta Neurol. Scand.* **2017**, *135*, 622–627. [CrossRef] [PubMed]
214. Corcoran, N.M.; Martin, D.; Hutter-Paier, B.; Windisch, M.; Nguyen, T.; Nheu, L.; Sundstrom, L.E.; Costello, A.J.; Hovens, C.M. Sodium selenate specifically activates PP2A phosphatase, dephosphorylates tau and reverses memory deficits in an Alzheimer's disease model. *J. Clin. Neurosci.* **2010**, *17*, 1025–1033. [CrossRef] [PubMed]
215. Kins, S.; Kurosinski, P.; Nitsch, R.M.; Gotz, J. Activation of the ERK and JNK signaling pathways caused by neuron-specific inhibition of PP2A in transgenic mice. *Am. J. Pathol.* **2003**, *163*, 833–843. [CrossRef]
216. Watkins, G.R.; Wang, N.; Mazalouskas, M.D.; Gomez, R.J.; Guthrie, C.R.; Kraemer, B.C.; Schweiger, S.; Spiller, B.W.; Wadzinski, B.E. Monoubiquitination promotes calpain cleavage of the protein phosphatase 2A (PP2A) regulatory subunit alpha4, altering PP2A stability and microtubule-associated protein phosphorylation. *J. Biol. Chem.* **2012**, *287*, 24207–24215. [CrossRef]
217. Liu, Y.; Guo, C.; Ding, Y.; Long, X.; Li, W.; Ke, D.; Wang, Q.; Liu, R.; Wang, J.Z.; Zhang, H.; et al. Blockage of AEP attenuates TBI-induced tau hyperphosphorylation and cognitive impairments in rats. *Aging (Albany NY)* **2020**, *12*, 19421–19439. [CrossRef]
218. Zhang, Z.; Song, M.; Liu, X.; Su Kang, S.; Duong, D.M.; Seyfried, N.T.; Cao, X.; Cheng, L.; Sun, Y.E.; Ping Yu, S.; et al. Delta-secretase cleaves amyloid precursor protein and regulates the pathogenesis in Alzheimer's disease. *Nat. Commun.* **2015**, *6*, 8762. [CrossRef]
219. Qian, W.; Yin, X.; Hu, W.; Shi, J.; Gu, J.; Grundke-Iqbal, I.; Iqbal, K.; Gong, C.X.; Liu, F. Activation of protein phosphatase 2B and hyperphosphorylation of Tau in Alzheimer's disease. *J. Alzheimers Dis.* **2011**, *23*, 617–627. [CrossRef]
220. Kurz, J.E.; Parsons, J.T.; Rana, A.; Gibson, C.J.; Hamm, R.J.; Churn, S.B. A significant increase in both basal and maximal calcineurin activity following fluid percussion injury in the rat. *J. Neurotrauma* **2005**, *22*, 476–490. [CrossRef]
221. Luo, J.; Ma, J.; Yu, D.Y.; Bu, F.; Zhang, W.; Tu, L.H.; Wei, Q. Infusion of FK506, a specific inhibitor of calcineurin, induces potent tau hyperphosphorylation in mouse brain. *Brain Res. Bull.* **2008**, *76*, 464–468. [CrossRef] [PubMed]
222. Reese, L.C.; Taglialatela, G. Neuroimmunomodulation by calcineurin in aging and Alzheimer's disease. *Aging Dis.* **2010**, *1*, 245–253.
223. Cicognola, C.; Brinkmalm, G.; Wahlgren, J.; Portelius, E.; Gobom, J.; Cullen, N.C.; Hansson, O.; Parnetti, L.; Constantinescu, R.; Wildsmith, K.; et al. Novel tau fragments in cerebrospinal fluid: Relation to tangle pathology and cognitive decline in Alzheimer's disease. *Acta Neuropathol.* **2019**, *137*, 279–296. [CrossRef] [PubMed]
224. Cicognola, C.; Satir, T.M.; Brinkmalm, G.; Matecko-Burmann, I.; Agholme, L.; Bergstrom, P.; Becker, B.; Zetterberg, H.; Blennow, K.; Hoglund, K. Tauopathy-Associated Tau Fragment Ending at Amino Acid 224 Is Generated by Calpain-2 Cleavage. *J. Alzheimers Dis.* **2020**, *74*, 1143–1156. [CrossRef]
225. Afreen, S.; Riherd Methner, D.N.; Ferreira, A. Tau(45-230) association with the cytoskeleton and membrane-bound organelles: Functional implications in neurodegeneration. *Neuroscience* **2017**, *362*, 104–117. [CrossRef] [PubMed]
226. Jiang, S.; Shao, C.; Tang, F.; Wang, W.; Zhu, X. Dynamin-like protein 1 cleavage by calpain in Alzheimer's disease. *Aging Cell* **2019**, *18*, e12912. [CrossRef]
227. Wang, C.Y.; Xie, J.W.; Wang, T.; Xu, Y.; Cai, J.H.; Wang, X.; Zhao, B.L.; An, L.; Wang, Z.Y. Hypoxia-triggered m-calpain activation evokes endoplasmic reticulum stress and neuropathogenesis in a transgenic mouse model of Alzheimer's disease. *CNS Neurosci. Ther.* **2013**, *19*, 820–833. [CrossRef]
228. Rao, M.V.; McBrayer, M.K.; Campbell, J.; Kumar, A.; Hashim, A.; Sershen, H.; Stavrides, P.H.; Ohno, M.; Hutton, M.; Nixon, R.A. Specific calpain inhibition by calpastatin prevents tauopathy and neurodegeneration and restores normal lifespan in tau P301L mice. *J. Neurosci.* **2014**, *34*, 9222–9234. [CrossRef] [PubMed]
229. Wang, Y.; Liu, Y.; Lopez, D.; Lee, M.; Dayal, S.; Hurtado, A.; Bi, X.; Baudry, M. Protection against TBI-Induced Neuronal Death with Post-Treatment with a Selective Calpain-2 Inhibitor in Mice. *J. Neurotrauma* **2018**, *35*, 105–117. [CrossRef]

230. Gamblin, T.C.; Chen, F.; Zambrano, A.; Abraha, A.; Lagalwar, S.; Guillozet, A.L.; Lu, M.; Fu, Y.; Garcia-Sierra, F.; LaPointe, N.; et al. Caspase cleavage of tau: Linking amyloid and neurofibrillary tangles in Alzheimer's disease. *Proc. Natl. Acad. Sci. USA* **2003**, *100*, 10032–10037. [CrossRef]
231. Rissman, R.A.; Poon, W.W.; Blurton-Jones, M.; Oddo, S.; Torp, R.; Vitek, M.P.; LaFerla, F.M.; Rohn, T.T.; Cotman, C.W. Caspase-cleavage of tau is an early event in Alzheimer disease tangle pathology. *J. Clin. Investig.* **2004**, *114*, 121–130. [CrossRef]
232. Basurto-Islas, G.; Luna-Munoz, J.; Guillozet-Bongaarts, A.L.; Binder, L.I.; Mena, R.; Garcia-Sierra, F. Accumulation of aspartic acid421- and glutamic acid391-cleaved tau in neurofibrillary tangles correlates with progression in Alzheimer disease. *J. Neuropathol. Exp. Neurol.* **2008**, *67*, 470–483. [CrossRef]
233. Fasulo, L.; Ugolini, G.; Cattaneo, A. Apoptotic effect of caspase-3 cleaved tau in hippocampal neurons and its potentiation by tau FTDP-mutation N279K. *J. Alzheimers Dis.* **2005**, *7*, 3–13. [CrossRef] [PubMed]
234. Spires-Jones, T.L.; de Calignon, A.; Matsui, T.; Zehr, C.; Pitstick, R.; Wu, H.Y.; Osetek, J.D.; Jones, P.B.; Bacskai, B.J.; Feany, M.B.; et al. In vivo imaging reveals dissociation between caspase activation and acute neuronal death in tangle-bearing neurons. *J. Neurosci.* **2008**, *28*, 862–867. [CrossRef]
235. DeTure, M.A.; Dickson, D.W. The neuropathological diagnosis of Alzheimer's disease. *Mol. Neurodegener.* **2019**, *14*, 32. [CrossRef]
236. Herring, B.E.; Nicoll, R.A. Long-Term Potentiation: From CaMKII to AMPA Receptor Trafficking. *Annu. Rev. Physiol.* **2016**, *78*, 351–365. [CrossRef] [PubMed]
237. Gascon, S.; Sobrado, M.; Roda, J.M.; Rodriguez-Pena, A.; Diaz-Guerra, M. Excitotoxicity and focal cerebral ischemia induce truncation of the NR2A and NR2B subunits of the NMDA receptor and cleavage of the scaffolding protein PSD-95. *Mol. Psychiatry* **2008**, *13*, 99–114. [CrossRef]
238. Bi, X.; Chen, J.; Baudry, M. Calpain-mediated proteolysis of GluR1 subunits in organotypic hippocampal cultures following kainic acid treatment. *Brain Res.* **1998**, *781*, 355–357. [CrossRef]
239. Riascos, D.; Nicholas, A.; Samaeekia, R.; Yukhananov, R.; Mesulam, M.M.; Bigio, E.H.; Weintraub, S.; Guo, L.; Geula, C. Alterations of Ca^{2+}-responsive proteins within cholinergic neurons in aging and Alzheimer's disease. *Neurobiol. Aging* **2014**, *35*, 1325–1333. [CrossRef]
240. Kelly, B.L.; Vassar, R.; Ferreira, A. Beta-amyloid-induced dynamin 1 depletion in hippocampal neurons. A potential mechanism for early cognitive decline in Alzheimer disease. *J. Biol. Chem.* **2005**, *280*, 31746–31753. [CrossRef]
241. Ferreira, A.P.A.; Casamento, A.; Carrillo Roas, S.; Halff, E.F.; Panambalana, J.; Subramaniam, S.; Schutzenhofer, K.; Chan Wah Hak, L.; McGourty, K.; Thalassinos, K.; et al. Cdk5 and GSK3beta inhibit fast endophilin-mediated endocytosis. *Nat. Commun.* **2021**, *12*, 2424. [CrossRef] [PubMed]
242. Walker, K.R.; Tesco, G. Molecular mechanisms of cognitive dysfunction following traumatic brain injury. *Front. Aging Neurosci.* **2013**, *5*, 29. [CrossRef] [PubMed]
243. Trinchese, F.; Fa, M.; Liu, S.; Zhang, H.; Hidalgo, A.; Schmidt, S.D.; Yamaguchi, H.; Yoshii, N.; Mathews, P.M.; Nixon, R.A.; et al. Inhibition of calpains improves memory and synaptic transmission in a mouse model of Alzheimer disease. *J. Clin. Investig.* **2008**, *118*, 2796–2807. [CrossRef] [PubMed]
244. Schoch, K.M.; Evans, H.N.; Brelsfoard, J.M.; Madathil, S.K.; Takano, J.; Saido, T.C.; Saatman, K.E. Calpastatin overexpression limits calpain-mediated proteolysis and behavioral deficits following traumatic brain injury. *Exp. Neurol.* **2012**, *236*, 371–382. [CrossRef]
245. Wang, Y.J.; Chen, G.H.; Hu, X.Y.; Lu, Y.P.; Zhou, J.N.; Liu, R.Y. The expression of calcium/calmodulin-dependent protein kinase II-alpha in the hippocampus of patients with Alzheimer's disease and its links with AD-related pathology. *Brain Res.* **2005**, *1031*, 101–108. [CrossRef]
246. Cho, H.; Hyeon, S.J.; Shin, J.Y.; Alvarez, V.E.; Stein, T.D.; Lee, J.; Kowall, N.W.; McKee, A.C.; Ryu, H.; Seo, J.S. Alterations of transcriptome signatures in head trauma-related neurodegenerative disorders. *Sci. Rep.* **2020**, *10*, 8811. [CrossRef] [PubMed]
247. Gu, Z.; Liu, W.; Yan, Z. beta-Amyloid impairs AMPA receptor trafficking and function by reducing Ca^{2+}/calmodulin-dependent protein kinase II synaptic distribution. *J. Biol. Chem.* **2009**, *284*, 10639–10649. [CrossRef]
248. Wang, X.L.; Deng, Y.X.; Gao, Y.M.; Dong, Y.T.; Wang, F.; Guan, Z.Z.; Hong, W.; Qi, X.L. Activation of alpha7 nAChR by PNU-282987 improves synaptic and cognitive functions through restoring the expression of synaptic-associated proteins and the CaM-CaMKII-CREB signaling pathway. *Aging (Albany NY)* **2020**, *12*, 543–570. [CrossRef]
249. Hooper, C.; Markevich, V.; Plattner, F.; Killick, R.; Schofield, E.; Engel, T.; Hernandez, F.; Anderton, B.; Rosenblum, K.; Bliss, T.; et al. Glycogen synthase kinase-3 inhibition is integral to long-term potentiation. *Eur. J. Neurosci.* **2007**, *25*, 81–86. [CrossRef]
250. Zhu, L.Q.; Wang, S.H.; Liu, D.; Yin, Y.Y.; Tian, Q.; Wang, X.C.; Wang, Q.; Chen, J.G.; Wang, J.Z. Activation of glycogen synthase kinase-3 inhibits long-term potentiation with synapse-associated impairments. *J. Neurosci.* **2007**, *27*, 12211–12220. [CrossRef]
251. DaRocha-Souto, B.; Coma, M.; Perez-Nievas, B.G.; Scotton, T.C.; Siao, M.; Sanchez-Ferrer, P.; Hashimoto, T.; Fan, Z.; Hudry, E.; Barroeta, I.; et al. Activation of glycogen synthase kinase-3 beta mediates beta-amyloid induced neuritic damage in Alzheimer's disease. *Neurobiol. Dis.* **2012**, *45*, 425–437. [CrossRef]

252. Cao, Q.; Meng, T.; Man, J.; Peng, D.; Chen, H.; Xiang, Q.; Su, Z.; Zhang, Q.; Huang, Y. aFGF Promotes Neurite Growth by Regulating GSK3beta-CRMP2 Signaling Pathway in Cortical Neurons Damaged by Amyloid-beta. *J. Alzheimers Dis.* **2019**, *72*, 97–109. [CrossRef]
253. Liao, Z.; Huang, Z.; Li, J.; Li, H.; Miao, L.; Liu, Y.; Zhang, J.; Xu, Y.; Li, Y. Regulation of CRMP2 by Cdk5 and GSK-3beta participates in sevoflurane-induced dendritic development abnormalities and cognitive dysfunction in developing rats. *Toxicol. Lett.* **2021**, *341*, 68–79. [CrossRef]
254. Pigino, G.; Morfini, G.; Pelsman, A.; Mattson, M.P.; Brady, S.T.; Busciglio, J. Alzheimer's presenilin 1 mutations impair kinesin-based axonal transport. *J. Neurosci.* **2003**, *23*, 4499–4508. [CrossRef]
255. Fu, K.; Lin, N.; Xu, Y.; Huang, E.; He, R.; Wu, Z.; Qu, D.; Chen, X.; Huang, T. CDK5-mediated hyperphosphorylation of Tau217 impairs neuronal synaptic structure and exacerbates cognitive impairment in Alzheimer's disease. *Transl. Psychiatry* **2025**, *15*, 302. [CrossRef] [PubMed]
256. Zohar, O.; Lavy, R.; Zi, X.; Nelson, T.J.; Hongpaisan, J.; Pick, C.G.; Alkon, D.L. PKC activator therapeutic for mild traumatic brain injury in mice. *Neurobiol. Dis.* **2011**, *41*, 329–337. [CrossRef] [PubMed]
257. Lorden, G.; Wozniak, J.M.; Dore, K.; Dozier, L.E.; Cates-Gatto, C.; Patrick, G.N.; Gonzalez, D.J.; Roberts, A.J.; Tanzi, R.E.; Newton, A.C. Enhanced activity of Alzheimer disease-associated variant of protein kinase Calpha drives cognitive decline in a mouse model. *Nat. Commun.* **2022**, *13*, 7200. [CrossRef] [PubMed]
258. Wu, H.Y.; Hudry, E.; Hashimoto, T.; Kuchibhotla, K.; Rozkalne, A.; Fan, Z.; Spires-Jones, T.; Xie, H.; Arbel-Ornath, M.; Grosskreutz, C.L.; et al. Amyloid beta induces the morphological neurodegenerative triad of spine loss, dendritic simplification, and neuritic dystrophies through calcineurin activation. *J. Neurosci.* **2010**, *30*, 2636–2649. [CrossRef]
259. Beattie, E.C.; Carroll, R.C.; Yu, X.; Morishita, W.; Yasuda, H.; von Zastrow, M.; Malenka, R.C. Regulation of AMPA receptor endocytosis by a signaling mechanism shared with LTD. *Nat. Neurosci.* **2000**, *3*, 1291–1300. [CrossRef]
260. Yin, Y.; Gao, D.; Wang, Y.; Wang, Z.H.; Wang, X.; Ye, J.; Wu, D.; Fang, L.; Pi, G.; Yang, Y.; et al. Tau accumulation induces synaptic impairment and memory deficit by calcineurin-mediated inactivation of nuclear CaMKIV/CREB signaling. *Proc. Natl. Acad. Sci. USA* **2016**, *113*, E3773–E3781. [CrossRef]
261. Puzzo, D.; Vitolo, O.; Trinchese, F.; Jacob, J.P.; Palmeri, A.; Arancio, O. Amyloid-beta peptide inhibits activation of the nitric oxide/cGMP/cAMP-responsive element-binding protein pathway during hippocampal synaptic plasticity. *J. Neurosci.* **2005**, *25*, 6887–6897. [CrossRef]
262. Campbell, J.N.; Register, D.; Churn, S.B. Traumatic brain injury causes an FK506-sensitive loss and an overgrowth of dendritic spines in rat forebrain. *J. Neurotrauma* **2012**, *29*, 201–217. [CrossRef]
263. Zeng, J.; Hu, X.F.; Sun, D.S.; Hong, X.Y.; Ma, J.Z.; Feng, Q. Alzheimer-like behavior and synaptic dysfunction in 3 x Tg-AD mice are reversed with calcineurin inhibition. *Exp. Brain Res.* **2024**, *242*, 1507–1515. [CrossRef]
264. Villegas, R.; Martinez, N.W.; Lillo, J.; Pihan, P.; Hernandez, D.; Twiss, J.L.; Court, F.A. Calcium release from intra-axonal endoplasmic reticulum leads to axon degeneration through mitochondrial dysfunction. *J. Neurosci.* **2014**, *34*, 7179–7189. [CrossRef]
265. Johnson, V.E.; Stewart, W.; Smith, D.H. Axonal pathology in traumatic brain injury. *Exp. Neurol.* **2013**, *246*, 35–43. [CrossRef]
266. Cai, Y.; Kanyo, J.; Wilson, R.; Bathla, S.; Cardozo, P.L.; Tong, L.; Qin, S.; Fuentes, L.A.; Pinheiro-de-Sousa, I.; Huynh, T.; et al. Subcellular proteomics and iPSC modeling uncover reversible mechanisms of axonal pathology in Alzheimer's disease. *Nat. Aging* **2025**, *5*, 504–527. [CrossRef]
267. Yan, X.X.; Jeromin, A.; Jeromin, A. Spectrin Breakdown Products (SBDPs) as Potential Biomarkers for Neurodegenerative Diseases. *Curr. Transl. Geriatr. Exp. Gerontol. Rep.* **2012**, *1*, 85–93. [CrossRef]
268. Knoferle, J.; Koch, J.C.; Ostendorf, T.; Michel, U.; Planchamp, V.; Vutova, P.; Tonges, L.; Stadelmann, C.; Bruck, W.; Bahr, M.; et al. Mechanisms of acute axonal degeneration in the optic nerve in vivo. *Proc. Natl. Acad. Sci. USA* **2010**, *107*, 6064–6069. [CrossRef] [PubMed]
269. Stys, P.K.; Jiang, Q. Calpain-dependent neurofilament breakdown in anoxic and ischemic rat central axons. *Neurosci. Lett.* **2002**, *328*, 150–154. [CrossRef]
270. Bian, F.; Nath, R.; Sobocinski, G.; Booher, R.N.; Lipinski, W.J.; Callahan, M.J.; Pack, A.; Wang, K.K.; Walker, L.C. Axonopathy, tau abnormalities, and dyskinesia, but no neurofibrillary tangles in p25-transgenic mice. *J. Comp. Neurol.* **2002**, *446*, 257–266. [CrossRef] [PubMed]
271. Sundaram, J.R.; Poore, C.P.; Sulaimee, N.H.; Pareek, T.; Asad, A.B.; Rajkumar, R.; Cheong, W.F.; Wenk, M.R.; Dawe, G.S.; Chuang, K.H.; et al. Specific inhibition of p25/Cdk5 activity by the Cdk5 inhibitory peptide reduces neurodegeneration in vivo. *J. Neurosci.* **2013**, *33*, 334–343. [CrossRef] [PubMed]
272. Reinhardt, L.; Kordes, S.; Reinhardt, P.; Glatza, M.; Baumann, M.; Drexler, H.C.A.; Menninger, S.; Zischinsky, G.; Eickhoff, J.; Frob, C.; et al. Dual Inhibition of GSK3beta and CDK5 Protects the Cytoskeleton of Neurons from Neuroinflammatory-Mediated Degeneration In Vitro and In Vivo. *Stem Cell Rep.* **2019**, *12*, 502–517. [CrossRef]

273. Touma, E.; Kato, S.; Fukui, K.; Koike, T. Calpain-mediated cleavage of collapsin response mediator protein(CRMP)-2 during neurite degeneration in mice. *Eur. J. Neurosci.* **2007**, *26*, 3368–3381. [CrossRef]
274. Stokin, G.B.; Lillo, C.; Falzone, T.L.; Brusch, R.G.; Rockenstein, E.; Mount, S.L.; Raman, R.; Davies, P.; Masliah, E.; Williams, D.S.; et al. Axonopathy and transport deficits early in the pathogenesis of Alzheimer's disease. *Science* **2005**, *307*, 1282–1288. [CrossRef]
275. Hou, S.T.; Jiang, S.X.; Aylsworth, A.; Ferguson, G.; Slinn, J.; Hu, H.; Leung, T.; Kappler, J.; Kaibuchi, K. CaMKII phosphorylates collapsin response mediator protein 2 and modulates axonal damage during glutamate excitotoxicity. *J. Neurochem.* **2009**, *111*, 870–881. [CrossRef]
276. Qu, J.; Matsouaka, R.; Betensky, R.A.; Hyman, B.T.; Grosskreutz, C.L. Calcineurin activation causes retinal ganglion cell degeneration. *Mol. Vis.* **2012**, *18*, 2828–2838. [PubMed]
277. Barrientos, S.A.; Martinez, N.W.; Yoo, S.; Jara, J.S.; Zamorano, S.; Hetz, C.; Twiss, J.L.; Alvarez, J.; Court, F.A. Axonal degeneration is mediated by the mitochondrial permeability transition pore. *J. Neurosci.* **2011**, *31*, 966–978. [CrossRef]
278. Okonkwo, D.O.; Buki, A.; Siman, R.; Povlishock, J.T. Cyclosporin A limits calcium-induced axonal damage following traumatic brain injury. *Neuroreport* **1999**, *10*, 353–358. [CrossRef]
279. Iwata, A.; Chen, X.H.; McIntosh, T.K.; Browne, K.D.; Smith, D.H. Long-term accumulation of amyloid-beta in axons following brain trauma without persistent upregulation of amyloid precursor protein genes. *J. Neuropathol. Exp. Neurol.* **2002**, *61*, 1056–1068. [CrossRef]
280. Cribbs, D.H.; Chen, L.S.; Cotman, C.W.; LaFerla, F.M. Injury induces presenilin-1 gene expression in mouse brain. *Neuroreport* **1996**, *7*, 1773–1776. [CrossRef] [PubMed]
281. Mandelkow, E.M.; Stamer, K.; Vogel, R.; Thies, E.; Mandelkow, E. Clogging of axons by tau, inhibition of axonal traffic and starvation of synapses. *Neurobiol. Aging* **2003**, *24*, 1079–1085. [CrossRef] [PubMed]
282. Rodrigues, E.M.; Weissmiller, A.M.; Goldstein, L.S. Enhanced beta-secretase processing alters APP axonal transport and leads to axonal defects. *Hum. Mol. Genet.* **2012**, *21*, 4587–4601. [CrossRef]
283. Tang, Y.; Scott, D.A.; Das, U.; Edland, S.D.; Radomski, K.; Koo, E.H.; Roy, S. Early and selective impairments in axonal transport kinetics of synaptic cargoes induced by soluble amyloid beta-protein oligomers. *Traffic* **2012**, *13*, 681–693. [CrossRef] [PubMed]
284. Vossel, K.A.; Zhang, K.; Brodbeck, J.; Daub, A.C.; Sharma, P.; Finkbeiner, S.; Cui, B.; Mucke, L. Tau reduction prevents Abeta-induced defects in axonal transport. *Science* **2010**, *330*, 198. [CrossRef] [PubMed]
285. Coleman, M. Molecular signaling how do axons die? *Adv. Genet.* **2011**, *73*, 185–217. [CrossRef] [PubMed]

International Journal of
Molecular Sciences

MDPI

Review

The Role of Oxygen Homeostasis and the HIF-1 Factor in the Development of Neurodegeneration

Elena V. Mitroshina * and Maria V. Vedunova

Institute of Biology and Biomedicine, Lobachevsky State University of Nizhny Novgorod, 23 Gagarin Avenue, 603022 Nizhny Novgorod, Russia; mvedunova@yandex.ru
* Correspondence: mitroshina@ibbm.unn.ru or helenmitroshina@gmail.com; Tel.: +7-950-604-5137

Abstract: Understanding the molecular underpinnings of neurodegeneration processes is a pressing challenge for medicine and neurobiology. Alzheimer's disease (AD) and Parkinson's disease (PD) represent the most prevalent forms of neurodegeneration. To date, a substantial body of experimental evidence has strongly implicated hypoxia in the pathogenesis of numerous neurological disorders, including AD, PD, and other age-related neurodegenerative conditions. Hypoxia-inducible factor (HIF) is a transcription factor that triggers a cell survival program in conditions of oxygen deprivation. The involvement of HIF-1α in neurodegenerative processes presents a complex and sometimes contradictory picture. This review aims to elucidate the current understanding of the interplay between hypoxia and the development of AD and PD, assess the involvement of HIF-1 in their pathogenesis, and summarize promising therapeutic approaches centered on modulating the activity of the HIF-1 complex.

Keywords: hypoxia; HIF-1; neurodegeneration; neuroprotection; PHD; Alzheimer's disease; Parkinson's disease

1. Introduction

The aging population in most developed nations underscores the critical importance of investigating the endogenous molecular mechanisms that may mitigate or decelerate the onset of age-related diseases, particularly neurodegenerative conditions, which are of paramount concern. Neurodegenerative diseases predominantly manifest as sporadic conditions, with fewer instances of hereditary transmission, affecting the central nervous system and resulting in the gradual and progressive deterioration of specific neuronal populations and their interconnections [1]. Given the ongoing aging trend and the extension of our life expectancy, the incidence of dementia is anticipated to rise significantly in the coming years [2,3].

According to the World Health Organization (WHO), Alzheimer's disease (AD) is the most prevalent form of dementia, accounting for an estimated 60–70% of cases globally and currently afflicting 50 million individuals worldwide [4]. Parkinson's disease (PD) ranks as the second most common neurodegenerative disorder [5], representing the foremost serious movement disorder globally. Importantly, PD surpasses other neurological disorders in terms of escalating rates of disability and mortality. Despite advancements, effective pharmacological interventions capable of preventing or halting the progression of neurodegenerative processes in PD, AD, and other chronic neurodegenerative disorders remain elusive. Therefore, identifying the key signaling pathways and genetic mechanisms amenable to pharmacological modulation to mitigate neurodegenerative changes in the brain represents a crucial yet unresolved challenge.

To date, a substantial body of experimental evidence has strongly implicated hypoxia in the pathogenesis of numerous neurological disorders, including Alzheimer's disease, Parkinson's disease, and other age-related neurodegenerative conditions [6]. For instance, several studies suggest that hypoxia during the perinatal period may serve as a precipitating factor for the onset of Alzheimer's disease later in adulthood [7]. Brain aging is linked with the emergence of chronic regions of microhypoxia within the nervous tissue. Prolonged exposure to hypoxia precipitates neuroinflammation by activating microglia, the resident immune cells in the brain. Chronic inflammation constitutes a fundamental component of neurodegenerative processes, leading to heightened levels of reactive oxygen species (ROS) and pro-inflammatory cytokines, which are hallmark features of various degenerative diseases of the central nervous system [8].

Consequently, there is a rationale for considering the intracellular systems responsible for adapting to hypoxic injury potential targets for preventing and treating neurodegenerative processes. The discovery of the transcription factor HIF-1 and its stability modulation using oxygen-dependent HIF prolyl hydroxylase started a new era in comprehending adaptive cell responses to hypoxic conditions. Greg Semenza, William Kaelin Jr., and Sir Piter Ratcliffe get the 2019 Nobel Prize in Physiology or Medicine for this research. HIF is a transcription factor that triggers a cell survival program in conditions of oxygen deprivation, activating over 100 genes that encompass various cellular processes involved in hypoxia adaptation, including angiogenesis, erythropoiesis, energy metabolism, cell proliferation, and cell cycle control [9]. Notably, the most renowned targets activated by HIF are the erythropoietin (EPO) and vascular endothelial growth factor (VEGF) genes. Specifically, the transcription factor HIF-1 can regulate the expression of numerous genes that mitigate neuronal cell death during neurodegeneration. For instance, the regulation of HIF-1 target genes involved in glycolysis or vascular regulation constitutes an early adaptive response to oxidative stress, potentially attenuating cognitive decline and delaying the progression to more severe stages of Alzheimer's disease [10].

The involvement of HIF-1α in neurodegenerative processes presents a complex and sometimes contradictory picture. While prolonged or excessive stressors, such as ischemia or oxidative stress, may transform HIF-1 into a trigger for neuronal cell death, potentially activating the Bnip3 factor and contributing to late-onset dementia in Alzheimer's disease [10], prolonged and severe hypoxia can induce HIF-1α-mediated stabilization of p53, subsequently triggering p53-dependent apoptosis [11]. Thus, the role of HIF-1 in the pathogenesis of neurodegenerative diseases necessitates careful examination. In this review, we initially examine the existing literature regarding the correlation between hypoxia and the onset of AD and PD. Subsequently, we delve into the role of HIF-1 in the pathophysiology of these conditions, followed by an exploration of the available data on the influence of modulating HIF-1 activity on the progression of neurodegenerative disorders.

2. The Role of Hypoxia in the Development of Neurodegenerative Diseases

2.1. The Role of Hypoxia in the Pathogenesis of Alzheimer's Disease

Emerging evidence suggests that hypoxia contributes to the pathogenesis of AD through various mechanisms, including the promotion of amyloid beta (Aβ) formation, stimulation of tau hyperphosphorylation, disruption of blood–brain barrier function, and the facilitation of neuronal degeneration (Figure 1).

2.1.1. Amyloid Processing

It is now known that hypoxia influences the formation and metabolism of beta-amyloid. Among the disintegrin and metalloproteinase domain-containing (ADAM) family of proteins, ADAM10 and ADAM17 are the principal amyloid precursor protein (APP)-cleaving enzymes acting at the α site of the protein [12,13]. ADAM10 predominantly functions as α-secretase in the brain, and recent identification of two rare ADAM10 mutations suggests their role as predisposing factors in early-onset AD [14]. Moreover, a single rare nonsynonymous variant rs142946965 [p.R215I] in ADAM17 was correlated with an autosomal dominant pattern of late-onset AD in one family [15]. Intriguingly, numerous studies have demonstrated that hypoxia markedly diminishes the expression and activity of the ADAM10 and ADAM17 proteins in the neurons, thereby reducing the cleavage of

APP via the non-amyloidogenic pathway [16–18]. Under the action of α-secretases, the extracellular fragment sAPPα is generated, which exerts several neuroprotective functions, including protection against ischemic damage and glutamate neurotoxicity. Furthermore, sAPPα exhibits numerous neurotrophic properties, such as promoting neuritogenesis, synaptogenesis, and memory retention. Additionally, the sAPPα peptide inhibits the activity of beta-site APP-cleaving enzyme 1 (BACE1), thereby impeding the amyloidogenic processing of APP. Hypoxia not only suppresses α-secretase activity but also enhances the diversion of APP processing toward the amyloidogenic pathway under hypoxic conditions [19].

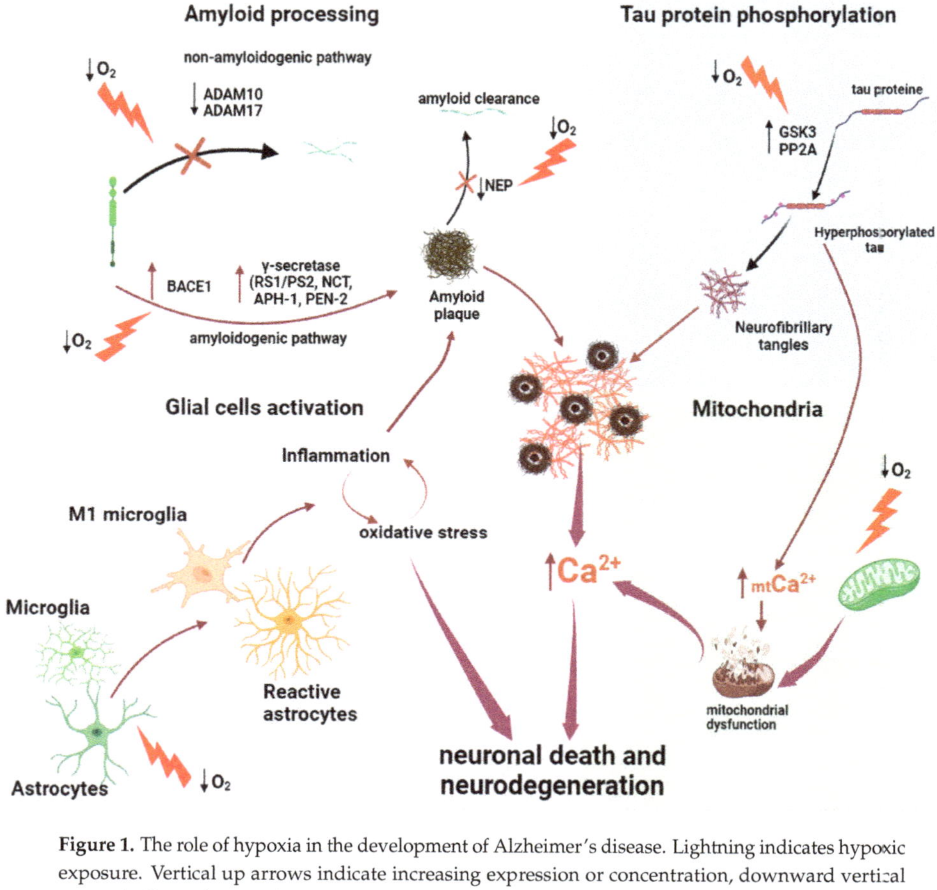

Figure 1. The role of hypoxia in the development of Alzheimer's disease. Lightning indicates hypoxic exposure. Vertical up arrows indicate increasing expression or concentration, downward vertical arrows indicate decreased expression or concentration. Curved arrows show the influence of some factors on others. If the curved arrow is crossed out, this indicates an inhibitory effect.

Research has demonstrated that chronic hypoxia upregulates the expression of β-secretase (BACE1), which facilitates amyloid precursor protein (APP) processing via the amyloidogenic pathway. The promoter region of the BACE1 gene contains a functional hypoxia-responsive element (HRE), with HIF-1α regulating BACE1 gene expression through binding to this HRE [19,20].

The γ-secretase complex, the second enzyme involved in the amyloidogenic processing pathway of APP, comprises four proteins: presenilin 1 or 2 (PS1 or PS2), nicastrin (NCT), anterior pharyngeal defect-1 (APH-1), and presenilin enhancer 2 (PEN2) [21]. Hypoxia has been found to elevate the expression of γ-secretase subunits [22]. Chronic hypoxia decreases the expression of DNA methyltransferase 3b (DNMT3β), leading to demethylation of the PEN2 CpG site and the promoters of the NCT and PS1 genes. Consequently, this results in transcriptional activation and increased protein levels of PEN2, NCT, and PS1, thereby enhancing γ-secretase activity. The promoter of the APH-1A gene contains HIF-1α-binding sites, leading to increased APH-1 expression under hypoxic conditions [23]. Therefore, studies suggest that hypoxia regulates the expression of γ-secretase complex subunits and subsequently augments its activity and amyloid beta production through epigenetic mechanisms [19,20,24].

Hypoxia diminishes the expression of Aβ-degrading enzymes, particularly the metalloprotease neprilysin (NEP), and involves the hypoxic induction of caspases [25]. The APP intracellular domain (AICD), a product of the amyloidogenic pathway, has been observed to bind to the NEP gene promoter, thereby activating its transcription. This mechanism potentially serves as a protective mechanism aimed at mitigating the neurotoxic effects of Aβ. However, AICD is also a caspase substrate. Consequently, the hypoxic induction of caspases can increase AICD cleavage, resulting in decreased NEP levels and amyloid accumulation [25].

Hypoxia can disrupt calcium homeostasis through the accumulation of Aβ, and subsequently, elevated levels of intracellular calcium can trigger cell death [26]. Ca^{2+} play a pivotal role in regulating various neuronal and astrocytic functions, including neurotransmitter release, synaptic plasticity, membrane excitability, gene transcription, cell proliferation, and cell death. The depolarization of senescent neurons induces an influx of Ca^{2+} from the extracellular space, leading to excitotoxicity [27] Elevated Ca^{2+} levels result in glutamate accumulation in the synaptic cleft, promoting the activation of postsynaptic AMPA/NMDA receptors, which further increases the neuronal calcium influx. Heightened cytosolic Ca^{2+} levels may facilitate Aβ production and its subsequent neurotoxic effects, while Aβ accumulation stimulates neuronal Ca^{2+} signaling. Consequently, a synergistic interplay between Ca^{2+} and Aβ may exacerbate neurodegeneration and cognitive impairment in individuals with Alzheimer's disease [19,28].

2.1.2. Mitochondria

Moreover, Aβ causes both cytosolic and mitochondrial Ca^{2+} overload both in vitro and in vivo. Misfolded and hyperphosphorylated tau protein also disrupts Ca^{2+} homeostasis in the mitochondria. This leads to mitochondrial dysfunction, the generation of ROS, and the activation of apoptosis (Figure 1). Mitochondrial dysfunction in AD is an important pathogenesis pathway and includes mitochondrial oxidative stress, disruption of cellular bioenergetics, and the induction of neuronal apoptosis [29].

However, it is important to note that although hypoxia promotes β-amyloid production, amyloidosis does not necessarily equate to the development of dementia or AD. The data on the β-amyloid content in the brain tissue of elderly individuals and centenarians vary considerably across different studies. For instance, in a study by Itoh et al., the density of senile amyloid plaques and neurofibrillary tangles in centenarians aged 101.5 ± 1.5 years was notably lower than in patients with AD (80.8 ± 3 years) but higher than in individuals without dementia symptoms at the age of 79.8 ± 3.2 years [30]. Similarly, in research conducted by Hauw et al., neuropathological examination of the brains of 20 centenarians over 100 years of age revealed persistent neurofibrillary tangles and diffuse Aβ deposits in 19 of the centenarians, with senile plaques being widespread. However, only five of them were diagnosed with dementia, including AD [31].

2.1.3. Tau Protein Hyperphosphorylation

Hypoxia contributes to tau protein hyperphosphorylation by disrupting the balance between kinases and phosphatases, notably glycogen synthase kinase-3β (GSK-3β), cyclin-dependent-like kinase-5 (CDK5), and protein phosphatase 2A (PP2A). Chronic hypoxia activates protein kinases like GSK-3β and CDK5 while inhibiting phosphatase PP2A, resulting in elevated tau phosphorylation levels [32,33].

2.1.4. Neuroinflammation

Neuroinflammation is a prominent consequence of cerebral hypoxia and plays a significant role in AD pathogenesis. Microglia, the resident macrophages of the central nervous system (CNS), are pivotal in mediating neuroinflammation. Hypoxia induces microglial overactivation, leading to the secretion of pro-inflammatory cytokines such as TNF-α, IL-1β, and interferon-γ. These cytokines can downregulate Aβ receptor expression and clearance while activating BACE1, ultimately causing neuronal damage and degeneration [34,35]. Furthermore, activated microglia trigger astrocyte activation, further amplifying the production of pro-inflammatory factors [36]. Furthermore, in addition to the activation of the microglia along the M1 inflammatory profile, hypoxia/ischemia likely upregulates the microglial expression of CD33 and its downstream target SHP-1, a tyrosine phosphatase that inhibits TREM2-driven phagocytosis. Consequently, the microglia fail to effectively phagocytose amyloid conglomerates, leading to the accumulation of plaques in the brain tissue. Additionally, hypoxia/ischemia promotes the deposition of sialylated gangliosides, including GM1, GM2, GM3, and GD1, which serve as ligands for inhibitory receptors such as CD33/Siglec-3 [20,37].

2.1.5. Sleep Apnea as a Risk Factor in the Development of Alzheimer's Diseases

Obstructive sleep apnea (OSA) syndrome is characterized by recurrent episodes of upper airway obstruction during sleep, leading to repetitive awakenings and episodes of hypoxia.

There is a growing body of research suggesting a correlation between sleep apnea, hypoxia, and the onset of Alzheimer's disease. Studies have reported that over 70% of individuals with dementia exhibit sleep-disordered breathing [38–40]. However, the precise mechanisms underlying how sleep disturbances, including OSA, may impact cognitive decline remain incompletely understood.

Recently, the glymphatic system of the brain was described, consisting of morphofunctional components such as Virchow–Robin spaces, spaces between the basement membrane, pericytes, astrocyte feet, the aquaporin receptor system of astrocytes, structures responsible for producing and resorbing cerebrospinal fluid, the interstitial space of the brain, and the space of liquor circulation. This system facilitates the removal of harmful substances and toxins, including Aβ42, accumulated during wakefulness from the brain. It has been demonstrated that this function is active only during sleep, particularly during slow-wave sleep. This may represent an additional mechanism by which sleep fragmentation, characteristic of OSA, may contribute to cognitive decline and the pathogenesis of Alzheimer's disease [41–43].

In light of these findings, a study involving 74 boys aged 15–17 years was conducted. Significantly elevated levels of the specific protein Aβ42, a key player in cognitive impairment, were observed in the plasma of adolescents with obstructive sleep apnea syndrome (OSAS) compared to their peers without OSAS [44]. It has been proposed that the disruption of the interaction between the cerebrospinal fluid and interstitial fluid in OSAS may contribute to this. The glymphatic system, responsible for exchanging solutes between these fluids, relies on the pressure cycles associated with respiration to facilitate flow from the interstitial fluid into the cerebrospinal fluid. During episodes of obstructive apnea, elevated intrathoracic and intracranial pressure occurs due to increased breathing efforts against a closed airway, followed by a sudden pressure reversal at the end of the apnea. Repeated fluctuations in pressure may disrupt the glymphatic flow, leading to the retention of metabolites such as Aβ and other neuron-derived substances in the interstitial fluid rather than their transport into the cerebrospinal fluid [45].

Neuroinflammation is a key component observed in both neurodegenerative diseases and sleep apnea. Intermittent hypoxia, characteristic of sleep-disordered breathing, also induces microglial activation and neuroinflammation. Accumulating experimental evidence suggests that chronic neuroinflammation resulting from sleep apnea may serve as a potential mechanism underlying an increased risk of Alzheimer's disease [46,47].

Intermittent hypoxia (IH) constitutes a significant component of obstructive sleep apnea, characterized by recurrent episodes of reduced blood oxygen levels [48].

In vivo studies have demonstrated that IH disrupts spatial learning and memory in rodents [49] while also adversely affecting hippocampal synaptic plasticity [50,51], likely mediated by heightened ROS production [52]. Increased expression of the HIF-1α protein has been observed in hippocampal neurons exposed to IH [53,54], with evidence suggesting a twofold rise in ROS expression and the impairment of N-methyl-D-aspartate receptor (NMDAr)-dependent long-term potentiation (LTP). Conversely, such effects were absent in heterozygous HIF-1α knockout animals [53].

Furthermore, IH induces increased amyloidogenic processing of APP and elevated production of beta-amyloid peptides. Short-term exposure to chronic IH for three days has been shown to significantly enhance beta-amyloid peptide production and the expression of BACE, presenilin, and HIF-1 [55].

Notably, studies indicate that modeling intermittent hypoxia in mice for 5 or 28 days induces alterations in genome methylation akin to aging, particularly affecting the metabolic cascades associated with neurogenesis, the cell cycle, and energy production, alongside an increase in phosphorylated tau protein [56].

However, some investigations have not observed elevated HIF-1 levels during IH development. For instance, in experiments modeling CIH over a prolonged period of four weeks in transgenic AD mice, it was noted that CIH significantly raised the Aβ42 levels in mice's brains without concurrent upregulation of HIF-1α expression [57]. There were no discernible differences in the levels of the BDNF, proBDNF, and HIF-1α proteins in the peripheral blood between the OSA patient group (n = 20) and the control group (n = 20), neither in the evening before bedtime nor in the morning after sleep [58].

This observation underscores the notion that the effects of HIF-1 depend on the duration and intensity of hypoxic exposure.

Hypoxia exerts multifaceted effects on various pathogenetic aspects, including autophagy, oxidative stress, mitochondrial and synaptic dysfunction, and endoplasmic reticulum stress, all of which can contribute directly or indirectly to neurodegeneration [34].

In summary, hypoxia can trigger several mechanisms implicated in the pathogenesis of Alzheimer's disease. Oxygen deprivation can induce the amyloidogenic cleavage of APP, leading to the accumulation of Aβ, which serves as the primary pathological trigger of AD. The pathological effects of Aβ on calcium homeostasis in the neurons and astrocytes contribute to neuronal cell death. Moreover, hypoxia not only promotes the formation and accumulation of Aβ but also independently disrupts the regulation of calcium homeostasis (in both the neurons and astrocytes) and induces oxidative stress, both of which contribute to neuronal cell death. Furthermore, Aβ accumulations may activate the microglia through a neurotoxic pathway, initiating a maladaptive neuroinflammatory response that exacerbates neurodegeneration. Additionally, hypoxia itself triggers neuroinflammation and neurodegeneration, which can be regarded as another initiating pathological trigger of AD.

2.2. The Influence of Hypoxia on the Development of Parkinson's Disease

Despite accumulating evidence suggesting the potential involvement of hypoxia and/or altered responses to hypoxia in the pathogenesis of Parkinson's disease, the precise nature of these relationships remains incompletely understood.

Several studies propose that hypoxia may serve as a trigger for the phosphorylation and aggregation of α-synucleins. Various post-translational modifications of α-synuclein are known, including phosphorylation, ubiquitination, glycosylation, and acetylation. Among these modifications, phosphorylation at the Ser129 site is considered a key event that promotes α-synuclein aggregation and subsequent pathology [59,60]. Phosphorylated α-synuclein (p-α-syn) represents the predominant pathological form of α-synuclein, with autopsy findings from PD patients indicating that 90% of the protein composition within Lewy bodies consists of p-α-syn. Under normal physiological conditions, α-synuclein primarily exists as a disordered monomer, with minimal modifications or aggregation observed in healthy individuals. However, p-α-syn facilitates the aggregation of α-synuclein into various toxic forms, including monomers, dimers, oligomers, fibrils, and fibrous aggregates, with α-synuclein oligomers considered particularly neurotoxic [61,62]. The accumulation of misfolded and aggregated forms of α-synuclein is observed to increase under hypoxic conditions [63]. Chronic short-term hypoxia has been shown to induce the conversion of α-synuclein into p-α-syn in mouse brain models, with p-α-syn aggregation occurring more rapidly and readily compared to α-syn aggregation, suggesting a potential mechanism for the propagation of α-synuclein pathology in PD. Hypoxia has also been found to promote the formation and spread of α-synuclein pathology in mouse models, accelerating the loss of dopaminergic neurons and the onset of behavioral abnormalities [64,65]. Additionally, rodent models of hypoxia, including those induced by middle cerebral artery occlusion (MCAO) and systemic hypoxia, have demonstrated increased expression of both total α-synuclein and p-α-syn [66–68]. Similar findings have been observed in nerve cell cultures. In an in vitro model, the total quantity of α-syn and p-α-syn and their oligomers demonstrated an increase after subjecting cells to a medium with oxygen levels of 0.5% for 24 h or 1% for 48 h [69].

Recent studies have also implicated OSA as a risk factor in Parkinson's disease. Patients with OSA exhibit significantly higher plasma levels of total α-synuclein and phosphorylated α-synuclein, which are positively correlated with disease severity and hypoxia levels. These findings suggest a potential involvement of hypoxia in Parkinson's disease pathogenesis [70]. Furthermore, some studies indicate that OSA is associated with an increased severity of cognitive dysfunction and motor symptoms in Parkinson's disease [71].

The association between Parkinson's disease and hypoxia may stem from the heightened susceptibility of the dopaminergic neurons located in the substantia nigra to oxygen deprivation. These neurons exhibit an elevated rate of energy metabolism compared to those in other regions and possess a basal respiration rate approximately threefold higher. Furthermore, they manifest increased levels of glycolysis, enhanced axonal branching, a higher mitochondrial density, and heightened vulnerability to cytotoxins [72]. These factors suggest that the dopaminergic neurons in the substantia nigra are particularly vulnerable and exhibit a degree of susceptibility to hypoxia [65].

Hypoxia-induced mitochondrial dysfunction may intertwine with α-syn pathology, and they may reciprocally influence each other (Figure 2). Anomalies in α-syn are believed to impact various mitochondrial functions, including the tricarboxylic acid cycle, oxidative phosphorylation, ROS and calcium homeostasis, and membrane potential. Mitochondrial dysfunction may arise from pathological interactions between α-syn and their membranes. Both soluble and aggregated forms of α-syn have been reported to engage with mitochondrial proteins, primarily facilitated by the C terminus of α-syn [73].

Multiple mitochondrial factors may influence the formation of alpha-synuclein pathology. The accumulation of ROS is associated with an increased loss of dopaminergic neurons in the substantia nigra in Parkinson's disease. Patients with PD demonstrate a widespread reduction in antioxidant defenses in the substantia nigra pars compacta, along with decreased levels of antioxidants in the peripheral blood [74]. Hypoxia can elevate mitochondrial ROS production primarily by limiting the oxygen availability to complex III and by modulating the expression of antioxidant genes [75].

The debate persists regarding whether α-syn pathology arises from mitochondrial dysfunction or whether mitochondrial damage is a consequence of α-syn aggregation and/or pathology formation. It is proposed that other factors, particularly hypoxia and the body's response to it, play a pivotal role in the onset of pathological processes affecting both α-syn and mitochondrial function in PD pathogenesis [73].

Furthermore, over 20 genetic mutations have been linked to PD, some of which are associated with the response to hypoxia. For instance, PARK1 (α-synuclein) is abundantly expressed in the neuronal cells, particularly at the presynaptic terminals, where it facilitates synaptic transport and neurotransmitter release. Mutations in this gene can lead to the formation of Lewy bodies, a process that may be exacerbated by hypoxia, ultimately contributing to neuronal dysfunction and death. ATP13A2 (PARK9) is a P-type ATPase involved in intracellular vesicle localization and cation homeostasis regulation. Disruptions in ATP13A2 activity due to mutations result in the development of an autosomal recessive form of PD. Given that the ATP13A2 gene promoter harbors an HRE, the stabilization of HIF-1α triggers ATP13A2 expression in dopaminergic neurons [64].

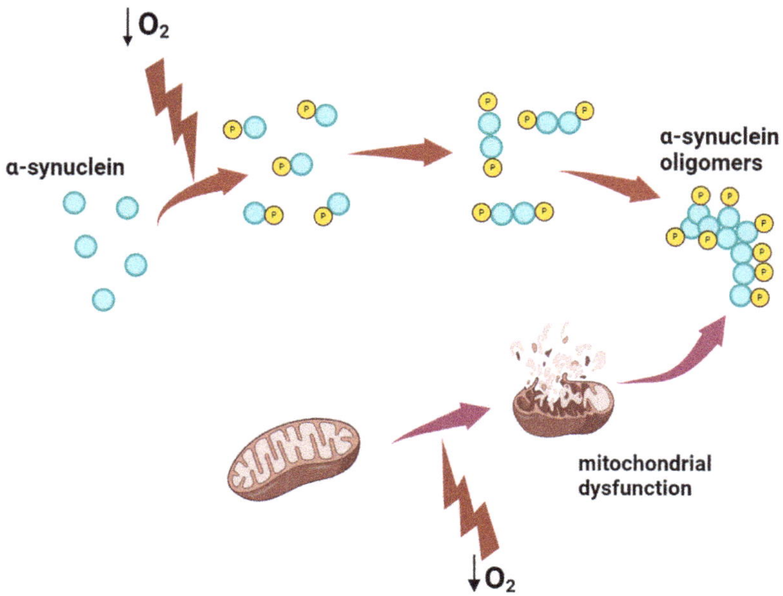

Figure 2. The role of hypoxia in α-syn aggregation and development of Parkinson's disease. Lightning indicates hypoxic exposure.

The available evidence strongly supports the notion that hypoxia represents a significant trigger for inducing neurodegenerative alterations. The primary mechanism facilitating the cellular adaptation of the brain to hypoxia is HIF-1. Consequently, the involvement of the HIF-1 factor in the pathogenesis of neurodegenerative diseases holds significant interest for researchers. Subsequently, we will explore the existing knowledge regarding the role of HIF-1 in the pathogenesis of AD and PD.

3. Hypoxia-Inducible Factor (HIF) in Neurodegenerative Diseases

3.1. Structure and Functions of HIF

The HIF family comprises heterodimeric transcription factors consisting of α and β subunits. The β subunit, also known as ARNT (aryl hydrocarbon receptor nuclear translocator), is constitutively expressed and resides within the cell nucleus continuously. In contrast, stabilization of the α subunit is governed by oxygen levels [76,77]. HIF-α encompasses three known isoforms: HIF-1α, HIF-2α, and HIF-3α (with its function still not fully elucidated). The expression patterns of HIF-1α and HIF-2α vary: HIF-1α is ubiquitously expressed in all organs, while HIF-2α expression is confined to specific tissues, such as the endothelium, kidneys, pancreas, liver, heart, lungs, intestine, and brain [78,79].

The protein structure of all HIF-α and ARNT isoforms exhibits similarity, featuring a bHLH (helix–loop–helix) domain facilitating binding to the HREs on DNA, and a PAS domain mediating subunit dimerization. HIF-α possesses a distinct oxygen-dependent degradation domain (ODDD), responsible for regulating subunit stability in an oxygen-dependent manner, along with two transactivation domains, namely the N-terminal (N-

TAD) and the C-terminal (C-TAD) (excluding HIF-3α). The C-TAD domain is accountable for interactions with coactivators such as p300 and CBP [79,80].

HIF-1α accumulation occurs in response to hypoxia and undergoes rapid degradation upon reoxygenation. Under normoxic conditions, HIF-1α undergoes hydroxylation at the Pro402 and Pro564 residues within the ODDD, facilitated by prolyl hydroxylase (PHD) enzymes [77,81,82]. This hydroxylation process necessitates molecular oxygen (O_2), iron in its ferrous state (Fe^{2+}), and 2-oxoglutarate as cofactors. In its hydroxylated form, HIF-1α is recognized by the von Hippel–Lindau protein (pVHL) and E3 ubiquitin ligase, leading to its targeting for rapid ubiquitin-mediated proteasomal degradation by the 26S proteasome [83,84]. A reduced flux of oxygen induces PHD hydroxylation, disrupting the interaction between HIF-1α and pVHL, resulting in the accumulation of HIF-1α. Subsequently, the accumulated HIF-1α translocates to the nucleus and forms a complex with HIF-1β to assemble transcriptionally active HIF-1 [85]. Under hypoxic conditions, HIF-1 functions as a transcription factor primarily by associating with the HREs in gene promoters, typically characterized by the 5′-(A/G) CGTG-3′ consensus sequence. This interaction with transcriptional coactivators such as CBP and p300 modulates the expression of various genes [86]. However, in the presence of oxygen, the HIF-1 inhibitory factor (FIH) hydroxylates the asparagine residues within the C-TAD of HIF-1α, blocking its interaction with coactivators and suppressing HIF-1 transcriptional activity. Although both FIH and PHD are deactivated by hypoxia, their activity can be inhibited by iron chelators, even during healthy development, as both enzymes harbor Fe^{2+} in their catalytic sites [75].

The induction of HIF-1α and the subsequent expression of its target genes activate a pivotal signaling pathway implicated in molecular and cellular adaptation to hypoxia. HIF-1α governs the expression of a diverse array of genes involved in vasomotor control, angiogenesis, erythropoiesis, iron metabolism, cell cycle regulation, cell proliferation and death, and energy metabolism. Notably, HIF-1α facilitates the efficient transport and delivery of nutrients and oxygen by upregulating the expression of erythropoietin (EPO), which stimulates red blood cell production, and vascular endothelial growth factor A (VEGFA), which fosters angiogenesis [79,86]. Additionally, HIF orchestrates cellular metabolic adjustments to hypoxia by favoring a shift from oxidative to glycolytic metabolism. Hypoxia profoundly impacts mitochondrial energy production, as reduced oxygen availability hinders ATP synthesis in the electron transport chain [75]. Consequently, HIF-1α suppresses aerobic oxidation and promotes glycolytic metabolism [87]. This is achieved by upregulating the expression of enzymes such as hexokinase (HK) and enolase 1 (ENO1), which accelerate pyruvate production, subsequently converted into lactate by HIF-1α-induced lactate dehydrogenase (LDHA) [88]. Furthermore, HIF-1α activates the gene encoding pyruvate dehydrogenase kinase (PDK1), which inactivates the TCA cycle enzyme pyruvate dehydrogenase (PDH), which converts pyruvate into acetyl-CoA [89]. As a result, the tricarboxylic acid cycle decelerates, and the production of the electron transport chain (ETC) substrates NADH and FADH2 is attenuated. To mitigate ETC activity and reduce ROS generation, HIF-1α induces the expression of NDUFA4L2, which dampens complex I activity. Moreover, the genes that regulate cytochrome c oxidase (the ETC Complex IV that reduces O_2 into H_2O) are also regulated by HIFs, including the COX4-2 isoform and the LON protease, which produces a switch from isoform COX4-1 to COX4-2 that optimizes respiration efficiency under hypoxic conditions [90]. Thus, HIF orchestrates metabolic adaptation to hypoxic environments.

Genes implicated in glucose transport or glycolysis are activated by HIF, including glucose transporter-1 and 3 (GLUT 1 and 3), pyruvate dehydrogenase kinase 1 (PDK1), and lactate dehydrogenase A (LDHA). In summary, HIF-1α-mediated gene expression enhances the protection of viable cells and facilitates the removal of cells damaged by hypoxia [75,86].

3.2. HIF for Alzheimer's Disease

The role of the HIF-1 factor in AD pathogenesis is dual. On the one hand, several studies indicate a decrease in HIF-1 expression in AD, likely contributing to disturbances in glucose metabolism. The evidence suggests impaired glucose uptake and metabolism in the brains of AD patients, which appear to be causes rather than consequences of neurodegeneration in AD [91]. The neurons in the brain lack the ability to synthesize or store glucose, relying entirely on the glucose transport across the blood–brain barrier facilitated by GLUT transporters, with GLUT1 and GLUT3 being predominant in the mammalian brain. GLUT1 is tasked with transporting glucose from the bloodstream into the extracellular space of the brain, while GLUT3 facilitates the transport of glucose from the extracellular space into the neurons. Reduced expression of GLUT1 and GLUT3 has been observed in AD brains [92–94], likely resulting from decreased levels of HIF-1α. The current data on the HIF-1 expression levels in AD patients are limited and conflicting. For instance, the brain levels of HIF-1 have been reported to be diminished in AD patients compared to age-matched healthy controls [95]. It is conceivable that reduced levels of HIF-1 in the brain are associated with increased oxidative stress, which is heightened in AD and may destabilize HIF-1 [96]. The induction of HIF-1 expression has been shown to increase GLUT1 expression, enhance glycolysis, and reduce oxidative stress [95,97]. These observations were corroborated in a study demonstrating that α-lipoic acid upregulates HIF-1α expression and restores glucose metabolism by stimulating the brain-derived neurotrophic factor (BDNF)/tropomyosin-linked kinase B (TrkB) pathway [98].

Impaired glucose uptake and metabolism have been shown to result in decreased levels of O-linked N-acetylglucosamine (O-GlcNAc) and subsequent hyperphosphorylation of tau protein, leading to the formation of neurofibrillary tangles and neuronal degeneration in the brains of AD patients. Under normal circumstances, O-glycosylation occurs, wherein β-N-acetylglucosamine (GlcNAc) attaches to the hydroxyl group of serine or threonine residues, regulating protein phosphorylation [96,99,100]. Additionally, elevated levels of HIF-1 have been reported to decrease tau protein phosphorylation and neuron inflammation, thereby offering cellular protection against apoptosis [101].

The levels of GLUT2, responsible for transporting glucose into the astrocytes, were found to be more than twice as high in AD brains compared to controls. Given that GLUT2 expression is mainly linked to the astrocytes in the brain, this observed increase in GLUT2 levels likely stems from astrocyte activation in AD [94]. Aβ-dependent activation of the astrocytes has been demonstrated to induce a prolonged reduction in HIF-1α expression and a decline in glycolytic rates. However, maintaining HIF-1α levels under conditions that prevent its proteasomal degradation can reverse glial activation and glycolytic alterations [102].

Several studies have highlighted the neuroprotective role of HIF-1 against beta-amyloid peptide toxicity [103,104], a topic that will be extensively discussed in subsequent sections. Therefore, HIF-1 contributes to neuronal survival during neurodegenerative processes.

An essential aspect concerning the involvement of HIF-1 in pathogenesis pertains to its impact on inflammatory mechanisms. In AD, the production of inflammatory cytokines such as TNF-α and IL-1β may activate HIF-1α, possibly by inhibiting PHD enzymes and enhancing HIF-1 stability in an oxygen-independent manner [105,106]. Subsequently, HIF-1 suppresses the expression of inflammatory cytokine receptors within the hippocampus, thereby mitigating the detrimental effects of excessive neuroinflammation [10,107]. Moreover, HIF-1α may diminish the production of pro-inflammatory cytokines like IL-6 and TNF-α by upregulating the expression of IL-10, which suppresses NLRP3 expression, thus modulating inflammasome activation and caspase-8 activation [108]. Alternatively, HIF-1α may exert its anti-inflammatory effects through the IL-10/JAK1/STAT3-mediated signaling pathway [109].

Contrary evidence suggests that HIF-1 may stimulate the production of inflammatory cytokines such as TNF-α, leading to the establishment of a positive feedback loop between hypoxia and inflammation [110]. Recent investigations by Jung et al. [111] on transgenic

3xTg AD mice revealed a reduction in capillary diameter alongside heightened expression of HIF-1 in the cerebral vascular endothelium. The authors proposed that HIF-1 in the vascular endothelial cells triggers the upregulation of the NADPH oxidase Nox4, instigating an inflammatory response. Additionally, the involvement of the NLRP1 inflammasome in this cascade has been underscored [112]. This line of inquiry aligns with the findings of P. Grammas, who demonstrated increased HIF-1α immunoreactivity in brain tissue from Tg (APPSWE) 2576-AD mice aged 50 to 60 weeks compared to Tg 2576 controls [112].

Further insights underscore that HIF-1 induces an augmentation in the expression of BACE1 and γ-secretase, pivotal factors in the amyloidogenic cleavage of APP, thereby indicating a direct positive association between HIF-1 and AD pathogenesis [113]. Moreover, HIF-1-deficient mice exhibit diminished BACE1 expression in select brain regions, including the hippocampus and cerebral cortex, highlighting the significant role of hypoxia/HIF-1 in modulating amyloidogenic APP processing. Guglielmotto et al. posited that reactive oxygen species (ROS) primarily drive the early activation of BACE1 following hypoxia, with subsequent activation of HIF-1α exacerbating BACE1 activity in later stages [114].

Moreover, cerebral ischemia and stroke have been shown to upregulate HIF-1α expression, which, via a pro-apoptotic pathway involving p53, augments the expression of BACE1, thereby contributing to the amyloidogenic process. While moderate hypoxia elicits the protective effects of HIF-1, severe hypoxia can lead to elevated levels of HIF-1α, activating the tumor suppressor gene p53 and subsequently inducing cellular apoptosis [103,115].

In work [116] observed that hypoxia induces cell death and autophagy in the microglia by elevating HIF-1α expression. In vitro studies revealed that the suppression of HIF-1α using pharmacological inhibitors (3-MA, Baf A1) or RNA interference mitigated microglial death and autophagy [116]. It is plausible that HIF-1α exacerbates the secretion of proinflammatory factors by the microglia while inhibiting phagocytosis.

Furthermore, the involvement of the HIF-1 factor in AD pathogenesis is intertwined with iron metabolism in brain tissue. Elevated iron levels may heighten toxic Aβ deposition and the hyperphosphorylation of tau protein [117,118]. A key regulator of the iron content and distribution in the brain is the TfR1 protein. Increasing evidence suggests that TfR gene transcription is modulated by hypoxia, with HIF-1α (the best studied transcriptional activator of hypoxia-responsive genes) playing a pivotal role in this regulation [119,120]. Recent investigations have demonstrated the upregulation of TfR1 protein expression in the cerebral cortex of 8-month-old female 5xFAD mice [121], which is associated with the activation of the HIF-1 signaling pathway, oxidative stress, and inflammation. Remarkably, this study revealed elevated HIF-1α mRNA expression in the cortical tissues of 5xFAD mice, alongside stabilization and nuclear translocation of the HIF-1 complex [121].

Thus, these findings delineate a dual role of HIF-1α in AD pathophysiology (Figure 3).

3.3. HIF-1 in Parkinson's Disease

While the precise molecular mechanisms of PD remain incompletely elucidated, considerable evidence implicates mitochondrial dysfunction and oxidative stress in its pathogenesis [122–124]. PD is characterized by mitochondrial deficits, including a reduced mitochondrial number and structural damage, alongside diminished activity of complex I of the mitochondrial respiratory chain, potentially mediated by α-synuclein [125–127].

The expression of HIF-1α and its downstream targets is downregulated in the neurons located within the substantia nigra of the brain in individuals with PD [128]. Notably, in a PD mouse model induced by 1-methyl-4-phenyl-1,2,3,6-tetrahydropyridine (MPTP), HIF-1α accumulation was inhibited in dopaminergic PC12 cell lines and mice [129].

An essential interplay between HIF-1 and Parkinson's disease involves tyrosine hydroxylase (TH). TH is responsible for the formation of L-DOPA, a pivotal step in dopamine biosynthesis, thus regulating the rate of the entire biosynthetic process. Parkinson's disease can be characterized as a syndrome marked by deficient striatal TH levels. Investigations into the activity of the tyrosine hydroxylase gene promoter reveal its regulation by the HRE. Hypoxia and agents that activate HIF-1 have been shown to enhance TH expression in the brain [130]. Additionally, studies suggest that HIF-1α plays a critical role in neuronal and dopaminergic differentiation in vitro. Conditional knockout of HIF-1α in young adult mice led to neuronal alterations and a 40% reduction in TH expression in the substantia nigra [131,132].

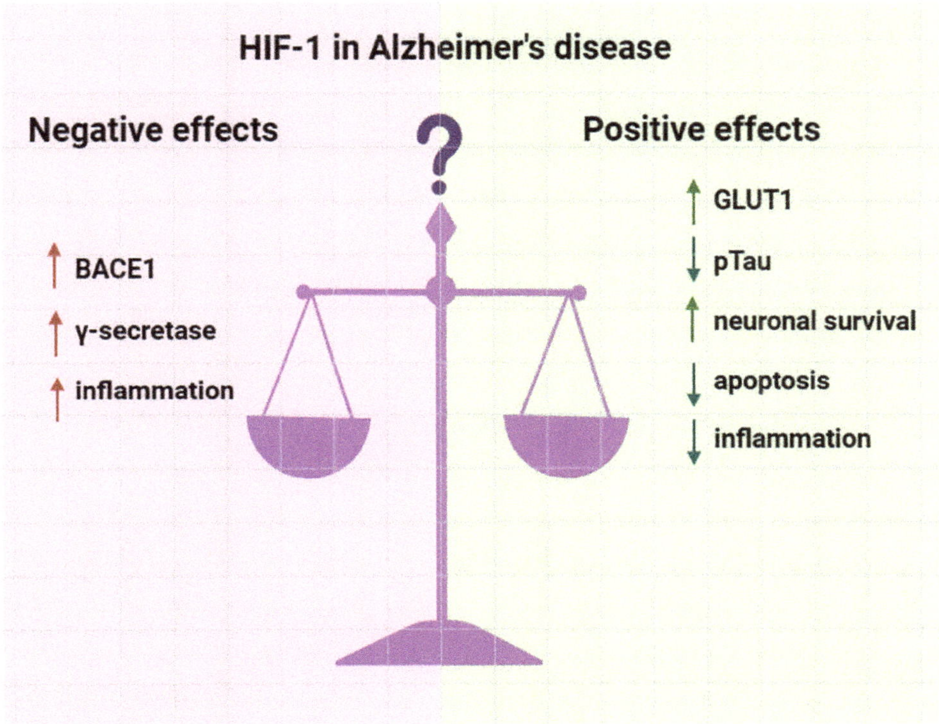

Figure 3. The double-edged sword of HIF-1 in Alzheimer's disease pathogenesis. Vertical up arrows indicate increasing expression or activity, downward vertical arrows indicate decreased expression or activity. Green arrows indicate positive effects, red arrows indicate negative ones.

Similar to various other neurodegenerative disorders, oxidative stress is a critical factor in the pathogenesis of PD. Dysregulation of ROS balance and the suppression of antioxidant defense mechanisms can lead to lipid peroxidation, protein oxidation, and DNA damage, contributing to the loss of dopaminergic cells in the substantia nigra associated with PD [133]. ROS have been implicated in stabilizing HIF-1α under certain conditions [75]. The mitochondria-targeted antioxidant mitoubiquinone (MitoQ) has been shown to abolish the hypoxia-induced production of ROS and decrease the level of the HIF-1α protein under low-oxygen conditions (3% O_2), while ROS-generating compounds increase HIF-1α stabilization, even under normoxia [134,135]. Conversely, HIF-1α plays a role in mitigating excessive ROS production by reorganizing mitochondrial respiratory chain complexes' oxygen consumption and suppressing the expression of mtDNA-encoded mRNAs [75]. The reduction in mitochondrial oxidative respiration and subsequent ROS production observed in PD compared to physiological levels may lead to HIF-1 destabilization.

DJ-1 (PARK7), the loss of which is associated with early PD onset [136], has been shown to negatively regulate the ubiquitination of the von Hippel–Lindau (VHL) α-subunit of HIF-1. DJ-1 modulates HIF-1α stability by influencing the HIF-VHL interaction, and DJ-1 deficiency results in reduced HIF-1α levels in both hypoxia and oxidative stress models. Conversely, elevated levels of HIF-1α and/or its targets have been implicated as a protective mechanism in certain in vitro PD models [137]. The decreased induction of vascular endothelial growth factor (VEGF) mRNA expression downstream of DJ-1 loss suggests the role of DJ-1 in mediating neuronal survival via the VHL-HIF-1α pathway [137].

An increase in iron content specifically within the substantia nigra pars compacta is linked to the biochemical pathology observed in PD. Iron is implicated in oxidative stress by facilitating the release of oxygen free radicals from H_2O_2. Studies propose that 6-hydroxydopamine (6-OHDA) induces lesions in the nigrostriatal dopaminergic neurons through metal-catalyzed free radical generation. To investigate this, the impact of the iron chelator deferoxamine (DFO) on 6-OHDA-induced dopaminergic neuron degeneration in rats was examined via intracerebroventricular injection of the inducer. The impact of the iron chelator deferoxamine (DFO), recognized as an inducer of HIF-1, on the 6-OHDA-induced degeneration of the dopaminergic neurons in rats was explored. These studies demonstrated that intracerebroventricular injection of DFO effectively shielded rats from neurodegeneration [138].

Furthermore, the HIF-1 target genes EPO and VEGF have been identified as contributors to neuronal protection in the context of PD pathogenesis. EPO exhibits neuroprotective effects against the dopaminergic neurotoxin 6-hydroxydopamine, while VEGF offers protection against 1-methyl-4-phenyl-1,2,3,6-tetrahydropyridine (MPTP) toxicity in the substantia nigra cells and may also facilitate dopaminergic neuron growth, proliferation, and differentiation. These findings suggest that HIF-1 may exert a neuroprotective influence on the brain in PD (Figure 4) [131]. This underscores the potential of HIF as a therapeutic target for neurodegenerative diseases.

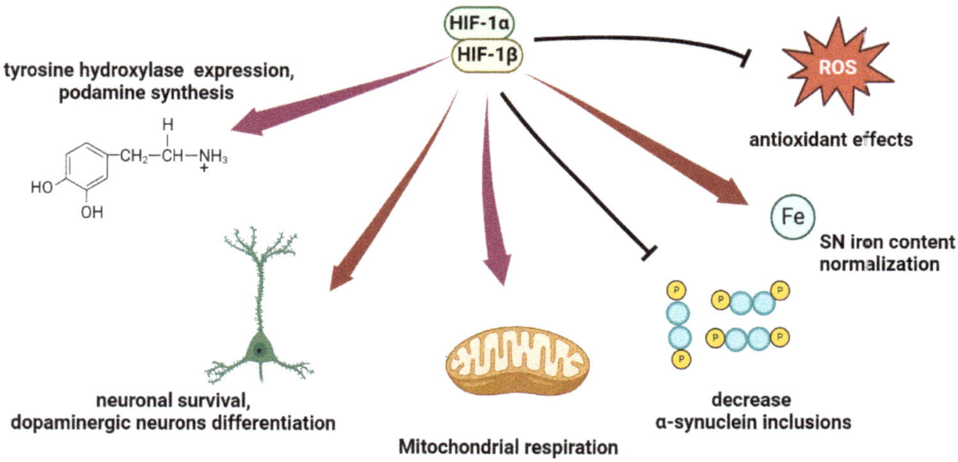

Figure 4. An overview of the potential of HIF-1α stabilization for neuroprotection in Parkinson's disease. Red and purple arrows indicate strengthening or support. Black lines indicate inhibition of ROS production or α-syn aggregation.

4. Modulation of HIF Activity as a Therapeutic Approach to Neurodegeneration

To modulate HIF activity, inhibitors of HIF hydroxylases are employed. These hydroxylases include prolyl hydroxylases (PHDs) and asparaginyl hydroxylases (FIHs). PHDs facilitate HIF-1α degradation, while FIH inhibits HIF-1 activity [139]. The enzymatic activity of PHDs relies on oxygen, iron, and 2-oxoglutarate [140]. Consequently, PHD activity can be hindered by small molecules either indirectly (noncompetitively), through reducing the cellular levels of oxygen, iron, or 2-oxoglutarate, or directly (competitively) via compounds binding to and obstructing the 2-oxoglutarate-binding site. The inhibition of hydroxylases has shown promise in conferring protective effects in neurodegenerative diseases [141].

Findings from both fundamental research and clinical trials have underscored that the activation of HIF-1 may represent a potent strategy for retarding symptom progression and ameliorating outcomes in AD [142]. For instance, iron chelators have the capacity to activate HIF due to iron serving as one of the cofactors for HIF PHDs. M30, a multitarget iron chelator, has been documented to elevate HIF-1 protein levels and trigger the expression of genes downstream of HIF-1 such as vascular endothelial growth factor and erythropoietin (VEGF and EPO) [104]. Simultaneously, M30 mitigates tau phosphorylation and shields the cortical neurons from amyloid-beta (Aβ) toxicity [143–145]. In addition, the M30 prolyl hydroxylase inhibitor, orally administered to mice for 14 days (2.5 mg/kg/day) after the MPTP neurotoxin, significantly increased striatal dopamine levels and increased tyrosine hydroxylase protein levels and activity [146]. These results have been confirmed by other studies [147,148]. Alejandra E Ramirez et al. demonstrated that M30 inhibited α-syn aggregation and decreased the formation of higher-molecular-weight species [149].

Deferoxamine (DFO), a widely utilized HIF-1 inducer, has been investigated in clinical trials involving AD patients and has demonstrated a deceleration of cognitive decline [142,150]. These investigations suggest that augmenting HIF-1 activity holds promise for forestalling neuronal demise and alleviating AD symptoms.

Lactoferrin (Lf) is an iron-binding glycoprotein renowned for its potent iron-chelating abilities, reminiscent of the synthetic chelator deferoxamine (DFO) [151]. These iron-binding properties enable Lf to stabilize the principal adaptive transcription factor, HIF-1α. The administration of Lf has been demonstrated to significantly elevate HIF-1α levels [152]. In murine models, Lf mitigates the dopaminergic neuron damage induced by 1-methyl-4-phenyl-1,2,3,6-tetrahydropyridine (MPTP) and subsequent dyskinesia. Furthermore, the stabilization of HIF-1α by Lf leads to the induction of tyrosine hydroxylase (TH) and several neuroprotective factors, thereby enhancing neuronal viability against MPP+ toxicity in Parkinson's disease [153]. Lf has also shown promise in ameliorating cognitive impairment in Alzheimer's disease mouse models through HIF-1α activation [115].

The presence of ferrous ions within the active sites is vital for the activity of prolyl hydroxylase domain-containing enzymes (PHDs). Cobalt can displace the sole free iron in these active sites, thereby deactivating these hydroxylases. Consequently, cobalt has the capacity to stabilize HIF-1α [154,155].

Moreover, in addition to cobalt and iron chelators, 2-oxoglutarate analogues have emerged as inhibitors of hydroxylation and inducers of HIF-1α protein. These analogues exhibit greater selectivity toward prolyl hydroxylases (PHDs) and asparaginyl hydroxylases (FIH) than simple iron chelators. Notable examples include 3,4-dihydroxybenzoate (3,4-DHB) and Roxadustat (FG-4592) [156,157]. For instance, 3,4-DHB has shown significant inhibition of FIH, with selectivity over PHDs [157,158]. In one study [159], subacute administration of 1-methyl-4-phenyl-1,2,3,6-tetrahydropyridine (MPTP) at doses of 2×20 mg/kg body weight 12 h apart resulted in a 30% loss of dopaminergic neurons in the substantia nigra pars compacta in mice. However, pretreatment with 3,4-DHB for 6 h before the initial administration of MPTP resulted in complete protection against dopaminergic cell loss, as well as preservation of tyrosine hydroxylase levels. Furthermore, DHB pretreatment led to increased levels of HIF-1α and increased expression of HIF-dependent genes such as vascular endothelial growth factor (VEGF) and heme oxygenase-1 (HO-1) [159]. Additionally, DHB demonstrated the ability to prevent microglial activation, resulting in decreased neuronal death both in vitro and in vivo in a PD model. These effects may be related to the ability of DHB to induce an increase in HO-1 levels, thereby eliciting both antioxidant and anti-inflammatory effects [160].

The regular use of iron-chelating HIF-PHDs may potentially lead to adverse effects. such as restless leg syndrome or anemia. Therefore, the exploration of non-iron-targeting HIF-PHDs holds promise in their therapeutic benefits. Roxadustat is an orally administered drug characterized by a short half-life, and when used in conjunction with an intermittent dosing schedule, it induces a transient rise in hypoxia-inducible factor (HIF) activity. It was jointly developed by FibroGen (San Francisco, CA, USA) and Astellas

(Tokyo, Japan)/AstraZeneca (Cambridge, UK) and was approved for the treatment of chronic kidney disease in China in December 2018 and in Japan in September 2019 for patients dependent on dialysis [79]. Recent studies have suggested that Roxadustat (FG-4592) may have the ability to partially penetrate the blood–brain barrier and induce the expression of HIF-1α in the brain tissue. Building upon this finding, researchers conducted an experiment on mice where MPTP (30 mg/kg/day) and Roxadustat (10 mg/kg/day) were intraperitoneally administered for 5 consecutive days. The results demonstrated that FG-4592 led to a dose-dependent increase in HIF-1α expression, accompanied by the induction of tyrosine hydroxylase (TH) and the activation of antioxidant enzymes such as Nrf2 and HO-1. Furthermore, a reduction in ROS production was observed, suggesting that these effects of HIF-1α activation may contribute to the protection of dopaminergic neurons against the cytotoxic effects of MPP+. Additionally, FG-4592 ameliorated the behavioral impairments induced by MPTP in mice [161]. However, further research is needed to elucidate the potential impact of FG-4592 on the development of Alzheimer's disease.

Orexin-A, a neuropeptide produced by the hypothalamic neurons, has garnered attention for its potential neuroprotective properties. Dysfunction in the orexinergic system has been linked to various neurodegenerative disorders, including AD and PD [162–165]. Orexins play pivotal roles in regulating feeding behavior, energy balance, sleep–wake cycles, autonomic nervous system function, and neuronal survival [166]. Multiple studies have demonstrated that Orexin-A can upregulate the expression of HIF-1 by inhibiting VHL or E3 ubiquitin ligase activity [167–169]. In PD pathogenesis, mitochondrial dysfunction may lead to reduced oxygen consumption, thereby activating prolyl hydroxylase and decreasing HIF-1α levels, which subsequently reduces the expression of several transcription factors implicated in protecting the brain against oxidative stress. Elevated levels of HIF-1 protein have been shown to mitigate toxin-induced neuronal damage, particularly by MPP+. Orexin-A not only induces HIF-1α expression but also activates downstream targets such as EPO and VEGF, exerting neuroprotective effects [170].

Furthermore, investigations into the genetic regulation of the HIF-1α gene have yielded promising insights. Chai et al. engineered an rAAV vector containing the human HIF-1α gene (rAAV-HIF-1α), which effectively expressed the HIF-1α protein in primary cultured hippocampal neurons and rat hippocampal tissue [171]. In vitro and in vivo findings suggest that rAAV-mediated HIF-1α gene delivery may ameliorate amyloid-β (Aβ)-induced cognitive deficits by reducing hippocampal neuron apoptosis in Alzheimer's disease models [171].

Furthermore, augmentation of HIF-1α expression through the utilization of a GV287-based lentiviral vector exhibited improvements in motor dysfunction and pathological alterations in an MPTP-induced PD model. This beneficial outcome was correlated with heightened expression of miR-128-3p, resulting in reduced Axin1 expression and subsequent activation of the Wnt/β-catenin signaling pathway [128].

However, contrasting findings have emerged regarding the neuroprotective role of HIF-1 inhibition. Recent research [172] published in 2024 introduced a novel class of 8-biaryl-2,2-dimethylbenzopyranamide derivatives, with compound D13 exhibiting promising efficacy in animal models associated with Alzheimer's disease and ischemic stroke. Experimental investigations revealed that compound D13 activates the PI3K/AKT/mTOR signaling pathway while concurrently suppressing HIF-1α and its downstream target genes [172]. In a separate study, knockdown and pharmacological inhibition of HIF-1α during a model of thiamine-deficiency-induced neurotoxicity markedly reduced the formation of BACE1, the C-terminal 99-amino-acid fragment (C99), and the development of AD-like pathology [173].

Moreover, the application of siHIF-1α lentivirus to induce HIF-1 deficiency led to a reduction in hypoxia-induced tau hyperphosphorylation and the restoration of diminished PP2Ac methylation levels. Additionally, this intervention resulted in the amelioration of learning and memory impairments observed in rats subjected to chronic hypoxia [174].

5. Conclusions

Hence, given the brain's heightened energy requirements, it is particularly susceptible to oxygen deprivation, and hypoxic injury can precipitate significant alterations in neuronal function, implicated in neurodegenerative conditions like Alzheimer's and Parkinson's disease. Numerous studies propose that modulating HIF-1α activity could mitigate cellular damage in these neurodegenerative disorders by orchestrating the transcription of a diverse array of genes, rendering the stimulation of the HIF-1α-mediated response a promising therapeutic avenue. Nonetheless, apprehension persists regarding the safety of eliciting the full transcriptional response mediated by HIF-1α given the potential for off-target effects. Hence, future research endeavors should prioritize comprehensive drug assessment and the development of safer therapeutic modalities.

The effects of HIF-1 are subject to significant debate. On the one hand, HIF-1 demonstrates the ability to prevent neuronal apoptosis and maintain neuronal differentiation. Additionally, its positive impact may correlate with the upregulation of GLUT1 and GLUT3, facilitating the maintenance of normal brain energy metabolism. Moreover, HIF-1α has been observed to normalize tau protein phosphorylation by activating GLUT1.

Conversely, overexpression of HIF-1α in the brains of individuals with Alzheimer's disease (AD) has been shown to inhibit the activation of insulin receptor substrate 1 in the neurons. This inhibition leads to insulin resistance and impaired glucose metabolism, contributing to neurodegeneration [175].

Additionally, heightened levels of HIF-1 have been associated with inflammation in both the brain and peripheral nervous system, as it directly activates the microglia and astrocytes, fostering neuroinflammation and the progression of various neurodegenerative conditions.

The role of HIF-1 in oxidative stress presents a similarly contentious issue. While HIF-1α can mitigate excessive ROS production by reorganizing oxygen consumption in mitochondrial respiratory chain complexes, several studies have indicated that heightened expression of HIF-1α typically results in increased ROS production by the mitochondria.

The imbalance in the HIF-1-mediated response to hypoxia may represent a novel pathological pathway for neurodegeneration. HIF-1α appears to serve a dual role in AD and PD, functioning as both a neuroprotective agent and a proinflammatory and neurotoxic factor. The specific role of HIF-1 likely depends on the degree and duration of hypoxia, as well as various associated factors.

Thus, in future therapeutic endeavors targeting HIF-1, it will be crucial to focus on maintaining appropriate levels of HIF-1α expression while maximizing its neuroprotective effects and minimizing its proinflammatory and neurotoxic impacts. In the early stages of neurodegeneration and under conditions of mild hypoxia, stimulating HIF-1α expression may decelerate disease progression by reducing inflammation and enhancing neuronal survival. Conversely, in cases of chronic hypoxia, particularly in advanced stages of AD, suppressing HIF-1α expression may prove to be a more efficacious approach.

Author Contributions: E.V.M. and M.V.V. wrote the manuscript. E.V.M. provided the idea. M.V.V. revised the manuscript. All authors have read and agreed to the published version of the manuscript.

Funding: This research was funded by a grant from the Russian Science Foundation (RSF), project no. 22-15-00178.

Conflicts of Interest: The authors declare no conflicts of interest.

List of Abbreviations

3,4-DHB—3,4-dihydroxybenzonate;
6-OHDA—6-hydroxydopamine;
α-syn—α-synuclein.
AAV—adeno-associated virus;
AD—Alzheimer's disease;

AD—Alzheimer's disease;
AICD—APP intracellular domain;
ApoE—apolipoprotein E;
APP—amyloid precursor protein;
ARNT—aryl hydrocarbon receptor nuclear translocator—HIF-β subunit;
Aβ—amyloid-β;
BACE-1—beta-site APP-cleaving enzyme 1;
DA—dopamine;
DFO—deferoxamine;
EPO—erythropoietin;
FIH—FIH—asparaginyl hydroxylase;
GLUT 1—glucose transporter-1;
GLUT 3—glucose transporter-3;
HIF—hypoxia-inducible factor;
HO-1—heme oxygenase-1;
HRE—hypoxia-responsive elements, short DNA sequences within the gene promoter;
IH—intermittent hypoxia;
LDHA—lactate dehydrogenase A;
Lf—lactoferrin;
MPP+—1-methyl-4-phenylpyridine;
MPTP—1-methyl-4-phenyl-1,2,3,6-tetrahydropyridine;
NFT—neurofibrillary tangles;
ODDD—oxygen-dependent degradation domain;
O-GlcNAc—O-linked N-acetylglucosamine;
OSA—obstructive sleep apnea;
PD—Parkinson's disease;
PDK1—pyruvate dehydrogenase 1;
PHD—prolyl hydroxylase;
pVHL—von Hippel–Lindau protein;
ROS—reactive oxygen species;
sAPPβ—soluble APPβ;
SNc—compact part of the substantia nigra;
TH—tyrosine hydroxylase;
VEGFA—vascular endothelial growth factor A

References

1. Reith, W. Neurodegenerative Erkrankungen. *Radiologe* **2018**, *58*, 241–258. [CrossRef] [PubMed]
2. Zhang, X.; Li, Q.; Cong, W.; Mu, S.; Zhan, R.; Zhong, S.; Zhao, M.; Zhao, C.; Kang, K.; Zhou, Z. Effect of Physical Activity on Risk of Alzheimer's Disease: A Systematic Review and Meta-Analysis of Twenty-Nine Prospective Cohort Studies. *Ageing Res. Rev.* **2023**, *92*, 102127. [CrossRef] [PubMed]
3. Yin, Y.; Tam, H.L.; Quint, J.; Chen, M.; Ding, R.; Zhang, X. Epidemiology of Dementia in China in 2010–2020: A Systematic Review and Meta-Analysis. *Healthcare* **2024**, *12*, 334. [CrossRef] [PubMed]
4. Wimo, A.; Seeher, K.; Cataldi, R.; Cyhlarova, E.; Dielemann, J.L.; Frisell, O.; Guerchet, M.; Jönsson, L.; Malaha, A.K.; Nichols, E.; et al. The Worldwide Costs of Dementia in 2019. *Alzheimer's Dement.* **2023**, *19*, 2865–2873. [CrossRef] [PubMed]
5. Tolosa, E.; Garrido, A.; Scholz, S.W.; Poewe, W. Challenges in the Diagnosis of Parkinson's Disease. *Lancet Neurol.* **2021**, *20*, 385–397. [CrossRef] [PubMed]
6. Burtscher, J.; Mallet, R.T.; Burtscher, M.; Millet, G.P. Hypoxia and Brain Aging: Neurodegeneration or Neuroprotection? *Ageing Res. Rev.* **2021**, *68*, 101343. [CrossRef] [PubMed]
7. Tarkowska, A. Hypoxic-Ischemic Brain Injury after Perinatal Asphyxia as a Possible Factor in the Pathology of Alzheimer's Disease. In *Cerebral Ischemia*; Exon Publications: Brisbane, QLD, Australia, 2021; pp. 45–60.
8. Hambali, A.; Kumar, J.; Hashim, N.F.M.; Maniam, S.; Mehat, M.Z.; Cheema, M.S.; Mustapha, M.; Adenan, M.I.; Stanslas, J.; Hamid, H.A. Hypoxia-Induced Neuroinflammation in Alzheimer's Disease: Potential Neuroprotective Effects of Centella Asiatica. *Front. Physiol.* **2021**, *12*, 712317. [CrossRef] [PubMed]
9. Ke, Q.; Costa, M. Hypoxia-Inducible Factor-1 (HIF-1). *Mol. Pharmacol.* **2006**, *70*, 1469–1480. [CrossRef] [PubMed]
10. Iyalomhe, O.; Swierczek, S.; Enwerem, N.; Chen, Y.; Adedeji, M.O.; Allard, J.; Ntekim, O.; Johnson, S.; Hughes, K.; Kurian, P.; et al. The Role of Hypoxia-Inducible Factor 1 in Mild Cognitive Impairment. *Cell. Mol. Neurobiol.* **2017**, *37*, 969–977. [CrossRef]
11. Lv, B.; Li, F.; Fang, J.; Xu, L.; Sun, C.; Han, J.; Hua, T.; Zhang, Z.; Feng, Z.; Jiang, X. Hypoxia Inducible Factor 1α Promotes Survival of Mesenchymal Stem Cells under Hypoxia. *Am. J. Transl. Res.* **2017**, *9*, 1521–1529.

12. Vingtdeux, V.; Marambaud, P. Identification and Biology of A-secretase. *J. Neurochem.* **2012**, *120*, 34–45. [CrossRef] [PubMed]
13. Chun, Y.S.; Cho, Y.Y.; Kwon, O.H.; Zhao, D.; Yang, H.O.; Chung, S. Substrate-Specific Activation of α-Secretase by 7-Deoxy-Trans-Dihydronarciclasine Increases Non-Amyloidogenic Processing of β-Amyloid Protein Precursor. *Molecules* **2020**, *25*, 646. [CrossRef] [PubMed]
14. Agüero, P.; Sainz, M.J.; García-Ayllón, M.-S.; Sáez-Valero, J.; Téllez, R.; Guerrero-López, R.; Pérez-Pérez, J.; Jiménez-Escrig, A.; Gómez-Tortosa, E. α-Secretase Nonsense Mutation (ADAM10 Tyr167*) in Familial Alzheimer's Disease. *Alzheimers Res. Ther.* **2020**, *12*, 139. [CrossRef] [PubMed]
15. Hartl, D.; May, P.; Gu, W.; Mayhaus, M.; Pichler, S.; Spaniol, C.; Glaab, E.; Bobbili, D.R.; Antony, P.; Koegelsberger, S.; et al. A Rare Loss-of-Function Variant of ADAM17 Is Associated with Late-Onset Familial Alzheimer Disease. *Mol. Psychiatry* **2020**, *25*, 629–639. [CrossRef] [PubMed]
16. Rybnikova, E.; Gluschenko, T.; Galeeva, A.; Tulkova, E.; Nalivaeva, N.N.; Makova, N.Z.; Turner, A.J.; Samoilov, M. Differential Expression of ADAM15 and ADAM17 Metalloproteases in the Rat Brain after Severe Hypobaric Hypoxia and Hypoxic Preconditioning. *Neurosci. Res.* **2012**, *72*, 364–373. [CrossRef] [PubMed]
17. Sun, M.; Zhou, T.; Zhou, L.; Chen, Q.; Yu, Y.; Yang, H.; Zhong, K.; Zhang, X.; Xu, F.; Cai, S.; et al. Formononetin Protects Neurons Against Hypoxia-Induced Cytotoxicity Through Upregulation of ADAM10 and SAβPPα. *J. Alzheimer's Dis.* **2012**, *28*, 795–808. [CrossRef] [PubMed]
18. Auerbach, I.D.; Vinters, H.V. Effects of Anoxia and Hypoxia on Amyloid Precursor Protein Processing in Cerebral Microvascular Smooth Muscle Cells. *J. Neuropathol. Exp. Neurol.* **2006**, *65*, 610–620. [CrossRef] [PubMed]
19. Lall, R.; Mohammed, R.; Ojha, U. What Are the Links between Hypoxia and Alzheimer's Disease? *Neuropsychiatr. Dis. Treat.* **2019**, *15*, 1343–1354. [CrossRef]
20. Salminen, A.; Kauppinen, A.; Kaarniranta, K. Hypoxia/Ischemia Activate Processing of Amyloid Precursor Protein: Impact of Vascular Dysfunction in the Pathogenesis of Alzheimer's Disease. *J. Neurochem.* **2017**, *140*, 536–549. [CrossRef]
21. Wong, E.; Frost, G.R.; Li, Y.-M. γ-Secretase Modulatory Proteins: The Guiding Hand Behind the Running Scissors. *Front. Aging Neurosci.* **2020**, *12*, 614690. [CrossRef]
22. Villa, J.C.; Chiu, D.; Brandes, A.H.; Escorcia, F.E.; Villa, C.H.; Maguire, W.F.; Hu, C.-J.; de Stanchina, E.; Simon, M.C.; Sisodia, S.S.; et al. Nontranscriptional Role of Hif-1α in Activation of γ-Secretase and Notch Signaling in Breast Cancer. *Cell Rep.* **2014**, *8*, 1077–1092. [CrossRef] [PubMed]
23. Li, L.; Zhang, X.; Yang, D.; Luo, G.; Chen, S.; Le, W. Hypoxia Increases Aβ Generation by Altering β- and γ-Cleavage of APP. *Neurobiol. Aging* **2009**, *30*, 1091–1098. [CrossRef]
24. Liu, H.; Qiu, H.; Yang, J.; Ni, J.; Le, W. Chronic Hypoxia Facilitates Alzheimer's Disease through Demethylation of Γ-secretase by Downregulating DNA Methyltransferase 3b. *Alzheimer's Dement.* **2016**, *12*, 130–143. [CrossRef] [PubMed]
25. Kerridge, C.; Kozlova, D.I.; Nalivaeva, N.N.; Turner, A.J. Hypoxia Affects Neprilysin Expression Through Caspase Activation and an APP Intracellular Domain-Dependent Mechanism. *Front. Neurosci.* **2015**, *9*, 163395. [CrossRef] [PubMed]
26. Verkhratsky, A.; Rodríguez-Arellano, J.J.; Parpura, V.; Zorec, R. Astroglial Calcium Signalling in Alzheimer's Disease. *Biochem. Biophys. Res. Commun.* **2017**, *483*, 1005–1012. [CrossRef] [PubMed]
27. Verkhratsky, A. Astroglial Calcium Signaling in Aging and Alzheimer's Disease. *Cold Spring Harb. Perspect. Biol.* **2019**, *11*, a035188. [CrossRef] [PubMed]
28. Cascella, R.; Cecchi, C. Calcium Dyshomeostasis in Alzheimer's Disease Pathogenesis. *Int. J. Mol. Sci.* **2021**, *22*, 4914. [CrossRef] [PubMed]
29. Calvo-Rodriguez, M.; Kharitonova, E.K.; Bacskai, B.J. In Vivo Brain Imaging of Mitochondrial Ca^{2+} in Neurodegenerative Diseases with Multiphoton Microscopy. *Biochim. Biophys. Acta Mol. Cell Res.* **2021**, *1868*, 118998. [CrossRef] [PubMed]
30. Itoh, Y.; Yamada, M.; Suematsu, N.; Matsushita, M.; Otomo, E. An Immunohistochemical Study of Centenarian Brains: A Comparison. *J. Neurol. Sci.* **1998**, *157*, 73–81. [CrossRef]
31. Hauw, J.J.; Zekry, D.; Seilhean, D.; Forette, B.; Gallinari, C.; Laurent, M.; Moulias, R.; Piette, F.; Sachet, A.; Duyckaerts, C. Neuropathology of the Cerebral Vessels of Centenarians. *J. Mal. Vasc.* **2002**, *27*, S13–S18.
32. Fang, B.; Zhao, Q.; Ling, W.; Zhang, Y.; Ou, M. Hypoxia Induces HT-22 Neuronal Cell Death via Orai1/CDK5 Pathway-Mediated Tau Hyperphosphorylation. *Am. J. Transl. Res.* **2019**, *11*, 7591–7603. [PubMed]
33. Pena, E.; San Martin-Salamanca, R.; El Alam, S.; Flores, K.; Arriaza, K. Tau Protein Alterations Induced by Hypobaric Hypoxia Exposure. *Int. J. Mol. Sci.* **2024**, *25*, 889. [CrossRef] [PubMed]
34. Zhang, F.; Niu, L.; Li, S.; Le, W. Pathological Impacts of Chronic Hypoxia on Alzheimer's Disease. *ACS Chem. Neurosci.* **2019**, *10*, 902–909. [CrossRef] [PubMed]
35. Zhang, X.; Le, W. Pathological Role of Hypoxia in Alzheimer's Disease. *Exp. Neurol.* **2010**, *223*, 299–303. [CrossRef] [PubMed]
36. Xie, L.; Zhang, N.; Zhang, Q.; Li, C.; Sandhu, A.F.; Williams, G., III; Lin, S.; Lv, P.; Liu, Y.; Wu, Q.; et al. Inflammatory Factors and Amyloid β-Induced Microglial Polarization Promote Inflammatory Crosstalk with Astrocytes. *Aging* **2020**, *12*, 22538. [CrossRef] [PubMed]
37. Salminen, A.; Kaarniranta, K.; Kauppinen, A. Hypoxia/Ischemia Impairs CD33 (Siglec-3)/TREM2 Signaling: Potential Role in Alzheimer's Pathogenesis. *Neurochem. Int.* **2021**, *150*, 105186. [CrossRef] [PubMed]
38. Osorio, R.S.; Gumb, T.; Pirraglia, E.; Varga, A.W.; Lu, S.; Lim, J.; Wohlleber, M.E.; Ducca, E.L.; Koushyk, V.; Glodzik, L.; et al. Sleep-Disordered Breathing Advances Cognitive Decline in the Elderly. *Neurology* **2015**, *84*, 1964–1971. [CrossRef] [PubMed]

39. Mullins, A.E.; Kam, K.; Parekh, A.; Bubu, O.M.; Osorio, R.S.; Varga, A.W. Obstructive Sleep Apnea and Its Treatment in Aging: Effects on Alzheimer's Disease Biomarkers, Cognition, Brain Structure and Neurophysiology. *Neurobiol. Dis.* **2020**, *145*, 105054. [CrossRef] [PubMed]
40. Lal, C.; Ayappa, I.; Ayas, N.; Beaudin, A.E.; Hoyos, C.; Kushida, C.A.; Kaminska, M.; Mullins, A.; Naismith, S.L.; Osorio, R.S.; et al. The Link between Obstructive Sleep Apnea and Neurocognitive Impairment: An Official American Thoracic Society Workshop Report. *Ann. Am. Thorac. Soc.* **2022**, *19*, 1245–1256. [CrossRef]
41. Astara, K.; Tsimpolis, A.; Kalafatakis, K.; Vavougios, G.D.; Xiromerisiou, G.; Dardiotis, E.; Christodoulou, N.G.; Samara, M.T.; Lappas, A.S. Sleep Disorders and Alzheimer's Disease Pathophysiology: The Role of the Glymphatic System. A Scoping Review. *Mech. Ageing Dev.* **2024**, *217*, 111899. [CrossRef]
42. Beschorner, N.; Nedergaard, M. Glymphatic System Dysfunction in Neurodegenerative Diseases. *Curr. Opin. Neurol.* **2024**, *37*, 182–188. [CrossRef] [PubMed]
43. Szlufik, S.; Kopeć, K.; Szleszkowski, S.; Koziorowski, D. Glymphatic System Pathology and Neuroinflammation as Two Risk Factors of Neurodegeneration. *Cells* **2024**, *13*, 286. [CrossRef] [PubMed]
44. Berdina, O.N.; Madaeva, I.M.; Bolshakova, S.E.; Sholokhov, L.F.; Rychkova, L.V. Obstructive Sleep Apnea and Amyloid-B42 in Adolescents: The Results of a Pilot Study. *Acta Biomed. Sci.* **2022**, *7*, 12–21. [CrossRef]
45. Ju, Y.S.; Finn, M.B.; Sutphen, C.L.; Herries, E.M.; Jerome, G.M.; Ladenson, J.H.; Crimmins, D.L.; Fagan, A.M.; Holtzman, D.M. Obstructive Sleep Apnea Decreases Central Nervous System–Derived Proteins in the Cerebrospinal Fluid. *Ann. Neurol.* **2016**, *80*, 154–159. [CrossRef] [PubMed]
46. Ulland, T.K.; Ewald, A.C.; Knutson, A.O.; Marino, K.M.; Smith, S.M.C.; Watters, J.J. Alzheimer's Disease, Sleep Disordered Breathing, and Microglia: Puzzling out a Common Link. *Cells* **2021**, *10*, 2907. [CrossRef] [PubMed]
47. Macheda, T.; Roberts, K.; Lyons, D.N.; Higgins, E.; Ritter, K.J.; Lin, A.; Alilain, W.J.; Bachstetter, A.D. Chronic Intermittent Hypoxia Induces Robust Astrogliosis in an Alzheimer's Disease-Relevant Mouse Model. *Neuroscience* **2019**, *398*, 55–63. [CrossRef] [PubMed]
48. Prabhakar, N.R.; Peng, Y.-J.; Nanduri, J. Hypoxia-Inducible Factors and Obstructive Sleep Apnea. *J. Clin. Investig.* **2020**, *130*, 5042–5051. [CrossRef] [PubMed]
49. Row, B.W.; Kheirandish, L.; Neville, J.J.; Gozal, D. Impaired Spatial Learning and Hyperactivity in Developing Rats Exposed to Intermittent Hypoxia. *Pediatr. Res.* **2002**, *52*, 449–453. [CrossRef] [PubMed]
50. Zhang, S.X.L.; Wang, Y.; Gozal, D. Pathological Consequences of Intermittent Hypoxia in the Central Nervous System. In *Comprehensive Physiology*; Wiley: Hoboken, NJ, USA, 2012; pp. 1767–1777.
51. Payne, R.S.; Goldbart, A.; Gozal, D.; Schurr, A. Effect of Intermittent Hypoxia on Long-Term Potentiation in Rat Hippocampal Slices. *Brain Res.* **2004**, *1029*, 195–199. [CrossRef]
52. Nair, D.; Dayyat, E.A.; Zhang, S.X.; Wang, Y.; Gozal, D. Intermittent Hypoxia-Induced Cognitive Deficits Are Mediated by NADPH Oxidase Activity in a Murine Model of Sleep Apnea. *PLoS ONE* **2011**, *6*, e19847. [CrossRef]
53. Arias-Cavieres, A.; Khuu, M.A.; Nwakudu, C.U.; Barnard, J.E.; Dalgin, G.; Garcia, A.J. A HIF1a-Dependent Pro-Oxidant State Disrupts Synaptic Plasticity and Impairs Spatial Memory in Response to Intermittent Hypoxia. *eNeuro* **2020**, *7*, ENEURO.0024-20.2020. [CrossRef] [PubMed]
54. Wall, A.M.; Corcoran, A.E.; O'Halloran, K.D.; O'Connor, J.J. Effects of Prolyl-Hydroxylase Inhibition and Chronic Intermittent Hypoxia on Synaptic Transmission and Plasticity in the Rat CA1 and Dentate Gyrus. *Neurobiol. Dis.* **2014**, *62*, 8–17. [CrossRef] [PubMed]
55. Ng, K.-M.; Lau, C.-F.; Fung, M.-L. Melatonin Reduces Hippocampal β-Amyloid Generation in Rats Exposed to Chronic Intermittent Hypoxia. *Brain Res.* **2010**, *1354*, 163–171. [CrossRef] [PubMed]
56. Yagishita, S.; Hirasawa, A. Intermittent Hypoxia Produces Alzheimer Disease? *Oncotarget* **2017**, *8*, 41786–41787. [CrossRef] [PubMed]
57. Shiota, S.; Takekawa, H.; Matsumoto, S.; Takeda, K.; Nurwidya, F.; Yoshioka, Y.; Takahashi, F.; Hattori, N.; Tabira, T.; Mochizuki, H.; et al. Chronic Intermittent Hypoxia/Reoxygenation Facilitate Amyloid-β Generation in Mice. *J. Alzheimer's Dis.* **2013**, *37*, 325–333. [CrossRef] [PubMed]
58. Gabryelska, A.; Sochal, M. Evaluation of HIF-1 Involvement in the BDNF and ProBDNF Signaling Pathways among Obstructive Sleep Apnea Patients. *Int. J. Mol. Sci.* **2022**, *23*, 14876. [CrossRef] [PubMed]
59. Hasegawa, M.; Fujiwara, H.; Nonaka, T.; Wakabayashi, K.; Takahashi, H.; Lee, V.M.-Y.; Trojanowski, J.Q.; Mann, D.; Iwatsubo, T. Phosphorylated α-Synuclein Is Ubiquitinated in α-Synucleinopathy Lesions. *J. Biol. Chem.* **2002**, *277*, 49071–49076. [CrossRef] [PubMed]
60. Chen, L.; Feany, M.B. α-Synuclein Phosphorylation Controls Neurotoxicity and Inclusion Formation in a Drosophila Model of Parkinson Disease. *Nat. Neurosci.* **2005**, *8*, 657–663. [CrossRef] [PubMed]
61. Bengoa-Vergniory, N.; Roberts, R.F.; Wade-Martins, R.; Alegre-Abarrategui, J. Alpha-Synuclein Oligomers: A New Hope. *Acta Neuropathol.* **2017**, *134*, 819–838. [CrossRef]
62. Winner, B.; Jappelli, R.; Maji, S.K.; Desplats, P.A.; Boyer, L.; Aigner, S.; Hetzer, C.; Loher, T.; Vilar, M.; Campioni, S.; et al. In Vivo Demonstration That α-Synuclein Oligomers Are Toxic. *Proc. Natl. Acad. Sci. USA* **2011**, *108*, 4194–4199. [CrossRef]
63. Muddapu, V.R.; Chakravarthy, V.S. Influence of Energy Deficiency on the Subcellular Processes of Substantia Nigra Pars Compacta Cell for Understanding Parkinsonian Neurodegeneration. *Sci. Rep.* **2021**, *11*, 1754. [CrossRef]

64. Guo, M.; Liu, W.; Luo, H.; Shao, Q.; Li, Y.; Gu, Y.; Guan, Y.; Ma, W.; Chen, M.; Yang, H.; et al. Hypoxic Stress Accelerates the Propagation of Pathological Alpha-synuclein and Degeneration of Dopaminergic Neurons. *CNS Neurosci. Ther.* **2023**, *29*, 544–558. [CrossRef] [PubMed]
65. Guo, M.; Ji, X.; Liu, J. Hypoxia and Alpha-Synuclein: Inextricable Link Underlying the Pathologic Progression of Parkinson's Disease. *Front. Aging Neurosci.* **2022**, *14*, 919343. [CrossRef]
66. Kim, T.; Mehta, S.L.; Kaimal, B.; Lyons, K.; Dempsey, R.J.; Vemuganti, R. Poststroke Induction of -Synuclein Mediates Ischemic Brain Damage. *J. Neurosci.* **2016**, *36*, 7055–7065. [CrossRef]
67. Unal-Cevik, I.; Gursoy-Ozdemir, Y.; Yemisci, M.; Lule, S.; Gurer, G.; Can, A.; Müller, V.; Kahle, P.J.; Dalkara, T. Alpha-Synuclein Aggregation Induced by Brief Ischemia Negatively Impacts Neuronal Survival in Vivo: A Study in [A30P]Alpha-Synuclein Transgenic Mouse. *J. Cereb. Blood Flow. Metab.* **2011**, *31*, 913–923. [CrossRef]
68. Hu, X.; Rea, H.C.; Wiktorowicz, J.E.; Perez-Polo, J.R. Proteomic Analysis of Hypoxia/Ischemia-Induced Alteration of Cortical Development and Dopamine Neurotransmission in Neonatal Rat. *J. Proteome Res.* **2006**, *5*, 2396–2404. [CrossRef] [PubMed]
69. Chen, T.; Wang, Q.; Chao, D.; Xia, T.C.; Sheng, S.; Li, Z.-R.; Zhao, J.-N.; Wen, G.-Q.; Ding, G.; Xia, Y. δ-Opioid Receptor Activation Attenuates the Oligomer Formation Induced by Hypoxia and/or α-Synuclein Overexpression/Mutation Through Dual Signaling Pathways. *Mol. Neurobiol.* **2019**, *56*, 3463–3475. [CrossRef] [PubMed]
70. Sun, H.; Sun, B.; Chen, D.; Chen, Y.; Li, W.; Xu, M.; Shen, Y.; Xu, Z.; Wang, Y.; Bu, X. Plasma α-synuclein Levels Are Increased in Patients with Obstructive Sleep Apnea Syndrome. *Ann. Clin. Transl. Neurol.* **2019**, *6*, 788–794. [CrossRef]
71. Elfil, M.; Bahbah, E.I.; Attia, M.M.; Eldokmak, M.; Koo, B.B. Impact of Obstructive Sleep Apnea on Cognitive and Motor Functions in Parkinson's Disease. *Mov. Disord.* **2021**, *36*, 570–580. [CrossRef]
72. Pacelli, C.; Giguère, N.; Bourque, M.-J.; Lévesque, M.; Slack, R.S.; Trudeau, L.-É. Elevated Mitochondrial Bioenergetics and Axonal Arborization Size Are Key Contributors to the Vulnerability of Dopamine Neurons. *Curr. Biol.* **2015**, *25*, 2349–2360. [CrossRef]
73. Burtscher, J.; Syed, M.M.K.; Keller, M.A.; Lashuel, H.A.; Millet, G.P. Fatal Attraction—The Role of Hypoxia When Alpha-Synuclein Gets Intimate with Mitochondria. *Neurobiol. Aging* **2021**, *107*, 128–141. [CrossRef] [PubMed]
74. Wei, Z.; Li, X.; Li, X.; Liu, Q.; Cheng, Y. Oxidative Stress in Parkinson's Disease: A Systematic Review and Meta-Analysis. *Front. Mol. Neurosci.* **2018**, *11*, 236. [CrossRef]
75. Lestón Pinilla, L.; Ugun-Klusek, A.; Rutella, S.; De Girolamo, L.A. Hypoxia Signaling in Parkinson's Disease: There Is Use in Asking "What HIF?". *Biology* **2021**, *10*, 723. [CrossRef] [PubMed]
76. Semenza, G.L. Hydroxylation of HIF-1: Oxygen Sensing at the Molecular Level. *Physiology* **2004**, *19*, 176–182. [CrossRef] [PubMed]
77. Semenza, G.L. Regulation of Oxygen Homeostasis by Hypoxia-Inducible Factor 1. *Physiology* **2009**, *24*, 97–106. [CrossRef] [PubMed]
78. Heikkilä, M.; Pasanen, A.; Kivirikko, K.I.; Myllyharju, J. Roles of the Human Hypoxia-Inducible Factor (HIF)-3α Variants in the Hypoxia Response. *Cell. Mol. Life Sci.* **2011**, *68*, 3885–3901. [CrossRef] [PubMed]
79. Leu, T.; Schützhold, V.; Fandrey, J.; Ferenz, K.B. When the Brain Yearns for Oxygen. *Neurosignals* **2019**, *27*, 50–61. [CrossRef] [PubMed]
80. Masoud, G.N.; Li, W. HIF-1α Pathway: Role, Regulation and Intervention for Cancer Therapy. *Acta Pharm. Sin. B* **2015**, *5*, 378–389. [CrossRef] [PubMed]
81. Bruick, R.K.; McKnight, S.L. A Conserved Family of Prolyl-4-Hydroxylases That Modify HIF. *Science* **2001**, *294*, 1337–1340. [CrossRef]
82. Ivan, M.; Kondo, K.; Yang, H.; Kim, W.; Valiando, J.; Ohh, M.; Salic, A.; Asara, J.M.; Lane, W.S.; Kaelin, W.G., Jr. HIFα Targeted for VHL-Mediated Destruction by Proline Hydroxylation: Implications for O_2 Sensing. *Science* **2001**, *292*, 464–468. [CrossRef]
83. Pereira, T.; Zheng, X.; Poellinger, L. Degradation of the Hypoxia-Inducible Factor 1α, Where Does It Happen? *Cell Cycle* **2006**, *5*, 2720–2722. [CrossRef] [PubMed]
84. Jaakkola, P.; Mole, D.R.; Tian, Y.-M.; Wilson, M.I.; Gielbert, J.; Gaskell, S.J.; von Kriegsheim, A.; Hebestreit, H.F.; Mukherji, M.; Schofield, C.J.; et al. Targeting of HIF-α to the von Hippel-Lindau Ubiquitylation Complex by O_2 -Regulated Prolyl Hydroxylation. *Science* **2001**, *292*, 468–472. [CrossRef] [PubMed]
85. Semenza, G.L. Oxygen Sensing, Hypoxia-Inducible Factors, and Disease Pathophysiology. *Annu. Rev. Pathol. Mech. Dis.* **2014**, *9*, 47–71. [CrossRef] [PubMed]
86. Dengler, V.L.; Galbraith, M.D.; Espinosa, J.M. Transcriptional Regulation by Hypoxia Inducible Factors. *Crit. Rev. Biochem. Mol. Biol.* **2014**, *49*, 1–15. [CrossRef] [PubMed]
87. Ishida, T.; Nakao, S.; Ueyama, T.; Harada, Y.; Kawamura, T. Metabolic Remodeling during Somatic Cell Reprogramming to Induced Pluripotent Stem Cells: Involvement of Hypoxia-Inducible Factor 1. *Inflamm. Regen.* **2020**, *40*, 8. [CrossRef] [PubMed]
88. Semenza, G.L.; Jiang, B.-H.; Leung, S.W.; Passantino, R.; Concordet, J.-P.; Maire, P.; Giallongo, A. Hypoxia Response Elements in the Aldolase A, Enolase 1, and Lactate Dehydrogenase A Gene Promoters Contain Essential Binding Sites for Hypoxia-Inducible Factor 1. *J. Biol. Chem.* **1996**, *271*, 32529–32537. [CrossRef]
89. Kim, J.; Tchernyshyov, I.; Semenza, G.L.; Dang, C.V. HIF-1-Mediated Expression of Pyruvate Dehydrogenase Kinase: A Metabolic Switch Required for Cellular Adaptation to Hypoxia. *Cell Metab.* **2006**, *3*, 177–185. [CrossRef] [PubMed]
90. Tello, D.; Balsa, E.; Acosta-Iborra, B.; Fuertes-Yebra, E.; Elorza, A.; Ordóñez, Á.; Corral-Escariz, M.; Soro, I.; López-Bernardo, E.; Perales-Clemente, E.; et al. Induction of the Mitochondrial NDUFA4L2 Protein by HIF-1α Decreases Oxygen Consumption by Inhibiting Complex I Activity. *Cell Metab.* **2011**, *14*, 768–779. [CrossRef]

91. Raut, S.; Bhalerao, A.; Powers, M.; Gonzalez, M.; Mancuso, S.; Cucullo, L. Hypometabolism, Alzheimer's Disease, and Possible Therapeutic Targets: An Overview. *Cells* **2023**, *12*, 2019. [CrossRef]
92. Peng, W.; Tan, C.; Mo, L.; Jiang, J.; Zhou, W.; Du, J.; Zhou, X.; Liu, X.; Chen, L. Glucose Transporter 3 in Neuronal Glucose Metabolism: Health and Diseases. *Metabolism* **2021**, *123*, 154869. [CrossRef]
93. Głuchowska, K.; Pliszka, M.; Szablewski, L. Expression of Glucose Transporters in Human Neurodegenerative Diseases. *Biochem. Biophys. Res. Commun.* **2021**, *540*, 8–15. [CrossRef] [PubMed]
94. Szablewski, L. Glucose Transporters in Brain: In Health and in Alzheimer's Disease. *J. Alzheimer's Dis.* **2016**, *55*, 1307–1320. [CrossRef] [PubMed]
95. Schubert, D. Glucose Metabolism and Alzheimer's Disease. *Ageing Res. Rev.* **2005**, *4*, 240–257. [CrossRef] [PubMed]
96. Liu, Y.; Liu, F.; Iqbal, K.; Grundke-Iqbal, I.; Gong, C.-X. Decreased Glucose Transporters Correlate to Abnormal Hyperphosphorylation of Tau in Alzheimer Disease. *FEBS Lett.* **2008**, *582*, 359–364. [CrossRef] [PubMed]
97. Soucek, T.; Cumming, R.; Dargusch, R.; Maher, P.; Schubert, D. The Regulation of Glucose Metabolism by HIF-1 Mediates a Neuroprotective Response to Amyloid Beta Peptide. *Neuron* **2003**, *39*, 43–56. [CrossRef] [PubMed]
98. Zhang, Y.; Yan, X.; Xu, S.; Pang, Z.; Li, L.; Yang, Y.; Fan, Y.; Wang, Z.; Yu, X.; Guo, C.; et al. α-Lipoic Acid Maintains Brain Glucose Metabolism via BDNF/TrkB/HIF-1α Signaling Pathway in P301S Mice. *Front. Aging Neurosci.* **2020**, *12*, 262. [CrossRef] [PubMed]
99. Park, J.; Lai, M.K.P.; Arumugam, T.V.; Jo, D.-G. O-GlcNAcylation as a Therapeutic Target for Alzheimer's Disease. *Neuromolecular Med.* **2020**, *22*, 171–193. [CrossRef] [PubMed]
100. Gupta, S.; Jinka, S.K.A.; Khanal, S.; Bhavnani, N.; Almashhori, F.; Lallo, J.; Mathias, A.; Al-Rhayyel, Y.; Herman, D.; Holden, J.G.; et al. Cognitive Dysfunction and Increased Phosphorylated Tau Are Associated with Reduced O-GlcNAc Signaling in an Aging Mouse Model of Metabolic Syndrome. *J. Neurosci. Res.* **2023**, *101*, 1324–1344. [CrossRef]
101. Wang, Y.-Y.; Huang, Z.-T.; Yuan, M.-H.; Jing, F.; Cai, R.-L.; Zou, Q.; Pu, Y.-S.; Wang, S.-Y.; Chen, F.; Yi, W.-M.; et al. Role of Hypoxia Inducible Factor-1α in Alzheimer's Disease. *J. Alzheimer's Dis.* **2021**, *80*, 949–961. [CrossRef]
102. Schubert, D.; Soucek, T.; Blouw, B. The Induction of HIF-1 Reduces Astrocyte Activation by Amyloid Beta Peptide. *Eur. J. Neurosci.* **2009**, *29*, 1323–1334. [CrossRef]
103. Merelli, A.; Rodríguez, J.C.G.; Folch, J.; Regueiro, M.R.; Camins, A.; Lazarowski, A. Understanding the Role of Hypoxia Inducible Factor During Neurodegeneration for New Therapeutics Opportunities. *Curr. Neuropharmacol.* **2018**, *16*, 1484–1498. [CrossRef] [PubMed]
104. Avramovich-Tirosh, Y.; Bar-Am, O.; Amit, T.; Youdim, M.B.H.; Weinreb, O. Up-Regulation of Hypoxia-Inducible Factor (HIF)-1α and HIF-Target Genes in Cortical Neurons by the Novel Multifunctional Iron Chelator Anti-Alzheimer Drug, M30. *Curr. Alzheimer Res.* **2010**, *7*, 300–306. [CrossRef] [PubMed]
105. Zhou, J.; Fandrey, J.; Schümann, J.; Tiegs, G.; Brüne, B. NO and TNF-α Released from Activated Macrophages Stabilize HIF-1α in Resting Tubular LLC-PK_1 Cells. *Am. J. Physiol. Cell Physiol.* **2003**, *284*, C439–C446. [CrossRef] [PubMed]
106. Hellwig-Bürgel, T.; Rutkowski, K.; Metzen, E.; Fandrey, J.; Jelkmann, W. Interleukin-1beta and Tumor Necrosis Factor-Alpha Stimulate DNA Binding of Hypoxia-Inducible Factor-1. *Blood* **1999**, *94*, 1561–1567. [CrossRef] [PubMed]
107. Xing, J.; Lu, J. HIF-1α Activation Attenuates IL-6 and TNF-α Pathways in Hippocampus of Rats Following Transient Global Ischemia. *Cell. Physiol. Biochem.* **2016**, *39*, 511–520. [CrossRef] [PubMed]
108. Gurung, P.; Li, B.; Subbarao Malireddi, R.K.; Lamkanfi, M.; Geiger, T.L.; Kanneganti, T.-D. Chronic TLR Stimulation Controls NLRP3 Inflammasome Activation through IL-10 Mediated Regulation of NLRP3 Expression and Caspase-8 Activation. *Sci. Rep.* **2015**, *5*, 14488. [CrossRef] [PubMed]
109. Sethuraman, A.; Rao, P.; Pranay, A.; Xu, K.; LaManna, J.C.; Puchowicz, M.A. Chronic Ketosis Modulates HIF1α-Mediated Inflammatory Response in Rat Brain. In *Oxygen Transport to Tissue XLII*; Springer: Berlin/Heidelberg, Germany, 2021; pp. 3–7.
110. Shin, D.H.; Li, S.H.; Yang, S.; Lee, B.L.; Lee, M.K.; Park, J. Inhibitor of Nuclear Factor-kappaB Alpha Derepresses Hypoxia-inducible Factor-1 during Moderate Hypoxia by Sequestering Factor Inhibiting Hypoxia-inducible Factor from Hypoxia-inducible Factor 1α. *FEBS J.* **2009**, *276*, 3470–3480. [CrossRef] [PubMed]
111. Jung, E.; Kim, Y.E.; Jeon, H.S.; Yoo, M.; Kim, M.; Kim, Y.-M.; Koh, S.-H.; Choi, Y.K. Chronic Hypoxia of Endothelial Cells Boosts HIF-1α-NLRP1 Circuit in Alzheimer's Disease. *Free Radic. Biol. Med.* **2023**, *204*, 385–393. [CrossRef] [PubMed]
112. Grammas, P.; Tripathy, D.; Sanchez, A.; Yin, X.; Luo, J. Brain Microvasculature and Hypoxia-Related Proteins in Alzheimer's Disease. *Int. J. Clin. Exp. Pathol.* **2011**, *4*, 616–627. [PubMed]
113. Alexander, C.; Li, T.; Hattori, Y.; Chiu, D.; Frost, G.R.; Jonas, L.; Liu, C.; Anderson, C.J.; Wong, E.; Park, L.; et al. Hypoxia Inducible Factor-1α Binds and Activates γ-Secretase for Aβ Production under Hypoxia and Cerebral Hypoperfusion. *Mol. Psychiatry* **2022**, *27*, 4264–4273. [CrossRef]
114. Guglielmotto, M.; Aragno, M.; Autelli, R.; Giliberto, L.; Novo, E.; Colombatto, S.; Danni, O.; Parola, M.; Smith, M.A.; Perry, G.; et al. The Up-regulation of BACE1 Mediated by Hypoxia and Ischemic Injury: Role of Oxidative Stress and HIF1α. *J. Neurochem.* **2009**, *108*, 1045–1056. [CrossRef]
115. Guo, C.; Yang, Z.-H.; Zhang, S.; Chai, R.; Xue, H.; Zhang, Y.-H.; Li, J.-Y.; Wang, Z.-Y. Intranasal Lactoferrin Enhances α-Secretase-Dependent Amyloid Precursor Protein Processing via the ERK1/2-CREB and HIF-1α Pathways in an Alzheimer's Disease Mouse Model. *Neuropsychopharmacology* **2017**, *42*, 2504–2515. [CrossRef] [PubMed]
116. Yang, Z.; Zhao, T.; Zou, Y.; Zhang, J.H.; Feng, H. Hypoxia Induces Autophagic Cell Death through Hypoxia-Inducible Factor 1α in Microglia. *PLoS ONE* **2014**, *9*, e96509. [CrossRef] [PubMed]

117. Solovyev, N.; El-Khatib, A.H.; Costas-Rodríguez, M.; Schwab, K.; Griffin, E.; Raab, A.; Platt, B.; Theuring, F.; Vogl, J.; Vanhaecke, F. Cu, Fe, and Zn Isotope Ratios in Murine Alzheimer's Disease Models Suggest Specific Signatures of Amyloidogenesis and Tauopathy. *J. Biol. Chem.* **2021**, *296*, 100292. [CrossRef] [PubMed]
118. Peters, D.G.; Connor, J.R.; Meadowcroft, M.D. The Relationship between Iron Dyshomeostasis and Amyloidogenesis in Alzheimer's Disease: Two Sides of the Same Coin. *Neurobiol. Dis.* **2015**, *81*, 49–65. [CrossRef] [PubMed]
119. Tacchini, L.; Bianchi, L.; Bernelli-Zazzera, A.; Cairo, G. Transferrin Receptor Induction by Hypoxia. *J. Biol. Chem.* **1999**, *274*, 24142–24146. [CrossRef] [PubMed]
120. Tacchini, L.; Gammella, E.; De Ponti, C.; Recalcati, S.; Cairo, G. Role of HIF-1 and NF-KB Transcription Factors in the Modulation of Transferrin Receptor by Inflammatory and Anti-Inflammatory Signals. *J. Biol. Chem.* **2008**, *283*, 20674–20686. [CrossRef] [PubMed]
121. Petralla, S.; Saveleva, L.; Kanninen, K.M.; Oster, J.S.; Panayotova, M.; Fricker, G.; Puris, E. Increased Expression of Transferrin Receptor 1 in the Brain Cortex of 5xFAD Mouse Model of Alzheimer's Disease Is Associated with Activation of HIF-1 Signaling Pathway. *Mol. Neurobiol.* **2024**. [CrossRef] [PubMed]
122. De Lau, L.M.; Breteler, M.M. Epidemiology of Parkinson's Disease. *Lancet Neurol.* **2006**, *5*, 525–535. [CrossRef]
123. Burbulla, L.F.; Song, P.; Mazzulli, J.R.; Zampese, E.; Wong, Y.C.; Jeon, S.; Santos, D.P.; Blanz, J.; Obermaier, C.D.; Strojny, C.; et al. Dopamine Oxidation Mediates Mitochondrial and Lysosomal Dysfunction in Parkinson's Disease. *Science* **2017**, *357*, 1255–1261. [CrossRef]
124. Hauser, D.N.; Hastings, T.G. Mitochondrial Dysfunction and Oxidative Stress in Parkinson's Disease and Monogenic Parkinsonism. *Neurobiol. Dis.* **2013**, *51*, 35–42. [CrossRef] [PubMed]
125. Di Maio, R.; Barrett, P.J.; Hoffman, E.K.; Barrett, C.W.; Zharikov, A.; Borah, A.; Hu, X.; McCoy, J.; Chu, C.T.; Burton, E.A.; et al. α-Synuclein Binds to TOM20 and Inhibits Mitochondrial Protein Import in Parkinson's Disease. *Sci. Transl. Med.* **2016**, *8*, 342ra78. [CrossRef] [PubMed]
126. Santos, D.; Esteves, A.R.; Silva, D.F.; Januário, C.; Cardoso, S.M. The Impact of Mitochondrial Fusion and Fission Modulation in Sporadic Parkinson's Disease. *Mol. Neurobiol.* **2015**, *52*, 573–586. [CrossRef] [PubMed]
127. Stefanis, L. α-Synuclein in Parkinson's Disease. *Cold Spring Harb. Perspect. Med.* **2012**, *2*, a009399. [CrossRef] [PubMed]
128. Zhang, G.; Chen, L.; Liu, J.; Jin, Y.; Lin, Z.; Du, S.; Fu, Z.; Chen, T.; Qin, Y.; Sui, F.; et al. HIF-1α/MicroRNA-128-3p Axis Protects Hippocampal Neurons from Apoptosis via the *Axin1* -Mediated Wnt/β-Catenin Signaling Pathway in Parkinson's Disease Models. *Aging* **2020**, *12*, 4067–4081. [CrossRef] [PubMed]
129. Agani, F.H.; Pichiule, P.; Chavez, J.C.; LaManna, J.C. The Role of Mitochondria in the Regulation of Hypoxia-Inducible Factor 1 Expression during Hypoxia. *J. Biol. Chem.* **2000**, *275*, 35863–35867. [CrossRef] [PubMed]
130. Czyzyk-Krzeska, M.F.; Furnari, B.A.; Lawson, E.E.; Millhorn, D.E. Hypoxia Increases Rate of Transcription and Stability of Tyrosine Hydroxylase MRNA in Pheochromocytoma (PC12) Cells. *J. Biol. Chem.* **1994**, *269*, 760–764. [CrossRef] [PubMed]
131. Zhang, Z.; Yan, J.; Chang, Y.; ShiDu Yan, S.; Shi, H. Hypoxia Inducible Factor-1 as a Target for Neurodegenerative Diseases. *Curr. Med. Chem.* **2011**, *18*, 4335–4343. [CrossRef]
132. Milosevic, J.; Maisel, M.; Wegner, F.; Leuchtenberger, J.; Wenger, R.H.; Gerlach, M.; Storch, A.; Schwarz, J. Lack of Hypoxia-Inducible Factor-1α Impairs Midbrain Neural Precursor Cells Involving Vascular Endothelial Growth Factor Signaling. *J. Neurosci.* **2007**, *27*, 412–421. [CrossRef]
133. Weng, M.; Xie, X.; Liu, C.; Lim, K.-L.; Zhang, C.; Li, L. The Sources of Reactive Oxygen Species and Its Possible Role in the Pathogenesis of Parkinson's Disease. *Park. Dis.* **2018**, *2018*, 9163040. [CrossRef]
134. Parandavar, E.; Yazdanparast, R. Differential Impact of Various Reactive Oxygen Species (ROS) on HIF-1α/P53 Direct Interaction in SK-N-MC Neuroblastoma Cells. *Cell Biosci.* **2017**, *7*, 52. [CrossRef] [PubMed]
135. Sanjuán-Pla, A.; Cervera, A.M.; Apostolova, N.; Garcia-Bou, R.; Víctor, V.M.; Murphy, M.P.; McCreath, K.J. A Targeted Antioxidant Reveals the Importance of Mitochondrial Reactive Oxygen Species in the Hypoxic Signaling of HIF-1α. *FEBS Lett.* **2005**, *579*, 2669–2674. [CrossRef] [PubMed]
136. Bonifati, V.; Rizzu, P.; van Baren, M.J.; Schaap, O.; Breedveld, G.J.; Krieger, E.; Dekker, M.C.J.; Squitieri, F.; Ibanez, P.; Joosse, M.; et al. Mutations in the *DJ-1* Gene Associated with Autosomal Recessive Early-Onset Parkinsonism. *Science* **2003**, *299*, 256–259. [CrossRef] [PubMed]
137. Parsanejad, M.; Zhang, Y.; Qu, D.; Irrcher, I.; Rousseaux, M.W.C.; Aleyasin, H.; Kamkar, F.; Callaghan, S.; Slack, R.S.; Mak, T.W.; et al. Regulation of the VHL/HIF-1 Pathway by DJ-1. *J. Neurosci.* **2014**, *34*, 8043–8050. [CrossRef] [PubMed]
138. Ben-Shachar, D.; Eshel, G.; Finberg, J.P.M.; Youdim, M.B.H. The Iron Chelator Desferrioxamine (Desferal) Retards 6-Hydroxydopamine-Induced Degeneration of Nigrostriatal Dopamine Neurons. *J. Neurochem.* **1991**, *56*, 1441–1444. [CrossRef] [PubMed]
139. Wohlrab, C.; Kuiper, C.; Vissers, M.C.; Phillips, E.; Robinson, B.A.; Dachs, G.U. Ascorbate Modulates the Hypoxic Pathway by Increasing Intracellular Activity of the HIF Hydroxylases in Renal Cell Carcinoma Cells. *Hypoxia* **2019**, *7*, 17–31. [CrossRef] [PubMed]
140. Flashman, E.; Davies, S.L.; Yeoh, K.K.; Schofield, C.J. Investigating the Dependence of the Hypoxia-Inducible Factor Hydroxylases (Factor Inhibiting HIF and Prolyl Hydroxylase Domain 2) on Ascorbate and Other Reducing Agents. *Biochem. J.* **2010**, *427*, 135–142. [CrossRef] [PubMed]

141. Kim, S.; Yang, E. Recent Advances in Developing Inhibitors for Hypoxia-Inducible Factor Prolyl Hydroxylases and Their Therapeutic Implications. *Molecules* **2015**, *20*, 20551–20568. [CrossRef]
142. Mitroshina, E.V.; Savyuk, M.O.; Ponimaskin, E.; Vedunova, M.V. Hypoxia-Inducible Factor (HIF) in Ischemic Stroke and Neurodegenerative Disease. *Front. Cell Dev. Biol.* **2021**, *9*, 703084. [CrossRef]
143. Kupershmidt, L.; Amit, T.; Bar-Am, O.; Youdim, M.B.H.; Weinreb, O. The Novel Multi-Target Iron Chelating-Radical Scavenging Compound M30 Possesses Beneficial Effects on Major Hallmarks of Alzheimer's Disease. *Antioxid. Redox Signal* **2012**, *17*, 860–877. [CrossRef]
144. Mechlovich, D.; Amit, T.; Bar-Am, O.; Mandel, S.; Youdim, M.; Weinreb, O. The Novel Multi-Target Iron Chelator, M30 Modulates HIF-1α-Related Glycolytic Genes and Insulin Signaling Pathway in the Frontal Cortex of APP/PS1 Alzheimer's Disease Mice. *Curr. Alzheimer Res.* **2014**, *11*, 119–127. [CrossRef]
145. Amit, T.; Bar-Am, O.; Mechlovich, D.; Kupershmidt, L.; Youdim, M.B.H.; Weinreb, O. The Novel Multitarget Iron Chelating and Propargylamine Drug M30 Affects APP Regulation and Processing Activities in Alzheimer's Disease Models. *Neuropharmacology* **2017**, *123*, 359–367. [CrossRef] [PubMed]
146. Gal, S.; Zheng, H.; Fridkin, M.; Youdim, M.B.H. Restoration of Nigrostriatal Dopamine Neurons in Post-MPTP Treatment by the Novel Multifunctional Brain-Permeable Iron Chelator-Monoamine Oxidase Inhibitor Drug, M30. *Neurotox. Res.* **2010**, *17*, 15–27. [CrossRef] [PubMed]
147. Youdim, M.B.H.; Kupershmidt, L.; Amit, T.; Weinreb, O. Promises of Novel Multi-Target Neuroprotective and Neurorestorative Drugs for Parkinson's Disease. *Park. Relat. Disord.* **2014**, *20*, S132–S136. [CrossRef] [PubMed]
148. Youdim, M.B.H. Multi Target Neuroprotective and Neurorestorative Anti-Parkinson and Anti-Alzheimer Drugs Ladostigil and M30 Derived from Rasagiline. *Exp. Neurobiol.* **2013**, *22*, 1–10. [CrossRef] [PubMed]
149. Ramirez, A.E.; Fernández-Pérez, E.J.; Olivos, N.; Burgos, C.F.; Boopathi, S.; Armijo-Weingart, L.; Pacheco, C.R.; González, W.; Aguayo, L.G. The Stimulatory Effects of Intracellular α-Synuclein on Synaptic Transmission Are Attenuated by 2-Octahydroisoquinolin-2(1H)-Ylethanamine. *Int. J. Mol. Sci.* **2021**, *22*, 13253. [CrossRef] [PubMed]
150. Farr, A.C.; Xiong, M.P. Challenges and Opportunities of Deferoxamine Delivery for Treatment of Alzheimer's Disease, Parkinson's Disease, and Intracerebral Hemorrhage. *Mol. Pharm.* **2021**, *18*, 593–609. [CrossRef] [PubMed]
151. Elzoghby, A.O.; Abdelmoneem, M.A.; Hassanin, I.A.; Abd Elwakil, M.M.; Elnaggar, M.A.; Mokhtar, S.; Fang, J.-Y.; Elkhodairy, K.A. Lactoferrin, a Multi-Functional Glycoprotein: Active Therapeutic, Drug Nanocarrier & Targeting Ligand. *Biomaterials* **2020**, *263*, 120355. [CrossRef]
152. Zakharova, E.T.; Kostevich, V.A.; Sokolov, A.V.; Vasilyev, V.B. Human Apo-Lactoferrin as a Physiological Mimetic of Hypoxia Stabilizes Hypoxia-Inducible Factor-1 Alpha. *BioMetals* **2012**, *25*, 1247–1259. [CrossRef]
153. Xu, S.-F.; Zhang, Y.-H.; Wang, S.; Pang, Z.-Q.; Fan, Y.-G.; Li, J.-Y.; Wang, Z.-Y.; Guo, C. Lactoferrin Ameliorates Dopaminergic Neurodegeneration and Motor Deficits in MPTP-Treated Mice. *Redox Biol.* **2019**, *21*, 101090. [CrossRef]
154. Nordquist, L.; Friederich-Persson, M.; Fasching, A.; Liss, P.; Shoji, K.; Nangaku, M.; Hansell, P.; Palm, F. Activation of Hypoxia-Inducible Factors Prevents Diabetic Nephropathy. *J. Am. Soc. Nephrol.* **2015**, *26*, 328–338. [CrossRef] [PubMed]
155. Xi, L.; Taher, M.; Yin, C.; Salloum, F.; Kukreja, R.C. Cobalt Chloride Induces Delayed Cardiac Preconditioning in Mice through Selective Activation of HIF-1α and AP-1 and INOS Signaling. *Am. J. Physiol. Heart Circ. Physiol.* **2004**, *287*, H2369–H2375. [CrossRef] [PubMed]
156. Li, Q.-Y.; Xiong, Q.-W.; Yao, X.; Liu, F.; Tang, X.; Fu, H.; Tong, T.; Mao, J.; Peng, W.-X. Roxadustat: Do We Know All the Answers? *Biomol. Biomed.* **2023**, *23*, 354. [CrossRef] [PubMed]
157. Ma, X.; Wang, X.; Cao, J.; Geng, Z.; Wang, Z. Effect of Proline Analogues on Activity of Human Prolyl Hydroxylase and the Regulation of HIF Signal Transduction Pathway. *PLoS ONE* **2014**, *9*, e95692. [CrossRef] [PubMed]
158. Selvaraju, V.; Parinandi, N.L.; Adluri, R.S.; Goldman, J.W.; Hussain, N.; Sanchez, J.A.; Maulik, N. Molecular Mechanisms of Action and Therapeutic Uses of Pharmacological Inhibitors of HIF–Prolyl 4-Hydroxylases for Treatment of Ischemic Diseases. *Antioxid. Redox Signal* **2014**, *20*, 2631–2665. [CrossRef] [PubMed]
159. Lee, D.W.; Rajagopalan, S.; Siddiq, A.; Gwiazda, R.; Yang, L.; Beal, M.F.; Ratan, R.R.; Andersen, J.K. Inhibition of Prolyl Hydroxylase Protects against 1-Methyl-4-Phenyl-1,2,3,6-Tetrahydropyridine-Induced Neurotoxicity. *J. Biol. Chem.* **2009**, *284*, 29065–29076. [CrossRef] [PubMed]
160. Chinta, S.J.; Rajagopalan, S.; Ganesan, A.; Andersen, J.K. A Possible Novel Anti-Inflammatory Mechanism for the Pharmacological Prolyl Hydroxylase Inhibitor 3,4-Dihydroxybenzoate: Implications for Use as a Therapeutic for Parkinson's Disease. *Park. Dis.* **2012**, *2012*, 364684. [CrossRef] [PubMed]
161. Li, X.; Cui, X.-X.; Chen, Y.-J.; Wu, T.-T.; Xu, H.; Yin, H.; Wu, Y.-C. Therapeutic Potential of a Prolyl Hydroxylase Inhibitor FG-4592 for Parkinson's Diseases in Vitro and in Vivo: Regulation of Redox Biology and Mitochondrial Function. *Front. Aging Neurosci.* **2018**, *10*, 121. [CrossRef] [PubMed]
162. Zhao, P.; You, Y.; Wang, Z.; Zhou, Y.; Chai, G.; Yan, G.; Jin, Z.; Wang, Q.; Sun, H. Orexin A Peptidergic System: Comparative Sleep Behavior, Morphology and Population in Brains between Wild Type and Alzheimer's Disease Mice. *Brain Struct. Funct.* **2022**, *227*, 1051–1065. [CrossRef]
163. Liu, M.-F.; Xue, Y.; Liu, C.; Liu, Y.-H.; Diao, H.-L.; Wang, Y.; Pan, Y.-P.; Chen, L. Orexin-A Exerts Neuroprotective Effects via OX1R in Parkinson's Disease. *Front. Neurosci.* **2018**, *12*, 415023. [CrossRef]

164. Shimizu, S.; Takenoshita, N.; Inagawa, Y.; Tsugawa, A.; Hirose, D.; Kaneko, Y.; Ogawa, Y.; Serisawa, S.; Sakurai, S.; Hirao, K.; et al. Positive Association Between Cognitive Function and Cerebrospinal Fluid Orexin A Levels in Alzheimer's Disease. *J. Alzheimer's Dis.* **2020**, *73*, 117–123. [CrossRef] [PubMed]
165. Fronczek, R.; van Geest, S.; Frölich, M.; Overeem, S.; Roelandse, F.W.C.; Lammers, G.J.; Swaab, D.F. Hypocretin (Orexin) Loss in Alzheimer's Disease. *Neurobiol. Aging* **2012**, *33*, 1642–1650. [CrossRef] [PubMed]
166. Villano, I.; La Marra, M.; Di Maio, G.; Monda, V.; Chieffi, S.; Guatteo, E.; Messina, G.; Moscatelli, F.; Monda, M.; Messina, A. Physiological Role of Orexinergic System for Health. *Int. J. Environ. Res. Public Health* **2022**, *19*, 8353. [CrossRef] [PubMed]
167. Sikder, D.; Kodadek, T. The Neurohormone Orexin Stimulates Hypoxia-Inducible Factor-1 Activity. *Genes Dev.* **2007**, *21*, 2995–3005. [CrossRef] [PubMed]
168. Gu, H.; Ru, Y.; Wang, W.; Cai, G.; Gu, L.; Ye, J.; Zhang, W.-B.; Wang, L. Orexin-A Reverse Bone Mass Loss Induced by Chronic Intermittent Hypoxia Through OX1R-Nrf2/HIF-1α Pathway. *Drug Des. Dev. Ther.* **2022**, *16*, 2145–2160. [CrossRef] [PubMed]
169. Wan, X.; Liu, Y.; Zhao, Y.; Sun, X.; Fan, D.; Guo, L. Orexin A Affects HepG2 Human Hepatocellular Carcinoma Cells Glucose Metabolism via HIF-1α-Dependent and -Independent Mechanism. *PLoS ONE* **2017**, *12*, e0184213. [CrossRef] [PubMed]
170. Feng, Y.; Liu, T.; Li, X.-Q.; Liu, Y.; Zhu, X.-Y.; Jankovic, J.; Pan, T.-H.; Wu, Y.-C. Neuroprotection by Orexin-A via HIF-1α Induction in a Cellular Model of Parkinson's Disease. *Neurosci. Lett.* **2014**, *579*, 35–40. [CrossRef] [PubMed]
171. Chai, X.; Kong, W.; Liu, L.; Yu, W.; Zhang, Z.; Sun, Y. A Viral Vector Expressing Hypoxia-Inducible Factor 1 Alpha Inhibits Hippocampal Neuronal Apoptosis. *Neural Regen. Res.* **2014**, *9*, 1145. [CrossRef] [PubMed]
172. Cao, R.; Du, F.; Cui, Y.; Qi, M.; Zhuang, J.; Wang, J.; Zhang, M.; Zhang, X.; Liu, Z.; Zou, L.; et al. Synthesis and Biological Evaluations of 8-Biaryl-2,2-Dimethylbenzopyranamide Derivatives against Alzheimer's Disease and Ischemic Stroke. *Bioorganic Chem.* **2024**, *143*, 107064. [CrossRef]
173. Hai, Y.; Ren, K.; Zhang, Y.; Yang, L.; Cao, H.; Yuan, X.; Su, L.; Li, H.; Feng, X.; Liu, D. HIF-1α Serves as a Co-Linker between AD and T2DM. *Biomed. Pharmacother.* **2024**, *171*, 116158. [CrossRef]
174. Lei, L.; Feng, J.; Wu, G.; Wei, Z.; Wang, J.-Z.; Zhang, B.; Liu, R.; Liu, F.; Wang, X.; Li, H.-L. HIF-1α Causes LCMT1/PP2A Deficiency and Mediates Tau Hyperphosphorylation and Cognitive Dysfunction during Chronic Hypoxia. *Int. J. Mol. Sci.* **2022**, *23*, 16140. [CrossRef] [PubMed]
175. Kapogiannis, D.; Boxer, A.; Schwartz, J.B.; Abner, E.L.; Biragyn, A.; Masharani, U.; Frassetto, L.; Petersen, R.C.; Miller, B.L.; Goetzl, E.J. Dysfunctionally Phosphorylated Type 1 Insulin Receptor Substrate in Neural-derived Blood Exosomes of Preclinical Alzheimer's Disease. *FASEB J.* **2015**, *29*, 589–596. [CrossRef] [PubMed]

International Journal of
Molecular Sciences

MDPI

Article

N-Stearidonoylethanolamine Restores CA1 Synaptic Integrity and Reduces Astrocytic Reactivity After Mild Traumatic Brain Injury

Anastasia Egoraeva, Igor Manzhulo, Darya Ivashkevich and Anna Tyrtyshnaia *

A.V. Zhirmunsky National Scientific Center of Marine Biology, Far Eastern Branch, Russian Academy of Sciences, 690041 Vladivostok, Russia; egoraeva_a@bk.ru (A.E.); i-manzhulo@bk.ru (I.M.); owncean@yandex.ru (D.I.)
* Correspondence: anna.tyrtyshnaia@bk.ru

Abstract

Mild traumatic brain injury (mTBI) disrupts hippocampal network function through coordinated alterations in glial reactivity, synaptic integrity, and adult neurogenesis. Effective therapeutic approaches targeting these interconnected processes remain limited. Lipid-derived molecules capable of modulating these mTBI-induced disturbances are emerging as promising neuroprotective candidates. Here, we investigated the effects of N-stearidonylethanolamine (SDEA), an ω-3 ethanolamide, in a mouse model of mTBI. SDEA treatment attenuated astrocytic reactivity, restored Arc expression, and improved dendritic spine density and morphology in the CA1 hippocampal area. In the dentate gyrus, mTBI reduced Ki-67-indexed proliferation while leaving DCX-positive immature neurons unchanged, and SDEA partially rescued proliferative activity. These effects were accompanied by improvements in anxiety-like behavior and working-memory performance. Together, these findings demonstrate that SDEA modulates several key components of the glia-synapse-neurogenesis axis and supports functional recovery of hippocampal circuits following mTBI. These results suggest that ω-3 ethanolamides may represent promising candidates for multi-target therapeutic strategies in mTBI.

Keywords: mild traumatic brain injury; N-stearidonoylethanolamine; astrocytic reactivity; neuroinflammation; dendritic spine plasticity; glia-synapse interactions; anxiety-like behavior

1. Introduction

Mild traumatic brain injury (mTBI) is a highly prevalent form of brain trauma and a major contributor to long-term neuropsychiatric complications, including anxiety, cognitive decline, and emotional dysregulation [1,2]. To model these processes, we employed a well-established closed-head mTBI paradigm that produces persistent functional and molecular alterations in the hippocampus without inducing gross tissue damage. This model is widely used to mimic key features of human mTBI, including long-term cognitive and emotional disturbances, glial reactivity, and synaptic dysfunction, while preserving overall brain architecture [3,4]. Despite its high incidence, the cellular and molecular mechanisms underlying persistent dysfunction after mTBI remain only partially understood, and no effective disease-modifying therapies are currently available. A growing body of evidence indicates that even mild injury triggers prolonged disturbances in hippocampal circuits, leading to alterations in glutamate homeostasis, synaptic plasticity, and neuroinflammatory signaling [3,4]. These changes are frequently dissociated from overt neuronal loss, suggesting that subtle structural and glial-mediated processes play a key role in the delayed emergence of behavioral deficits.

Astrocytes represent a central hub in this cascade. Following mTBI, astrocytes undergo pronounced reactive remodeling characterized by GFAP upregulation, NF-κB activation, cytokine release, and disruption of metabolic and glutamate-buffering functions [5–8]. This reactive state contributes to synaptic vulnerability by amplifying inflammatory signaling and impairing the stability of excitatory synapses, particularly in CA1, which is known to be highly sensitive to metabolic and inflammatory stress [9,10]. Activation of astrocytes and engagement of inflammasome pathways (multiprotein innate immune complexes that regulate maturation of pro-inflammatory cytokines) are critical mechanisms in the pathophysiology of TBI, influencing the development of secondary injury and determining long-term neurological outcomes [11]. Astrocyte-derived IL-6 can further modify PSD-95 palmitoylation (a post-translational lipid modification that regulates membrane anchoring and synaptic retention of PSD-95) and enhance the synaptic recruitment of GluR1-containing AMPA receptors, demonstrating a direct glia-to-synapse control of AMPAR localization [12]. In neuronal cultures, pro-inflammatory cytokines such as IL-1β decrease dendritic spine density and alter PSD-95 distribution without inducing neuronal death, indicating a primarily synaptic level of injury, and inflammatory molecules released from mechanically injured astrocytes similarly induce presynaptic loss in cortical networks [9,13]. Together, these studies support the view that astrocyte-driven inflammation can disrupt synaptic scaffolds, AMPA receptor organization, and activity-dependent plasticity programs such as Arc (activity-regulated cytoskeleton-associated protein, an immediate early gene critically involved in AMPA receptor trafficking and activity-dependent synaptic plasticity), thereby weakening hippocampal network connectivity even when gross neuronal survival is preserved. Parallel alterations in hippocampal neurogenesis, including reduced proliferation of dentate gyrus progenitors, further constrain recovery of circuit function after injury [14].

Against this background, lipid-derived signaling molecules capable of modulating glial reactivity and synaptic stability have attracted increasing attention as potential therapeutic candidates after mild traumatic brain injury. In particular, ethanolamides derived from ω-3 polyunsaturated fatty acids represent a biologically active class of endogenous lipid mediators with demonstrated neuroprotective and anti-inflammatory properties. Among them, ω-3 ethanolamides such as N-docosahexaenoylethanolamine (synaptamide, DHEA) and eicosapentaenoylethanolamide (EPEA) have attracted particular interest due to their anti-inflammatory, neurotrophic, and synapse-stabilizing properties [15–19]. Several studies also indicate that ω-3 PUFAs and their ethanolamide derivatives, such as DHEA and EPEA, can modulate neurogenic processes by improving the inflammatory and trophic milieu within the dentate gyrus [16,20], although their role in mTBI has not yet been systematically examined. To date, it remains unclear whether ω-3 ethanolamides can simultaneously modulate astrocytic reactivity, synaptic integrity, neurogenic responses, and behavioral outcomes after mTBI.

In this study we examine how N-stearidonoylethanolamine (SDEA, an ethanolamide of stearidonic acid, 18:4, n-3) influences hippocampal function after mTBI by assessing multiple levels of circuit regulation. SDEA is a less studied member of the ω-3 ethanolamide family. Although direct data on SDEA biological activity are limited, its structural similarity to other ω-3 ethanolamides suggests that it may share key functional properties, including modulation of neuroinflammatory signaling, regulation of glial reactivity, and stabilization of synaptic architecture [18,19]. Stearidonic acid serves as a metabolic precursor of longer-chain ω-3 fatty acids, and its ethanolamide derivative may act as a bioactive lipid mediator influencing glia-neuron interactions [21]. Based on the established actions of DHEA and

https://doi.org/10.3390/ijms27010471

EPEA, we hypothesized that SDEA could attenuate astrocytic activation, preserve synaptic integrity, and support activity-dependent plasticity following mTBI.

We evaluated the effects of SDEA on glial reactivity, immediate early gene activity, synaptic organization within the CA1 region and dentate gyrus, proliferative activity within the subgranular zone (SGZ) of the dentate gyrus (DG), a region associated with adult hippocampal neurogenesis, as well as behavioral performance in recognition memory, working memory, and anxiety-related paradigms. This integrative analysis combines molecular, structural, and behavioral indicators to characterize the glia-synapse-neurogenesis axis under injury conditions and identifies SDEA as a potential multi-target modulator of hippocampal recovery.

2. Results

2.1. Behavioral Outcomes: Cognitive and Emotional Consequences of mTBI and Treatment

2.1.1. Elevated Plus Maze

Given that mild traumatic brain injury often produces subtle behavioral alterations without overt cognitive deficits, we first examined anxiety-related behavior to establish functional consequences of injury and treatment. The effects of mTBI and treatment with SDEA on anxiety-related behavior were assessed in the elevated plus maze (EPM) [22] (Figure 1a). A two-way ANOVA was performed separately for the closed arms, open arms, and central zone.

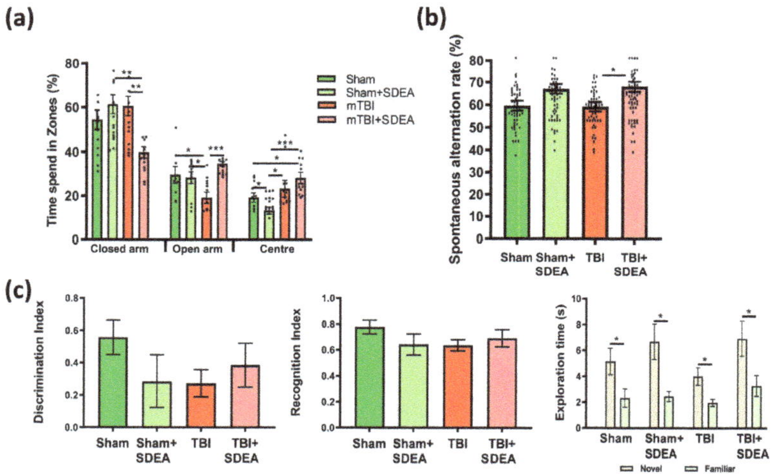

Figure 1. Effects of mild traumatic brain injury (mTBI) and SDEA treatment on anxiety-like behavior, working memory, and recognition memory in mice. (**a**) Elevated Plus Maze. The percentage of time spent in the closed arms, open arms, and center of the maze is shown for the four experimental groups: "Sham", "Sham + SDEA", "mTBI", and "mTBI + SDEA" (n = 14 per group). Individual data points are overlaid on the box plot. Data are expressed as mean ± SEM (%). Two-way ANOVA. Post hoc Tukey test: * $p < 0.05$, ** $p < 0.01$, *** $p < 0.001$. (**b**) Y-maze spontaneous alternation test. The percentage of spontaneous alternations (%). Because Y-maze alternation was measured across multiple sessions (4 trials × 14 animals), n was 56 per group. Values are shown as mean ± SEM. Two-way ANOVA. Post hoc Tukey test: * $p < 0.05$. (**c**) Novel Object Recognition (NOR) test. The graphs show the Discrimination Index, the Recognition Index, and the exploration time (s) for novel and familiar objects in the four groups. Within-group comparisons revealed significant differences between novel and familiar object exploration in all experimental groups (* $p < 0.05$), n = 14 per group. Data are expressed as mean ± SEM (%). Two-way ANOVA. Post hoc Tukey test.

Closed Arms

The percentage of time spent in the closed arms did not significantly differ between groups "Sham" (54.42 ± 4.58%) and "mTBI" (60.61 ± 4.40%) groups, indicating that mTBI alone did not substantially alter avoidance behavior ($F_{1, 52} = 3.89$, $p = 0.054$). SDEA administration produced a statistically significant decrease in closed-arm occupancy in injured animals ("mTBI": 60.61 ± 4.40% vs. "mTBI + SDEA": 39.89 ± 2.56%, $p = 0.003$). Although the main effect of treatment did not reach statistical significance ($F_{1, 52} = 3.26$, $p = 0.078$), a strong interaction was detected ($F_{1, 52} = 12.77$, $p = 0.0008$), indicating that TBI-induced behavioral deficits were restored in SDEA-treated animals.

Open Arms

mTBI animals exhibited a clear behavioral tendency toward reduced open-arm exploration compared with sham controls ("Sham": 29.63 ± 3.95%; "mTBI": 19.13 ± 2.73%). SDEA treatment increased open-arm time in mTBI animals ("mTBI + SDEA": 34.66 ± 1.15%), restoring exploration after trauma. Two-way ANOVA revealed a significant main effect of treatment ($F_{1, 52} = 7.30$, $p = 0.0096$) and interaction between trauma and treatment ($F_{1, 52} = 10.06$, $p = 0.0027$). These results demonstrate that SDEA exerts an anxiolytic-like effect, counteracting the trauma-induced reduction in open-arm exploration.

Central Zone

In the central zone, mTBI animals did not differ significantly from sham controls, indicating that traumatic injury alone did not alter exploratory behavior in this area. The mean percentage of time spent in the center was 19.21 ± 2.23% in Sham and 23.31 ± 3.81% in mTBI animals ($p > 0.05$). SDEA exerted bidirectional, trauma-dependent effects: it slightly reduced central exploration in sham mice ("Sham + SDEA": 13.24 ± 1.59%) but increased it in injured animals ("mTBI + SDEA": 28.21 ± 2.62%) compared to controls. Two-way ANOVA confirmed a significant main effect of trauma ($F_{1, 52} = 12.94$, $p = 0.001$) and a significant interaction between trauma and treatment ($F_{1, 52} = 4.21$, $p = 0.0458$), while the main effect of treatment alone was not significant ($F_{1, 52} = 0.036$, $p = 0.85$). These data suggest that SDEA selectively increased central-zone exploration in injured animals while slightly suppressing it in intact mice, indicating a state-dependent, bidirectional modulation of risk-assessment behavior rather than a uniform anxiolytic action.

Together, these findings demonstrate that mTBI induces an anxiety-like phenotype, characterized by reduced exploration of open zones. SDEA treatment attenuated these alterations, restoring the balance between risk-taking and avoidance behavior. This normalization of exploratory activity supports the neuroprotective and anxiolytic potential of SDEA following mild traumatic brain injury.

2.1.2. Y-Maze

The Y-maze spontaneous alternation test measures working memory and exploratory strategy, relying on the mouse's innate tendency to alternate between arms [23] (Figure 1b). It is a low-stress task sensitive to hippocampal-prefrontal coordination and well suited for mild TBI models, which may show subtle network inefficiency rather than overt cognitive loss.

mTBI did not significantly affect spontaneous alternation performance, whereas SDEA treatment significantly increased alternation rate in injured animals. Mean values were 59.56 ± 2.15% for "Sham", 66.27 ± 2.28% for "Sham-SDEA", 59.28 ± 2.01% "mTBI", and 67.30 ± 2.35% "mTBI-SDEA". Thus, SDEA improved working-memory-related behavior independent of trauma status ($F_{1, 220}$) = 11.12, $p = 0.001$.

These findings show that mild injury alone did not impair working memory, consistent with literature reporting preserved Y-maze alternation after mTBI, while SDEA enhanced hippocampal network efficiency.

https://doi.org/10.3390/ijms27010471

2.1.3. Novel Object Recognition

The novel object recognition (NOR) test was employed to assess long-term recognition memory, a form of recognition-based memory that depends on hippocampal-cortical interactions, including the perirhinal, entorhinal, and hippocampal regions [24] (Figure 1c).

Two-way ANOVA revealed no significant main effects of Trauma ($F_{1,52} = 0.52$, $p = 0.47$) or Treatment ($F_{1,52} = 0.40$, $p = 0.53$), and no factors' interaction ($F_{1,52} = 2.32$, $p = 0.13$) for either the discrimination or recognition indices. Likewise, exploration times for the familiar and novel objects did not differ significantly between groups ($p > 0.22$). Within-group analysis demonstrated that mice in all experimental groups spent significantly more time exploring the novel object compared to the familiar one ($p < 0.05$), indicating preserved object recognition performance across groups.

The absence of changes in NOR performance indicates that recognition memory was largely unaffected by mild traumatic brain injury in this model. These findings highlight a dissociation between affective and cognitive outcomes of mTBI: while emotional regulation and synaptic plasticity are selectively vulnerable, baseline recognition memory and neuronal differentiation are resilient.

As expected for a mild TBI model, behavioral alterations were subtle and did not manifest as severe deficits across all parameters.

2.2. *Effect of mTBI and SDEA Treatment on Hippocampal Neurogenesis*

Because behavioral changes may reflect underlying alterations in hippocampal cellular plasticity, we next examined markers of adult neurogenesis in the dentate gyrus.

Ki-67-positive proliferating cells (Figure 2a) and DCX-positive cells (Figure 2b), representing immature neuronal populations, were analyzed within the SGZ of the DG, a region associated with adult hippocampal neurogenesis. Ki-67 is a marker of proliferating cells across all active phases of the cell cycle, whereas DCX labels immature neurons and reflects early neuronal differentiation within the adult neurogenic niche [25,26]. These markers allow phase-specific assessment of the neurogenic process.

For ki-67, two-way ANOVA revealed a strong main effect of trauma ($F_{1,96} = 12.35$, $p = 0.0007$) and a significant interaction between trauma and treatment ($F_{1,96} = 6.41$, $p = 0.013$). mTBI markedly reduced the number of Ki-67-positive cells compared with sham controls ("mTBI": 317.3 $\pm$ 57.0 vs. "Sham": 1288.4 $\pm$ 275.5, cells/mm^3, mean $\pm$ SEM), indicating post-traumatic suppression of cell proliferation in the subgranular zone (SGZ). SDEA treatment partially restored proliferative activity in injured mice (599.5 $\pm$ 87.5 cells/mm^3), while not affecting baseline proliferation in sham animals (757.1 $\pm$ 127.8 cells/mm^3). Tukey's post hoc test confirmed a significant difference between "Sham" and "mTBI" ($p = 0.0003$) and between "Sham" and "mTBI + SDEA" ($p = 0.0163$) (Figure 2c). These data demonstrate that mTBI strongly inhibits hippocampal cell proliferation, whereas SDEA exerts a restorative effect, counteracting trauma-induced reduction in Ki-67 expression and suggesting protection of the neurogenic niche.

Analysis of DCX-positive cells revealed a different pattern. Two-way ANOVA showed no significant main effect of trauma ($F_{1,96} = 1.34$, $p = 0.25$) and no factors' interaction ($F_{1,96} = 0.06$, $p = 0.80$), though there was a near-significant trend for a treatment effect ($F_{1,96} = 2.97$, $p = 0.088$). Post hoc analysis did not reveal statistically significant group differences ($p > 0.05$) (Figure 2d). Thus, DCX marks late progenitors and immature neurons, which remained stable across groups, despite the pronounced reduction in Ki-67-positive proliferating cells.

The divergent responses of Ki-67 and DCX indicate a phase-dependent modulation of hippocampal neurogenesis following mTBI. Because the assessment was performed 11 weeks after injury, the acute inflammatory effects of mTBI likely exerted their strongest

 https://doi.org/10.3390/ijms27010471

impact on the early proliferative phase, reducing Ki-67⁺ cell numbers, whereas the pool of DCX⁺ immature neurons, which reflects a later, post-mitotic stage of neurogenesis, appears largely preserved at this chronic time point. This dissociation likely reflects the temporal delay between cell proliferation and neuronal maturation [25,26]. Together, these results support a phasic model in which mTBI transiently disrupts the earliest stage of neurogenesis, whereas SDEA may exert stage-specific effects by partially restoring proliferation while preserving immature neuron numbers required for hippocampal plasticity and cognitive recovery.

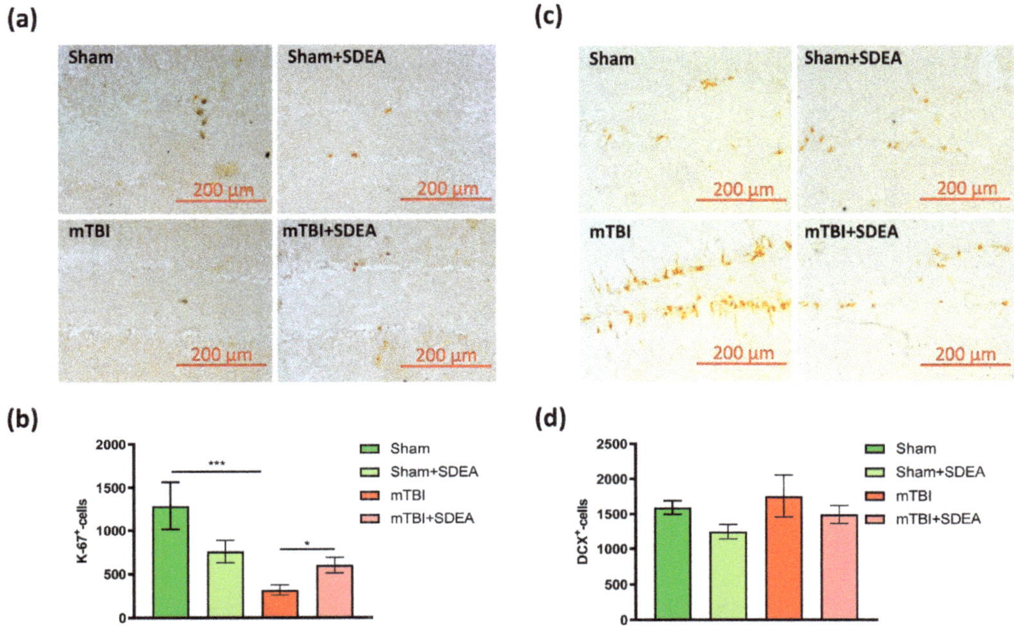

Figure 2. Immunohistochemical analysis of hippocampal neurogenesis markers Ki-67 and DCX after mTBI and SDEA treatment. (**a**,**b**) Ki-67 immunoreactivity in the dentate gyrus. Representative micrographs (**a**) show Ki-67⁺ proliferating cells (brown nuclei) in the subgranular zone of the dentate gyrus. Scale bar = 200 µm. Quantitative analysis (**b**) shows the number of Ki-67⁺ cells/mm^3 (n = 25 slices per group). Data are expressed as mean ± SEM. Two-way ANOVA, Tukey's post hoc test (* $p < 0.05$, *** $p < 0.001$). (**c**,**d**) DCX immunoreactivity in the dentate gyrus. Representative images (**c**) show DCX⁺ immature neurons in the subgranular zone of the dentate gyrus. Scale bar = 200 µm. Quantification (**d**) shows the number of DCX⁺ cells/mm^3 (n = 25 slices per group). Data are expressed as mean ± SEM. Two-way ANOVA, Tukey's post hoc test.

2.3. *Effect of mTBI and SDEA Treatment on Arc Production*

Beyond cell proliferation and differentiation, successful neurogenesis requires the functional integration of newborn neurons into existing hippocampal networks. Although functional integration of newborn neurons is a component of adult neurogenesis, immediate-early gene Arc (activity-regulated cytoskeleton-associated protein) expression was assessed here as a general marker of activity-dependent plasticity within the dentate gyrus, without implying cell-type specificity. This integration depends on activity-regulated synaptic mechanisms that enable newly formed granule cells to participate in memory encoding [27,28]. Therefore, after characterizing proliferation and maturation markers (Ki-67, DCX), we next evaluated Arc expression in the dentate gyrus to assess whether the observed changes in neurogenesis are accompanied by alterations in activity-dependent plasticity.

 https://doi.org/10.3390/ijms27010471

We measured the number of Arc-positive cells in the dentate gyrus across four groups (Figure 3a). After mTBI, Arc expression markedly decreased compared to sham animals (1948.8 ± 294.4 vs. 4009.8 ± 521.4, cells/mm^3, p = 0.0004). Treatment with SDEA restored Arc levels in the injured group (3219.1 ± 345.0 cells/mm^3), which were significantly higher than in "mTBI" mice (p = 0.0078) and became comparable to sham controls (p = 0.400). In contrast, SDEA had little effect in sham animals (3558.6 ± 199.4 vs. 4009.8 ± 521.4, cells/mm^3, p = 0.426). Overall, these findings show that mTBI strongly suppresses Arc expression in the DG, and SDEA selectively restores Arc after injury, without changing baseline Arc levels in uninjured mice. The two-way ANOVA revealed a significant main effect of Trauma ($F_{1, 96}$ =11.69, p = 0.00095), no significant main effect of Treatment ($F_{1, 96}$ =1.12, p = 0.293), and a significant factors' interaction ($F_{1, 96}$ =5.96, p = 0.0166) (Figure 3b).

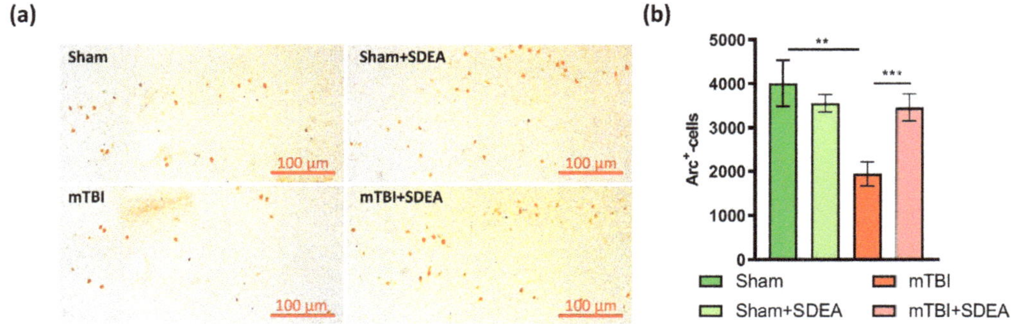

Figure 3. Immunohistochemical detection of Arc$^+$ cells in the dentate gyrus after mTBI and SDEA treatment. (**a**) Representative micrographs showing Arc$^+$-immunopositive cells (brown nuclei) in the DG. Scale bar = 100 μm. (**b**) Quantification of Arc$^+$ cells/mm^3 (n = 25 slices per group). Data are shown as mean ± SEM (cells/mm^3). Two-way ANOVA, Tukey's post hoc test (** $p < 0.01$, *** $p < 0.001$).

These findings show that mTBI strongly suppresses Arc expression, consistent with evidence that reduced Arc reflects weakened excitatory drive and impaired experience-dependent synaptic remodeling [27]. SDEA selectively reversed this deficit in injured animals, suggesting that it enhances synaptic activation and promotes re-engagement of dentate granule cells in hippocampal processing. The pattern in which Arc expression decreases in parallel with reduced proliferation (Ki-67) and recovers after SDEA is consistent with proposals that Arc interacts with activity-dependent processes that may indirectly influence neurogenic integration [28]. Together, these results indicate that SDEA restores activity-dependent signaling pathways that are crucial for functional recruitment of hippocampal circuits after mild brain injury.

2.4. Spine Remodeling After mTBI and SDEA Treatment

Since neurogenic and synaptic processes are tightly coupled within hippocampal networks, we subsequently assessed dendritic spine architecture and synaptic protein expression. Dendritic spine morphology reflects the structural basis of synaptic plasticity and functional connectivity in the hippocampus. Mature mushroom spines are associated with stable, high-conductance excitatory synapses, whereas thin spines and filopodia represent dynamic, plastic structures involved in learning-related remodeling [29,30]. mTBI often disrupts this balance, reducing mature synapses and promoting unstable spine types, which contribute to long-term synaptic inefficiency. To assess whether SDEA treatment could counteract these chronic morphological alterations, we analyzed dendritic spine density in hippocampal neurons 11 weeks after injury.

https://doi.org/10.3390/ijms27010471

2.4.1. CA1 Apical Dendrites

Two-way ANOVA revealed that both trauma and treatment significantly affected the total density of dendritic spines in CA1 apical dendrites (Trauma: $F_{1,56} = 12.45$, $p = 0.0010$; Treatment: $F_{1,56} = 5.60$, $p = 0.0225$), with no interaction between the factors ($p = 0.39$). TBI significantly reduced total spine density ("Sham": 2.99 ± 0.09 vs. "mTBI": 2.2 ± 0.12 spines/µm; $p = 0.004$), whereas treatment with SDEA partially restored it ("mTBI + SDEA": 2.72 ± 0.15; $p = 0.047$ vs. "mTBI") reaching levels not different from controls ($p = 0.61$) (Figure 4a).

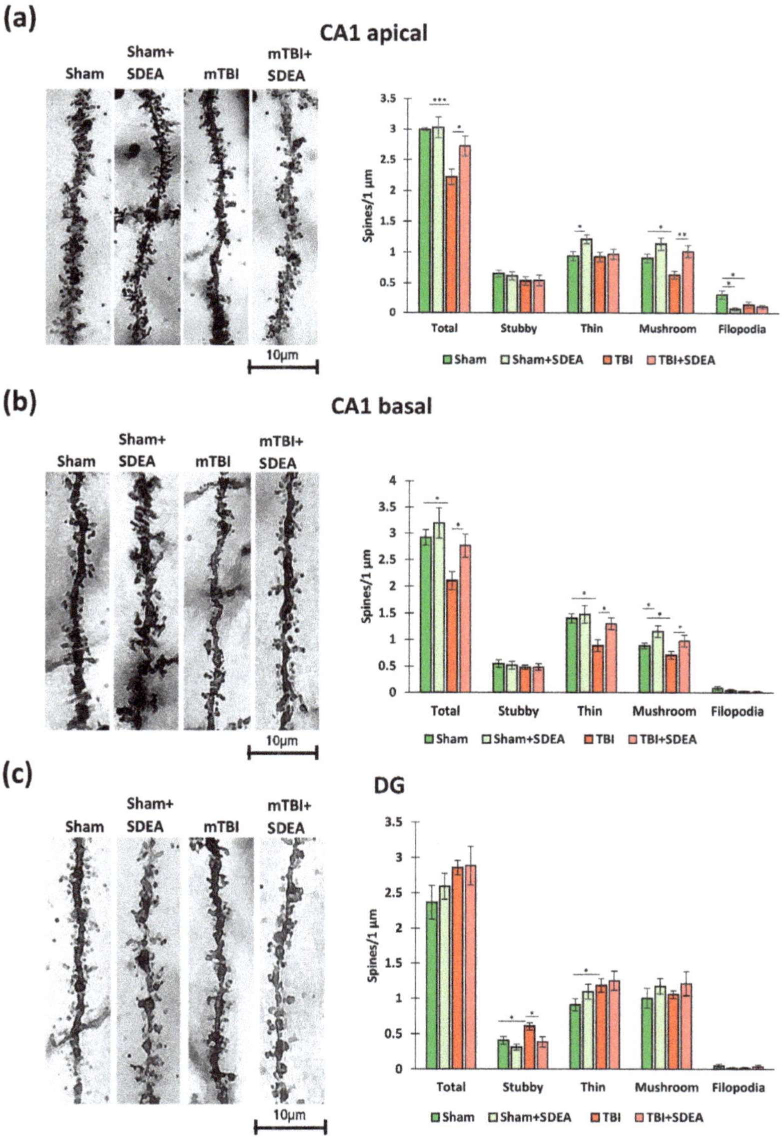

Figure 4. Dendritic spine morphology in apical and basal dendrites of CA1 pyramidal neurons and dendrites of dentate gyrus granule cells after mTBI and SDEA treatment. (**a**) CA1 apical dendrites.

https://doi.org/10.3390/ijms27010471

Representative Golgi-stained apical dendritic segments from the four experimental groups (**left**). Scale bar = 10 µm. Quantification of total spine density and spine subtypes (stubby, thin, mushroom, filopodia) is shown as spines/µm (n = 15 dendrites per group from 3 mice) (**right**). Data are expressed as mean ± SEM. Two-way ANOVA, Tukey's post hoc test (* $p < 0.05$, ** $p < 0.01$, *** $p < 0.001$). (**b**) CA1 basal dendrites. Representative Golgi-stained basal dendritic segments from the four experimental groups (**left**). Scale bar = 10 µm. Quantification of total spine density and spine subtypes (stubby, thin, mushroom, filopodia) is shown as spines/µm (n = 15 dendrites per group from 3 mice) (**right**). Data are expressed as mean ± SEM. Two-way ANOVA, Tukey's post hoc test (* $p < 0.05$). (**c**) Dentate gyrus (DG) dendrites. Representative Golgi-stained dendritic segments from DG granule neurons (**left**). Scale bar = 10 µm. Quantification of total spine density and spine subtypes (stubby, thin, mushroom, filopodia) is shown as spines/µm (n = 15 dendrites per group from 3 mice) (**right**). Data are expressed as mean ± SEM. Two-way ANOVA, Tukey's post hoc test (* $p < 0.05$).

For mushroom dendritic spines, both trauma ($F_{1,56} = 5.90$, $p = 0.019$) and treatment ($F_{1,56} = 14.74$, $p = 0.0004$) had strong effects. Mushroom spine density dropped after trauma ("Sham": 0.91 ± 0.07 vs. "mTBI": 0.64 ± 0.06; spines/µm, $p = 0.003$) and significantly increased under SDEA ("TBI + SDEA": 1.02 ± 0.09; $p = 0.012$ vs. "mTBI"), suggesting partial restoration of mature synapses.

For thin dendritic spines, treatment exerted a significant effect ($F_{1,56} = 4.85$, $p = 0.033$). SDEA elevated thin spine density in predominantly in sham animals ("Sham + SDEA": 1.21 ± 0.07 vs. "Sham": 0.94 ± 0.07; spines/µm, $p = 0.041$). These changes may reflect enhanced structural plasticity and turnover of excitatory synapses.

For stubby dendritic spines, no significant effects of trauma or treatment were found. The mean values remained stable (0.53–0.65 spines/µm), indicating that transitional spine forms are relatively resistant to both injury and pharmacological modulation.

For filopodia dendritic spines, a significant treatment effect ($F_{1,56} = 9.88$, $p = 0.003$) and Trauma × Treatment interaction ($F_{1,56} = 5.82$, $p = 0.020$) were observed. Filopodia density was highest in "Sham" (0.31 ± 0.06, spines/µm) and decreased under SDEA (0.07 ± 0.02, spines/µm; $p = 0.001$). This reduction may reflect stabilization of dendritic contacts and suppression of excessive spine turnover. Thus, mTBI caused a significant loss of mature mushroom spines and reduced overall spine density, while SDEA reversed these deficits and limited immature protrusions, suggesting structural stabilization and synaptic recovery in CA1 apical dendrites.

2.4.2. CA1 Basal Dendrites

In CA1 basal dendrites, two-way ANOVA revealed significant main effects of both trauma and treatment on the total dendritic spine density in CA1 basal dendrites (Trauma: $F_{1,56} = 9.36$, $p = 0.0042$; Treatment: $F_{1,56} = 5.27$, $p = 0.0276$), with no significant interaction ($p = 0.35$). Following trauma, total spine density decreased from 2.92 ± 0.14 spines/µm in "Sham" to 2.10 ± 0.16 spines/µm in "mTBI" ($p = 0.006$). SDEA administration partially restored spine density to 2.76 ± 0.21 spines/µm ($p = 0.041$ vs. "mTBI"), approaching control levels ("Sham + SDEA": 3.19 ± 0.28) (Figure 4b).

For mushroom dendritic spines, significant main effects of both trauma ($F_{1,56} = 4.28$, $p = 0.046$) and treatment ($F_{1,56} = 9.83$, $p = 0.0034$) were detected. Mushroom spine density dropped after trauma ("mTBI": 0.710 ± 0.069 spines/µm vs. "Sham": 0.885 ± 0.051 spines/µm; $p = 0.028$), while SDEA increased this parameter both in injured ("mTBI + SDEA": 0.976 ± 0.105 spines/µm; $p = 0.039$ vs. "mTBI") and sham animals ("Sham + SDEA": 1.158 ± 0.105 spines/µm; $p = 0.021$ vs. "Sham"). These results indicate that SDEA enhances maintenance and restoration of mature, mushroom-type synapses.

Both trauma ($F_{1,56} = 8.95$, $p = 0.0050$) and treatment ($F_{1,56} = 4.52$, $p = 0.0404$) significantly affected thin spines. Trauma reduced their density from 1.40 ± 0.08

https://doi.org/10.3390/ijms27010471

to 0.89 ± 0.10 spines/μm (p = 0.008), while treatment promoted partial recovery ("mTBI + SDEA": 1.30 ± 0.10; p = 0.041 vs. "mTBI"). Thin spines, being structurally plastic, appear highly responsive to both injury and pharmacological stimulation.

Neither trauma nor treatment significantly influenced stubby spine density (Trauma: $F_{1,56}$ = 0.68, p = 0.42; Treatment: $F_{1,56}$ = 0.04, p = 0.84). Mean values remained within a narrow range (0.47–0.54 spines/μm) across all groups, indicating relative structural stability of this intermediate spine form under both conditions.

Filopodia density was significantly affected by trauma ($F_{1,56}$ = 4.87, p = 0.034) but not by treatment (p = 0.24). Post hoc tests showed a reduction in "mTBI" animals (0.020 ± 0.014) compared to "Sham" (0.085 ± 0.031; p = 0.044), consistent with trauma-induced suppression of nascent spine formation. Treatment did not further alter this parameter ("mTBI + SDEA": 0.014 ± 0.014; p = 0.61 vs. "mTBI").

Thus, in basal dendrites of CA1 pyramidal neurons, mild traumatic brain injury caused a substantial reduction in total spine density, driven primarily by the loss of mushroom and thin spines. mSDEA treatment significantly mitigated these effects, restoring both mature (mushroom) and plastic (thin) spine populations. Stubby spines remained unaffected, while filopodia were reduced following trauma, suggesting suppressed synaptogenesis. Together, these data indicate that SDEA preserves dendritic integrity and promotes synaptic stability in basal dendrites after mTBI.

2.4.3. Dentate Gyrus Granule Cells

In DG granule neurons, two-way ANOVA showed a trend of increased total spine density after trauma ($F_{1,56}$ = 3.90, p = 0.056), without significant treatment effects (p = 0.52). Trauma increased mean density from 2.37 ± 0.23 ("Sham") to 2.86 ± 0.10 spines/μm ("TBI"; p = 0.059) insignificantly, consistent with reactive synaptogenesis in the DG (Figure 4c).

For mushroom dendritic spines, no significant differences were detected (p > 0.18). Densities remained stable across all groups (1.00–1.21 spines/μm), suggesting that mature synaptic contacts in the DG are less affected by trauma or treatment than in CA1.

Trauma significantly increased thin spine density ($F_{1,56}$ = 4.69, p = 0.037), whereas treatment had no independent effect (p = 0.21). TBI animals exhibited higher thin spine numbers (1.19 ± 0.09 vs. 0.91 ± 0.08 spines/μm; p = 0.049), and treatment further elevated them (1.26 ± 0.13 spines/μm; p = 0.041 vs. "TBI"). This enhancement may reflect adaptive plasticity and structural compensation in granule cells.

For stubby dendritic spines, both Trauma ($F_{1,56}$ = 6.62, p = 0.014) and Treatment ($F_{1,56}$ = 8.88, p = 0.005) were significant. Stubby spine density rose after trauma (0.605 ± 0.047 vs. 0.407 ± 0.051 spines/μm; p = 0.013) and normalized with SDEA (0.386 ± 0.068 spines/μm; p = 0.031 vs. "TBI"), suggesting reduced structural immaturity.

For filopodia dendritic spines, no significant changes were found (p > 0.15). Filopodia remained rare (0.01–0.05 spines/μm), consistent with limited de novo spine formation in adult granule cells. Thus, DG neurons exhibited a trauma-induced increase in thin spines, coupled with elevated stubby forms, suggesting reactive plasticity. SDEA modulated this response by normalizing stubby spines, thus favoring a more flexible yet stable synaptic architecture.

2.5. SDEA Suppresses mTBI-Induced Astroglial Activation

Astrocytic activation is a hallmark of post-traumatic neuroinflammation and is characterized by increased GFAP expression, cellular hypertrophy, and altered metabolic and glutamatergic support [31,32]. Persistent GFAP elevation reflects chronic reactivity and is widely used as a biomarker of astroglial injury in both tissue and serum studies. To determine whether SDEA treatment modulates astroglial activation in the hippocampus

 https://doi.org/10.3390/ijms27010471

during the chronic phase after mild TBI, we quantified the GFAP-positive area in the DG (Figure 5a) and CA1 subregions (Figure 5b) 11 weeks post-injury.

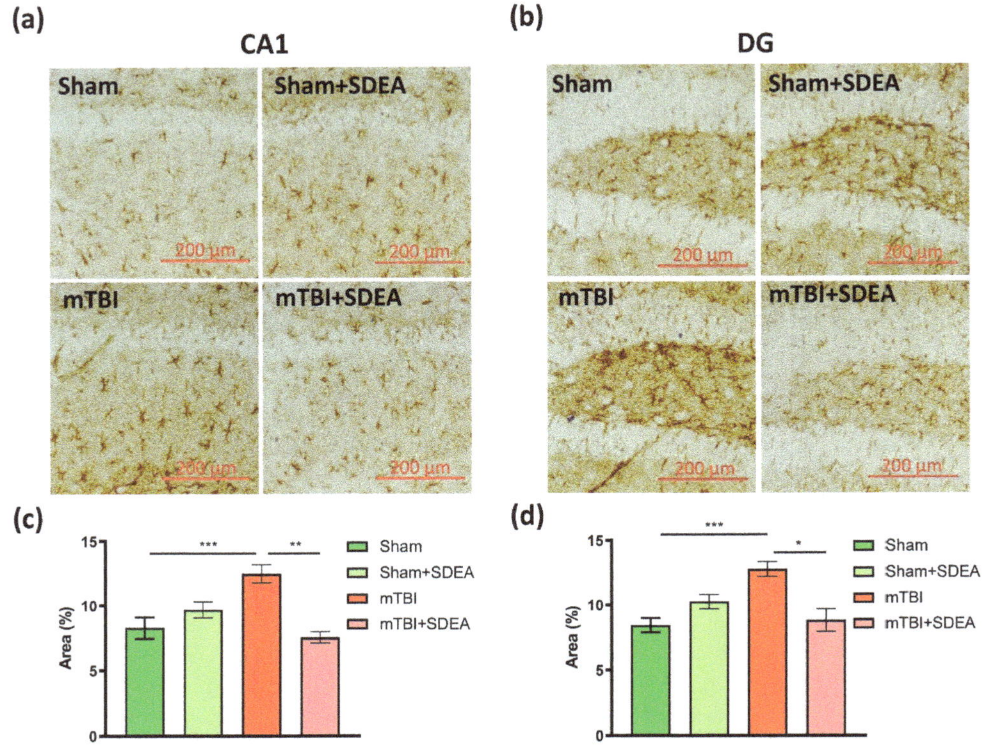

Figure 5. GFAP immunoreactivity in the CA1 region and dentate gyrus (DG) after mTBI and SDEA treatment. (**a**) Representative GFAP-stained astrocytes in the CA1 region of the hippocampus. Scale bar = 200 µm. (**b**) Representative GFAP immunostaining in the DG across the same experimental groups. Scale bar = 200 µm. (**c**) Quantification of GFAP-positive area (%) in CA1 (n = 25 slices per group). Data are presented as mean ± SEM. Two-way ANOVA. Tukey post hoc test (** $p < 0.01$, *** $p < 0.001$). (**d**) Quantification of GFAP-positive area (%) in DG (n = 25 slices per group). Data are presented as mean ± SEM. Two-way ANOVA. Tukey post hoc test (* $p < 0.05$, *** $p < 0.001$).

Quantitative assessment of GFAP-positive area in the CA1 region demonstrated a significant effect of treatment ($F_{1,96} = 7.27$, $p = 0.0084$) and a strong interaction between trauma and treatment ($F_{1,96} = 23.53$, $p < 0.00001$), indicating that SDEA efficiently modulated astroglial activation following mild TBI. The main effect of trauma was not significant ($p = 0.12$). In the "mTBI" group, the GFAP-positive area was markedly elevated (12.52 ± 0.68%), reflecting persistent astrocytic reactivity 11 weeks post-injury. SDEA treatment significantly reduced this value to 7.53 ± 0.43%, restoring it close to control levels. In sham-operated animals, SDEA caused no significant change compared with vehicle (9.69 ± 0.63% vs. 8.26 ± 0.83%), confirming that its effect was specific to post-traumatic conditions (Figure 5c).

In the DG region, two-way ANOVA showed a significant main effect of trauma ($F_{1,96} = 5.06$, $p = 0.027$), a tendency for treatment ($p = 0.10$), and a strong interaction between trauma and treatment ($F_{1,96} = 20.12$, $p < 0.0001$), indicating that SDEA significantly modulated astroglial reactivity in injured animals. In the "mTBI" group, the GFAP-positive area was markedly elevated (12.81 ± 0.56%), indicating persistent astrocytic activation

 https://doi.org/10.3390/ijms27010471

11 weeks after injury. Treatment with SDEA reduced this value to 8.81 ± 0.87%, restoring it toward control levels. In sham-operated mice, SDEA produced no significant change compared with vehicle (10.27 ± 0.57% vs. 8.42 ± 0.55%), confirming that its effect was specific to post-traumatic conditions (Figure 5d).

These results indicate that SDEA effectively attenuates long-term astroglial activation across the hippocampus, thereby supporting recovery of glia–neuron homeostasis and potentially facilitating synaptic and neurogenic repair after mild traumatic brain injury.

2.6. SDEA Normalizes BIN1 Expression and Restores Synaptic Marker Balance After mTBI

The structural and functional integrity of synapses critically depends on the surrounding glial environment, particularly astrocytes, which form the "tripartite synapse" and regulate glutamate homeostasis, stabilization of postsynaptic structures, and AMPA-mediated transmission. In the preceding section, we showed that mTBI induces pronounced astrocytic activation and an increase in GFAP-positive area in both CA1 and DG, indicating reactive remodeling of the glial component of the hippocampal network. Such glial changes can directly influence synaptic protein composition, as reactive astrocytes disrupt glutamate uptake, release pro-inflammatory mediators, and alter local metabolic conditions, processes that collectively lead to reductions in key structural proteins (Synaptophysin, PSD-95) and disturbances in AMPA receptor subunit expression [33]. AMPA receptor subunits GluR1 and GluR2 determine synaptic strength and Ca^{2+} permeability, with GluR2 conferring Ca^{2+}-impermeability and stabilizing mature synaptic contacts [34–36]. Disruption of these proteins after TBI reflects impaired structural maintenance of excitatory synapses. Therefore, the next step of our analysis was to quantify presynaptic (Synaptophysin) and postsynaptic density (PSD-95) proteins, as well as AMPA receptor subunits (GluR1/2), to determine how astrocytic reactivity in mTBI relates to alterations in synaptic architecture and excitatory transmission, and to assess the extent to which SDEA can restore these molecular indices.

2.6.1. Synaptophysin

Synaptophysin levels were significantly increased after mTBI compared to sham animals ("Sham": 1.00 ± 0.01; "mTBI": 2.14 ± 0.02, $p < 0.001$), indicating increased presynaptic vesicle density, potentially reflecting compensatory presynaptic remodeling. Importantly, SDEA treatment further increased synaptophysin expression in mTBI mice compared to untreated mTBI animals ($p < 0.001$), suggesting that SDEA potentiates injury-associated presynaptic remodeling (Figure 6a,b).

2.6.2. PSD-95

mTBI animals exhibited a significant reduction in PSD-95 protein levels compared to sham controls ("Sham": 1.00 ± 0.01; "mTBI": 0.66 ± 0.01, $p < 0.001$). SDEA treatment significantly increased PSD-95 expression in mTBI mice compared to untreated mTBI animals ($p < 0.001$), indicating attenuation of trauma-associated postsynaptic alterations (Figure 6a,b).

2.6.3. GluR1

mTBI was associated with a reduction in GluR1 protein expression relative to sham controls ("Sham": 1.27 ± 0.03; "mTBI": 0.81 ± 0.05, $p < 0.001$). SDEA treatment led to a further decrease in GluR1 levels compared to the "mTBI" group, suggesting additional modulation of AMPA receptor subunit composition during the post-injury period (Figure 6a,b). This additional reduction in GluR1 may reflect adaptive rebalancing of AMPA receptor composition rather than further synaptic impairment.

https://doi.org/10.3390/ijms27010471

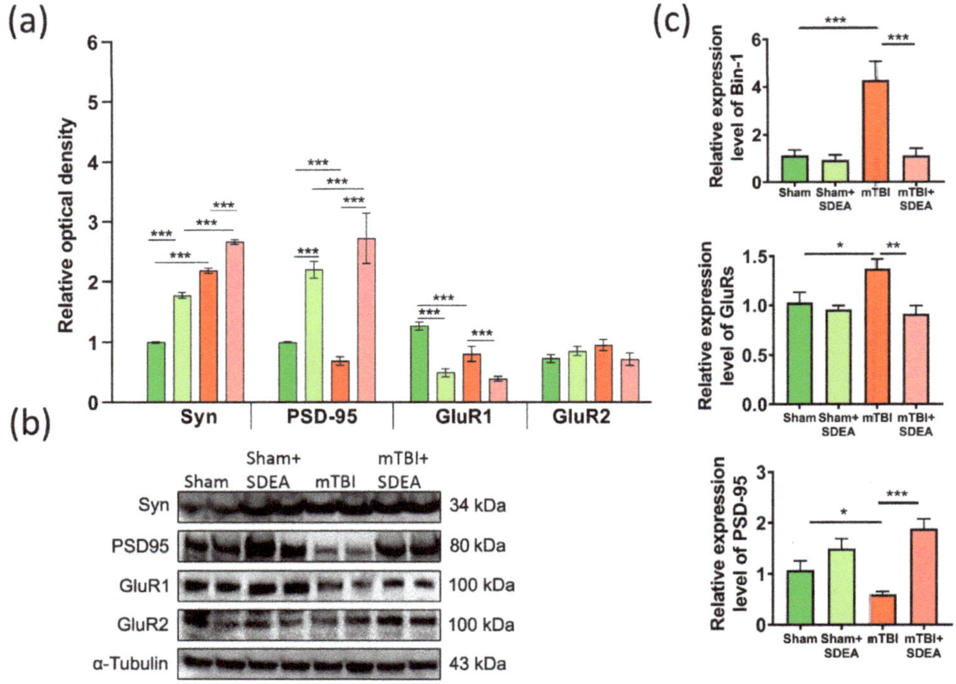

Figure 6. Expression of synaptic proteins and AMPA receptor subunits after mTBI and SDEA treatment. (**a**) Quantification of relative optical density of synaptic proteins Synaptophysin (Syn), PSD-95, and AMPA receptor subunits GluR1 and GluR2 across the four experimental groups. Data are shown as mean ± SEM (relative optical density, normalized to α-tubulin); $n = 6$ (animals per group). Two-way ANOVA. Tukey post hoc: *** $p < 0.001$. (**b**) Representative Western blot bands showing protein levels of Syn (~34 kDa), PSD-95 (~80 kDa), GluR1 (~100 kDa), GluR2 (~100 kDa), and loading control α-Tubulin (~43 kDa) from hippocampal homogenates of all experimental groups. (**c**) Relative mRNA expression of Bin1, GluRs (combined AMPA receptor subunits), and PSD-95 measured by qPCR. Values are presented as mean ± SEM (relative expression normalized to housekeeping genes); $n = 3$ (animals per group). Two-way ANOVA. Tukey post hoc: * $p < 0.05$, ** $p < 0.01$, *** $p < 0.001$.

2.6.4. GluR2

No significant differences in GluR2 protein levels were detected between "Sham", "mTBI", and "mTBI + SDEA" groups (Figure 6a,b), indicating that neither mild traumatic brain injury nor SDEA treatment significantly affected GluR2 expression at the protein level (Figure 6a,b).

Collectively, these data indicate that mTBI induces alterations in synaptic protein composition, including reduced PSD-95 and changes in AMPA receptor subunits. SDEA treatment attenuated several trauma-associated synaptic alterations, particularly at the level of postsynaptic scaffolding proteins and receptor composition, consistent with partial stabilization of excitatory synapses.

To evaluate transcriptional changes accompanying the synaptic protein alterations, we analyzed mRNA expression of Psd95, GluR subunits and Bin1 in the hippocampus by quantitative PCR. BIN1 (Bridging Integrator 1) protein was included in this study as a key regulator of membrane curvature and endocytic trafficking, processes that are crucial for maintaining synaptic structure and receptor dynamics. BIN1 has been repeatedly identified as one of the major genetic risk factors for Alzheimer's disease, where its dysregulation contributes to abnormal membrane remodeling, Tau pathology, and synaptic loss [37–39].

https://doi.org/10.3390/ijms27010471

Given the growing evidence that TBI accelerates neurodegenerative processes and shares common molecular pathways with Alzheimer's disease, assessing BIN1 expression provides important insight into how mild TBI affects neuronal membrane homeostasis and synaptic stability. By analyzing BIN1 together with synaptic markers (Synaptophysin, PSD-95, GluR), we aimed to determine whether the observed synaptic alterations after mTBI are linked to BIN1-dependent mechanisms and whether SDEA treatment can normalize this Alzheimer-like molecular signature.

2.6.5. Bin-1 mRNA

Two-way ANOVA revealed significant main effects of trauma ($F_{1,8} = 13.34$, $p = 0.0016$), treatment ($F_{1,8} = 13.53$, $p = 0.0015$), and a significant interaction ($F_{1,8} = 10.50$, $p = 0.0041$). Bin-1 expression was markedly increased in "mTBI" animals (4.28 ± 0.80, SEM) compared to sham controls (1.13 ± 0.23, $p < 0.001$), while SDEA reduced Bin-1 mRNA to near-control levels (1.12 ± 0.31, $p < 0.001$ vs. "mTBI") (Figure 6c).

2.6.6. Psd-95 mRNA

A strong treatment effect ($F_{1,8} = 23.58$, $p = 0.0013$) and a significant trauma $\times$ treatment interaction ($F_{1,8} = 6.03$, $p = 0.039$) were observed. Psd-95 mRNA was significantly downregulated after mTBI (0.69 ± 0.12) compared to sham (1.14 ± 0.15, $p = 0.002$), while SDEA treatment significantly increased Psd-95 mRNA expression in mTBI mice compared to untreated mTBI animals ($p < 0.001$), indicating reversal of trauma-associated transcriptional downregulation (Figure 6c).

2.6.7. GluR Subunits mRNA

Analysis of combined GluR1/2 transcripts showed significant effects of treatment ($F_{1,8} = 9.21$, $p = 0.0162$) and interaction ($F_{1,8} = 5.01$, $p = 0.043$). mTBI led to a decrease in AMPA receptor mRNA (0.79 ± 0.11 vs. sham 1.02 ± 0.08, $p < 0.05$), indicating impaired transcription of excitatory receptor subunits. SDEA partially restored GluR expression (0.96 ± 0.10, $p < 0.05$ vs. "mTBI") (Figure 6c).

Thus, at the transcriptional level, mTBI induced a characteristic pattern of Bin-1/GluR upregulation and Psd95 downregulation, reflecting disturbed membrane trafficking and synaptic gene expression. The concordance between PSD-95 protein and mRNA levels contrasts with the lack of direct correspondence observed for AMPA receptor subunits. This dissociation likely reflects the distinct regulatory mechanisms governing these proteins. While PSD-95 expression is largely transcriptionally regulated, AMPA receptor abundance at synapses is dynamically controlled by activity-dependent trafficking, endocytosis, and recycling, which can uncouple protein levels from transcript abundance [33,34]. SDEA administration attenuated these trauma-associated transcriptional changes, consistent with modulation of synaptic gene expression programs following injury.

2.7. Serum Neuroglial Markers Following mTBI and SDEA Treatment

Serum levels of neuroglial proteins (MBP, S100B, GFAP) and cytokines (IL-6) are widely used biomarkers of CNS injury and systemic neuroglial stress [40–42]. Two-way ANOVA revealed significant main effects of both Trauma and Treatment, as well as factors' interactions for all markers ($p < 0.01$). Serum levels of MBP ($F_{1,20} = 40.8$, $p < 0.0001$), S100 ($F_{1,20} = 42.3$, $p < 0.0001$), GFAP ($F_{1,20} = 17.8$, $p = 0.0004$), and IL-6 ($F_{1,20} = 55.6$, $p < 0.0001$) were all markedly increased in the "TBI" group compared to controls (all Tukey $p < 0.001$). SDEA treatment significantly attenuated these elevations, reducing MBP to 1.31 ± 0.11 vs. 2.36 ± 0.24 ($p = 0.0002$), S100 to 1.19 ± 0.12 vs. 2.98 ± 0.28 ($p < 0.0001$), GFAP to 1.09 ± 0.10 vs. 2.10 ± 0.19 ($p < 0.0001$), and IL-6 to 2.46 ± 0.22 vs. 5.10 ± 0.45 ($p < 0.0001$).

https://doi.org/10.3390/ijms27010471

No significant differences were observed between "Sham" and "Sham-SDEA" groups, indicating that SDEA alone did not alter baseline serum profiles (Figure 7a,b).

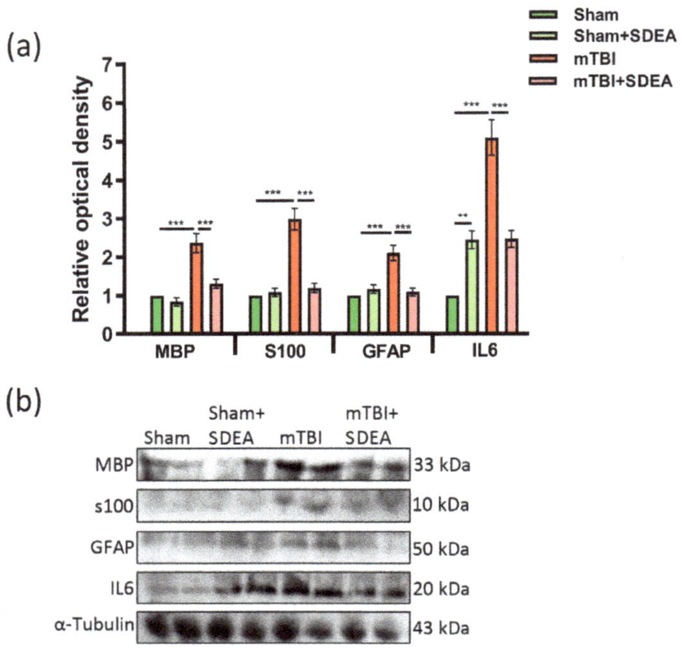

Figure 7. Serum levels of glial and inflammatory markers after mTBI and SDEA treatment. (**a**) Quantification of relative optical density of MBP, S100, GFAP, and IL-6 in serum samples from the four experimental groups. Data are presented as mean ± SEM (relative optical density normalized to α-tubulin); n = 6 (animals per group). Two-way ANOVA. Tukey post hoc test (** $p < 0.01$, *** $p < 0.001$). (**b**) Representative Western blot bands for MBP (~33 kDa), S100 (~10 kDa), GFAP (~50 kDa), IL-6 (~20 kDa), and loading control α-Tubulin (~43 kDa) from serum samples of the four experimental groups.

The coordinated increase in MBP, S100, GFAP, and IL-6 after mTBI reflects axonal/glial injury and systemic inflammation, consistent with increased neuroglial stress and systemic responses associated with CNS injury. SDEA treatment significantly reduced serum levels of all markers elevated by mTBI.

3. Discussion

Mild traumatic brain injury is increasingly recognized as a condition that can produce long-lasting glial reactivity and synaptic dysfunction even in the absence of overt neuronal loss or severe cognitive deficits [43–46]. In our study, mTBI elicited a coordinated set of molecular, structural, and mild behavioral changes in the hippocampus, including astrocytic activation, synaptic protein loss, dendritic spine alterations, reduced activity-dependent Arc expression, and impaired ki-67-positive proliferation within the DG SGZ. Importantly, treatment with stearidonic acid ethanolamide (SDEA) attenuated many of these alterations and ameliorated the anxiety-like behavior induced by injury. Given the mild nature of the injury, behavioral alterations were subtle, and SDEA effects should be interpreted as modulation of behavioral responses under injury conditions rather than full normalization of overt deficits.

A key finding of our study was the elevation of GFAP within hippocampal tissue, which indicates local reactive astrogliosis. Reactive astrocytes undergo hypertrophy, cy-

https://doi.org/10.3390/ijms27010471

toskeletal remodeling, and altered regulation of glutamate and cytokine signaling [47–50], changes that can profoundly influence neuronal excitability and synaptic stability. In parallel, we observed increases in serum GFAP, S100B, MBP, and IL-6. These circulating proteins represent a distinct biological dimension: they reflect glial stress and axonal strain. Because blood–brain barrier integrity was not directly assessed in the present study, changes in circulating GFAP, S100B, MBP, and IL-6 should be interpreted as indirect indicators rather than definitive evidence of BBB disruption [51–54]. Thus, tissue and serum markers reflect complementary but different aspects of the post-traumatic response.

In line with this glial activation, we also detected coordinated alterations in synaptic structure and plasticity-related signaling. Synaptic alterations were heterogeneous, with decreases in PSD-95 and GluR1 but increases in synaptophysin. Although these measurements reflect signals averaged across subregions, the magnitude and pattern of the changes align with the dendritic spine deficits we quantified specifically in CA1. PSD-95 is essential for anchoring and stabilizing AMPA receptors at postsynaptic sites, and its loss is a hallmark of impaired synaptic plasticity [55,56]. AMPA receptor dynamics are further compromised by inflammatory cytokines such as IL-6 and TNF-α, which promote GluR1 internalization and reduce synaptic incorporation [57,58]. The reduction in GluR1 that we observed is therefore consistent with the inflammatory milieu of mTBI and with the GFAP-based evidence of astrocytic activation. Together, these findings suggest that synaptic weakening after injury arises from convergent astrocytic and neuronal mechanisms.

Morphological analyses further revealed a clear regional specificity: dendritic spine density was markedly reduced in CA1 pyramidal neurons, affecting both apical and basal dendrites. DG granule cells showed fewer structural alterations than CA1, suggesting partial resistance in this mild-injury model. This pattern is strongly supported by previous work demonstrating that CA1 neurons are disproportionately vulnerable to metabolic, inflammatory, and mechanical stress due to their high excitatory drive, dense NMDA receptor expression, and selective susceptibility to glutamate-mediated toxicity [59,60]. Numerous studies have shown that CA1 pyramidal neurons exhibit pronounced synaptic and structural deficits even after mild injuries [4]. In contrast, DG granule neurons exhibited only modest structural changes. This pattern is compatible with the notion that the dentate gyrus possesses a unique neurogenic niche and high structural plasticity, which may provide some buffering capacity against injury-induced loss [61,62]. At the same time, several studies have shown that, although DG neurons often survive after traumatic brain injury, their dendrites and synapses can still undergo substantial degeneration [63], indicating that such resilience is relative rather than absolute. In this context, our finding of preserved DG spine morphology but marked CA1 spine loss suggests that mild trauma in our model primarily destabilizes CA1 networks, while DG granule cells either remain structurally intact or are better compensated by ongoing neurogenesis.

This regional dissociation, marked spine loss in CA1 with relative preservation of DG morphology, suggests that not only structural connectivity but also activity-dependent plasticity may be differentially affected across hippocampal subfields. Consistent with this idea, we next examined the immediate early gene Arc, a key marker of experience- and activity-dependent synaptic plasticity in the dentate gyrus, to determine whether mTBI and SDEA modulate hippocampal network engagement at the cellular level. Arc, an immediate-early gene essential for synaptic plasticity and AMPA receptor trafficking [64], was reduced in the DG following mTBI. While Arc is classically described as an immediate early gene, long-lasting changes in its expression have been observed in conditions involving chronic synaptic remodeling and persistent network dysfunction. Therefore, the altered Arc levels detected at 11 weeks post-mTBI likely reflect sustained activity-dependent plasticity rather than acute gene induction. Reduced Arc is typically associated with impaired memory

 https://doi.org/10.3390/ijms27010471

encoding and diminished hippocampal network engagement [65,66]. Yet, in contrast to what might be predicted from Arc downregulation alone, the NOR test revealed no deficits. This apparent discrepancy likely reflects the multilayered nature of DG function. However, recognition memory relies not only on Arc-mediated plasticity but also on intact structural connectivity and pattern separation processes [67]. Because DG dendritic spines and DCX-positive cells were largely preserved in our study, the reduction in Arc may represent a functional but subthreshold impairment, insufficient to compromise NOR performance. These findings support the notion that molecular changes may precede or occur without measurable behavioral deficits [68]. Interestingly, while Heim et al. reported cognitive impairments despite minimal detectable anatomical changes, our animals exhibited the opposite dissociation: pronounced molecular and synaptic abnormalities, particularly in CA1, without measurable deficits in recognition or working memory. This suggests an even more subtle form of "silent" hippocampal dysfunction, in which injury-induced alterations in Arc expression, synaptic proteins, and dendritic spine architecture are effectively compensated at the behavioral level, at least in tasks primarily relying on DG-dependent pattern separation.

By contrast, CA1-dependent functions appear much more sensitive to the structural and molecular disruptions we observed. CA1 plays a central role in emotional regulation and stress reactivity [69,70], and CA1-selective synaptic weakening can increase anxiety-like behavior even in the absence of cognitive decline [59,71,72]. This matches our behavioral results: anxiety-like behavior increased after mTBI, whereas NOR and Y-maze performance remained intact. The preserved architecture of the DG together with only mild Arc and Ki-67 suppression suggests that DG networks could compensate for injury-related changes, supporting normal recognition memory performance. At the same time, the combination of CA1 spine loss, reduced PSD-95/GluR1, and strong astrocytic activation offers a clear mechanism for the heightened anxiety-like behavior.

Finally, the restorative actions of SDEA align with the broader biology of ω-3 polyunsaturated fatty acid-derived ethanolamides, which function as lipid mediators coordinating glial reactivity and synaptic stability. Synaptamide and related ethanolamides such as EPEA engage GPR110-dependent cAMP/PKA signaling, attenuate NF-κB-driven astrocytic activation, and support neurite and axonal growth [15,16,73,74]. Because astrocytic NF-κB activation contributes substantially to excitatory synapse vulnerability after mTBI [5,9], the reduction in GFAP-associated reactivity observed after SDEA likely reflects improved glial-synaptic homeostasis. ω-3-derived ethanolamides may also influence activity-dependent plasticity through their capacity to reduce neuroinflammatory signaling and thereby preserve the molecular environment required for experience-driven synaptic responses. Synaptamide has been shown to attenuate microglial activation and improve hippocampal plasticity in inflammatory conditions [75], suggesting that ω-3 ethanolamides can indirectly support plasticity-related gene expression by stabilizing glial-synaptic interactions. This is relevant given the central role of Arc in structural and functional synaptic adaptations [76]. Recent work demonstrates that astrocyte-dependent glutamate clearance regulates Arc translation and supports hippocampal spatial memory [77], illustrating how glial metabolic signaling can shape Arc-mediated plasticity. In this context, the normalization of Arc expression following SDEA treatment is consistent with a broader framework in which ω-3 ethanolamides restore the permissive environment for experience-dependent plasticity by dampening glial cytokine output and maintaining synaptic homeostasis. In parallel, several N-acylethanolamides including synaptamide, EPEA, and PEA are known to influence neurogenic pathways by reducing inflammatory stress and supporting the cellular environment required for progenitor proliferation. PEA has been shown to restore hippocampal neurogenesis under inflammatory and neuropathic conditions by attenuating

https://doi.org/10.3390/ijms27010471

glial activation [78,79], while DHA and its derivatives, including synaptamide, promote neuronal differentiation and neuritogenesis through GPR110-mediated cAMP/PKA signaling [20,80,81]. These mechanisms provide a plausible basis for the increase in Ki-67-positive cells observed after SDEA treatment, suggesting improved DG niche conditions through combined anti-inflammatory and pro-neurogenic actions. These convergent mechanisms provide a cohesive explanation for the observed normalization of synaptic protein expression, restoration of Arc-dependent plasticity, rescue of CA1 dendritic spines, and stimulation of Ki-67-positive proliferative activity. The accompanying reduction in anxiety-like behavior is consistent with the particular sensitivity of CA1 circuits to glial dysregulation and excitatory synaptic loss, supporting the interpretation that SDEA acts as a multi-target modulator capable of re-establishing hippocampal circuit stability after mTBI.

The observation that SDEA exerted both injury-independent effects and injury × treatment interactions suggests the involvement of multiple, context-dependent mechanisms of action. Some effects likely reflect baseline modulatory properties of SDEA on glial reactivity and synaptic organization under physiological conditions, whereas other effects become evident only in the presence of trauma-induced network dysfunction. Such dual action is consistent with a compound that both stabilizes homeostatic processes and counteracts pathological alterations triggered by mTBI.

In summary, our findings demonstrate that mTBI induces coordinated glial, synaptic, and neurogenic disturbances that are anatomically and functionally selective, disproportionately affecting CA1 while sparing DG-dependent recognition memory. The partial mismatch between Arc downregulation in DG and preserved NOR performance likely reflects subthreshold functional impairment insufficient to disrupt behavior. SDEA effectively mitigated astrocytic activation, preserved synaptic structure, restored molecular plasticity markers, and reduced anxiety-like behavior, highlighting its therapeutic potential as a multi-target modulator of post-traumatic hippocampal plasticity.

4. Materials and Methods

4.1. Isolation and Characterization of Stearidonic Acid Ethanolamide (SDEA)

The stearidonic acid ethanolamide (N-stearidonylethanolamine) used in this study was produced from lipid extracts of *Sardinops melanostictus*. To obtain the target N-acylethanolamide, polyunsaturated fatty acids were first converted to their ethyl ester forms, after which the esterified fraction was reacted with ethanolamine to generate PUFA ethanolamides. Following synthesis, the reaction products were extracted with chloroform and washed repeatedly with water to remove polar contaminants.

The crude mixture of ethanolamides was initially purified by silica gel column chromatography. Subsequent separation of individual PUFA ethanolamides was achieved by preparative HPLC using a Shimadzu LC-8A system equipped with a UV/VIS SPD-20A detector (λ = 205 nm) (Shimadzu, Kyoto, Japan). Chromatographic separation was performed on a Supelco Discovery HS C18 column (250 × 50 mm, 10 μm particle size) (Supelco, Bellefonte, PA, USA) under isocratic elution with ethanol/water (70:30, *v*/*v*) at a flow rate of 50 mL/min. Fractions containing N-acylethanolamides were collected, evaporated under reduced pressure, and subjected to GC and GC–MS analysis for final identification. The obtained SDEA was a pale yellow, slightly viscous oil with a purity of 94.8%. To determine the composition of the ethanolamide fraction, samples were derivatized to their trimethylsilyl (TMS) forms. Briefly, 1 mg of fatty acid ethanolamides was mixed with 50 μL of BSTFA (N,O-bis(trimethylsilyl)trifluoroacetamide) and incubated at 60 °C for 1 h under an argon atmosphere. After cooling, 1 mL of hexane was added, and 1 μL of the silylated solution was injected into a Shimadzu GC-2010 Plus system equipped with a

https://doi.org/10.3390/ijms27010471

flame ionization detector (Shimadzu, Kyoto, Japan) and a Supelco SLB™-5 ms capillary column (30 m × 0.25 mm) (Supelco, Bellefonte, PA, USA).

Gas chromatographic analysis was performed using the following temperature program: initial temperature 180 °C; ramp 2 °C/min to 260 °C; hold at 260 °C for 35 min. Injector and detector temperatures were both set to 260 °C. Structural confirmation of TMS-NAE derivatives was carried out by GC–MS using a Shimadzu TQ-8040 mass spectrometer (Shimadzu, Kyoto, Japan) with electron impact ionization (70 eV) and the same chromatographic conditions. Mass spectral analysis yielded a molecular ion at m/z 391 for TMS-N-SDEA, with a prominent fragment at m/z 376 corresponding to loss of a methyl group. Total ion chromatograms and representative mass spectra are shown in Supplementary Figure S1a,b.

4.2. Animals

Male C57BL/6J mice (8–10 weeks old) were bred and housed in the animal facility (vivarium) of the A.V. Zhirmunsky National Scientific Center of Marine Biology, Far Eastern Branch of the Russian Academy of Sciences (Vladivostok, Russia). All experimental groups were derived from the same local breeding colony, ensuring internal genetic consistency across comparisons. Adult male mice (3 months old; 28–30 g) were used in all experiments. Animals were housed in groups of four under standard laboratory conditions (22 ± 1 °C; 12:12 light/dark cycle; food and water ad libitum). Only male mice were used in this study to reduce biological variability associated with estrous cycle-dependent hormonal fluctuations and because the work was designed as an exploratory mechanistic analysis. A total of 68 mice were used (n = 17 per group: "Sham", "Sham + SDEA", "mTBI", "mTBI + SDEA"). All procedures adhered to the EU Directive 2010/63/EU [82] and were approved by the Animal Ethics Committee of the National Scientific Center for Marine Biology, Far East Branch of the Russian Academy of Sciences (No. 2/2025, 10 February 2025).

4.3. mTBI Model

Mild traumatic brain injury was modeled using a standardized closed-head impact procedure. Mice were first placed in an induction chamber and anesthetized with 4.5% isoflurane (Laboratories Karizoo, SA, Barcelona, Spain) delivered in 100% oxygen using a rodent anesthetic vaporizer (VetFloTM, Kent Scientific Corporation, Torrington, CT, USA). After 1–3 min, when a stable level of anesthesia was confirmed, the scalp was shaved and a thin metallic head shield (1.25 mm) was fixed over the parietal area to distribute the mechanical load during impact.

For injury induction, animals were positioned on a flexible polyurethane support mounted on the platform of a vertical impact apparatus (IMP-1020, Shanghai TOW Intelligent Technology, Shanghai, China). The head shield was centered beneath the guide tube to ensure reproducibility of the impact coordinates. A 100 g weight was then elevated to 40 cm and released through the tube by an electromagnet, producing a controlled mild impact to the skull without craniotomy.

Immediately after the procedure, animals were examined for acute adverse events such as convulsions or external cranial deformation; neither was detected in any experimental group. Mice were then transferred to a warmed recovery surface and monitored until spontaneous movement, normal posture, and thermoregulation were restored. No mortality occurred throughout the study. Sham-operated animals underwent the same preparation steps, including anesthesia and placement under the device, but the weight was not released.

 https://doi.org/10.3390/ijms27010471

4.4. SDEA Preparation and Administration

The test compound (SDEA) was prepared freshly before injections. The emulsion was generated by dispersing the compound in water with continuous vortexing at high speed (3000 rpm for 5 min). Ethanol (1.5% of total injection volume) served as an emulsion stabilizer. An identical ethanol concentration was included in the vehicle solution for control groups. The compound was administered subcutaneously at 10 mg/kg, following a dosing regimen consisting of two pre-treatment days, administration on the day of the injury, and daily injections for the subsequent seven days. Vehicle-treated mice received matched injections of the carrier solution. Four experimental groups were formed: "Sham"—mice underwent all preparatory steps, including anesthesia and positioning under the impact apparatus, but the impact was not delivered. These animals received the vehicle solution containing 1.5% ethanol, serving as baseline controls for both the surgical manipulation and solvent administration ($n = 17$); "Sham + SDEA"—animals received the same sham procedure as described above but were treated with SDEA at 10 mg/kg. This group controlled for the pharmacological effects of the compound in the absence of traumatic injury ($n = 17$); "mTBI"—mice in this group were subjected to a single mild closed-head impact, modeling mTBI, and subsequently received the vehicle solution ($n = 17$); "mTBI + SDEA"—these animals experienced mTBI and were treated with SDEA at 10 mg/kg according to the full dosing schedule (two pre-injury injections, one injection on the day of injury, and daily injections for seven days after) ($n = 17$). This group was used to determine the therapeutic efficacy of SDEA in mitigating mTBI-induced deficits. Behavioral testing was performed in a dedicated cohort ($n = 14$ per group). For gene expression analysis (qPCR), an independent cohort was used to avoid potential confounding effects of behavioral testing on transcriptional readouts. All procedures (mTBI induction, treatment regimen, timing of tissue collection) were identical to the behavioral cohort. The experimental design is schematically summarized in Figure 8.

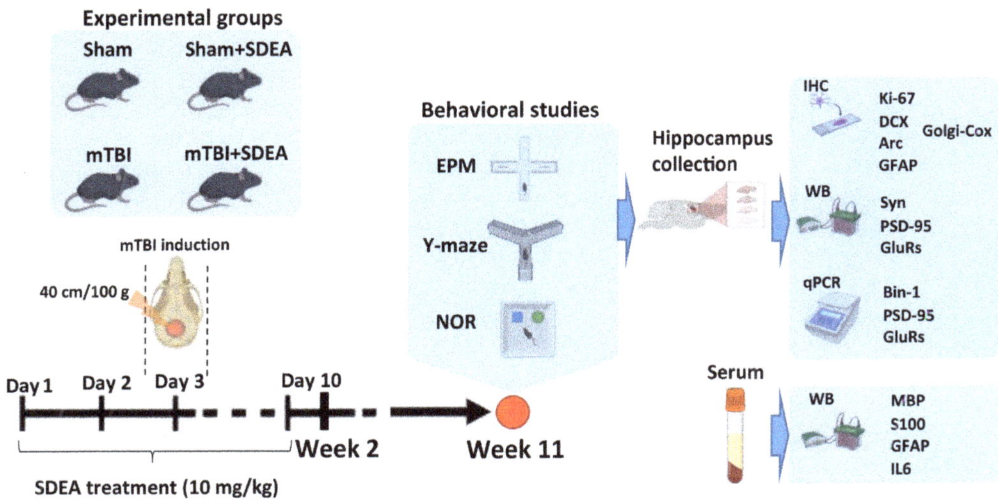

Figure 8. Experimental design and workflow of the study. Mice were divided into four experimental groups: "Sham", "Sham + SDEA", "mTBI", and "mTBI + SDEA". Mild traumatic brain injury (mTBI) was induced using a weight-drop model (40 cm/100 g). SDEA treatment (10 mg/kg) was administered as indicated in the timeline. Behavioral assessments, including the elevated plus maze (EPM), Y-maze, and novel object recognition (NOR) tests, were performed at Week 11. At the end of the experiment (Week 11), animals were euthanized, and the hippocampus was dissected for subsequent molecular, immunohistochemical, and morphological analyses (Golgi-Cox staining).

https://doi.org/10.3390/ijms27010471

4.5. Behavioral Testing

All behavioral assessments were performed between 10:00 and 16:00 under dim ambient lighting (25–30 lux). Mice were habituated to the testing room for at least 30 min prior to each procedure. All apparatuses were cleaned with 70% ethanol between trials to eliminate olfactory cues.

4.5.1. Y-Maze Spontaneous Alternation Test

Working memory performance was assessed using the spontaneous alternation paradigm in a Y-shaped acrylic maze (30 × 10 × 20 cm), consisting of three identical arms arranged at 120°. All three arms were identical in appearance and odor, and no external cues or bedding differences were introduced, as spontaneous alternation relies on the animal's intrinsic exploratory behavior rather than explicit arm discrimination. Each mouse was placed at the junction of the three arms and allowed to explore freely for 5 min. An arm entry was counted when the animal completely crossed into an arm with all four paws. The sequence of arm visits was recorded to evaluate spontaneous alternation, defined as consecutive entries into three distinct arms. The test was administered four times at weeks 10–11.

The spontaneous alternation rate (SAR) was calculated as:

$$SAR = N/A \times 100\% \quad (1)$$

where SAR—the spontaneous alternation rate, N—the quantity of consecutive entries into the three nonrepeating arms, A—total number of possible alternations (total arm entries—2).

This parameter was used as the primary measure of working memory performance in the Y-maze test.

4.5.2. Novel Object Recognition (NOR) Test

Recognition memory was examined using a standard NOR protocol adapted from Bevins and Besheer [83]. Animals underwent habituation to the testing arena one day prior to the experiment. Training phase: each mouse was placed in the center of the arena containing two identical objects and allowed 10 min of unrestricted exploration. Object interaction was defined as active investigation, including sniffing or touching the object at a distance of <1–2 cm. Testing phase: after a 24 h retention interval, one of the familiar objects was replaced with a novel one. Mice were again placed in the arena for 5 min, and exploration was video-recorded with an overhead camera. To quantify recognition memory, the recognition index (*RI*) and discrimination index (*DI*) were calculated as:

$$RI = T_N/(T_N + T_F) \quad (2)$$

$$DI = (T_N - T_F)/(T_N + T_F) \quad (3)$$

where T_N—time for novel object (s), T_F—time for familiar object (s). Objects and the arena were thoroughly cleaned with 70% ethanol between trials.

These indices were used for subsequent statistical comparisons between experimental groups in the NOR test.

4.5.3. Elevated Plus Maze (EPM)

Anxiety-related behavior was evaluated using a standard elevated plus maze (Panlab, Harvard Apparatus, Holliston, MA, USA) consisting of two open arms (30 × 5 cm, 1 cm ledge) and two closed arms (30 × 5 cm, 15 cm walls), elevated 50 cm above the floor. Each mouse was placed on the central platform facing an open arm and allowed to explore

 https://doi.org/10.3390/ijms27010471

freely for 5 min. Behavior was recorded using a SMART 3.0 video-tracking system (Panlab, Harvard Apparatus, Holliston, MA, USA). The following parameters were quantified: time spent in open arms, time spent in closed arms, and time spent in the center. Increased exploration of open arms was interpreted as reduced anxiety-like behavior, whereas avoidance of open arms indicated heightened anxiety [21].

4.5.4. Tissue Collection

The animals were euthanized under deep anesthesia 11 weeks after surgery, immediately after behavioral testing. Mice were first placed in an induction chamber and anesthetized with 4.5% isoflurane (Laboratories Karizoo, SA, Barcelona, Spain) delivered in 100% oxygen using a rodent anesthetic vaporizer (VetFloTM, Kent Scientific Corporation, Torrington, CT, USA). Brains designated for immunohistochemistry were perfused with PBS followed by 4% paraformaldehyde and postfixed in 4% paraformaldehyde for 24 h. Brains for biochemical analysis were rapidly dissected; hippocampi were isolated on ice, snap-frozen in liquid nitrogen, and stored at −80 °C. Blood was collected via cardiac puncture, centrifuged (3000× g, 10 min), and serum was stored at −80 °C.

4.5.5. Immunohistochemistry

After fixation, brains were rinsed in PBS, processed through a standard dehydration and clearing series, and embedded in paraffin blocks. Serial coronal sections (7 μm) were cut using a Leica RM2245 rotary microtome (Leica, Wetzlar, Germany). For immunohistochemical analysis, brain sections were first deparaffinized through a graded xylene and ethanol series and then incubated in 3% hydrogen peroxide for 10 min to suppress endogenous peroxidase activity. After rinsing in 0.1 M phosphate buffer (pH 7.2), sections were transferred into a blocking solution containing 2% bovine serum albumin (Sigma-Aldrich, Darmstadt, Germany) and 0.01% Triton X-100 for 1 h at room temperature to reduce nonspecific antibody binding. Following the blocking step, sections were incubated with primary antibodies for 24 h at 4 °C. The following Abcam primary antibodies were used depending on the target: GFAP (astrocytes): ab7260, 1:1000; Arc (activity-dependent immediate early gene): ab183183, 1:1000; Ki-67 (proliferating cells): ab16667, 1:500; DCX (immature neurons): ab18723, 1:500. Negative controls were processed identically but without addition of primary antibodies. After overnight incubation, sections were washed and incubated with the appropriate HRP-conjugated secondary antibody (Abcam, Cambridge, UK) for 15 min, followed by incubation with streptavidin-HRP (ab64269, Abcam) for 10 min. Visualization was performed using DAB chromogen (Abcam), producing a brown reaction product. Finally, the tissue was rinsed in distilled water, dehydrated through ascending ethanol series, cleared in xylene, and mounted using Dako mounting medium (CS705, Denver, CO, USA).

4.5.6. Image Acquisition and Quantification

Microscopic evaluation was carried out on a Zeiss Axio Imager system equipped with an AxioCam 503 color camera and controlled through ZEN Blue (Carl Zeiss, Oberkochen, Germany) v.3.4.. Digital images were further processed and quantified in ImageJ 1.53t (NIH, Bethesda, MD, USA). Prior to analysis, each micrograph was converted to 8-bit format, and uneven background illumination was corrected using the rolling-ball subtraction algorithm (radius = 50). A region of interest (ROI) corresponding to the CA1 or dentate gyrus was manually outlined based on anatomical landmarks. GFAP immunoreactivity was segmented using a fixed intensity threshold, which was applied uniformly to all images. The percentage of GFAP-positive area was calculated as the fraction of thresholded pixels relative to the total ROI area. The sections were imaged using identical microscope settings (objective, exposure time, and illumination) for all experimental groups. Quantitative analy-

https://doi.org/10.3390/ijms27010471

sis was performed using ImageJ software by measuring the GFAP-positive area within CA1 and DG hippocampal regions. All image acquisition and quantification procedures were conducted by an experimenter blinded to group assignment. For quantitative assessment of immunolabeling, every sixth section throughout the hippocampus was selected. All image analyses were performed by an investigator blinded to group assignments to prevent bias.

4.5.7. Golgi-Cox Staining and Dendritic Spine Analysis

Structural plasticity was assessed using the FD Rapid GolgiStain Kit (FD NeuroTechnologies, Columbia, MD, USA). Brains were impregnated for 14 days, cryosectioned at 100 μm, and mounted on gelatin-coated slides. Fully impregnated CA1 pyramidal neurons (apical and basal dendrites) and DG granule neurons were selected based on standard criteria. Dendritic segments (20–40 μm) were imaged at 100× oil immersion, and spine density and morphology (thin, stubby, mushroom, filopodia) were quantified in ImageJ.

4.5.8. Western Blotting

Western blotting was performed to quantify serum levels of GFAP, S100B, MBP, and IL-6, as well as hippocampal expression of Synaptophysin, PSD-95, GluR1, and GluR2. For each experimental group, hippocampal tissue or serum was collected from six biologically independent animals (n = 6 per group). Each biological sample was processed individually without pooling. For every sample, two technical replicates (duplicate lanes) were run on the same gel to ensure reliability of signal detection. Serum proteins were precipitated and resuspended in lysis buffer, whereas hippocampal tissue was homogenized manually using a micropestle in ice-cold RIPA buffer supplemented with protease inhibitors (cOmplete™, Sigma-Aldrich, St. Louis, MI, USA). Lysates were centrifuged at 12,000× g for 15 min at 4 °C, and the supernatants were collected. Protein concentration was determined with a standard colorimetric assay, after which all samples were normalized to 2 mg/mL. Each lysate was mixed 1:1 with 1× Laemmli sample buffer (Biorad, Hercules, CA, USA) containing 5% β-mercaptoethanol, heated at 94 °C for 5 min, and briefly centrifuged before loading. Equal amounts of protein (60 μg per lane) were separated on Any kDa Mini-PROTEAN precast gels (Biorad, Hercules, CA, USA) alongside a molecular weight ladder (Thermo Fisher Scientific, Waltham, MA, USA). Gels were run at a constant current of 15 mA per gel. Proteins were transferred to PVDF membranes using the Trans-Blot Turbo transfer system (Biorad, Hercules, CA, USA) and the RTA transfer kit under manufacturer-recommended settings. Membranes were incubated in 2% BSA in PBS for 1 h at room temperature and then washed in PBS containing 0.1% Tween-20 (PBS-T). Primary antibodies were applied overnight at 4 °C: GFAP (1:1000, ab7260), S100B (1:1000, ab52642), MBP (1:1000, ab7349), IL-6 (1:1000, ab290735), Synaptophysin (1:500, ab8049), PSD-95 (1:500, ab2723), GluR1 (1:1000, ab31232), GluR2 (1:1000, ab133477), α-tubulin (ab7291, 1:5000, loading control) all from Abcam (Cambridge, UK). After washing (3 × 5 min in PBS-T), membranes were incubated with secondary HRP-conjugated antibodies Anti-Rabbit (Abcam) and Anti-Mouse (Abcam, Cambridge, UK). for 1 h at room temperature, followed by three additional washes. Chemiluminescent substrate (Biorad, Hercules, CA, USA) was applied for 5 min, and signal was visualized with the ChemiDoc imaging system (Biorad, Hercules, CA, USA). Band intensities were quantified in ImageJ (NIH, USA) and normalized to the corresponding loading control. For GluR2 quantification, densitometric analysis was performed only for the band corresponding to the expected molecular weight of GluR2, as indicated by the antibody datasheet. Additional bands were excluded from analysis.

4.5.9. Real-Time PCR

RT-PCR was used to measure hippocampal transcriptional levels of BIN1, GluRs, and PSD95. Because tissue from the primary experimental cohort was fully allocated to histo-

https://doi.org/10.3390/ijms27010471

logical and protein analyses, qPCR was performed using an additional independent cohort of mice (three animals per group). Hippocampal samples were homogenized using metal beads in a TissueLyser LT system (Qiagen, Hilden, Germany). Total RNA was isolated with the ExtractRNA kit (Evrogen, Moscow, Russia) and further purified using CleanRNA Standard columns (Evrogen, Moscow, Russia). RNA purity and concentration were assessed spectrophotometrically (Thermo Fisher, Waltham, MA, USA). Reverse transcription was performed using the RNAscribe RT kit (Biolambix, Novosibirsk, Russia) according to the manufacturer's protocol. Real-time PCR was conducted on a CFX96 system (Bio-Rad Laboratories, Hercules, CA, USA) using the SYBR Green I qPCR mix (Evrogen, Moscow, Russia). Reactions (20 μL total volume) contained: 1 μL of cDNA, 0.5 μM forward and reverse primers, SYBR Green master mix. Thermal profile: 95 °C for 10 min, 40 cycles of: 95 °C for 15 s, 67 °C for 40 s, 72 °C for 10 s, melt-curve analysis to confirm single-product amplification. Primers for BIN1, GluRs, and PSD95 were designed in Primer Premier 5 and synthesized by Evrogen (Moscow, Russia). GAPDH and β-actin served as internal reference genes. Relative expression was calculated using the ΔΔCt method, with normalization to the geometric mean of both housekeeping genes.

4.5.10. Statistical Analysis

Data distribution was assessed with Shapiro–Wilk test. Behavioral, biochemical, and histological data were analyzed using two-way ANOVA (factors: Trauma × Treatment), followed by Tukey's post hoc test. Spine-type-specific datasets were analyzed separately for each dendritic compartment (CA1 apical, CA1 basal, DG) using the same statistical model. Values are presented as mean ± SEM. Statistical significance was set at $p < 0.05$. Graphs were prepared in GraphPad Prism 10 (San Diego, CA, USA).

5. Conclusions

N-stearidonylethanolamine effectively counteracts key pathological consequences of mTBI. Treatment attenuated circulating injury biomarkers and astrocytic reactivity, restored PSD-95, AMPA receptor subunits, Arc expression, and rescued CA1 dendritic spine loss while preserving DG structure. These molecular and structural improvements were accompanied by normalization of anxiety-like behavior, whereas recognition and working memory remained intact following mTBI. Together, the findings indicate that SDEA acts as a multi-target modulator of glia-synapse interactions, selectively stabilizing vulnerable CA1 circuits and promoting hippocampal resilience after mild brain trauma.

Limitations: A limitation of the present study is the exclusive use of male animals. Sex-dependent differences in neuroinflammatory responses and synaptic plasticity after traumatic brain injury have been reported, and future studies will be required to assess whether the effects of SDEA extend to female animals.

Although the SGZ of the dentate gyrus represents a neurogenic niche in the adult brain, the present study assessed proliferative activity using Ki-67 without cell-type-specific markers. Therefore, the observed changes should be interpreted as modulation of proliferation within a neurogenesis-associated region rather than direct evidence of altered adult neurogenesis.

Although N-acylethanolamides are highly lipophilic and have been reported to access the central nervous system, direct measurements of SDEA brain penetration were not performed in this study. Therefore, the present findings support functional CNS effects of SDEA but do not directly demonstrate its concentration or structural integrity within brain tissue.

Supplementary Materials: The following supporting information can be downloaded at: https://www.mdpi.com/article/10.3390/ijms27010471/s1.

https://doi.org/10.3390/ijms27010471

Author Contributions: Conceptualization, A.T. and I.M.; methodology, A.T., A.E., I.M. and D.I.; validation, A.T. and I.M.; investigation, A.T., A.E., I.M. and D.I.; resources, A.T.; data curation, A.T. and I.M.; writing—original draft preparation, A.T.; writing—review and editing, A.T.; visualization, A.T.; project administration, A.T. and I.M.; funding acquisition, A.T. All authors have read and agreed to the published version of the manuscript.

Funding: The study was supported by the grant of the Russian Science Foundation No. 25-75-10017.

Institutional Review Board Statement: The study was conducted in accordance with the EU Directive 2010/63/EU and was approved by the Animal Ethics Committee of the National Scientific Center for Marine Biology, Far East Branch of the Russian Academy of Sciences (No. 2/2025, 10 February 2025).

Informed Consent Statement: Not applicable.

Data Availability Statement: The raw data supporting the conclusions of this article will be made available by the authors on request.

Conflicts of Interest: The authors declare no conflicts of interest.

Abbreviations

The following abbreviations are used in this manuscript:

mTBI	mild traumatic brain injury
CNS	central nervous system
DG	dentate gyrus
CA1	cornu ammonis area 1
PUFA	polyunsaturated fatty acid
NAE	N-acylethanolamide
SDEA	stearidonic acid ethanolamide
DHA	docosahexaenoic acid
EPA	eicosapentaenoic acid
DHEA (synaptamide)	N-docosahexaenoylethanolamide
EPEA	N-eicosapentaenoylethanolamide
PEA	palmitoylethanolamide
GFAP	glial fibrillary acidic protein
S100B	S100 calcium-binding protein B
MBP	myelin basic protein
IL-6	interleukin-6
IL-1β	interleukin-1 beta
TNF-α	tumor necrosis factor alpha
PSD-95	postsynaptic density protein 95
AMPAR	AMPA-type glutamate receptor
GluR1/GluR2	AMPA receptor subunits
NMDAR1/NMDAR2	NMDA receptor subunits
Arc	activity-regulated cytoskeleton-associated protein
BIN1	bridging integrator 1
EPM	elevated plus maze
NOR	novel object recognition
DI	discrimination index
RI	recognition index
SAR	spontaneous alternation rate (Y-maze)
PVDF	polyvinylidene difluoride
ECL	enhanced chemiluminescence
RT-PCR	real-time polymerase chain reaction
SEM	standard error of the mean
ANOVA	analysis of variance

 https://doi.org/10.3390/ijms27010471

References

1. Haarbauer-Krupa, J.; Pugh, M.J.; Prager, E.M.; Harmon, N.; Wolfe, J.; Yaffe, K. Epidemiology of chronic effects of traumatic brain injury. *J. Neurotrauma* **2021**, *38*, 3235–3247. [CrossRef] [PubMed]
2. Gardner, A.J.; Tonks, J.; Potter, S.; Yates, P.J.; Reuben, A.; Ryland, H.; Williams, H. Neuropsychological assessment of mTBI in adults. In *Traumatic Brain Injury: A Clinician's Guide to Diagnosis, Management, and Rehabilitation*; Springer International Publishing: Cham, Switzerland, 2019; pp. 57–73.
3. Prins, M.L.; Alexander, D.; Giza, C.C.; Hovda, D.A. Repeated mild traumatic brain injury: Mechanisms of cerebral vulnerability. *J. Neurotrauma* **2013**, *30*, 30–38. [CrossRef] [PubMed]
4. Langlois, L.D.; Selvaraj, P.; Simmons, S.C.; Gouty, S.; Zhang, Y.; Nugent, F.S. Repetitive mild traumatic brain injury induces persistent alterations in spontaneous synaptic activity of hippocampal CA1 pyramidal neurons. *IBRO Neurosci. Rep.* **2022**, *12*, 157–162. [CrossRef] [PubMed]
5. Burda, J.E.; Sofroniew, M.V. Reactive gliosis and the multicellular response to CNS damage and disease. *Neuron* **2014**, *81*, 229–248. [CrossRef] [PubMed]
6. Jayakumar, A.R.; Tong, X.Y.; Ruiz-Cordero, R.; Bregy, A.; Bethea, J.R.; Bramlett, H.M.; Norenberg, M.D. Activation of NF-κB mediates astrocyte swelling and brain edema in traumatic brain injury. *J. Neurotrauma* **2014**, *31*, 14. [CrossRef]
7. Michinaga, S.; Koyama, Y. Pathophysiological responses and roles of astrocytes in traumatic brain injury. *Int. J. Mol. Sci.* **2021**, *22*, 6418. [CrossRef] [PubMed]
8. Mira, R.G.; Lira, M.; Cerpa, W. Traumatic brain injury: Mechanisms of glial response. *Front. Physiol.* **2021**, *12*, 740939. [CrossRef] [PubMed]
9. Lantoine, J.; Procès, A.; Villers, A.; Halliez, S.; Buée, L.; Ris, L.; Gabriele, S. Inflammatory molecules released by mechanically injured astrocytes trigger presynaptic loss in cortical neuronal networks. *ACS Chem. Neurosci.* **2021**, *12*, 3885–3897. [CrossRef] [PubMed]
10. Zhuang, Y.; Xu, X.; Li, H.; Niu, F.; Yang, M.; Ge, Q.; Lu, S.; Deng, Y.; Wu, H.; Zhang, B.; et al. Megf10-related engulfment of excitatory postsynapses by astrocytes following severe brain injury. *CNS Neurosci. Ther.* **2023**, *29*, 2873–2883. [CrossRef] [PubMed]
11. Zhang, H.; Zhang, X.; Chai, Y.; Wang, Y.; Zhang, J.; Chen, X. Astrocyte-mediated inflammatory responses in traumatic brain injury: Mechanisms and potential interventions. *Front. Immunol.* **2025**, *16*, 1584577. [CrossRef]
12. Liu, L.; Dai, L.; Xu, D.; Wang, Y.; Bai, L.; Chen, X.; Li, M.; Yang, S.; Tang, Y. Astrocyte secretes IL-6 to modulate PSD-95 palmitoylation in basolateral amygdala and depression-like behaviors induced by peripheral nerve injury. *Brain Behav. Immun.* **2022**, *104*, 139–154. [CrossRef] [PubMed]
13. Mishra, A.; Kim, H.J.; Shin, A.H.; Thayer, S.A. Synapse loss induced by interleukin-1β requires pre- and post-synaptic mechanisms. *J. Neuroimmune Pharmacol.* **2012**, *7*, 571–578. [CrossRef] [PubMed]
14. Ngwenya, L.B.; Danzer, S.C. Impact of traumatic brain injury on neurogenesis. *Front. Neurosci.* **2019**, *12*, 1014. [CrossRef] [PubMed]
15. Kim, H.Y.; Spector, A.A. Synaptamide, endocannabinoid-like derivative of docosahexaenoic acid with cannabinoid-independent function. *Prostaglandins Leukot. Essent. Fatty Acids* **2013**, *88*, 121–125. [CrossRef] [PubMed]
16. Pellitteri, R.; La Cognata, V.; Russo, C.; Patti, A.; Sanfilippo, C. Protective role of eicosapentaenoic and docosahexaenoic and their N-ethanolamide derivatives in olfactory glial cells affected by lipopolysaccharide-induced neuroinflammation. *Molecules* **2024**, *29*, 4821. [CrossRef] [PubMed]
17. Mandal, G.; Alboni, S.; Cattane, N.; Marizzoni, M.; Saleri, S.; Arslanovski, N.; Mariani, N.; Kirkpatrick, M.; Cattaneo, A.; Pariante, C.M.; et al. The dietary ligands, omega-3 fatty acid endocannabinoids and short-chain fatty acids prevent cytokine-induced reduction of human hippocampal neurogenesis and alter the expression of genes involved in neuroinflammation and neuroplasticity. *Mol. Psychiatry* **2025**, *30*, 5338–5355. [CrossRef] [PubMed]
18. Kim, H.Y.; Spector, A.A. N-Docosahexaenoylethanolamine: A neurotrophic and neuroprotective metabolite of docosahexaenoic acid. *Mol. Asp. Med.* **2018**, *64*, 34–44. [CrossRef] [PubMed]
19. Park, T.; Chen, H.; Kim, H.Y. GPR110 (ADGRF1) mediates anti-inflammatory effects of N-docosahexaenoylethanolamine. *J. Neuroinflamm.* **2019**, *16*, 225. [CrossRef] [PubMed]
20. Kim, H.Y.; Spector, A.A.; Xiong, Z.M. A synaptogenic amide N-docosahexaenoylethanolamide promotes hippocampal development. *Prostaglandins Other Lipid Mediat.* **2011**, *96*, 114–120. [CrossRef]
21. Calder, P.C. Mechanisms of action of (n-3) fatty acids. *J. Nutr.* **2012**, *142*, 592S–599S. [CrossRef] [PubMed]
22. Walf, A.A.; Frye, C.A. The use of the elevated plus maze as an assay of anxiety-related behavior in rodents. *Nat. Protoc.* **2007**, *2*, 322–328. [CrossRef] [PubMed]
23. Prieur, E.A.; Jadavji, N.M. Assessing spatial working memory using the spontaneous alternation Y-maze test in aged male mice. *Bio-Protocol* **2019**, *9*, e3162. [CrossRef]
24. Lueptow, L.M. Novel object recognition test for the investigation of learning and memory in mice. *J. Vis. Exp. JoVE* **2017**, *126*, 55718. [CrossRef]

 https://doi.org/10.3390/ijms27010471

25. Zhao, X.; van Praag, H. Steps towards standardized quantification of adult neurogenesis. *Nat. Commun.* **2020**, *11*, 4275. [CrossRef] [PubMed]
26. Gebara, E.; Sultan, S.; Kocher-Braissant, J.; Toni, N. Adult hippocampal neurogenesis inversely correlates with microglia in conditions of voluntary running and aging. *Front. Neurosci.* **2013**, *7*, 145. [CrossRef] [PubMed]
27. Bramham, C.R.; Worley, P.F.; Moore, M.J.; Guzowski, J.F. The immediate early gene arc/arg3.1: Regulation, mechanisms, and function. *J. Neurosci.* **2008**, *28*, 11760–11767. [CrossRef] [PubMed]
28. Penrod, R.D.; Kumar, J.; Smith, L.N.; McCalley, D.; Nentwig, T.B.; Hughes, B.W.; Barry, G.M.; Glover, K.; Taniguchi, M.; Cowan, C.W. Activity-regulated cytoskeleton-associated protein (Arc/Arg3.1) regulates anxiety- and novelty-related behaviors. *Genes Brain Behav.* **2019**, *18*, e12561. [CrossRef] [PubMed]
29. Runge, K.; Cardoso, C.; de Chevigny, A. Dendritic spine plasticity: Function and mechanisms. *Front. Synaptic Neurosci.* **2020**, *12*, 36. [CrossRef] [PubMed]
30. Bourne, J.; Harris, K.M. Do thin spines learn to be mushroom spines that remember? *Curr. Opin. Neurobiol.* **2007**, *17*, 381–386. [CrossRef] [PubMed]
31. Sacco, M.A.; Gualtieri, S.; Tarallo, A.P.; Verrina, M.C.; Calafiore, J.; Princi, A.; Lombardo, S.; Ranno, F.; Di Cello, A.; Gratteri, S.; et al. The Role of GFAP in Post-Mortem Analysis of Traumatic Brain Injury: A Systematic Review. *Int. J. Mol. Sci.* **2024**, *26*, 185. [CrossRef] [PubMed]
32. Janigro, D.; Mondello, S.; Posti, J.P.; Unden, J. GFAP and S100B: What you always wanted to know and never dared to ask. *Front. Neurol.* **2022**, *13*, 835597. [CrossRef] [PubMed]
33. Glantz, L.A.; Gilmore, J.H.; Hamer, R.M.; Lieberman, J.A.; Jarskog, L.F. Synaptophysin and postsynaptic density protein 95 in the human prefrontal cortex from mid-gestation into early adulthood. *Neuroscience* **2007**, *149*, 582–591. [CrossRef] [PubMed]
34. Hollmann, M.; Hartley, M.; Heinemann, S. Ca^{2+} permeability of KA-AMPA--gated glutamate receptor channels depends on subunit composition. *Science* **1991**, *252*, 851–853. [CrossRef] [PubMed]
35. He, K.; Song, L.; Cummings, L.W.; Goldman, J.; Huganir, R.L.; Lee, H.K. Stabilization of Ca^{2+}-permeable AMPA receptors at perisynaptic sites by GluR1-S845 phosphorylation. *Proc. Natl. Acad. Sci. USA* **2009**, *106*, 20033–20038. [CrossRef] [PubMed]
36. Hagihara, H.; Ohira, K.; Toyama, K.; Miyakawa, T. Expression of the AMPA receptor subunits GluR1 and GluR2 is associated with granule cell maturation in the dentate gyrus. *Front. Neurosci.* **2011**, *5*, 100. [CrossRef] [PubMed]
37. Glennon, E.B.; Lau, D.H.; Gabriele, R.M.C.; Taylor, M.F.; Troakes, C.; Opie-Martin, S.; Elliott, C.; Killick, R.; Hanger, D.P.; Perez-Nievas, B.G.; et al. Bridging Integrator-1 protein loss in Alzheimer's disease promotes synaptic tau accumulation and disrupts tau release. *Brain Commun.* **2020**, *2*, fcaa011. [CrossRef] [PubMed]
38. Dourlen, P.; Kilinc, D.; Landrieu, I.; Chapuis, J.; Lambert, J.C. BIN1 and Alzheimer's disease: The tau connection. *Trends Neurosci.* **2025**, *48*, 349–361. [CrossRef] [PubMed]
39. Saha, O.; Melo de Farias, A.R.; Pelletier, A.; Siedlecki-Wullich, D.; Landeira, B.S.; Gadaut, J.; Carrier, A.; Vreulx, A.C.; Guyot, K.; Shen, Y.; et al. The Alzheimer's disease risk gene BIN1 regulates activity-dependent gene expression in human-induced glutamatergic neurons. *Mol. Psychiatry* **2024**, *29*, 2634–2646. [CrossRef] [PubMed]
40. Kim, H.J.; Tsao, J.W.; Stanfill, A.G. The current state of biomarkers of mild traumatic brain injury. *JCI Insight* **2018**, *3*, e97105. [CrossRef] [PubMed]
41. Oris, C.; Kahouadji, S.; Durif, J.; Bouvier, D.; Sapin, V. S100B, actor and biomarker of mild traumatic brain injury. *Int. J. Mol. Sci.* **2023**, *24*, 6602. [CrossRef] [PubMed]
42. Hossain, I.; Marklund, N.; Czeiter, E.; Hutchinson, P.; Buki, A. Blood biomarkers for traumatic brain injury: A narrative review of current evidence. *Brain Spine* **2023**, *4*, 102735. [CrossRef] [PubMed]
43. Sofroniew, M.V.; Vinters, H.V. Astrocytes: Biology and pathology. *Acta Neuropathol.* **2010**, *119*, 7–35. [CrossRef] [PubMed]
44. Loane, D.J.; Kumar, A. Microglia in the TBI brain: The good, the bad, and the dysregulated. *Exp. Neurol.* **2016**, *275*, 316–327. [CrossRef] [PubMed]
45. Simon, D.W.; McGeachy, M.J.; Bayır, H.; Clark, R.S.; Loane, D.J.; Kochanek, P.M. The far-reaching scope of neuroinflammation after traumatic brain injury. *Nat. Rev. Neurol.* **2017**, *13*, 171–191. Erratum in *Nat. Rev. Neurol.* **2017**, *13*, 572. [CrossRef] [CrossRef] [PubMed]
46. Johnson, V.E.; Stewart, J.E.; Begbie, F.D.; Trojanowski, J.Q.; Smith, D.H.; Stewart, W. Inflammation and white matter degeneration persist for years after a single traumatic brain injury. *Brain* **2013**, *136*, 28–42. [CrossRef] [PubMed]
47. Shandra, O.; Winemiller, A.R.; Heithoff, B.P.; Munoz-Ballester, C.; George, K.K.; Benko, M.J.; Zuidhoek, I.A.; Besser, M.N.; Curley, D.E.; Edwards, G.F., III; et al. Repetitive diffuse mild traumatic brain injury causes an atypical astrocyte response and spontaneous recurrent seizures. *J. Neurosci.* **2019**, *39*, 1944–1963. [CrossRef] [PubMed]
48. Moretti, L.; Cristofori, I.; Weaver, S.M.; Chau, A.; Portelli, J.N.; Grafman, J. Cognitive decline in older adults with a history of traumatic brain injury. *Lancet Neurol.* **2012**, *11*, 1103–1112. [CrossRef] [PubMed]
49. Khakh, B.S.; Sofroniew, M.V. Diversity of astrocyte functions and phenotypes in neural circuits. *Nat. Neurosci.* **2015**, *18*, 942–952. [CrossRef] [PubMed]

 https://doi.org/10.3390/ijms27010471

50. Chen, Y.; Swanson, R.A. Astrocytes and brain injury. *J. Cereb. Blood Flow Metab.* **2003**, *23*, 137–149. [CrossRef] [PubMed]
51. Papa, L.; Brophy, G.M.; Welch, R.D.; Lewis, L.M.; Braga, C.F.; Tan, C.N.; Ameli, N.J.; Lopez, M.A.; Haeussler, C.A.; Mendez Giordano, D.I.; et al. Time course and diagnostic accuracy of glial and neuronal blood biomarkers GFAP and UCH-L1 in a large cohort of trauma patients with and without mild traumatic brain injury. *JAMA Neurol.* **2016**, *73*, 551–560. [CrossRef] [PubMed]
52. Marchi, N.; Cavaglia, M.; Fazio, V.; Bhudia, S.; Hallene, K.; Janigro, D. Peripheral markers of blood-brain barrier damage. *Clin. Chim. Acta* **2004**, *342*, 1–12. [CrossRef] [PubMed]
53. Berger, R.P.; Adelson, P.D.; Pierce, M.C.; Dulani, T.; Cassidy, L.D.; Kochanek, P.M. Serum neuron-specific enolase, S100B, and myelin basic protein concentrations after inflicted and noninflicted traumatic brain injury in children. *J. Neurosurg.* **2005**, *103*, 61–68. [CrossRef] [PubMed]
54. Sun, Y.; Bai, L.; Niu, X.; Wang, Z.; Yin, B.; Bai, G.; Zhang, D.; Gan, S.; Sun, C.; Wang, S.; et al. Elevated serum levels of inflammation-related cytokines in mild traumatic brain injury are associated with cognitive performance. *Front. Neurol.* **2019**, *10*, 1120. [CrossRef] [PubMed]
55. Chen, X.; Nelson, C.D.; Li, X.; Winters, C.A.; Azzam, R.; Sousa, A.A.; Leapman, R.D.; Gainer, H.; Sheng, M.; Reese, T.S. PSD-95 is required to sustain the molecular organization of the postsynaptic density. *J. Neurosci.* **2011**, *31*, 6329–6338. [CrossRef] [PubMed]
56. Opazo, P.; Sainlos, M.; Choquet, D. Regulation of AMPA receptor surface diffusion by PSD-95 slots. *Curr. Opin. Neurobiol.* **2012**, *22*, 453–460. [CrossRef] [PubMed]
57. Beattie, E.C.; Stellwagen, D.; Morishita, W.; Bresnahan, J.C.; Ha, B.K.; Von Zastrow, M.; Beattie, M.S.; Malenka, R.C. Control of synaptic strength by glial TNFα. *Science* **2002**, *295*, 2282–2285. [CrossRef] [PubMed]
58. Bernardino, L.; Xapelli, S.; Silva, A.P.; Jakobsen, B.; Poulsen, F.R.; Oliveira, C.R.; Vezzani, A.; Malva, J.O.; Zimmer, J. Modulator effects of interleukin-1β and tumor necrosis factor-α on AMPA-induced excitotoxicity in mouse organotypic hippocampal slice cultures. *J. Neurosci.* **2005**, *25*, 6734–6744. [CrossRef]
59. Michaelis, E.K. Selective neuronal vulnerability in the hippocampus: Relationship to neurological diseases and mechanisms for differential sensitivity of neurons to stress. In *The Clinical Neurobiology of the Hippocampus*; Oxford University Press: Oxford, UK, 2012; pp. 54–76. [CrossRef]
60. Lana, D.; Ugolini, F.; Giovannini, M.G. An overview on the differential interplay among neurons–astrocytes–microglia in CA1 and CA3 hippocampus in hypoxia/ischemia. *Front. Cell. Neurosci.* **2020**, *14*, 585833. [CrossRef] [PubMed]
61. Alkadhi, K.A. Cellular and molecular differences between area CA1 and the dentate gyrus of the hippocampus. *Mol. Neurobiol.* **2019**, *56*, 6566–6580. [CrossRef] [PubMed]
62. Vasic, V.; Schmidt, M.H.H. Resilience and vulnerability to pain and inflammation in the hippocampus. *Int. J. Mol. Sci.* **2017**, *18*, 739. [CrossRef] [PubMed]
63. Gao, X.; Chen, J. Moderate traumatic brain injury promotes neural precursor proliferation without increasing neurogenesis in the adult hippocampus. *Exp. Neurol.* **2013**, *239*, 38–48. [CrossRef] [PubMed]
64. Bramham, C.R.; Alme, M.N.; Bittins, M.; Kuipers, S.D.; Nair, R.R.; Pai, B.; Panja, D.; Schubert, M.; Soule, J.; Tiron, A.; et al. The Arc of synaptic memory. *Exp. Brain Res.* **2010**, *200*, 125–140. Erratum in *Exp. Brain Res.* **2011**, *209*, 317. [CrossRef] [CrossRef] [PubMed]
65. Haley, M.; Bertrand, J.; Anderson, V.T.; Fuad, M.; Frenguelli, B.G.; Corrêa, S.A.L.; Wall, M.J. Arc expression regulates long-term potentiation magnitude and metaplasticity in area CA1 of the hippocampus in ArcKR mice. *Eur. J. Neurosci.* **2023**, *58*, 4166–4180. [CrossRef] [PubMed]
66. Khan, Z.U.; Carretero-Rey, M.; de León-López, C.A.M.; Navarro-Lobato, I. Memory-associated immediate early genes: Roles in synaptic function, memory processes, and neurological diseases. *Mol. Neurobiol.* **2025**, *62*, 15885–15915. [CrossRef] [PubMed]
67. Stevenson, R.F.; Reagh, Z.M.; Chun, A.P.; Murray, E.A.; Yassa, M.A. Pattern separation and source memory engage distinct hippocampal and neocortical regions during retrieval. *J. Neurosci.* **2020**, *40*, 843–851. [CrossRef] [PubMed]
68. Heim, L.R.; Bader, M.; Edut, S.; Rachmany, L.; Baratz-Goldstein, R.; Lin, R.; Elpaz, A.; Qubty, D.; Bikovski, L.; Rubovitch, V.; et al. The invisibility of mild traumatic brain injury: Impaired cognitive performance as a silent symptom. *J. Neurotrauma* **2017**, *34*, 2518–2528. [CrossRef] [PubMed]
69. Tomar, A.; Polygalov, D.; McHugh, T.J. Differential impact of acute and chronic stress on CA1 spatial coding and gamma oscillations. *Front. Behav. Neurosci.* **2021**, *15*, 710725. [CrossRef] [PubMed]
70. Ghasemi, M.; Navidhamidi, M.; Rezaei, F.; Azizikia, A.; Mehranfard, N. Anxiety and hippocampal neuronal activity: Relationship and potential mechanisms. *Cogn. Affect. Behav. Neurosci.* **2022**, *22*, 431–449. [CrossRef] [PubMed]
71. Bartsch, T.; Döhring, J.; Reuter, S.; Finke, C.; Rohr, A.; Brauer, H.; Deuschl, G.; Jansen, O. Selective neuronal vulnerability of human hippocampal CA1 neurons: Lesion evolution, temporal course, and pattern of hippocampal damage in diffusion-weighted MR imaging. *J. Cereb. Blood Flow Metab.* **2015**, *35*, 1836–1845. [CrossRef] [PubMed]
72. Chen, L.W.; Sun, D.; Davis, S.L.; Haswell, C.C.; Dennis, E.L.; Swanson, C.A.; Whelan, C.D.; Gutman, B.; Jahanshad, N.; Iglesias, J.E.; et al. Smaller hippocampal CA1 subfield volume in posttraumatic stress disorder. *Depress. Anxiety* **2018**, *35*, 1018–1029. [CrossRef] [PubMed]

73. Balvers, M.G.; Verhoeckx, K.C.; Plastina, P.; Wortelboer, H.M.; Meijerink, J.; Witkamp, R.F. Docosahexaenoic acid and eicosapentaenoic acid are converted by 3T3-L1 adipocytes to N-acyl ethanolamines with anti-inflammatory properties. *Biochim. Biophys. Acta* **2010**, *1801*, 1107–1114. [CrossRef] [PubMed]
74. Bazan, N.G. Cell survival matters: Docosahexaenoic acid signaling, neuroprotection and photoreceptors. *Trends Neurosci.* **2006**, *29*, 263–271. [CrossRef] [PubMed]
75. Tyrtyshnaia, A.; Bondar, A.; Konovalova, S.; Sultanov, R.; Manzhulo, I. N-Docosahexanoylethanolamine reduces microglial activation and improves hippocampal plasticity in a murine model of neuroinflammation. *Int. J. Mol. Sci.* **2020**, *21*, 9703. [CrossRef] [PubMed]
76. Ringsevjen, H.; Egbenya, D.L.; Bieler, M.; Davanger, S.; Hussain, S. Activity-regulated cytoskeletal-associated protein (Arc) in presynaptic terminals and extracellular vesicles in hippocampal synapses. *Front. Mol. Neurosci.* **2023**, *16*, 1225533. [CrossRef] [PubMed]
77. Riboldi, J.G.; Correa, J.; Renfijes, M.M.; Tintorelli, R.; Viola, H. Arc and BDNF mediated effects of hippocampal astrocytic glutamate uptake blockade on spatial memory stages. *Commun. Biol.* **2024**, *7*, 1032. [CrossRef]
78. Lama, A.; Pirozzi, C.; Annunziata, C.; Morgese, M.G.; Senzacqua, M.; Severi, I.; Calignano, A.; Trabace, L.; Giordano, A.; Meli, R.; et al. Palmitoylethanolamide counteracts brain fog improving depressive-like behaviour in obese mice: Possible role of synaptic plasticity and neurogenesis. *Br. J. Pharmacol.* **2021**, *178*, 845–859. [CrossRef] [PubMed]
79. Boccella, S.; Cristiano, C.; Romano, R.; Iannotta, M.; Belardo, C.; Farina, A.; Guida, F.; Piscitelli, F.; Palazzo, E.; Mazzitelli, M. Ultra-micronized palmitoylethanolamide rescues the cognitive decline-associated loss of neural plasticity in the neuropathic mouse entorhinal cortex-dentate gyrus pathway. *Neurobiol. Dis.* **2019**, *121*, 106–119. [CrossRef]
80. Rashid, M.A.; Katakura, M.; Kharebava, G.; Kevala, K.; Kim, H.Y. N-Docosahexaenoylethanolamine is a potent neurogenic factor for neural stem cell differentiation. *J. Neurochem.* **2013**, *125*, 869–884. [CrossRef]
81. Dyall, S.C. Long-chain omega-3 fatty acids and the brain: A review of the independent and shared effects of EPA, DPA and DHA. *Front. Aging Neurosci.* **2015**, *7*, 52. [CrossRef] [PubMed]
82. *Directive 2010/63/EU of the European Parliament and of the Council of 22 September 2010 on the protection of animals used for scientific purposes*; European Union: Brussels, Belgium, 2010.
83. Bevins, R.A.; Besheer, J. Object recognition in rats and mice: A one-trial non-matching-to-sample learning task to study 'recognition memory'. *Nat. Protoc.* **2006**, *1*, 1306–1311. [CrossRef] [PubMed]

 https://doi.org/10.3390/ijms27010471

International Journal of
Molecular Sciences

MDPI

Review

Chronic Traumatic Encephalopathy as the Course of Alzheimer's Disease

Magdalena Pszczołowska [1], Kamil Walczak [1], Weronika Miśków [1], Katarzyna Antosz [1], Joanna Batko [1], Donata Kurpas [2] and Jerzy Leszek [3,*]

1 Faculty of Medicine, Wroclaw Medical University, Ludwika Pasteura 1, 50-367 Wrocław, Poland; magdalena.pszczolowska@gmail.com (M.P.)
2 Faculty of Health Sciences, Wroclaw Medical University, Ul. Kazimierza Bartla 5, 51-618 Wrocław, Poland
3 Clinic of Psychiatry, Department of Psychiatry, Wroclaw Medical University, Ludwika Pasteura 10, 50-367 Wrocław, Poland
* Correspondence: jerzy.leszek@umw.edu.pl

Abstract: This editorial investigates chronic traumatic encephalopathy (CTE) as a course of Alzheimer's disease (AD). CTE is a debilitating neurodegenerative disease that is the result of repeated mild traumatic brain injury (TBI). Many epidemiological studies show that experiencing a TBI in early or middle life is associated with an increased risk of dementia later in life. Chronic traumatic encephalopathy (CTE) and Alzheimer's disease (AD) present a series of similar neuropathological features that were investigated in this work like recombinant tau into filaments or the accumulation and aggregation of Aβ protein. However, these two conditions differ from each other in brain–blood barrier damage. The purpose of this review was to evaluate information about CTE and AD from various articles, focusing especially on new therapeutic possibilities for the improvement in cognitive skills.

Keywords: chronic traumatic encephalopathy; traumatic brain injury; Alzheimer's Disease; oxidative stress; dementia; cognitive skills

1. Methodology Based on Preferred Presenting Items for Systematic Reviews and Meta-Analyses (PRISMA)

In this research study, the electronic database searches were focused on chronic traumatic encephalopathy and traumatic brain injury, as well as their connections and common points with Alzheimer's Disease. This article is based on original articles, meta-analyses, characteristic cases, and systematic reviews that describe the above issues. Starting with studies from PubMed, the searched keywords were "chronic traumatic encephalopathy", "traumatic brain injury", "Alzheimer's Disease", "dementia", and "cognitive skills", mostly in a period from 2010 to February 2024. Additionally, articles related to the topics discussed and important publications from the past were included. As for the excluded articles, those that did not directly refer to the topic discussed, were outdated or too limited, and did not typically fit the context of the work were rejected.

2. Introduction

Chronic traumatic encephalopathy (CTE) is a debilitating neurodegenerative disease that results from repeated mild traumatic brain injury (TBI). It is increasingly seen in athletes [1]. Many epidemiological studies show that experiencing a TBI in early or middle life is associated with an increased risk of dementia later in life. Multiple mild TBIs in professional boxers have been observed to be associated with a high risk of chronic traumatic encephalopathy (CTE), a type of dementia with distinctive clinical and pathological features [2]. Chronic traumatic encephalopathy (CTE) and Alzheimer's disease (AD) share several similar neuropathological features. These are primarily neurofibrillary tangles and hyperphosphorylated tau, but these entities have been described separately. New research shows that CTE and AD can occur in the same patients [3]. The neurofibrillary tangles found in the cortex in AD and CTE have different distributions. In AD, tangles in layers V and VI are visible in the deeper cortex with a laminar distribution.

In contrast, tangles may be perivascular in CTE and occur mainly in the more superficial layers II and III. Neurofibrillary tangles occur in the substantia nigra of people with CTE but are rare in AD. In AD, confusion appears most frequently in CA1; in CTE, confusion is seen in the horn of Ammon (CA1-CA4) [3].

Cases of CTE have been reported to coexist with Alzheimer's disease (AD) pathology. In the cohort of CTE patients described by Stein and colleagues, amyloid β peptide (Aβ) deposition in the form of diffuse or neuritic plaque was found in 52% of cases [4]. The diagnosis of CTE is made post-mortem, but specific medical tools can identify characteristic features of CTE based on serum, clinical history, cerebrospinal fluid, and neuroimaging biomarkers. Knowledge of tau isoform processing mechanisms may lead to new treatments targeting proteases. Their goal is to prevent the formation of toxic tau fragments in neurodegenerative diseases of tauopathy [5]. CTE is compared to other known neurodegenerative entities, including Alzheimer's disease and dementia with Lewy bodies [5]. Understanding the physiological processes in CTE and AD will allow us to compare these diseases and learn about their possible impact on each other.

3. Alzheimer's Disease—Basic Facts

Alzheimer's Disease (AD) belongs to the group of neurodegenerative diseases of the central nervous system. It is the most common cause of dementia (60-80% of cases), and it is expected that the number of patients with this diagnosis will increase due to the growing proportion of people who live beyond the age of sixty-five [6,7]. In the United States, Alzheimer's remains the fifth-leading cause of death among people aged 65 and older [6].

There are many risk factors for AD and, among others, they are age, genetics (APOE-e4, Down syndrome, genetic mutations of the APP gene and genes for presenilin-1 and presenilin-2, and a first-degree relative with AD), smoking, obesity, hypertension, high cholesterol levels, diabetes, diet, low physical activity, lower socioeconomic status, and poor sleep quality. There have also been protective factors found, such as more years of formal education and remaining socially and mentally active throughout life, a heart-healthy diet, for example Mediterranean, Dietary Approaches to Stop Hypertension (DASH), or Mediterranean–DASH Intervention for Neurodegenerative Delay (MIND), and physical activity [6].

AD pathogenesis is based on the accumulation of the amyloid-beta (Aβ) outside of the cells and hyperphosphorylated tau proteins inside them. Those processes interfere with transmission between synapses and block the transportation of nutrients and molecules nourishing neurons. Another essential pathology is the degeneration of the cholinergic neurons in the nucleus basalis of Meynert. Amyloid deposits activate proinflammatory cytokines, which lead to chronic inflammation and exacerbation of the neurodegeneration process [8]. Many studies have found that vascular pathology and age-related changes in the blood vessels, such as brain–blood barrier (BBB) dysfunction or hypoperfusion and hypoxia, may play their part in AD's pathogenesis [9,10].

Primary symptoms of AD are memory, language, and thinking problems. Personality, mood, and behavior changes, such as wandering, are also observed. With time, the impairment progresses, affecting other functions, including walking or swallowing. Patients with AD live on an average for four to eight years after diagnosis [6].

The diagnosis of AD is made based on DSM-5 criteria, which include meeting the requirements of major or mild neurocognitive disorder, insidious onset, and gradual progression of at least one cognitive disorder. Probable AD is diagnosed if there is evidence of a causative AD genetic mutation or if all of the following are present: evidence of a decline in memory and learning and at least one other cognitive domain, gradual decline in cognition, and no evidence of mixed etiology [11]. ICD-11 criteria are also used in the diagnosis. They include meeting all diagnostic requirements for dementia, the presumption of AD based on clinical assessment, standardized cognitive and neuropsychological testing, neuroimaging, genetic and medical tests, family history, and gradual onset with progressing memory and word-finding difficulties and mild functional impairment [12]. Additional tests can be used, such as TK, MRI, FDG amyloid PET, or CSF examination. TK and MRI can show hippocampal and cortical atrophy in temporal and parietal regions, whereas the changes in CSF include low Aβ and high tau and phospho-tau levels [7,13]. An essential part of the diagnostic process is also excluding other probable causes of dementia, such as vitamin B12 deficiency, hypothyroidism, HIV, or depression [13]. Current strategies in the pharmacological treatment of AD include acetylcholinesterase inhibitors, donepezil, rivastigmine, and galantamine, which improve cognition, and an NMDA receptor antagonist—memantine, which is used in moderate and severe forms of AD [8]. New therapies involve monoclonal antibodies such as aducanumab and gantenerumab, which target Aβ [14].

4. Chronic Traumatic Encephalopathy

Chronic traumatic encephalopathy (CTE) is a progressive, neurodegenerative condition that can be revealed after prolonged exposure to repeated episodes of blunt head trauma. CTE manifests itself as persistent cognitive and neuropsychiatric symptoms [15–17].

The discourse of CTE was first started in 1928 by Harrison Stanford Martland using the term "punch-drunk". His observation was related to the characteristic neuropsychiatric syndrome, which some boxers presented after being exposed to repetitive traumatic brain injuries (TBI) [17–19]. From that point, more and more cases and studies were described. These studies featured scrupulously detailed variations, such as cortical atrophy, ventricular enlargement, fenestrated septum pellucidum, flattening of parts of the nervous system (e.g., the corpus callosum), substantia nigra depigmentation, cerebellar fibrosis, neuronal loss, and gliosis. There were also findings in neurons of the medial temporal lobe, cortex, and brainstem of argyrophilic neurofibrillary tangles (NFTs) [20]. As a result of all of these explorations and analyses, the neuropathological criteria for CTE were defined by the NINDS/NIBIB in 2016, with diagnostic refinements made in 2021 [21].

In Martland's works, "punch-drunk" referred to specific symptoms, mood issues, and forms of behavior (described, e.g., as "goofy" or "slug nutty") [21,22]. However, nowadays, CTE refers precisely to the neuropathological diagnosis in tissues and can be definitively diagnosed only by post-mortem neuropathological examination [21].

Starting from epidemiology, CTE is mainly associated with professional and amateur athletes, especially American football players and boxers, but it should not only be. Plenty of different groups of people are exposed to repetitive head impacts (RHIs), also known as repetitive traumatic brain injuries (TBIs), defined as cumulative exposure to recurrent concussive and subconcussive events. These injuries can be caused through contact and collision sports at any age, military service, severe epilepsy attacks, or being the victim of domestic violence [16,17,23,24].

Molecular Mechanisms of Tau Protein in CTE and AD

From a molecular point of view, CTE is a tauopathy with progressive accumulation of the tau protein observed even in middle-aged patients (the first significant findings were made in American football players in 2005) [19,23,25]. The characteristic pathognomonic lesions of a perivascular accumulation of neuronal phosphorylated tau (p-Tau) are described. These neuropathological findings can initially localize in mechanical trauma regions and then spread into the whole structure through an inflammatory response. The pathology of this process is exceptional, as these lesions are unlike any observed tauopathy [16,21,23,24] and will be precisely described in the subsequent paragraph.

Referring to TBIs mentioned at the beginning, it is worth highlighting that a traumatic brain injury (TBI) increases the risk of developing neurodegenerative diseases (NDs) and neurocognitive disorders. A TBI triggers extensive epigenetic modifications somehow. However, the precise molecular pathways still need to be fully elucidated and require more research. While taking this approach, other NDs (such as Alzheimer's or Parkinson's Diseases (AD and PD)) are worth mentioning, as they are strongly connected with the epigenetic changes of the comprehensively described mechanisms [23,26]. The present data suggest that a TBI is a potent risk factor for these NDs (CTE, AD, and PD) due to, e.g., persistent neuroinflammation, but most theories are hypothetical theses [23]. In conclusion, TBIs are being considered a cause of significant preventable neurodegeneration [16].

Lastly, there is a controversial side to CTE. As featured in Randolph's publication in 2018, there are also individual opinions that CTE is not an actual disease. Some scientists point out the lack of specific, verifiable data and accept it on the case study level with doubtful pathomechanisms. Additionally, the all-cause mortality rate as well as the suicide rate are about less than half of those expected in the presumptions of CTE [16,27]. To summarize, this area requires lots of confirmed and solid data. Nevertheless, there is a connection between TBIs and CTE and NDs [16,23].

5. Assembly of Recombinant Tau into Filaments (CTE and AD)

AD is a neurodegenerative disease, with intracellular neurofibrillary tangles of hyperphosphorylated tau and extracellular amyloid plaques composed of amyloid-β (Aβ) being characteristic in this condition [28]. Most age-related neurodegenerative diseases involve assembling small numbers of soluble proteins into insoluble amyloid fibrils. Of these proteins, tau is the most commonly affected. Tau is made into fibers in many diseases, including Alzheimer's disease (AD), chronic traumatic encephalopathy (CTE), globular glial tauopathy (GGT), corticobasal degeneration (CBD), or Pick's disease (PID) [29].

5.1. *Tau Protein Structure and Isoforms*

Tau is expressed mainly in the central and peripheral nervous system and is most abundant in the axons of nerve cells. It is divided into an N-terminal projection domain, a proline-rich region, a repetitive region, and a C-terminal domain [30]. Research suggests that tau in the form of oligomers, or more specifically in the form of filaments, is considered a neurotoxin. Before medical symptoms appear, tau oligomers are sequestered in the early stages of the disease. Morphologically, tau oligomers take the form of a sphere using atomic force microscopy. They fuse with two or more tau molecules between 6 and 20 nm in size [31].

In the human brain, six different isoforms are generated by alternative mRNA splicing mechanisms, which range from 45 to 65 kDa in length according to SDS gels. The isoforms differ in the presence or absence of the R2 domain, one of four partially repeating regions of the microtubule-binding domain (MTBR) designated R1, R2, R3, and R4, and the presence or absence of two 29 aa amino-terminal inserts encoded by exons 2 and 3 that are referred to as 0N, 1N, or 2N tau [32].

5.2. *Pathological Tau Phosphorylation*

In CTE, the deposition of hyperphosphorylated tau usually begins in the perivascular furrow area in the cortex and then spreads unevenly in the cortex. In contrast, AD shows a diffuse distribution of hyperphosphorylated tau in cortical regions [33]. Three tau phosphorylation sites, Thr231, Thr181, and Ser199, serve as biomarkers of AD. Hyperphosphorylation of tau at Thr212/Thr231/Ser262 or Ser199/Ser202/Thr205 can cause cell death because it leads to microtubule instability [34]. Phosphorylation at Thr175, Tor 231, Ser199, and Ser422 has been reported in CTE [35]. Phosphorylation of tau at Thr231 induces a cis–trans conformational change in p-tau. Within hours after a TBI, the cis p-tau conformation appears in neurons before tau oligomers, prefibrillar tangles, and NFTs are formed and cause damage to axons [36]. We can infer that cis p-tau has a function that drives neurodegeneration.

The assembly of tau into filaments can be initiated or accelerated by adding granules. Human tauopathies may spread in the brain. It has been studied that tau assemblies, when applied extracellularly, can cause aggregates to form, and then tau aggregates are laid [30].

The spread of the tau protein requires seeding and the capture and release of aggregates. Expressed tau can only be buried if it is capable of aggregation. Aggregation inhibitors may be responsible for reducing tau-induced seeding and spreading. Tau induces them. Many studies believe that monomeric tau cannot form aggregates [30].

Individual diseases differ in tau folds, but tau filaments from individuals with a given disease have the same structures. Identical protofilaments can arrange themselves differently to form distinct filaments, so-called ultrastructural polymorphs. In AD, two protofilaments join symmetrically to form paired helical filaments (PHFs) or asymmetrically to form straight filaments (SFs). Alzheimer's and CTE folds consist virtually entirely of R3 and R4. The distribution of protofilament fold cores explains the composition of tau filament isoforms [37]. Recombinant full-length tau is highly soluble [38]. Tau in Alzheimer's Disease also differs from normal tau by its very high degree of phosphorylation. The high degree of phosphorylation predisposes tau to aggregation in PHFs, but no conclusive confirmatory studies exist. Phosphorylation, particularly of the repeat domain, clearly regulates tau–microtubule interactions [39]. Moreover, while tau filaments exhibit diverse configurations across individual diseases, the involvement of Aβ protein emerges as a crucial factor in the pathogenesis of Alzheimer's disease (AD) and chronic traumatic encephalopathy (CTE). Understanding the structural intricacies of tau filaments sheds light on the interplay between tau pathology and Aβ accumulation, elucidating the complex mechanisms underlying neurodegeneration.

6. Aβ pathology in AD and CTE

The Aβ protein plays a crucial part in AD pathogenesis. Its accumulation and aggregation outside the cells trigger AD progression [8]. It consists of 39–40 amino acids, and its biogenesis involves proteolysis of the APP protein, i.e., a transmembrane protein. There are two paths of APP degradation: amyloidogenic and nonamyloidogenic. In the nonamyloidogenic pathway, secretases of alpha and gamma are engaged, and they lead to the formation of soluble products. In the pathogenesis of AD, APP is degraded through the cleavage of beta- and gamma-secretase, leading to the generation of insoluble Aβ, which aggregates into plaques [40], as illustrated in Figure 1.

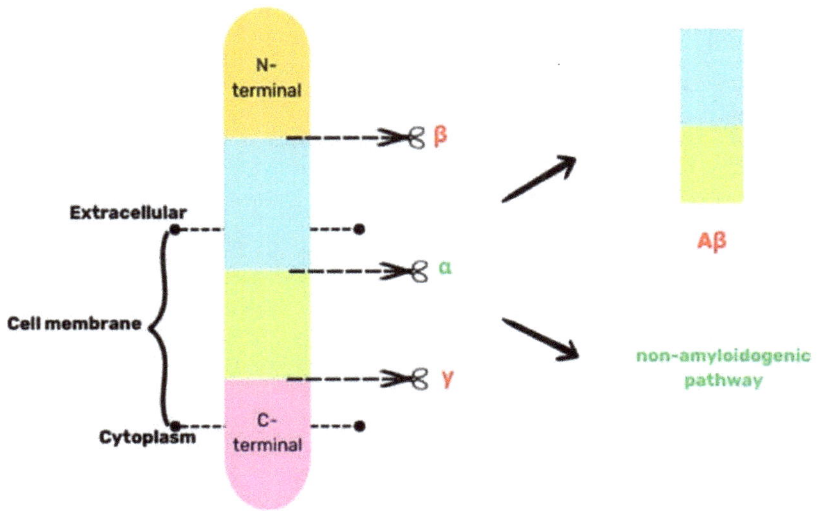

Figure 1. A simplified scheme of the biogenesis of Aβ based on [40].

Aβ oligomers have many neurotoxic effects. Aβ deposits activate various Toll-like receptors (TLRs) and coreceptors and cytokines of IL-1β family production, leading to chronic inflammation and consequently to an impairment in dendritic spines and disruption in the microglial clearance of Aβ. The synthesis of nitric oxide (NO) and the activation of CDKs is also enhanced, which further increases the peptide's ability to aggregate and suppress synaptic plasticity, as well as increasing the hyperphosphorylation of tau [8]. The aggregation of Aβ also promotes free radicals such as reactive oxidative species (ROS) through binding with metal ions, such as zinc and copper [41]. Oxidative changes in Aβ caused by ROS make its clearance even less effective. Aβ-induced oxidative stress also leads to the oxidation of proteins and lipids, causing cell membrane defects of DNA, resulting in its damage [40]. All of the effects are presented in Figure 2.

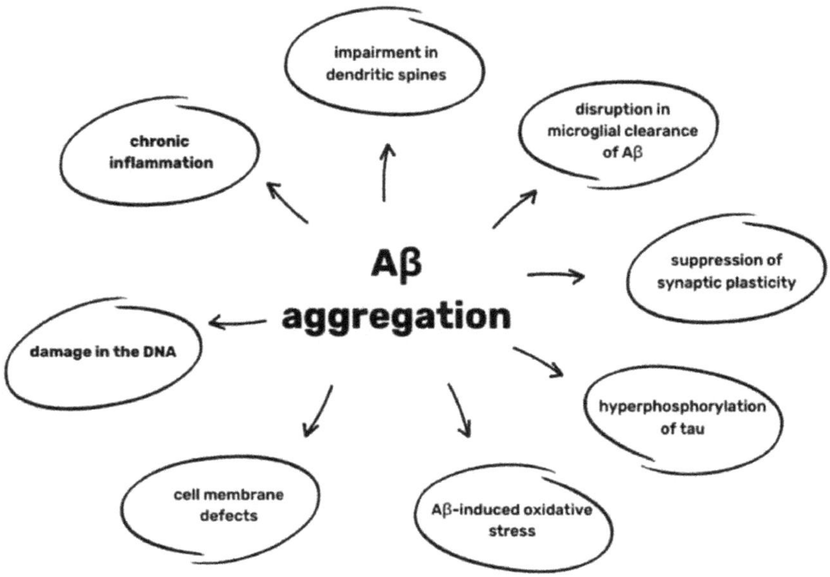

Figure 2. Effects of Aβ aggregation based on [8,40,41].

Aβ also aggregates in cerebral and leptomeningeal blood vessels, leading to cerebral amyloid angiopathy (CAA). The result of this pathology is the destruction of blood vessel walls, which increases the risk of either microhemorrhages or extensive lobar hemorrhages [42,43]. Another consequence of CAA can be leukoencephalopathy and CAA-related inflammation or angiitis. The described pathologies can cause a stroke and lead to further progression of dementia and cognitive impairment [43].

When it comes to chronic traumatic encephalopathy (CTE), Aβ can be found in the brains of about half of diagnosed patients, but the role of its accumulation in CTE is still not well understood [44]. It appears that in CTE patients, age-dependent Aβ deposition is accelerated compared to the general population, and they develop Aβ accumulation at a younger age [45,46]. Furthermore, in CTE, damage to cerebral blood vessels is observed, and it is associated with CAA [7]. Aβ presence is also connected to CTE progression. Studies show that Aβ is not present in stage I of CTE; in stage II, it is already observed in 19% of patients older than 50, and the number grows significantly in stages III and IV [8]. It was also observed that it increased the stage of CTE tauopathy. It also accelerated the development of dementia. The amyloid plaque localization is also different from that in AD—it is more peripheral, suggesting that its accumulation's pathogenesis is based mainly on impaired CSF clearance of Aβ [7]. It was also found that the presence of APOE ε4, one of the significant risk factors for AD, leads to greater accumulation of Aβ after a traumatic brain injury (TBI) and further increases the likelihood of its deposition in CTE [6].

The role of Aβ in CTE is still not fully known. Therefore, studies on this topic are needed to find more possible connections between CTE and AD in this field, which could lead to a better understanding of both diseases.

7. Endothelial Cell and BBB Damage in CTE and AD

The BBB is a physical barrier consisting of endothelial cells (ECs), which are connected tightly and communicate with themselves through specific junctional proteins. These cells act as the proper guards of the barrier; they maintain the integrity and regulate the movement of plasma proteins, ions, and all nutrients and toxins in the system. ECs can manage the expression of various transporters capable of different types of molecular transport (e.g., solute carrier-mediated, receptor-mediated, active efflux, and ion) [47–49]. The construction of the BBB is presented in Figure 3.

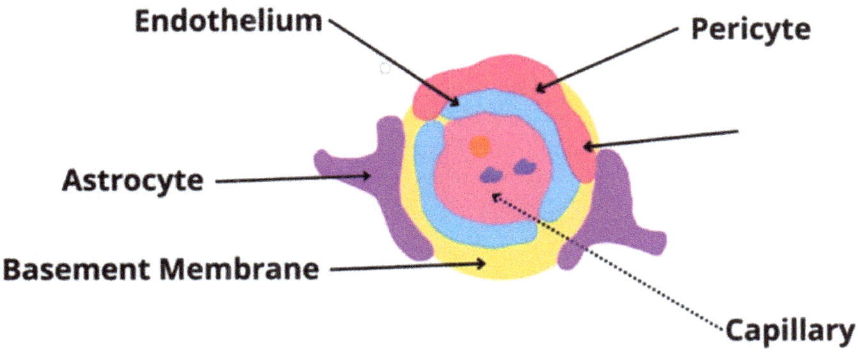

Figure 3. The blood–brain barrier—a representation image of the most crucial structure [1–3].

Regarding BBB construction, ECs are attached to a basement membrane of specific collagen (type IV), fibronectin, and laminin. Not only are ECs surrounded by a basement membrane, but pericytes also are. Pericytes can express specialized particles or partitions in signaling pathways as well, and thanks to their perivascular location, pericytes are an essential part of the immune response of the nervous system [47–49].

Lastly, astrocytes, the characteristic star-shaped glial cells, must be mentioned. By connecting with their endfeet wrapping around the capillary walls, astrocytes connect and communicate with the brain's capillaries and neurons, constituting a linkage between the blood and the brain [47–49].

All of these things considered, the small arterial cross-sections are surrounded by these three different cell types, which determine the pivotal role of the BBB [48,50]. From this point, the image of the solid structure that surrounds the minor arterials is made. However, besides the isolation and the protection of the brain from different agents such as toxins or microorganisms, the BBB plays a vital role in the communication between these two worlds. EDc are strictly connected by specialized tight junctions, and even though the substance transport is minimal, it is worth emphasizing that oxygen and carbon dioxide can diffuse across the barrier without hindrance [48,49]. Additionally, special transport is performed for a whole variety of required agents, e.g., transmembrane diffusion for small lipid-soluble molecules, low-weight compounds (<400 Da), or ethanol (due to its low-hydrogen bond structure); solute carrier-mediated transport (CMT) for compounds such as carbohydrates, amino acids, fatty acids, monocarboxylic acids, nucleotides, hormones, vitamins, and organic ions; receptor-mediated transcytosis (RMT) for peptides, polypeptides, or even proteins; and the sodium pump (Na+ and K+-ATPase) for sodium and potassium flux [48,51,52].

Knowing the BBB's major components and functions, it is perfectly understandable that any BBB dysfunction can disrupt the complexity of brain homeostasis and reduce its protective effect on the central nervous system (CNS). Initially, these changes were attributed to vascular dementia. However, a new approach has appeared with the growing amount of data [53]. These changes can be detrimental to a variety of NDs. The ongoing endothelial pathology can contribute to the damage of the nervous structure and to Aβ and tau pathology, as described above [48,50,53].

This paragraph briefly specifies the BBB damage in NDs, such as AD and CTE. In the following parts, mitochondrial dysfunction in the vasculature and cerebrovascular inflammation will be discussed.

We start with the most apparent and visible way to observe BBB damage—the section. Described analyses of post-mortem tissues from patients with AD and other NDs demonstrate brain capillary leakages. Additionally, the accumulation of white or red blood cells from invalid infiltration and the degeneration of pericytes and EDc are observed with various molecular changes in the BBB and all CNS structures [48]. Moreover, when comparing AD with CTE, the results are adequate to confirm their dissimilar pathophysiology. In AD, the most prominent changes are the accumulation of fibrinogen, thrombin, albumin, IgG, and hemosiderin in the cortex and hippocampus. Mentioned proteins are often found to be characteristic of Aβ. In CTE, the perivascular macrophages and histiocytes with hemosiderin can be noticed [48,54,55].

While focusing on the post-mortem studies mentioned and characterized earlier, Aβ is a noticeable factor of BBB damage. The accumulation of Aβ in the cerebral meninges and smaller cerebral vessels, called cerebral amyloid angiopathy (CAA), is featured as the major BBB disruptor and a significant hallmark of AD. Moreover, even on the AD prediction level, vascular biomarkers can be detected before the noticeable CAA and Aβ deposits, leading to a cause-and-effect sequence conclusion [56].

Successively, introduction to the two-hit vascular hypothesis of AD is required. In this approach, the initial step of the pathological pathway, following genetic predisposition, is specifically cerebral blood vessel damage. From this point, BBB dysfunction starts and leads to all of the neurodegenerative processes described in the previous paragraphs. In addition to the first step, lifestyle and environment are mentioned, where variable factors can also influence Aβ propagation, which determines the second step and the development of AD [51,57]. This hypothesis is illustrated in Figure 4.

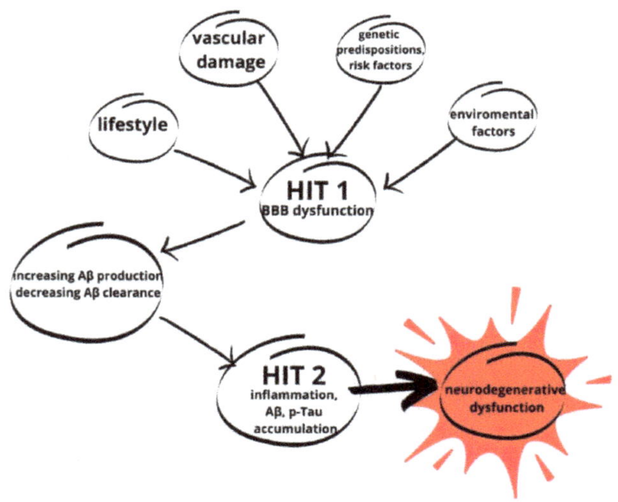

Figure 4. The two-hit hypothesis of AD. The first hit (HIT 1) is due to variables such as lifestyle, genetic predispositions and risk factors, environmental factors, and vascular damage described in this paper. The variables initiate a cascade of neurodegenerative changes. With BBB damage, the accumulation of Aβ starts, which strongly contributes to HIT 2. From HIT 2, through inflammation and Aβ and p-Tau accumulation, neurodegenerative changes lead to NDs (based on [51]).

Returning to other NDs, CTE mainly, BBB disruptions are also found. In the injury-exposed regions of the brain, a high density of perivascular p-Tau is described, with immunoreactivity signs of interruptions in the tight junctions (especially its components such as claudin-5 and zonula occludens-1) and with enhanced extravasation of endogenous blood components such as fibrinogen and IgG [58,59].

Damage of the BBB is a complex issue. It is worth remembering that these neurodegenerative changes limit or demolish the physiological processes and impact the pathophysiological ones. The damage also results in toxins and all kinds of xenobiotic accumulation. These xenobiotics, such as food additives, allergens, pollution, and drugs, also react with neurogenic brain functions and structures, contributing to disease development [48]. Moreover, the disturbed mechanism of cleansing the cerebral vascular space leads to accelerated Aβ accumulation connected with AD. From this point, another chain of events is made, where the reduced blood flow and increased Aβ levels are described due to the BBB damage. These changes successively promote pathological p-Tau pathways [60].

Approaching the end of this paragraph, it is important to mention that another step of xenobiotic accumulation is a valid aspect of neurodegeneration. Apart from p-Tau formation, an increasing inflammatory response due to the activation of astrocytes and microglia is also observed. The stormy secretion of cytokines and chemokines occurs in vascular and cerebral tissues, contributing to progressive neurodegenerative changes. This cerebrovascular inflammation will be discussed later [48].

8. Mitochondrial Dysfunction in the Vasculature in CTE and AD

Transitioning from the discussion of xenobiotic accumulation and inflammatory responses in neurodegeneration, we now turn our focus to another critical aspect: mitochondrial dysfunction in the vasculature in both CTE and AD. Traumatic brain injury can instigate chronic traumatic encephalopathy by disrupting blood flow within the nervous system, leading to ischemia and subsequent elevation of reactive oxygen species (ROS) levels. These molecules are produced in small amounts in healthy mitochondria; however, due to stress conditions, their level is increased. This may lead to many threatening actions in the cell, such as the induction of apoptosis through lipid peroxidation. Interestingly, ROS may release cytochrome C (CytC), which also may start apoptosis [61] or activate Nuclear factor-kappaB (NF-κB), which regulates the immune response. However, increased levels of NF-κB were discovered in neurons that were dying [61]. In a study on rodents, cerebral ischemia resulted in the activation of the NF-κB/MAPK/JNK pathway [62]. Moreover, elevated ROS levels may lead to DNA destruction like base damage and single-strand breaks (SSBs) [63] and mitochondrial permeabilization.

While numerous studies prove that ischemic brain damage upregulates the protection against antioxidants [64], some research indicates the downregulation of this defense. Kaur et al.'s study described reduced antioxidant defense in the early hours of post-stroke [6]. Antioxidant protection mechanisms are considered inefficient in the defense against elevated ROS levels, leading to cell death [65].

What is more, mitochondrial dysfunction due to brain injury is associated with an increased level of nitric oxide (NO). NO may regulate the functioning of this organelle through S-nitrosylation of mitochondrial proteins, like Receptor-interacting protein 3 (RIP3), a molecule sensor regulating cell apoptosis and necrosis [64]. This molecule can modulate apoptosis at several levels, such as blocking the apoptosome [66], inhibiting caspases [67], or influencing other cellular targets [68]. It was also proven that NO can cause mitochondria to commence cell death [69].

The ratio of NAD/NADH is critical when regulating many processes in the cell. As a result of severe trauma, poly (ADP-ribose) polymerase-1 (PARP-1) is the most active enzyme because it can promote DNA repair [66]. Activated PARP-1 leads to a significant decrease in intracellular NAD, especially under conditions of metabolic stress, while the production of NAD is limited due to the reduction in intracellular ATP [67].

Mitochondrial fragmentation and dendritic damage may be observed as a response to severe injury. Mitochondria may also change their structure—from a thin and long morphology to being swollen and fragmented. Ischemic stroke or brain trauma may affect mitochondria via two different mechanisms—they can protect neurons by adsorbing excessive Ca2+ and increasing levels of ATP or they can discharge proapoptotic factors that lead to cell death [70]. Proteins that mediate mitochondrial fission and fusion, such as Drp1, Opa1, and Mfn1/, are necessary for ischemic neuronal death [71,72].

Additionally, previous studies demonstrate that amyloid-β plaques enhance mitochondrial fission by increasing the generation of NO, which promotes S-nitrosylation of Drp1 [73,74]. These findings suggest that disturbances in mitochondrial activity are similar to traumatic brain injuries and AD. Furthermore, mitochondria play a crucial role in various astrocyte processes, emphasizing the importance of balanced mitochondrial dynamics in maintaining brain health [75,76].

9. Cerebrovascular Inflammation in CTE and AD

Building upon the link between traumatic brain injury (TBI) and Alzheimer's disease (AD), recent research by Tajiri et al. [77] confirms a pathological connection between the two conditions. In their experimental model, TBI significantly impacted the cognitive performance of AD-transgenic mice and nontransgenic (NT) mice trained in spatial memory tasks. Two and six weeks post-TBI, AD mice exhibited worsened performance compared to AD mice without a TBI and NT mice, regardless of TBI exposure [78].

The aftermath of a TBI involves damage to various brain structures, including neurons, axons, glia, and blood vessels, initiating numerous biochemical pathways that lead to cell death and neurodegeneration [79]. This primary injury not only affects the surrounding tissues directly but also triggers a cascade of neurodegeneration that can spread throughout the brain. The immediate inflammatory response, characterized by microglial activation, persists for at least one year post-injury, as evidenced by studies in rats and rhesus monkeys [80,81]. These findings suggest that the onset of a progressive degenerative process is associated with TBIs, potentially contributing to the later development of AD [81]. Microglial response to the injury is immediate [82], lasting for at least 1 year, which is proven by a study on rats [77]. There was significant tissue loss and ventriculomegaly in the hemisphere ipsilateral to the injured brain. The results of this study suggest the commencement of a progressive degenerative process connected to the occurrence of a TBI. Similar conclusions are drawn in a study on the rhesus monkey, whose microglia and macrophages were still activated one year after TBI [80].

A secondary cell death mechanism resulting from TBI is the secretion by injured cells of a neurotransmitter called intracellular glutamate. It is secreted into extracellular spaces [83], leading to cell necrosis and apoptosis by activating protein phosphatases, phospholipases, and endonucleases to fragment DNA [84].

Secondary cell death following a TBI may also be caused by neuroinflammation [85]. Some research claims that this process may last for many years and predispose patients to the development of AD [82]. Immune cells, microglia, cytokines, chemokines, and other inflammatory molecules are responsible for immunological responses [86]. A recent study has revealed increased levels of cytokine CCL11 in the frontal cortex in the brains of CTE patients as opposed to patients with AD [87]. Excessive activation leads to the upregulation of proinflammatory factors and increases the amount of neurotoxic molecules and free radicals, activating additional cell death mechanisms [78]. What is more, excessive expression of major histocompatibility complex class II (MHCII+) also leads to neurodegeneration [88].

Intriguingly, microglia exhibit two distinct phenotypes: M1, which increases the production of proinflammatory cytokines while decreasing anti-inflammatory cytokines, and M2, which stimulates inflammation by reducing proinflammatory cytokines and promoting anti-inflammatory molecules [89].

Similarly, astrocytes play a dual role in inflammation, capable of both increasing and decreasing inflammatory responses. Surface glycoproteins like the scavenger receptor activate pathways involved in neuroinflammation and the development of neurodegenerative diseases [90].

10. Treatment Methods of CTE

As primary injury can only be prevented by avoiding traumatic brain injuries (TBIs), the therapeutic focus for chronic traumatic encephalopathy (CTE) shifts toward mitigating events occurring during secondary cell death. The primary goal is to delay the onset of neuroinflammation [91]. One promising avenue in this regard is minocycline, a tetracycline derivative known to pass through the blood–brain barrier. Research indicates that minocycline reduces the concentration of proinflammatory cytokines and chemokines by modulating inflammatory regulators. Animal studies have shown normalization of serum and tissue levels of various inflammatory, vascular, neuronal, and glial markers, with animals treated with minocycline demonstrating improved neurological outcomes [92,93]. Nonetheless, further investigations are warranted to ascertain its suitability for humans with TBIs.

Another promising drug is melatonin, a hormone from the pineal gland. This hormone may inhibit microglial activation and reduce proinflammatory cytokine concentrations, limiting microglial activation. A study on mice revealed that the nuclear factor erythroid 2-related factor 2 and antioxidant-responsive element (Nrf2-ARE), an antioxidant, is transferred from the cytoplasm to the nucleus. It activates the Nrf2-ARE signaling pathway, which protects the cell from many pathological processes, including oxidative stress [94].

Another potential treatment option is statins, which have limited microglial and astrocyte activation. They reduce proinflammatory cytokines IL-1 β and TNF-α and intracellular adhesion molecules. Drugs like rosuvastatin limit brain superoxide production and inhibit factor-kappa B (NF-κB) activation. This substance has also promoted the activation of microglial cells after SAH [95]. Preclinical studies have proven its effectiveness in animals and humans in treating TBIs [96].

Stem cells are a hope for patients with TBIs due to their high degree of proliferation. This is the reason why the administration of stem cells improves cognitive skills, limits inflammation, and decreases cell death [97,98]. A special type called mesenchymal stem cells (MSCs) is used especially for TBI treatment because it strengthens the BBB. It does so by limiting the production of several chemokines, which results in decreasing inflammation. Decreasing the level of CXCL2 occurs by limiting neutrophil infiltration. Limiting CCL2 stops immune cell relocation, and the blocking of RANTES stops T-cell activation [99,100].

Interestingly, stem cells may be administrated not only to replace the damaged tissue but also to modulate the inflammatory response [101]. Stem cells may release a paracrine set of regulating molecules like cytokines, chemokines, and growth factors [102]. Those molecules may be released through exocytosis or by transport proteins and extracellular vesicles [103]. This is a new treatment possibility for neurodegenerative diseases, for example AD. There was a study conducted in which MSCs were intravenously injected into a transgenic mouse model of AD. After six weeks of injections, mice improved spatial learning, and there was a decrease in Aβ plaques in the brain [104]. MSCs may also be activated to express TSG-6, a multifunctional protein that modulates inflammation. This molecule built of a hyaluronan-binding Link domain and CUB module may inhibit neutrophil migration, enhancing the inhibiting activity of some enzymes that disturb the inflammation process [105–107].

The administration of granulocyte colony-stimulating factor (G-CSF) along with stem cells improves the treatment outcome due to the improvement in the survival rate of the stem cells. G-CSF can cross the BBB. This combined therapy increases angiogenesis in the brain and decreases inflammation and apoptosis [81].

Due to the mitochondrial imbalance which appears as a result of a TBI, some antioxidants have been proposed as a treatment option. Molecules like CoQ10 or MitoQ presented positive trial results [108,109]. Other therapeutic options are nicotinamide riboside (NR) and nicotinamide mononucleotide (NMN), which are NAD+ precursors [110].

One of the new ideas for treating AD is to use inhibiting substances like capsaicin, coenzyme Q2, mucidin, and stigmatellin, the largest multi-subunit complex that contributes to creating reactive oxygen species (ROS) [111].

Pharmacological therapy may be divided into the areas of cognition that it aims to improve. For memory impairment, the most suitable therapy is cholinesterase inhibitors [112,113]. Apathy, which is often a symptom for patients with dementia, may be treated with levodopa formulations, dopamine agonists, and stimulants [114]. Stimulants may be also useful for impaired attention treatment, which is common in CTE patients. The most commonly used drug is methylphenidate. This drug was found to decrease mental fatigue and improve mental processing speed, attention, and working memory [115]. Depression and anxiety are common in CTE patients. Using antidepressants in treating those conditions may also be beneficial in improving cognitive skills [116,117]. Agitation, which also occurs in CTE patients, is treated with different drugs depending on the cause of this pathology. Anxiety caused by agitation may be treated with sertraline or escitalopram. Night-time agitation, which is caused by sleep disturbances, is primarily treated with sleep hygiene, followed by low doses of risperidone [118].

It is very important to note that nonpharmacologic or combination therapy may be the first-line treatment for CTE. One of the disease treatments is also exercising. Improving blood circulation in the brain thanks to aerobic exercise has been described in some studies [19,20]. The positive influence of aerobic exercise on neuroplasticity and the subsequent increase in neurogenesis was also investigated [21].

Another option for patients suffering from CTE is a Mediterranean diet. This type of diet is said to be balanced in the best possible way, providing the patient with the highest amount of nutrients and, at the same time, the lowest dose of saturated fats [22].

Cognitive rehabilitation is said to improve daily life quality, but not to treat the disease. This kind of therapy is a set of interventions that help the patient's cognitive skills improve. It is beneficial at any time after the injury, in both mild and severe injuries. It is very important that at the beginning of the treatment process, there is a precise examination of cognitive skills performed to obtain the most accurate treatment plan [23]. This form of treatment was proposed in the last century; however, new inventions in technology have allowed us to obtain better results [24].

As a support for the main therapy, there may also be mood/behavior therapy administered. This would be beneficial for patients presenting depression or anxiety symptoms. Cognitive behavioral therapy (CBT) helps to create balanced cognitive patterns to avoid the stress and anxiety associated with psychological problems. CBT aims to teach functional and helpful behaviors. It is also believed that practicing mindfulness is helpful for some patients [25].

Some sources claim that occupational–ocular therapy (OOT) may improve cognitive skills in patients suffering from CTE which limits blurred or double vision. The help from certified ocular therapists may be beneficial in limiting other visual problems and improving visual processing and perceptual skills. Also, vestibular therapy may improve cognitive skills for patients whose hearing is impaired from a TBI [19].

Due to physical therapy, patients with impaired motor systems influence improvements, even in neurological functioning. Transcranial magnetic stimulation is believed to cause many beneficial results, the same as hyperbaric oxygen therapy [19]. For patients who suffer from cognitive impairment associated with neuroendocrine dysfunction, a new possibility for improvement is endocrine therapy. The hormonal problems caused by a TBI may be gonadotropin deficiency, growth hormone deficiency, secondary adrenal insufficiency, diabetes insipidus, inappropriate antidiuretic hormone secretion (SIADH), or thyroid-stimulating hormone deficiency [21]. Sometimes, this kind of hormonal disturbance may be only slightly intense and health care providers may not recognize the problem because of the brain injury itself. It is crucial to treat these conditions with proper hormone substitution.

As we look ahead to the future of CTE treatment, the focus lies on two main directions: better prevention of traumatic brain injuries (TBIs) and more effective strategies for reducing neuroinflammation. The ideal treatment would involve controlling the cells and molecular processes contributing to secondary cell death, including immune cells, astrocytes, cytokines, and chemokines. A deeper understanding of CTE pathology will undoubtedly inform future treatment modalities.

11. Imagining of CTE

Transitioning from the prospects of treatment, a critical aspect of diagnosing CTE involves neuropathological examination. This diagnostic approach allows for the visualization of aggregations of the hyperphosphorylated tau protein, which can be observed in both neurons and glial cells [119,120]. Specifically, the identification of cis p-Tau, a distinct isomer believed to be generated following traumatic brain injury, holds diagnostic significance [121]. Additionally, according to the Preliminary National Institute of Neurological Disorders and Stroke criteria, the presence of disordered neurites around small vessels in an irregular pattern at the depths of the cortical sulci is required for the pathological diagnosis of CTE [122].

Mild cases may not present any macroscopic changes but advanced stages of CTE are characterized by decreased brain weight, extension of the ventricles, and atrophy of the neuronal tissue, especially in the frontal and temporal lobes. In some of the most severe cases, atrophy of the thalamus and hypothalamus may be observed. In most cases, changes like septal fenestrations or depigmentation of the substantia nigra may be observed [15].

Together with the severity of p-tau neurodegeneration, the degree of axonal disintegrity increases [123]. There may also be abnormally phosphorylated TDP-43 proteins observed. This protein may be found in the cerebral cortex, the medial temporal lobe, the diencephalon, the brainstem, and even the spinal cord [124].

Positron emission tomography (PET) allows us to see changes in the brain taking place during a TBI. Post-mortem studies have revealed microglial activation using the ligand 3HPK11195, which was present even many years after the traumatic injury. In the analysis of astrocyte activity, which also helps to characterize the range of the brain damage, there may also be fluorescence microscopes [76]. Among the PET ligands used in CTE neuroimaging may also be Fluorodeoxyglucose (FDG), which measures brain metabolism and glucose uptake level. Florbetapir (18F-AV-45) indicates fibrillar beta-amyloid deposits, while Flortaucipir (T807, 18F-AV1451) binds to the paired helical filament (PHF) tau. 18F-FMZ is a ligand that is helpful in GABA metabolism observation and 11C-DPA-713 facilitates neuroinflammation level observation [125].

Magnetic resonance spectroscopy (MRS) is also useful in CTE imaging. It reflects on metabolite concentration, which allows us to observe brain metabolism. High lactate levels indicate hypoxia and impairment of perfusion, while choline (Cho) increases points at diffuse axonal injury [125].

Magnetic resonance imaging (MRI) delivers detailed brain morphology information. It is beneficial in quantifying volumetric changes in the patient's brain. Conventional MR cannot visualize the microscopic damage, but those pathologies may be detected with diffusion tensor imaging (DTI), a noninvasive MRI method. This helps discover changes in the brain's white matter, relying on the water's diffusion properties [126]. A functional MRI (fMRI), which reflects cerebral blood flow in the brain, helps diagnose and observe the disease course [127].

Differentiating CTE from other neurodegenerative diseases, especially AD, is very important to ensure the best treatment options are chosen. Crucial to distinguishing CTE from AD is histopathological examination. AD shows a common and highly consistent pattern of distribution of neurofibrillary tangles (NFTs); on the contrary, the distribution of NFTs is more sporadic in CTE [128]. More characteristic of CTE is the presence of NFTS at the depths of sulci and the perivascular clustering of NFTs [129]. Interestingly, Tau distribution in CTE may reflect head injury patterns since the depths of the sulci are areas of stress concentration due to forces during a brain injury. Studies on contact-sport athletes revealed that pathological findings in the superior and dorsal–lateral frontal lobes in their brains matched frequent high-intensity impacts on the crown of their helmets [130]. The Aβ pathology in CTE patients is more significant in the depths of the sulci, contrary to AD patients, which had similar levels of Aβ plaque deposition both in the sulci and gyri [44].

12. Conclusions

The transition from discussions on neurobiological mechanisms to therapeutic possibilities underlines the interconnected nature of AD and CTE, two neurodegenerative diseases that exhibit both similarities and differences. A contradiction of the thesis stated in the title may be the fact that the β-helix region is differently confirmed in CTE patients' brains compared to those of people suffering from AD. This different conformation is responsible for creating a hydrophobic cavity in tau filaments. What is more, there are varied protofilament interfaces in both conditions [131]. In the work that used multiplex ELISA, there were specific clusters for each tauopathy identified. There were different proteins characteristic of CTE and AD. CCL21 was the most specific for CTE and FLT3L for AD [132]. These facts are, based on some researchers' arguments, contrary to the common pathology of the diseases.

By delving into topics such as cerebral vasculitis, Aβ pathology, and mitochondrial dysfunction in both conditions, this study highlights the intricate relationship between them. Moreover, it elucidates how neuropathological features characteristic of both diseases can coexist within a single individual, emphasizing the importance of dual diagnosis.

Understanding this interplay between neurotrauma and neurodegeneration is crucial as it offers insights into the impact of injury and age on disease development. Furthermore, by presenting treatment methods for CTE and exploring better biomarkers for detecting damage, this study paves the way for improved diagnosis and management strategies.

Moving forward, continued scientific inquiry into the relationship between CTE and AD holds promise for advancing our understanding of these overlapping human diseases. By uncovering new insights, researchers may ultimately contribute to the development of more effective diagnostic tools and therapeutic interventions for individuals affected by these debilitating conditions.

Author Contributions: Conceptualization, J.L. and M.P.; methodology, J.L.; validation, J.L., M.P. and D.K.; formal analysis, M.P., K.W., W.M., K.A. and J.B.; investigation, M.P., K.W., W.M., K.A. and J.B.; resources, M.P., K.W., W.M., K.A. and J.B.; data curation, M.P., K.W., W.M., K.A. and J.B.; writing—original draft preparation, M.P., K.W., W.M., K.A. and J.B.; writing—review and editing, J.L. and D.K.; visualization, M.P. and K.A.; supervision, J.L. and D.K.; project administration, J.L. All authors have read and agreed to the published version of the manuscript.

Funding: There was no funding for this study.

Institutional Review Board Statement: Not applicable.

Informed Consent Statement: Not applicable.

Data Availability Statement: Not applicable.

Conflicts of Interest: All authors declare no conflicts of interest.

References

1. Fesharaki-Zadeh, A. Chronic Traumatic Encephalopathy: A Brief Overview. *Front. Neurol.* **2019**, *10*, 713. [CrossRef]
2. Shively, S.; Scher, A.I.; Perl, D.P.; Diaz-Arrastia, R. Dementia Resulting from Traumatic Brain Injury. *Arch. Neurol.* **2012**, *69*, 1245–1251. [CrossRef]
3. Turner, R.C.; Lucke-Wold, B.P.; Robson, M.J.; Lee, J.M.; Bailes, J.E. Alzheimer's disease and chronic traumatic encephalopathy: Distinct but possibly overlapping disease entities. *Brain Inj.* **2016**, *30*, 1279–1292. [CrossRef]
4. Fesharaki-Zadeh, A. A Case of Possible Chronic Traumatic Encephalopathy and Alzheimer's Disease in an Ex-Football Player. *Neurologist* **2022**, *27*, 249–252. [CrossRef]

5. Brett, B.L.; Wilmoth, K.; Cummings, P.; Solomon, G.S.; McCrea, M.A.; Zuckerman, S.L. The Neuropathological and Clinical Diagnostic Criteria of Chronic Traumatic Encephalopathy: A Critical Examination in Relation to Other Neurodegenerative Diseases. *J. Alzheimers Dis.* **2019**, *68*, 591–608. [CrossRef] [PubMed]
6. 2023 Alzheimer's disease facts and figures. *Alzheimer's Dement.* **2023**, *19*, 1598–1695. [CrossRef] [PubMed]
7. Apostolova, L.G. Alzheimer Disease. *Contin. Lifelong Learn. Neurol.* **2016**, *22*, 419–434. [CrossRef]
8. Khan, S.; Barve, K.H.; Kumar, M.S. Recent Advancements in Pathogenesis, Diagnostics and Treatment of Alzheimer's Disease. *Curr. Neuropharmacol.* **2020**, *18*, 1106–1125. [CrossRef]
9. Kapasi, A.; Schneider, J.A. Vascular contributions to cognitive impairment, clinical Alzheimer's disease, and dementia in older persons. *Biochim. Biophys. Acta (BBA)—Mol. Basis Dis.* **2016**, *1862*, 878–886. [CrossRef] [PubMed]
10. Zlokovic, B.V. Neurovascular pathways to neurodegeneration in Alzheimer's disease and other disorders. *Nat. Rev. Neurosci.* **2011**, *12*, 723–738. [CrossRef] [PubMed]
11. American Psychiatric Association. *American Psychiatric Association: Diagnostic and Statistical Manual of Mental Disorders*, 5th ed.; American Psychiatric Association: Arlington, TX, USA, 2013.
12. World Health Organization (WHO). *International Classification of Diseases*; WHO: Geneva, Switzerland, 2021.
13. Atri, A. The Alzheimer's Disease Clinical Spectrum. *Med. Clin. N. Am.* **2019**, *103*, 263–293. [CrossRef] [PubMed]
14. Scheltens, P.; De Strooper, B.; Kivipelto, M.; Holstege, H.; Chételat, G.; Teunissen, C.E.; Cummings, J.; van der Flier, W.M. Alzheimer's disease. *Lancet* **2021**, *397*, 1577–1590. [CrossRef] [PubMed]
15. Mckee, A.C.; Abdolmohammadi, B.; Stein, T.D. *The Neuropathology of Chronic Traumatic Encephalopathy*; Elsevier: Amsterdam, The Netherlands, 2018; pp. 297–307. [CrossRef]
16. Byard, R.; Tiemensma, M.; Buckland, M.E.; Vink, R. Chronic traumatic encephalopathy (CTE)—Features and forensic considerations. *Forensic Sci. Med. Pathol.* **2023**, *19*, 620–624. [CrossRef] [PubMed]
17. McKee, A.C.; Mez, J.; Abdolmohammadi, B.; Butler, M.; Huber, B.R.; Uretsky, M.; Babcock, K.; Cherry, J.D.; Alvarez, V.E.; Martin, B.; et al. Neuropathologic and Clinical Findings in Young Contact Sport Athletes Exposed to Repetitive Head Impacts. *JAMA Neurol.* **2023**, *80*, 1037. [CrossRef] [PubMed]
18. Martland, H.S. Punch Drunk. *J. Am. Med. Assoc.* **1928**, *91*, 1103. [CrossRef]
19. Bieniek, K.F.; Cairns, N.J.; Crary, J.F.; Dickson, D.W.; Folkerth, R.D.; Keene, C.D.; Litvan, I.; Perl, D.P.; Stein, T.D.; Vonsattel, J.-P.; et al. The Second NINDS/NIBIB Consensus Meeting to Define Neuropathological Criteria for the Diagnosis of Chronic Traumatic Encephalopathy. *J. Neuropathol. Exp. Neurol.* **2021**, *80*, 210–219. [CrossRef]
20. McKee, A.C.; Cairns, N.J.; Dickson, D.W.; Folkerth, R.D.; Keene, C.D.; Litvan, I.; Perl, D.P.; Stein, T.D.; Vonsattel, J.-P.; Stewart, W.; et al. The first NINDS/NIBIB consensus meeting to define neuropathological criteria for the diagnosis of chronic traumatic encephalopathy. *Acta Neuropathol.* **2016**, *131*, 75–86. [CrossRef] [PubMed]
21. McKee, A.C.; Stein, T.D.; Huber, B.R.; Crary, J.F.; Bieniek, K.; Dickson, D.; Alvarez, V.E.; Cherry, J.D.; Farrell, K.; Butler, M.; et al. Chronic traumatic encephalopathy (CTE): Criteria for neuropathological diagnosis and relationship to repetitive head impacts. *Acta Neuropathol.* **2023**, *145*, 371–394. [CrossRef] [PubMed]
22. Casper, S.T. Punch-Drunk Slugnuts: Violence and the Vernacular History of Disease. *Isis* **2022**, *113*, 266–288. [CrossRef]
23. Smolen, P.; Dash, P.K.; Redell, J.B. Traumatic brain injury-associated epigenetic changes and the risk for neurodegenerative diseases. *Front. Neurosci.* **2023**, *17*, 1259405. [CrossRef]
24. Nowinski, C.J.; Bureau, S.C.; Buckland, M.E.; Curtis, M.A.; Daneshvar, D.H.; Faull, R.L.M.; Grinberg, L.T.; Hill-Yardin, E.L.; Murray, H.C.; Pearce, A.J.; et al. Applying the Bradford Hill Criteria for Causation to Repetitive Head Impacts and Chronic Traumatic Encephalopathy. *Front. Neurol.* **2022**, *13*, 938163. [CrossRef] [PubMed]
25. Omalu, B.I.; DeKosky, S.T.; Minster, R.L.; Kamboh, M.I.; Hamilton, R.L.; Wecht, C.H. Chronic Traumatic Encephalopathy in a National Football League Player. *Neurosurgery* **2005**, *57*, 128–134. [CrossRef]
26. Berson, A.; Nativio, R.; Berger, S.L.; Bonini, N.M. Epigenetic Regulation in Neurodegenerative Diseases. *Trends Neurosci.* **2018**, *41*, 587–598. [CrossRef]
27. Randolph, C. Chronic traumatic encephalopathy is not a real disease. *Arch. Clin. Neuropsychol.* **2018**, *33*, 644–648. [CrossRef] [PubMed]
28. Muralidar, S.; Ambi, S.V.; Sekaran, S.; Thirumalai, D.; Palaniappan, B. Role of tau protein in Alzheimer's disease: The prime pathological player. *Int. J. Biol. Macromol.* **2020**, *163*, 1599–1617. [CrossRef] [PubMed]
29. Shi, Y.; Murzin, A.G.; Falcon, B.; Epstein, A.; Machin, J.; Tempest, P.; Newell, K.L.; Vidal, R.; Garringer, H.J.; Sahara, N.; et al. Cryo-EM structures of tau filaments from Alzheimer's disease with PET ligand APN-1607. *Acta Neuropathol.* **2021**, *141*, 697–708. [CrossRef]
30. Goedert, M.; Eisenberg, D.S.; Crowther, R.A. Propagation of Tau Aggregates and Neurodegeneration. *Annu. Rev. Neurosci.* **2017**, *40*, 189–210. [CrossRef]
31. Mamun, A.; Uddin, M.; Mathew, B.; Ashraf, G. Toxic tau: Structural origins of tau aggregation in Alzheimer's disease. *Neural Regen. Res.* **2020**, *15*, 1417. [CrossRef]
32. Boyarko, B.; Hook, V. Human Tau Isoforms and Proteolysis for Production of Toxic Tau Fragments in Neurodegeneration. *Front. Neurosci.* **2021**, *15*, 702788. [CrossRef]
33. Katsumoto, A.; Takeuchi, H.; Tanaka, F. Tau Pathology in Chronic Traumatic Encephalopathy and Alzheimer's Disease: Similarities and Differences. *Front. Neurol.* **2019**, *10*, 980. [CrossRef]

34. Shahpasand, K.; Uemura, I.; Saito, T.; Asano, T.; Hata, K.; Shibata, K.; Toyoshima, Y.; Hasegawa, M.; Hisanaga, S. Regulation of Mitochondrial Transport and Inter-Microtubule Spacing by Tau Phosphorylation at the Sites Hyperphosphorylated in Alzheimer's Disease. *J. Neurosci.* **2012**, *32*, 2430–2441. [CrossRef]
35. Moszczynski, A.J.; Strong, W.; Xu, K.; McKee, A.; Brown, A.; Strong, M.J. Pathologic Thr 175 tau phosphorylation in CTE and CTE with ALS. *Neurology* **2018**, *90*, e380–e387. [CrossRef] [PubMed]
36. Kondo, A.; Shahpasand, K.; Mannix, R.; Qiu, J.; Moncaster, J.; Chen, C.-H.; Yao, Y.; Lin, Y.-M.; Driver, J.A.; Sun, Y.; et al. Antibody against early driver of neurodegeneration cis P-tau blocks brain injury and tauopathy. *Nature* **2015**, *523*, 431–436. [CrossRef] [PubMed]
37. Lövestam, S.; Koh, F.A.; van Knippenberg, B.; Kotecha, A.; Murzin, A.G.; Goedert, M.; Scheres, S.H. Assembly of recombinant tau into filaments identical to those of Alzheimer's disease and chronic traumatic encephalopathy. *eLife* **2022**, *11*, e76494. [CrossRef] [PubMed]
38. Pérez, M.; Valpuesta, J.M.; Medina, M.; de Garcini, E.M.; Avila, J. Polymerization of τ into Filaments in the Presence of Heparin: The Minimal Sequence Required for τ-τ Interaction. *J. Neurochem.* **1996**, *67*, 1183–1190. [CrossRef] [PubMed]
39. Kampers, T.; Friedhoff, P.; Biernat, J.; Mandelkow, E.-M.; Mandelkow, E. RNA stimulates aggregation of microtubule-associated protein tau into Alzheimer-like paired helical filaments. *FEBS Lett.* **1996**, *399*, 344–349. [CrossRef] [PubMed]
40. Sun, X.; Chen, W.-D.; Wang, Y.-D. β-Amyloid: The key peptide in the pathogenesis of Alzheimer's disease. *Front. Pharmacol.* **2015**, *6*, 221. [CrossRef] [PubMed]
41. Cheignon, C.; Tomas, M.; Bonnefont-Rousselot, D.; Faller, P.; Hureau, C.; Collin, F. Oxidative stress and the amyloid beta peptide in Alzheimer's disease. *Redox Biol.* **2018**, *14*, 450–464. [CrossRef] [PubMed]
42. Trejo-Lopez, J.A.; Yachnis, A.T.; Prokop, S. Neuropathology of Alzheimer's Disease. *Neurotherapeutics* **2022**, *19*, 173–185. [CrossRef]
43. Yamada, M.; Naiki, H. *Cerebral Amyloid Angiopathy*; Elsevier: Amsterdam, The Netherlands, 2012; pp. 41–78. [CrossRef]
44. Stein, T.D.; Montenigro, P.H.; Alvarez, V.E.; Xia, W.; Crary, J.F.; Tripodis, Y.; Daneshvar, D.H.; Mez, J.; Solomon, T.; Meng, G.; et al. Beta-amyloid deposition in chronic traumatic encephalopathy. *Acta Neuropathol.* **2015**, *130*, 21–34. [CrossRef]
45. Turk, K.W.; Geada, A.; Alvarez, V.E.; Xia, W.; Cherry, J.D.; Nicks, R.; Meng, G.; Daley, S.; Tripodis, Y.; Huber, B.R.; et al. A comparison between tau and amyloid-β cerebrospinal fluid biomarkers in chronic traumatic encephalopathy and Alzheimer disease. *Alzheimers Res. Ther.* **2022**, *14*, 28. [CrossRef]
46. McKee, A.C.; Stein, T.D.; Kiernan, P.T.; Alvarez, V.E. The Neuropathology of Chronic Traumatic Encephalopathy. *Brain Pathol.* **2015**, *25*, 350–364. [CrossRef] [PubMed]
47. Abrahamson, E.E.; Ikonomovic, M.D. Brain injury-induced dysfunction of the blood brain barrier as a risk for dementia. *Exp. Neurol.* **2020**, *328*, 113257. [CrossRef] [PubMed]
48. Sweeney, M.D.; Sagare, A.P.; Zlokovic, B.V. Blood–brain barrier breakdown in Alzheimer disease and other neurodegenerative disorders. *Nat. Rev. Neurol.* **2018**, *14*, 133–150. [CrossRef] [PubMed]
49. Zhao, W.; Xu, Z.; Cao, J.; Fu, Q.; Wu, Y.; Zhang, X.; Long, Y.; Zhang, X.; Yang, Y.; Li, Y.; et al. Elamipretide (SS-31) improves mitochondrial dysfunction, synaptic and memory impairment induced by lipopolysaccharide in mice. *J. Neuroinflamm.* **2019**, *16*, 230. [CrossRef] [PubMed]
50. Archie, S.R.; Al Shoyaib, A.; Cucullo, L. Blood-Brain Barrier Dysfunction in CNS Disorders and Putative Therapeutic Targets: An Overview. *Pharmaceutics* **2021**, *13*, 1779. [CrossRef]
51. Zhao, Z.; Nelson, A.R.; Betsholtz, C.; Zlokovic, B.V. Establishment and Dysfunction of the Blood-Brain Barrier. *Cell* **2015**, *163*, 1064–1078. [CrossRef] [PubMed]
52. Pardridge, W. Targeted delivery of protein and gene medicines through the blood–brain barrier. *Clin. Pharmacol. Ther.* **2015**, *97*, 347–361. [CrossRef] [PubMed]
53. Toledo, J.B.; Arnold, S.E.; Raible, K.; Brettschneider, J.; Xie, S.X.; Grossman, M.; Monsell, S.E.; Kukull, W.A.; Trojanowski, J.Q. Contribution of cerebrovascular disease in autopsy confirmed neurodegenerative disease cases in the National Alzheimer's Coordinating Centre. *Brain* **2013**, *136*, 2697–2706. [CrossRef]
54. Zipser, B.D.; Johanson, C.E.; Gonzalez, L.; Berzin, T.M.; Tavares, R.; Hulette, C.M.; Vitek, M.P.; Hovanesian, V.; Stopa, E.G. Microvascular injury and blood–brain barrier leakage in Alzheimer's disease. *Neurobiol. Aging* **2007**, *28*, 977–986. [CrossRef]
55. Omalu, B.I.; Fitzsimmons, R.P.; Hammers, J.; Bailes, J. Chronic traumatic encephalopathy in a professional American wrestler. *J. Forensic Nurs.* **2010**, *6*, 130–136. [CrossRef]
56. Kuhn, J.; Sharman, T. Cerebral Amyloid Angiopathy. [Updated 2023 Jun 5]. In *StatPearls [Internet]*; StatPearls Publishing: Treasure Island, FL, USA, 2024. Available online: https://www.ncbi.nlm.nih.gov/books/NBK556105/ (accessed on 18 March 2024).
57. Nelson, A.R.; Sweeney, M.D.; Sagare, A.P.; Zlokovic, B.V. Neurovascular dysfunction and neurodegeneration in dementia and Alzheimer's disease. *Biochim. Biophys. Acta (BBA) Mol. Basis Dis.* **2016**, *1862*, 887–900. [CrossRef] [PubMed]
58. Doherty, C.P.; O'Keefe, E.; Wallace, E.; Loftus, T.; Keaney, J.; Kealy, J.; Humphries, M.M.; Molloy, M.G.; Meaney, J.F.; Farrell, M.; et al. Blood–Brain Barrier Dysfunction as a Hallmark Pathology in Chronic Traumatic Encephalopathy. *J. Neuropathol. Exp. Neurol.* **2016**, *75*, 656–662. [CrossRef] [PubMed]
59. Farrell, M.; Aherne, S.; O'Riordan, S.; O'Keeffe, E.; Greene, C.; Campbell, M. Blood-brain barrier dysfunction in a boxer with chronic traumatic encephalopathy and schizophrenia. *Clin. Neuropathol.* **2019**, *38*, 51–58. [CrossRef] [PubMed]
60. Kisler, K.; Nelson, A.R.; Montagne, A.; Zlokovic, B.V. Cerebral blood flow regulation and neurovascular dysfunction in Alzheimer disease. *Nat. Rev. Neurosci.* **2017**, *18*, 419–434. [CrossRef] [PubMed]

61. Ueno, T.; Sawa, Y.; Kitagawa-Sakakida, S.; Nishimura, M.; Morishita, R.; Kaneda, Y.; Kohmura, E.; Yoshimine, T.; Matsuda, H. Nuclear factor-κB decoy attenuates neuronal damage after global brain ischemia: A future strategy for brain protection during circulatory arrest. *J. Thorac. Cardiovasc. Surg.* **2001**, *122*, 720–727. [CrossRef] [PubMed]
62. Kim, E.J.; Raval, A.P.; Hirsch, N.; Perez-Pinzon, M.A. Ischemic Preconditioning Mediates Cyclooxygenase-2 Expression Via Nuclear Factor-Kappa B Activation in Mixed Cortical Neuronal Cultures. *Transl. Stroke Res.* **2010**, *1*, 40–47. [CrossRef]
63. Manzanero, S.; Santro, T.; Arumugam, T.V. Neuronal oxidative stress in acute ischemic stroke: Sources and contribution to cell injury. *Neurochem. Int.* **2013**, *62*, 712–718. [CrossRef] [PubMed]
64. Dahm, C.C.; Moore, K.; Murphy, M.P. Persistent S-Nitrosation of Complex I and Other Mitochondrial Membrane Proteins by S-Nitrosothiols but Not Nitric Oxide or Peroxynitrite. *J. Biol. Chem.* **2006**, *281*, 10056–10065. [CrossRef]
65. Singh, R.; Lakhanpal, D.; Kumar, S.; Sharma, S.; Kataria, H.; Kaur, M.; Kaur, G. Late-onset intermittent fasting dietary restriction as a potential intervention to retard age-associated brain function impairments in male rats. *Age* **2012**, *34*, 917–933. [CrossRef]
66. Kosenko, E.; Montoliu, C.; Giordano, G.; Kaminsky, Y.; Venediktova, N.; Buryanov, Y.; Felipo, V. Acute ammonia intoxication induces an NMDA receptor-mediated increase in poly(ADP-ribose) polymerase level and NAD + metabolism in nuclei of rat brain cells. *J. Neurochem.* **2004**, *89*, 1101–1110. [CrossRef] [PubMed]
67. Soane, L.; Kahraman, S.; Kristian, T.; Fiskum, G. Mechanisms of impaired mitochondrial energy metabolism in acute and chronic neurodegenerative disorders. *J. Neurosci. Res.* **2007**, *85*, 3407–3415. [CrossRef] [PubMed]
68. Bernassola, F.; Catani, M.V.; Corazzari, M.; Rossi, A.; Melino, G. Inactivation of multiple targets by nitric oxide in CD95-triggered apoptosis. *J. Cell Biochem.* **2001**, *82*, 123–133. [CrossRef] [PubMed]
69. Vieira, H.; Kroemer, G. Mitochondria as Targets of Apoptosis Regulation by Nitric Oxide. *IUBMB Life* **2003**, *55*, 613–616. [CrossRef] [PubMed]
70. Kislin, M.; Sword, J.; Fomitcheva, I.V.; Croom, D.; Pryazhnikov, E.; Lihavainen, E.; Toptunov, D.; Rauvala, H.; Ribeiro, A.S.; Khiroug, L.; et al. Reversible Disruption of Neuronal Mitochondria by Ischemic and Traumatic Injury Revealed by Quantitative Two-Photon Imaging in the Neocortex of Anesthetized Mice. *J. Neurosci.* **2017**, *37*, 333–348. [CrossRef]
71. Owens, K.; Park, J.H.; Gourley, S.; Jones, H.; Kristian, T. Mitochondrial dynamics: Cell-type and hippocampal region specific changes following global cerebral ischemia. *J. Bioenerg. Biomembr.* **2015**, *47*, 13–31. [CrossRef]
72. Peng, C.; Rao, W.; Zhang, L.; Wang, K.; Hui, H.; Wang, L.; Su, N.; Luo, P.; Hao, Y.; Tu, Y.; et al. Mitofusin 2 ameliorates hypoxia-induced apoptosis via mitochondrial function and signaling pathways. *Int. J. Biochem. Cell Biol.* **2015**, *69*, 29–40. [CrossRef]
73. Cho, D.-H.; Nakamura, T.; Fang, J.; Cieplak, P.; Godzik, A.; Gu, Z.; Lipton, S.A. S-Nitrosylation of Drp1 Mediates β-Amyloid-Related Mitochondrial Fission and Neuronal Injury. *Science (1979)* **2009**, *324*, 102–105. [CrossRef]
74. Lu, M.-H.; Zhao, X.-Y.; Yao, P.-P.; Xu, D.-E.; Ma, Q.-H. The Mitochondrion: A Potential Therapeutic Target for Alzheimer's Disease. *Neurosci. Bull.* **2018**, *34*, 1127–1130. [CrossRef]
75. Sarkar, S.; Malovic, E.; Harischandra, D.S.; Ngwa, H.A.; Ghosh, A.; Hogan, C.; Rokad, D.; Zenitsky, G.; Jin, H.; Anantharam, V.; et al. Manganese exposure induces neuroinflammation by impairing mitochondrial dynamics in astrocytes. *Neurotoxicology* **2018**, *64*, 204–218. [CrossRef]
76. Guilhaume-Correa, F.; Pickrell, A.M.; VandeVord, P.J. The Imbalance of Astrocytic Mitochondrial Dynamics Following Blast-Induced Traumatic Brain Injury. *Biomedicines* **2023**, *11*, 329. [CrossRef] [PubMed]
77. Smith, D.H.; Chen, X.-H.; Pierce, J.E.S.; Wolf, J.A.; Trojanowski, J.Q.; Graham, D.I.; Mcintosh, T.K. Progressive Atrophy and Neuron Death for One Year Following Brain Trauma in the Rat. *J. Neurotrauma* **1997**, *14*, 715–727. [CrossRef]
78. Kabadi, S.; Faden, A. Neuroprotective Strategies for Traumatic Brain Injury: Improving Clinical Translation. *Int. J. Mol. Sci.* **2014**, *15*, 1216–1236. [CrossRef] [PubMed]
79. Sullivan, P.G.; Thompson, M.B.; Scheff, S.W. Cyclosporin A Attenuates Acute Mitochondrial Dysfunction Following Traumatic Brain Injury. *Exp. Neurol.* **1999**, *160*, 226–234. [CrossRef] [PubMed]
80. Nagamoto-Combs, K.; McNeal, D.W.; Morecraft, R.J.; Combs, C.K. Prolonged Microgliosis in the Rhesus Monkey Central Nervous System after Traumatic Brain Injury. *J. Neurotrauma* **2007**, *24*, 1719–1742. [CrossRef]
81. Acosta, S.A.; Tajiri, N.; Shinozuka, K.; Ishikawa, H.; Sanberg, P.R.; Sanchez-Ramos, J.; Song, S.; Kaneko, Y.; Borlongan, C.V. Combination Therapy of Human Umbilical Cord Blood Cells and Granulocyte Colony Stimulating Factor Reduces Histopathological and Motor Impairments in an Experimental Model of Chronic Traumatic Brain Injury. *PLoS ONE* **2014**, *9*, e90953. [CrossRef]
82. Giunta, B.; Obregon, D.; Velisetty, R.; Sanberg, P.R.; Borlongan, C.V.; Tan, J. The immunology of traumatic brain injury: A prime target for Alzheimer's disease prevention. *J. Neuroinflamm.* **2012**, *9*, 678. [CrossRef]
83. Mehta, A.; Prabhakar, M.; Kumar, P.; Deshmukh, R.; Sharma, P.L. Excitotoxicity: Bridge to various triggers in neurodegenerative disorders. *Eur. J. Pharmacol.* **2013**, *698*, 6–18. [CrossRef]
84. Martin, L.J.; Al-Abdulla, N.A.; Brambrink, A.M.; Kirsch, J.R.; Sieber, F.E.; Portera-Cailliau, C. Neurodegeneration in Excitotoxicity, Global Cerebral Ischemia, and Target Deprivation: A Perspective on the Contributions of Apoptosis and Necrosis. *Brain Res. Bull.* **1998**, *46*, 281–309. [CrossRef]
85. Tajiri, N.; Kellogg, S.L.; Shimizu, T.; Arendash, G.W.; Borlongan, C.V. Traumatic Brain Injury Precipitates Cognitive Impairment and Extracellular Aβ Aggregation in Alzheimer's Disease Transgenic Mice. *PLoS ONE* **2013**, *8*, e78851. [CrossRef]

86. Tajiri, N.; Hernandez, D.; Acosta, S.; Shinozuka, K.; Ishikawa, H.; Ehrhart, J.; Diamandis, T.; Gonzales-Portillo, C.; Borlongan, M.C.; Tan, J.; et al. Suppressed cytokine expression immediatey following traumatic brain injury in neonatal rats indicates an expeditious endogenous anti-inflammatory response. *Brain Res.* **2014**, *1559*, 65–71. [CrossRef]
87. Cherry, J.D.; Stein, T.D.; Tripodis, Y.; Alvarez, V.E.; Huber, B.R.; Au, R.; Kiernan, P.T.; Daneshvar, D.H.; Mez, J.; Solomon, T.M.; et al. CCL11 is increased in the CNS in chronic traumatic encephalopathy but not in Alzheimer's disease. *PLoS ONE* **2017**, *12*, e0185541. [CrossRef]
88. Hernandez-Ontiveros, D.G.; Tajiri, N.; Acosta, S.; Giunta, B.; Tan, J.; Borlongan, C.V. Microglia Activation as a Biomarker for Traumatic Brain Injury. *Front. Neurol.* **2013**, *4*, 30. [CrossRef]
89. Chhor, V.; Le Charpentier, T.; Lebon, S.; Oré, M.-V.; Celador, I.L.; Josserand, J.; Degos, V.; Jacotot, E.; Hagberg, H.; Sävman, K.; et al. Characterization of phenotype markers and neuronotoxic potential of polarised primary microglia in vitro. *Brain Behav. Immun.* **2013**, *32*, 70–85. [CrossRef] [PubMed]
90. Wu, L.-Y.; Bao, X.-Q.; Sun, H.; Zhang, D. Scavenger receptor on astrocytes and its relationship with neuroinflammation. *Zhongguo Yi Xue Ke Xue Yuan Xue Bao* **2014**, *36*, 330–335. [CrossRef] [PubMed]
91. Szmydynger-Chodobska, J.; Gandy, J.R.; Varone, A.; Shan, R.; Chodobski, A. Synergistic Interactions between Cytokines and AVP at the Blood-CSF Barrier Result in Increased Chemokine Production and Augmented Influx of Leukocytes after Brain Injury. *PLoS ONE* **2013**, *8*, e79328. [CrossRef]
92. Kovesdi, E.; Kamnaksh, A.; Wingo, D.; Ahmed, F.; Grunberg, N.E.; Long, J.B.; Kasper, C.E.; Agoston, D.V. Acute Minocycline Treatment Mitigates the Symptoms of Mild Blast-Induced Traumatic Brain Injury. *Front. Neurol.* **2012**, *3*, 111. [CrossRef]
93. Xu, L.; Fagan, S.C.; Waller, J.L.; Edwards, D.; Borlongan, C.V.; Zheng, J.; Hill, W.D.; Feuerstein, G.; Hess, D.C. Low dose intravenous minocycline is neuroprotective after middle cerebral artery occlusion-reperfusion in rats. *BMC Neurol.* **2004**, *4*, 7. [CrossRef] [PubMed]
94. Ding, K.; Wang, H.; Xu, J.; Li, T.; Zhang, L.; Ding, Y.; Zhu, L.; He, J.; Zhou, M. Melatonin stimulates antioxidant enzymes and reduces oxidative stress in experimental traumatic brain injury: The Nrf2–ARE signaling pathway as a potential mechanism. *Free Radic. Biol. Med.* **2014**, *73*, 1–11. [CrossRef]
95. Uekawa, K.; Hasegawa, Y.; Ma, M.; Nakagawa, T.; Katayama, T.; Sueta, D.; Toyama, K.; Kataoka, K.; Koibuchi, N.; Kawano, T.; et al. Rosuvastatin Ameliorates Early Brain Injury after Subarachnoid Hemorrhage via Suppression of Superoxide Formation and Nuclear Factor-Kappa B Activation in Rats. *J. Stroke Cerebrovasc. Dis.* **2014**, *23*, 1429–1439. [CrossRef]
96. Loane, D.J.; Faden, A.I. Neuroprotection for traumatic brain injury: Translational challenges and emerging therapeutic strategies. *Trends Pharmacol. Sci.* **2010**, *31*, 596–604. [CrossRef] [PubMed]
97. Tajiri, N.; Duncan, K.; Antoine, A.; Pabon, M.; Acosta, S.A.; de la Pena, I.; Hernadez-Ontiveros, D.G.; Shinozuka, K.; Ishikawa, H.; Kaneko, Y.; et al. Stem cell-paved biobridge facilitates neural repair in traumatic brain injury. *Front. Syst. Neurosci.* **2014**, *8*, 116. [CrossRef] [PubMed]
98. Antonucci, I.; Pantalone, A.; Tete, S.; Salini, V.; Borlongan, C.V.; Hess, D.; Stuppia, L. Amniotic Fluid Stem Cells: A Promising Therapeutic Resource for Cell-Based Regenerative Therapy. *Curr. Pharm. Des.* **2012**, *18*, 1846–1863. [CrossRef] [PubMed]
99. Yoo, S.-W.; Chang, D.-Y.; Lee, H.-S.; Kim, G.-H.; Park, J.-S.; Ryu, B.-Y.; Joe, E.-H.; Lee, Y.-D.; Kim, S.-S.; Suh-Kim, H. Immune following suppression mesenchymal stem cell transplantation in the ischemic brain is mediated by TGF-β. *Neurobiol. Dis.* **2013**, *58*, 249–257. [CrossRef] [PubMed]
100. Zhang, R.; Liu, Y.; Yan, K.; Chen, L.; Chen, X.-R.; Li, P.; Chen, F.-F.; Jiang, X.-D. Anti-inflammatory and immunomodulatory mechanisms of mesenchymal stem cell transplantation in experimental traumatic brain injury. *J. Neuroinflamm.* **2013**, *10*, 871. [CrossRef] [PubMed]
101. Bonaventura, G.; Munafò, A.; Bellanca, C.M.; La Cognata, V.; Iemmolo, R.; Attaguile, G.A.; Di Mauro, R.; Di Benedetto, G.; Cantarella, G.; Barcellona, M.L.; et al. Stem Cells: Innovative Therapeutic Options for Neurodegenerative Diseases? *Cells* **2021**, *10*, 1992. [CrossRef] [PubMed]
102. Vizoso, F.J.; Eiro, N.; Cid, S.; Schneider, J.; Perez-Fernandez, R. Mesenchymal Stem Cell Secretome: Toward Cell-Free Therapeutic Strategies in Regenerative Medicine. *Int. J. Mol. Sci.* **2017**, *18*, 1852. [CrossRef] [PubMed]
103. Nawaz, M.; Camussi, G.; Valadi, H.; Nazarenko, I.; Ekström, K.; Wang, X.; Principe, S.; Shah, N.; Ashraf, N.M.; Fatima, F.; et al. The emerging role of extracellular vesicles as biomarkers for urogenital cancers. *Nat. Rev. Urol.* **2014**, *11*, 688–701. [CrossRef]
104. Kim, K.-S.; Kim, H.S.; Park, J.-M.; Kim, H.W.; Park, M.-K.; Lee, H.-S.; Lim, D.S.; Lee, T.H.; Chopp, M.; Moon, J. Long-term immunomodulatory effect of amniotic stem cells in an Alzheimer's disease model. *Neurobiol. Aging* **2013**, *34*, 2408–2420. [CrossRef]
105. Watanabe, J.; Shetty, A.K.; Hattiangady, B.; Kim, D.-K.; Foraker, J.E.; Nishida, H.; Prockop, D.J. Administration of TSG-6 improves memory after traumatic brain injury in mice. *Neurobiol. Dis.* **2013**, *59*, 86–99. [CrossRef]
106. Milner, C.M.; Higman, V.A.; Day, A.J. TSG-6: A pluripotent inflammatory mediator? *Biochem. Soc. Trans.* **2006**, *34*, 446–450. [CrossRef] [PubMed]
107. Mahoney, D.J.; Mulloy, B.; Forster, M.J.; Blundell, C.D.; Fries, E.; Milner, C.M.; Day, A.J. Characterization of the Interaction between Tumor Necrosis Factor-stimulated Gene-6 and Heparin. *J. Biol. Chem.* **2005**, *280*, 27044–27055. [CrossRef] [PubMed]
108. McManus, M.J.; Murphy, M.P.; Franklin, J.L. The Mitochondria-Targeted Antioxidant MitoQ Prevents Loss of Spatial Memory Retention and Early Neuropathology in a Transgenic Mouse Model of Alzheimer's Disease. *J. Neurosci.* **2011**, *31*, 15703–15715. [CrossRef] [PubMed]

109. Muthukumaran, K.; Kanwar, A.; Vegh, C.; Marginean, A.; Elliott, A.; Guilbeault, N.; Badour, A.; Sikorska, M.; Cohen, J.; Pandey, S. Ubisol-Q10 (a Nanomicellar Water-Soluble Formulation of CoQ10) Treatment Inhibits Alzheimer-Type Behavioral and Pathological Symptoms in a Double Transgenic Mouse (TgAPEswe, PSEN1dE9) Model of Alzheimer's Disease. *J. Alzheimer's Dis.* **2017**, *61*, 221–236. [CrossRef] [PubMed]
110. Hou, Y.; Lautrup, S.; Cordonnier, S.; Wang, Y.; Croteau, D.L.; Zavala, E.; Zhang, Y.; Moritoh, K.; O'Connell, J.F.; Baptiste, B.A.; et al. NAD + supplementation normalizes key Alzheimer's features and DNA damage responses in a new AD mouse model with introduced DNA repair deficiency. *Proc. Natl. Acad. Sci. USA* **2018**, *115*, E1876–E1885. [CrossRef]
111. Fato, R.; Bergamini, C.; Bortolus, M.; Maniero, A.L.; Leoni, S.; Ohnishi, T.; Lenaz, G. Differential effects of mitochondrial Complex I inhibitors on production of reactive oxygen species. *Biochim. Biophys. Acta (BBA)-Bioenerg.* **2009**, *1787*, 384–392. [CrossRef]
112. Polich, G.; Iaccarino, M.A.; Zafonte, R. *Psychopharmacology of Traumatic Brain Injury*; Elsevier: Amsterdam, The Netherlands, 2019; pp. 253–267. [CrossRef]
113. Anderson, N.D. State of the science on mild cognitive impairment (MCI). *CNS Spectr.* **2019**, *24*, 78–87. [CrossRef]
114. Cantu, R.; Budson, A. Management of chronic traumatic encephalopathy. *Expert Rev. Neurother.* **2019**, *19*, 1015–1023. [CrossRef]
115. Kakehi, S.; Tompkins, D.M. A Review of Pharmacologic Neurostimulant Use During Rehabilitation and Recovery After Brain Injury. *Ann. Pharmacother.* **2021**, *55*, 1254–1266. [CrossRef]
116. Salter, K.L.; McClure, J.A.; Foley, N.C.; Sequeira, K.; Teasell, R.W. Pharmacotherapy for Depression Posttraumatic Brain Injury: A Meta-analysis. *J. Head Trauma Rehabil.* **2016**, *31*, E21–E32. [CrossRef]
117. Silverberg, N.D.; Panenka, W.J. Antidepressants for depression after concussion and traumatic brain injury are still best practice. *BMC Psychiatry* **2019**, *19*, 100. [CrossRef] [PubMed]
118. Rahmani, E.; Lemelle, T.M.; Samarbafzadeh, E.; Kablinger, A.S. Pharmacological Treatment of Agitation and/or Aggression in Patients with Traumatic Brain Injury: A Systematic Review of Reviews. *J. Head Trauma Rehabil.* **2021**, *36*, E262–E283. [CrossRef] [PubMed]
119. McKee, A.C. The Neuropathology of Chronic Traumatic Encephalopathy: The Status of the Literature. *Semin. Neurol.* **2020**, *40*, 359–369. [CrossRef] [PubMed]
120. Smith, D.H.; Johnson, V.E.; Trojanowski, J.Q.; Stewart, W. Chronic traumatic encephalopathy—Confusion and controversies. *Nat. Rev. Neurol.* **2019**, *15*, 179–183. [CrossRef] [PubMed]
121. VanItallie, T.B. Traumatic brain injury (TBI) in collision sports: Possible mechanisms of transformation into chronic traumatic encephalopathy (CTE). *Metabolism* **2019**, *100*, 153943. [CrossRef] [PubMed]
122. Katz, D.I.; Bernick, C.; Dodick, D.W.; Mez, J.; Mariani, M.L.; Adler, C.H.; Alosco, M.L.; Balcer, L.J.; Banks, S.J.; Barr, W.B.; et al. National Institute of Neurological Disorders and Stroke Consensus Diagnostic Criteria for Traumatic Encephalopathy Syndrome. *Neurology* **2021**, *96*, 848–863. [CrossRef] [PubMed]
123. Holleran, L.; Kim, J.H.; Gangolli, M.; Stein, T.; Alvarez, V.; McKee, A.; Brody, D.L. Axonal disruption in white matter underlying cortical sulcus tau pathology in chronic traumatic encephalopathy. *Acta Neuropathol.* **2017**, *133*, 367–380. [CrossRef]
124. McKee, A.C.; Gavett, B.E.; Stern, R.A.; Nowinski, C.J.; Cantu, R.C.; Kowall, N.W.; Perl, D.P.; Hedley-Whyte, E.T.; Price, B.; Sullivan, C.; et al. TDP-43 Proteinopathy and Motor Neuron Disease in Chronic Traumatic Encephalopathy. *J. Neuropathol. Exp. Neurol.* **2010**, *69*, 918–929. [CrossRef] [PubMed]
125. Lin, A.; Charney, M.; Shenton, M.E.; Koerte, I.K. *Chronic Traumatic Encephalopathy: Neuroimaging Biomarkers*; Elsevier: Amsterdam, The Netherlands, 2018; pp. 309–322. [CrossRef]
126. Sundman, M.; Doraiswamy, P.M.; Morey, R.A. Neuroimaging assessment of early and late neurobiological sequelae of traumatic brain injury: Implications for CTE. *Front. Neurosci.* **2015**, *9*, 334. [CrossRef]
127. Logothetis, N.K. What we can do and what we cannot do with fMRI. *Nature* **2008**, *453*, 869–878. [CrossRef]
128. McKee, A.C.; Cantu, R.C.; Nowinski, C.J.; Hedley-Whyte, E.T.; Gavett, B.E.; Budson, A.E.; Santini, V.E.; Lee, H.-S.; Kubilus, C.A.; Stern, R.A. Chronic Traumatic Encephalopathy in Athletes: Progressive Tauopathy After Repetitive Head Injury. *J. Neuropathol. Exp. Neurol.* **2009**, *68*, 709–735. [CrossRef] [PubMed]
129. McKee, A.C.; Stein, T.D.; Nowinski, C.J.; Stern, R.A.; Daneshvar, D.H.; Alvarez, V.E.; Lee, H.-S.; Hall, G.; Wojtowicz, S.M.; Baugh, C.M.; et al. The spectrum of disease in chronic traumatic encephalopathy. *Brain* **2013**, *136*, 43–64. [CrossRef] [PubMed]
130. McKee, A.C.; Daneshvar, D.H.; Alvarez, V.E.; Stein, T.D. The neuropathology of sport. *Acta Neuropathol.* **2014**, *127*, 29–51. [CrossRef] [PubMed]
131. Falcon, B.; Zivanov, J.; Zhang, W.; Murzin, A.G.; Garringer, H.J.; Vidal, R.; Crowther, R.A.; Newell, K.L.; Ghetti, B.; Goedert, M.; et al. Novel tau filament fold in chronic traumatic encephalopathy encloses hydrophobic molecules. *Nature* **2019**, *568*, 420–423. [CrossRef]
132. Cherry, J.D.; Baucom, Z.H.; Eppich, K.G.; Kirsch, D.; Dixon, E.R.; Tripodis, Y.; Bieniek, K.F.; Farrell, K.; Whitney, K.; Uretsky, M.; et al. Neuroimmune proteins can differentiate between tauopathies. *J. Neuroinflamm.* **2022**, *19*, 278. [CrossRef]

International Journal of
Molecular Sciences

MDPI

Perspective

Alzheimer's Is a Multiform Disease of Sustained Neuronal Integrated Stress Response Driven by the C99 Fragment Generated Independently of AβPP; Proteolytic Production of Aβ Is Suppressed in AD-Affected Neurons: Evolution of a Theory

Vladimir Volloch [1,*] and Sophia Rits-Volloch [2,3,*]

1 Department of Developmental Biology, Harvard School of Dental Medicine, Boston, MA 02115, USA
2 Division of Molecular Medicine, Children's Hospital, Boston, MA 02115, USA
3 Department of Biological Chemistry and Molecular Pharmacology, Harvard Medical School, Boston, MA 02115, USA
* Correspondence: vladimir.volloch@gmail.com (V.V.); rits@crystal.harvard.edu (S.R.-V.)

Abstract: The present *Perspective* analyzes the remarkable evolution of the Amyloid Cascade Hypothesis 2.0 (ACH2.0) theory of Alzheimer's disease (AD) since its inception a few years ago, as reflected in the diminishing role of amyloid-beta (Aβ) in the disease. In the initial iteration of the ACH2.0, Aβ-protein-precursor (AβPP)-derived intraneuronal Aβ (iAβ), accumulated to neuronal integrated stress response (ISR)-eliciting levels, triggers AD. The neuronal ISR, in turn, activates the AβPP-independent production of its C99 fragment that is processed into iAβ, which drives the disease. The second iteration of the ACH2.0 stemmed from the realization that AD is, in fact, a disease of the sustained neuronal ISR. It introduced two categories of AD—conventional and unconventional—differing mainly in the manner of their causation. The former is caused by the neuronal ISR triggered by AβPP-derived iAβ, whereas in the latter, the neuronal ISR is elicited by stressors distinct from AβPP-derived iAβ and arising from brain trauma, viral and bacterial infections, and various types of inflammation. Moreover, conventional AD always contains an unconventional component, and in both forms, the disease is driven by iAβ generated independently of AβPP. In its third, the current, iteration, the ACH2.0 posits that proteolytic production of Aβ is suppressed in AD-affected neurons and that the disease is driven by C99 generated independently of AβPP. Suppression of Aβ production in AD seems an oxymoron: Aβ is equated with AD, and the later is inconceivable without the former in an ingrained Amyloid Cascade Hypothesis (ACH)-based notion. But suppression of Aβ production in AD-affected neurons is where the logic leads, and to follow it we only need to overcome the inertia of the preexisting assumptions. Moreover, not only is the generation of Aβ suppressed, so is the production of all components of the AβPP proteolytic pathway. This assertion is not a quantum leap (unless overcoming the inertia counts as such): the global cellular protein synthesis is severely suppressed under the neuronal ISR conditions, and there is no reason for constituents of the AβPP proteolytic pathway to be exempted, and they, apparently, are not, as indicated by the empirical data. In contrast, tau protein translation persists in AD-affected neurons under ISR conditions because the human tau mRNA contains an internal ribosomal entry site in its 5′UTR. In current mouse models, iAβ derived from AβPP expressed exogenously from human transgenes elicits the neuronal ISR and thus suppresses its own production. Its levels cannot principally reach AD pathology-causing levels regardless of the number of transgenes or the types of FAD mutations that they (or additional transgenes) carry. Since the AβPP-independent C99 production pathway is inoperative in mice, the current transgenic models have no potential for developing the full spectrum of AD pathology. What they display are only effects of the AβPP-derived iAβ-elicited neuronal ISR. The paper describes strategies to construct adequate transgenic AD models. It also details the utilization of human neuronal cells as the only adequate model system currently available for conventional and unconventional AD. The final alteration of the ACH2.0, introduced in the present *Perspective*, is that AβPP, which supports neuronal functionality and viability, is, after all, potentially produced in AD-affected neurons, albeit not conventionally but in an ISR-driven and -compatible process. Thus, the present narrative begins with the "omnipotent" Aβ capable of both triggering and driving the disease and ends up with this peptide largely dislodged from its pedestal and retaining its central role in triggering the disease in only one, although prevalent (conventional), category of AD (and driving it in none). Among interesting inferences of the present *Perspective* is the determination that "sporadic AD" is not sporadic at all ("non-familial" would be a much better designation). The term has fatalistic connotations, implying that the disease can strike at random. This is patently not the case: The conventional disease affects a distinct subpopulation, and the basis for unconventional AD is well understood. Another conclusion is that, unless prevented, the occurrence of conventional AD is inevitable given a sufficiently long lifespan. This *Perspective* also defines therapeutic directions not to be taken as well as auspicious ways forward. The former category includes ACH-based drugs (those interfering with the proteolytic production of Aβ and/or depleting extracellular Aβ). They are legitimate (albeit inefficient) preventive agents for conventional AD. There is, however, a proverbial snowball's chance in hell of them being effective in symptomatic AD, lecanemab, donanemab, and any other "...mab" or "...stat" notwithstanding. They comprise Aβ-specific antibodies, inhibitors of beta- and gamma-secretase, and modulators of the latter. In the latter category, among ways to go are the following: (1) Depletion of iAβ, which, if sufficiently "deep", opens up a tantalizing possibility of once-in-a-lifetime preventive transient treatment for conventional AD and aging-associated cognitive decline, AACD. (2) Composite therapy comprising the degradation of C99/iAβ and concurrent inhibition of the neuronal ISR. A single transient treatment could be sufficient to arrest the progression of conventional AD and prevent its recurrence for life. Multiple recurrent treatments would achieve the same outcome in unconventional AD. Alternatively, the sustained reduction/removal of unconventional neuronal ISR-eliciting stressors through the elimination of their source would convert unconventional AD into conventional one, preventable/treatable by a single transient administration of the composite C99/iAβ depletion/ISR suppression therapy. Efficient and suitable ISR inhibitors are available, and it is explicitly clear where to look for C99/iAβ-specific targeted degradation agents—activators of BACE1 and, especially, BACE2. Directly acting C99/iAβ-specific degradation agents such as proteolysis-targeting chimeras (PROTACs) and molecular-glue degraders (MGDs) are also viable options. (3) A circumscribed shift (either upstream or downstream) of the position of transcription start site (TSS) of the human AβPP gene, or, alternatively, a gene editing-mediated excision or replacement of a small, defined segment of its portion encoding 5′-untranslated region of AβPP mRNA; targeting AβPP RNA with anti-antisense oligonucleotides is another possibility. If properly executed, these RNA-based strategies would not interfere with the protein-coding potential of AβPP mRNA, and each would be capable of both preventing and stopping the AβPP-independent generation of C99 and thus of either preventing AD or arresting the progression of the disease in its conventional and unconventional forms. The paper is interspersed with "validation" sections: every conceptually significant notion is either validated by the existing data or an experimental procedure validating it is proposed.

Keywords: Amyloid Cascade Hypothesis 2.0 (ACH2.0); conventional and unconventional Alzheimer's disease (AD); AβPP-independent generation of the C99 fragment; neuronal integrated stress response (ISR); AD as the disease of the neuronal ISR; ISR-mediated suppression of the AβPP proteolytic pathway in AD-affected neurons; C99 as the driver of AD; inhibition of the neuronal ISR as AD therapy; concurrent inhibition of the neuronal ISR and activation of BACE1 and BACE2 as composite AD therapy; RNA-dependent amplification of human AβPP mRNA; RNA-based AD therapies

1. Introduction

The present *Perspective* outlines and analyzes the remarkable evolution of a theory of Alzheimer's disease (AD), designated the Amyloid Cascade Hypothesis 2.0 (ACH2.0), since its formulation a few years ago [1–10]. The ACH2.0 shares a portion of its name with the preceding theory of AD, the Amyloid Cascade Hypothesis (ACH). This overlap is intentional. In the ACH, amyloid-beta (Aβ) is both the causative agent and the driver of the disease. At the time of the inception of the ACH2.0, it was assumed that the same is conceptually true for this theory of AD as well. With time, our understanding of AD evolved, and the ACH2.0 evolved accordingly. If anything, the evolution of the ACH2.0 is a story of the diminishing role of Aβ in the disease. In the initial iteration of the ACH2.0, Aβ-protein-precursor (AβPP)-derived intraneuronal Aβ (iAβ) accumulated to neuronal integrated stress response (ISR)-eliciting levels triggers AD, and iAβ produced independently of AβPP drives the disease. The realization that AD is actually a disease of the sustained neuronal ISR (which activates AβPP-independent production of C99 and, potentially, iAβ) prompted the second iteration of the ACH2.0. It introduced two categories of AD, conventional and unconventional, differing mainly in the manner of their causation. The former is caused by the neuronal ISR triggered by AβPP-derived iAβ, whereas in the latter, stressors that are distinct from AβPP-derived iAβ and arise from traumatic brain injury, bacterial and viral infections, and various forms of inflammation elicit the neuronal ISR. In its both forms, the disease is driven by iAβ generated independently of AβPP. The third iteration of the ACH2.0 posits that proteolytic production of Aβ is suppressed in AD-affected neurons and that the C99 fragment of AβPP generated independently of AβPP drives the disease. Moreover, not only is the production of Aβ suppressed, so is the production of all components of the AβPP proteolytic pathway, including AβPP and beta- and gamma-secretases. The final modification of the ACH2.0, introduced in the present *Perspective*, is that AβPP, which is required for the neuronal functionality and viability, is, after all, produced in AD-affected neurons, albeit not conventionally but in an ISR-driven and -compatible process. Thus, the story we are about to relay begins with an "omnipotent" Aβ capable of both triggering and driving the disease and ends with this peptide largely dislodged from its pedestal and retaining its central role in merely triggering the disease in only one, albeit prevalent, (conventional) category of AD (and driving it in none; thus, "central but not causative" in George Perry's words). To present the complete picture, we begin the story with the birth of AD as a distinct research field.

2. A Brief Historical Background

The field of Alzheimer's disease has a defined birthday: 3 November 1906. On that day, Dr. Alois Alzheimer presented a case study of his patient, Auguste D., at the 37th annual meeting of the German regional psychiatric association in Tubingen. The disorder that he described was yet to be called "Alzheimer's disease" but was referred to as "a peculiar severe disease of the cerebral cortex" marked by distinctive neuritic plaques and neurofibrillary tangles as its main histological attributes. The presentation elicited little enthusiasm or even interest; no post-presentation discussion, customary at that time, followed. In fact, not even questions were asked, and the session progressed promptly to more fashionable psychoanalytical cases. Nevertheless, the abstract of this presentation was included in the proceedings of the conference published in 1906 [11], and one year later, in 1907, Dr. Alzheimer succeeded in publishing his complete lecture [12]. The disease acquired its by now familiar name only in 1910, when Dr. Kraepelin, a colleague and mentor of Dr. Alzheimer, described it in the eighth edition of his textbook "Psychiatrie" [13,14] and referred to it as "Alzheimer's disease". Initially, the field developed quite slowly, in part due to the rarity of the disease (the life expectancy in Germany at the time was 48 years, insufficient for sporadic AD to develop; Dr. Alzheimer himself, for example, died in 1915 aged only 51); in his remaining lifetime, Dr. Alzheimer described only four additional clinical cases [15,16]. A noticeable milestone occurred in 1968, when Blessed and co-workers described what they interpreted as a quantitative association between cerebral plaques in brains of elderly subjects and senile changes in their behavior [17]. The next milestone took place in 1984, almost 80 years after the introduction of AD. This is the year when Glenner and Wong identified and sequenced the main component of neuritic plaques (Dr. Alzheimer referred to plaques as "miliary foci", which he described as dystrophic neuronal processes around a "special substance in the cortex"; what Glenner and Wong isolated and characterized was this "special substance"); they designated it as amyloid-beta (Aβ) peptide and speculated that it was derived from a larger precursor [18]. Since this seminal event, the development of the field rapidly accelerated. In 1987 three research groups, using the amino acid sequence of Aβ elucidated in [17], confirmed the prediction of Glenner and Wong [18]: they identified, independently and nearly simultaneously, the gene encoding the protein precursor of human Aβ (AβPP) and obtained a complete sequence of its cDNA [19–21]. Finally, in 1991, Goate and co-workers detected the first (of many to follow) AD-causing AβPP mutation [22]. This burst of developments in the field culminated, in 1992, in the formulation, by Hardy and Higgins, of a major theory of AD called the Amyloid Cascade Hypothesis, ACH [23]. This theory was destined to guide the field for the following three decades.

3. The Amyloid Cascade Hypothesis: From Domination to Unsustainability

The essence of the ACH is best summarized in the words of its inventors, Hardy and Higgins: "Our hypothesis is that the extracellular deposition of amyloid β protein, the main component of the plaques, is the causative agent of Alzheimer's pathology and that the neurofibrillary tangles, cell loss, vascular damage, and dementia follow as the direct result of this deposition" [23]. It should be emphasized that in 1992, the time of formulation of the ACH, cerebral plaques ostensibly associated with AD were known for decades; it was, in fact Dr. Alzheimer who established their occurrence in the disease [11,12]. The identity of the main component of the plaques was also known since 1984 [18], for a better part of a decade. This knowledge, however, was insufficient to anoint Aβ as the causative agent of AD. What allowed it was the discovery, in 1991, of an AβPP mutation affecting the production of Aβ in the AβPP proteolytic pathway and segregating with the early-onset disease (familiar AD, FAD) [22].

At the time of its introduction, the ACH was consistent with the accumulated body of empirical data. It also defined the path forward and was, therefore, enthusiastically accepted by the scientific community. The ACH guided the design and construction of transgenic AD models and the development of AD drugs. The logic of these endeavors was straightforward: overexpress human AβPP in mice and the disease will follow; deplete extracellular Aβ and reduce the amyloid plaques load and the disease would be stopped and possibly cured. Under this guidance numerous AD drug candidates were developed that either deplete extracellular Aβ (e.g., various Aβ-specific monoclonal antibodies) or suppress the production of Aβ in the AβPP proteolytic pathway (e.g., inhibitors of the beta-site AβPP cleaving enzyme, BACE1, also known as beta-secretase). These types of AD drugs are referred to henceforth as "ACH-based drugs". Many ACH-based drugs were spectacularly successful not only in reducing but also in reversing the symptoms in transgenic mouse models overexpressing Aβ from numerous human AβPP transgenes [24–26].

But this is where the first cracks in the ACH appeared. Transgenic, human Aβ-overexpressing mice indeed deposit excessive amyloid plaques and exhibit some neurodegeneration and certain cognitive impairments, such as defects of the neuronal plasticity, learning, and memory formation. However, these mouse models appear unable to develop the full spectrum of AD pathology. Tellingly, none of the ubiquitous current transgenic mouse models develops neurofibrillary tangles (NFT), a major hallmark of AD. Therefore, it can be argued, and, indeed, it was argued by us [7,10] that transgenic mice overexpressing human AβPP are not AD models (further discussed in Section 19 below). Whatever symptoms and pathology these models develop are apparently due to the overexpression of human Aβ. It is, therefore, no surprise that ACH-based drugs targeting overexpressed Aβ provide relief from those symptoms and pathology.

The cracks in the ACH theory became chasms and crevasses when numerous human trials of candidate ACH-based drugs in symptomatic AD were conducted. In all such trials ACH-based drugs failed as spectacularly as they succeeded in transgenic mouse models; no efficacy whatsoever was seen, with some trials terminating prematurely [27,28] (the observed marginal effects of lecanemab and donanemab in very early AD [29–33] do not contradict this statement; see Section 39 below). Importantly, it became apparent that the drugs failed not because they underperformed mechanistically in humans (versus mice). To the contrary, the results of the trials indicated that they performed remarkably well in humans. Thus, verubecestat, an inhibitor of BACE1, substantially suppressed the production of Aβ in the AβPP proteolytic pathway and cleared extracellular Aβ, resulting in 80% drop of levels of Aβ in CSF of AD patients [27,28]. The findings that the drugs accomplished their mechanistic mission but did not provide any relief to the patients suggested that the underlying theory is incorrect.

Furthermore, as PET scan methods were developed that allow the evaluation of the amyloid plaques load in the living brain, it became apparent that there is no good correlation between the amount of amyloid deposition and the occurrence of AD. In a large fraction, about one third, of the aged general population the degree of extracellular Aβ deposition is comparable with or even exceeds that seen in AD patients. These individuals, however, develop neither cognitive impairment nor AD pathology in their lifetimes [34–40]. The reverse is also true: some AD patients exhibit no excessive levels of extracellular Aβ deposition [41]. Considered cumulatively, the observations described above provide a convincing indication that extracellular Aβ neither causes nor drives AD pathology and, consequently, that the ACH is unsustainable.

4. The Central Role of Aβ in Conventional Alzheimer's Disease Is Incontestable

The conclusion of the preceding section notwithstanding, the centrality of Aβ in conventional Alzheimer's disease is indisputable. The basis for this assertion is the nature of mutations that either cause familiar AD or protect from the disease. As mentioned above, Goate and co-workers discovered the first FAD-causing mutation in 1991 [22]. Since then, numerous mutations within either AβPP of presenilins (PSENs, components of the gamma-secretase complex) were discovered [42]. With one exception, all these mutations cause early-onset AD. The exception is the Icelandic mutation within the Aβ segment of AβPP, which confers on its carriers protection not only from AD but also from aging-associated cognitive decline, AACD [43,44]. All these mutations have one common attribute: they affect only the structure, production, and cleavages of Aβ. Hardy and Higgins deemed a single AβPP AD-causing mutation sufficient to assign to Aβ the central role in the disease [23], a notion readily accepted by the AD research field. The discovery of dozens of such mutations affecting, in one way or another, solely Aβ strongly affirm the centrality of Aβ in AD. Below, in Sections 32–35, we describe the mechanistic underlying of these mutations. The mechanisms involved not only explain how these mutations exert their effect (i.e., the causation of or protection from AD and AACD) but also suggest effective approaches to intervene with the disease. Total cerebral Aβ population consists of two subclasses: extracellular Aβ and intraneuronal Aβ (iAβ), both accumulating during the lifetime (physiologically occurring processes underlying accumulation of iAβ are described in Section 24 below). Since Aβ is apparently central in Alzheimer's disease and because extracellular Aβ can be ruled out as the causative and driving agent of AD, it logically follows that this distinction belongs to intraneuronal Aβ. This conclusion is supported by numerous observations that AD symptoms correlate much better with the levels of iAβ than with those of extracellular Aβ [45–57]. Co-incidentally, as described in the following section, the same conclusion is derived from the results of clinical trials of ACH-based candidate drugs in symptomatic AD.

5. The ACH2.0, Initial Version: Conventional Alzheimer's Disease Is Triggered by AβPP-Derived Intraneuronal Aβ (iAβ) and Driven by iAβ Generated Independently of AβPP

The major principles of the initial (Variant One) formulation of the ACH2.0, an AD theory replacing the ACH, stem from two sources. One is the certitude, articulated above, that conventional AD is triggered and driven by Aβ. Another source is the results of multiple clinical trials of candidate ACH-based drugs in symptomatic AD. Of those, especially important are trials of verubecestat, an effective inhibitor of BACE1 [27,28]. This is because whereas drugs like Aβ-specific antibodies deplete extracellular Aβ, verubecestat (and other BACE1 inhibitors) both clears extracellular Aβ and suppresses the production of Aβ in the AβPP proteolytic pathway.

The logic applied to the analysis of the outcomes of clinical trials is straightforward. (1) A substantial removal of extracellular Aβ has no positive effect whatsoever. Therefore, it is intraneuronal Aβ, iAβ, which triggers and drives AD. (2) A substantial suppression of the production of Aβ in the AβPP proteolytic pathway has no positive effect whatsoever. Therefore, the intraneuronal Aβ that drives AD is generated independently of AβPP and retained within the disease-affected neurons. These two notions constitute the principal attributes of the ACH2.0 in its initial iteration [1], which posits the following: Alzheimer's disease is triggered by AβPP-derived intraneuronal Aβ, iAβ, accumulated over certain critical threshold, and is driven by Aβ produced independently of Aβ-protein-precursor (AβPP) and retained intraneuronally as iAβ [1–10]. Whereas the notion of intraneuronal Aβ, iAβ, is, as described below, rather conventional and its origins are well understood, the concept of the generation of Aβ independently of its genome-encoded precursor, AβPP, is anything but (this is the reason why one of our initial publications on the subject is entitled "News from Mars" [58]). It may appear outlandish and even farfetched. It is neither. In fact, this concept, which defines the active core, the "engine" of AD in the ACH2.0 paradigm, has been already formulated, twice, a couple of decades previously but was deemed unnecessary and unfounded, and was "waiting in the wings" for a considerable time. Presently, it is strongly supported by the empirical data, and its time has apparently arrived. Familiarity with it is necessary in order to continue the present narrative. The following is its brief history (note that the present narrative resumes in Section 24 below).

6. Singular Attributes of the AUG Codon Encoding Methionine 671 of Human AβPP

The present section picks up the AD story in 1987. This is the year when several laboratories, independently and nearly simultaneously, cloned and sequenced cDNA encoding human AβPP [19–21]. As a reminder, it is 770 amino acid residues long (albeit there are shorter and, actually, some longer versions) and Aβ is embedded within its C-terminal segment and released by two proteolytic cleavages. One is enacted by beta-secretase, known also as BACE (Beta-site AβPP-Cleaving Enzyme). The "beta-site" of the above designation is localized between amino acid residues 671 and 672 and the cleavage by BACE at this site produces the 99 amino acid residues long C-terminal fragment (CTF) of AβPP designated rather imaginatively C99. C99 contains Aβ as its N-terminal segment and Asp672 of AβPP forms the N-end of both C99 and the future Aβ. This is the immediate precursor of Aβ. The release of Aβ is completed by gamma-secretase that cleaves C99 within a narrow spatial window and generates peptides 39–43 (Aβ39-Aβ43) amino acid residues long (with prevailing species Aβ40 and Aβ42).

The same year, 1987, subsequent to publication of the nucleotide sequence of human AβPP cDNA, two researchers, Breimer and Denny, noticed that the amino acid residue in position 671 of AβPP (i.e., the position contiguously preceding the C99 segment of AβPP) is methionine [59], referred to henceforth as Met671. They also noted that the AUG codon encoding Met671 of human AβPP is positioned within the optimal translation initiation nucleotide context (known as Kozak motif or Kozak consensus sequence, after Merilyn Kozak who discovered it). In its own right, such localization is remarkable due to its rarity. Even more remarkable is the fact, also noticed by Breimer and Denny [59], that the AUG encoding Met671 of human AβPP is the only AUG codon in AβPP mRNA that is situated within the optimal translation initiation nucleotide context. Human AβPP mRNA contains twenty in-frame, methionine-encoding AUG codons, and not even the AUG encoding translation-initiating methionine is embedded within the optimal translation initiation nucleotide context. Breimer and Denny's paper [59] happened to be published on April 1st. However, it was anything but a joke. The singularity of the location of the AUG encoding Met671 of human AβPP prompted Breimer and Denny to posit that such a propitious positioning is not random but potentially underlies a physiological function, namely the internal initiation of translation of intact AβPP mRNA, a process, they proposed, possibly induced in Alzheimer's disease [59]. This would result in C99 (presuming the translation-initiating methionine 671 is cleaved off co-translationally by N-terminal methionine aminopeptidase, MAP, as typically occurs in translation; further discussed in Section 23 below) as the primary translation product. Thus, the C99 fragment of AβPP would be generated independently of AβPP and could, upon gamma-secretase cleavage, yield Aβ, also produced in the AβPP-independent manner [59].

7. Ruling out the Internal Initiation of Translation of the Intact Human AβPP mRNA

Shortly after Breimer and Denny published their proposal [59], two research groups attempted to test it [60,61]. Both employed the same rationale. If the coding region of AβPP mRNA is altered well upstream of the AUG encoding Met671, the alteration would affect the conventional translation but would not interfere with translation initiated internally from the AUG codon in question. To implement this approach, one research group introduced frame-shifting mutations upstream from the AUG codon encoding Met671 of human AβPP [60]. Another research group inserted a translation stop codon, also well upstream of the AUG encoding Met671 of human AβPP [61]. In both approaches, alteration of the human AβPP gene was expected to stop the conventional production of AβPP and, consequently, of C99 and Aβ, but not to interfere with the production of C99 (and Aβ) initiated from the AUG encoding Met671, provided it occurs. In both approaches, therefore, the production of C99 and Aβ from altered AβPP mRNA would report on the occurrence of the unconventional internal initiation of translation at the AUG encoding Met671 of AβPP. In both approaches no production of C99 and/or Aβ from the altered AβPP expression vectors was observed, and both groups declared the possibility of the internal initiation of translation from the AUG codon in question as "ruled out" [60,61] (further discussed below; see Section 22).

8. Initiation of Translation from the AUG Encoding Met671 of Human AβPP Remains a Viable Possibility

Conceptually, however, the occurrence of the in-frame AUG codon singularly positioned within the optimal translation initiation nucleotide context and contiguously preceding the C99-encoding segment of human AβPP mRNA opens up another possibility of generating C99 (and, subsequently, Aβ) independently of AβPP. The solution is rather simple: generate a 5′-truncated AβPP mRNA where the AUG codon encoding Met671 of human AβPP is the first, 5′-most translation initiation codon, and translation of C99 independently of AβPP would occur conventionally. The key question in this regard is: Does such a process occur in Alzheimer's disease? The answer is, apparently, affirmative, and the discovery of a mechanism capable of accomplishing this constitutes a part of our story. This mechanism has been described in detail elsewhere [58] and is briefly summarized in Sections 10–14 below. However, to keep the narrative in chronological order, we first describe, in the following section, a puzzling observation that was not understood at the time it was made, but later played a pivotal role in the presently narrated story.

9. 3′-Extended Human AβPP cDNA: An Artifact or a Clue?

The next event in our narrative occurs in 1988. This is the year when a paper by Mita and co-workers was published [62] describing human AβPP cDNA that differed in its nucleotide sequence from human AβPP cDNAs described initially by other three laboratories and referred to henceforth as "conventional" human AβPP cDNA [19–21]. It is, in fact, identical to the conventional cDNAs except it contains a substantial extension at its 3′ end [62]. At the time of its introduction the genomic nucleotide sequence upstream from the conventionally defined human AβPP gene was not yet known, and the authors of [62] suggested that the 3′-extended human AβPP cDNA originated as a complement of corresponding human AβPP mRNA initiated from a transcription start site (TSS) positioned well upstream from the "conventional" TSS of the human AβPP gene [62]. Later the same year, however, the nucleotide sequence upstream of the conventional TSS of the human AβPP gene was determined and published [63], and it ruled out the proposed origin of the observed 3′-extended cDNA as a transcript of human AβPP mRNA initiated upstream from the conventional TSS; no such mRNA species occurs. Lacking a plausible explanation of their results, the authors of [62] declared their observation an artifact and corrected their paper accordingly [62]. Several years later, however, when a phenomenon of physiologically occurring mammalian RNA-dependent mRNA was discovered and characterized (see the following section), this observation provided a clue and, potentially, a proof of concept, for the cellular mechanism capable of producing 5′-truncated human AβPP mRNA where the AUG codon encoding Met671 of AβPP is the first, 5′-most translation initiation codon (see Section 13 below). The general outlines of mammalian RNA-dependent mRNA amplification are sketched out in the following section.

10. Mammalian RNA-Dependent mRNA Amplification: General Outline

The next development pertinent to our story takes place in 1996. This is the year when the physiological occurrence of the RNA-dependent amplification of mammalian mRNA was posited [64] and further developed in subsequent publications [58,65–73]. As follows from its designation, this process is capable of amplifying mRNA molecules, i.e., of utilizing existing mRNA molecules as templates to generate new ones. It is of the utmost relevance to AD because, potentially, it is capable of generating a version of the original gene-transcribed mRNA molecule (referred to henceforth as mRNA progenitor) 5′-truncated within its coding region and thus encoding only a CTF of the original polypeptide; in other words, this mechanism is conceptually capable of generating a variant of AβPP mRNA that, given the availability of in-frame AUG contiguously preceding the C99-encoding segment of AβPP mRNA and provided this AUG is the first translation initiation codon, can be translated into C99 thus producing this polypeptide independently of AβPP.

Mammalian RNA-dependent mRNA amplification was first discovered in differentiating erythropoietic cells where it is responsible for the large-scale production of globin polypeptides [65,66]. Subsequently, this process was shown to underlie the generation of vast amounts of matrix proteins during their extracellular deposition [67]. In this process every eligible gene-transcribed mRNA molecule ("mRNA progenitor") is utilized repeatedly as a template for generation of additional mRNA molecules. Since mRNA progenitors can number hundreds, in some cases thousands of molecules per cell, RNA-dependent mRNA amplification is equivalent to a massive gene amplification; its utilization in production of specific proteins could be orders of magnitude more efficient than conventional cellular protein synthesis. This process was shown to occur in two distinct phases. The first phase, referred to as "chimeric" for the reasons articulated below, is of especial interest in terms of its relevance for Alzheimer disease because it can result in a severely 5′-truncated mRNA end product encoding only a CTF portion of the original protein. This phase is outlined in the present and subsequent sections. Upon completion of the first phase of mRNA amplification, one of the two resulting end products can, if properly modified, serve as a progenitor in a polymerase chain reaction (PCR) process that always produces mRNA molecules retaining the entire coding capacity of the mRNA progenitor and thus encoding the intact original polypeptide. This phase of RNA-dependent mRNA amplification is described, and its relevance to Alzheimer's disease explained, in Section 79 below.

Stages of the chimeric phase of mammalian RNA-dependent mRNA amplification are outlined in Figure 1. For the purposes of orientation, the progenitor mRNA (also referred to as conventional mRNA) is shown in the top panel; its elements of importance are its termini and the translation-initiating AUG codon (marked "AUG"). In the first stage of the chimeric phase of mRNA amplification (marked "1" in the middle panel of Figure 1) RNA-dependent RNA polymerase (RdRp) initiates transcription of the progenitor mRNA molecule within its 3′-terminal poly(A) segment and completes it with transcribing the 5′-terminal cap-G into "C", which is not encoded in the genome.

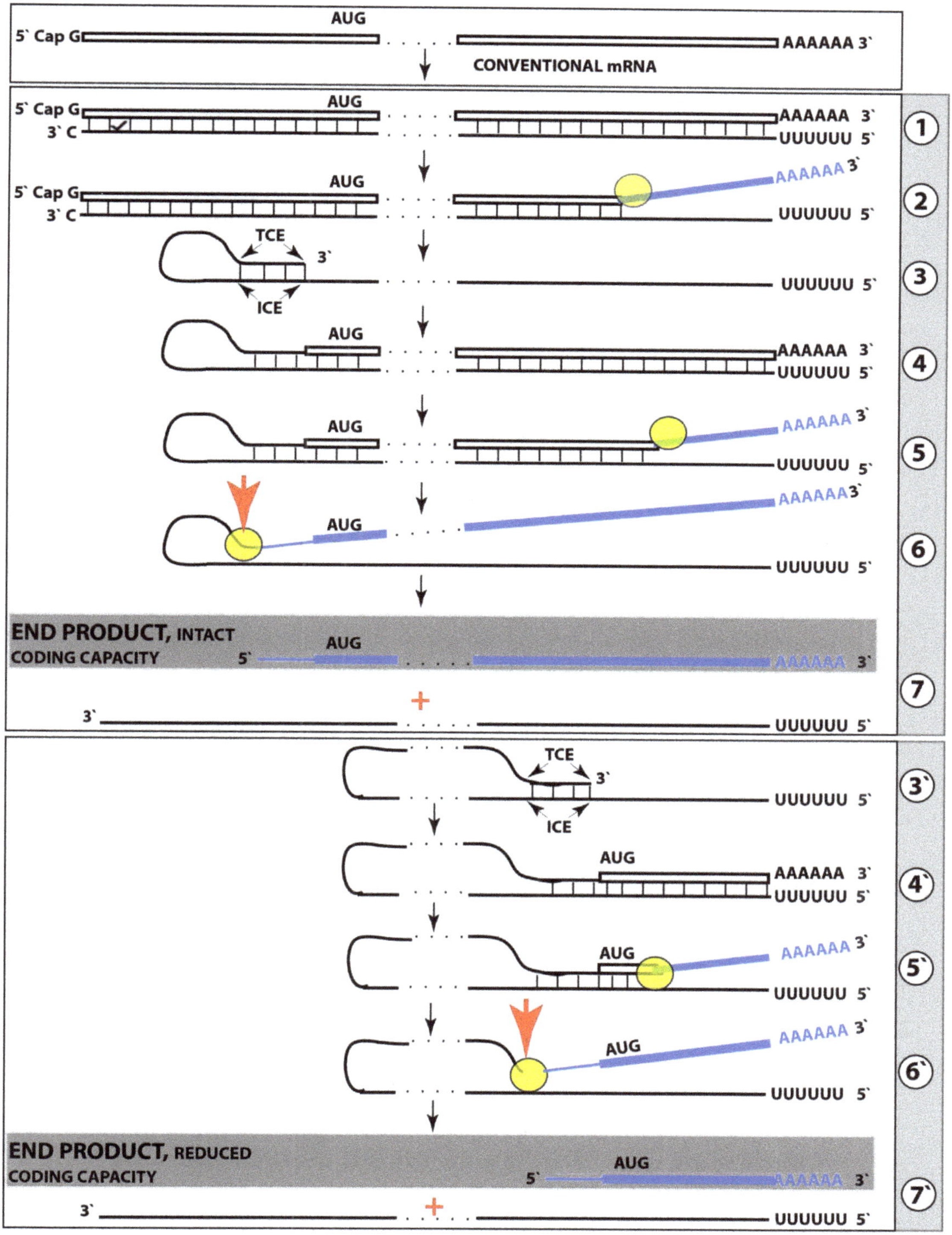

Figure 1. Mammalian RNA-dependent mRNA amplification: Principal stages. *Single lines*: Antisense RNA. *Boxed lines*: Sense RNA. *Blue boxed lines*: Single-stranded RNA separated from its complementary

RNA strand by helicase. *Yellow circles*: Helicase complex containing helicase strand-separating activity, nucleotide modifying activity and RNA cleaving activity. *Red arrows*: positions of the cleavage of the intermediate generating RNA end products of the amplification process. *RdRp*: RNA-dependent RNA polymerase. *AUG*: Translation initiation codon. *TCE*: 3′ Terminal Complementary Element of the antisense RNA strands. *ICE*: Internal Complementary Element of the antisense RNA strands. *Top panel*: Conventionally transcribed mRNA molecule referred to as progenitor mRNA. *Middle panel*: Stages of the chimeric pathway of the RNA-dependent mRNA amplification; the ICE is located within a portion of antisense RNA corresponding to the 5′ UTR of the progenitor mRNA molecule. (**1**): The progenitor mRNA is transcribed by RdRp into antisense RNA. (**2**): Complementary RNA strands are separated by the helicase activity. Helicase complex mounts the 3′ poly(A) segment and moves along the sense RNA strand modifying on average ever fifth nucleotide. (**3**): Folding of the antisense RNA into self-priming configuration is guided by the interaction of its TCE and ICE elements. (**4**): RdRp extends the 3′ terminus of self-primed antisense RNA. This produces a hairpin-like molecule containing both sense and antisense RNA components and referred to as the chimeric RNA intermediate. (**5**): Complementary strands of the chimeric RNA intermediate are separated by the helicase activity. Nucleotide modifications introduced during the separation prevent the re-annealing of the sense and antisense RNA strands. (**6**): When the helicase reaches the single-stranded portion of he chimeric intermediate (either the 5′ end of the TCE or a TCE/ICE mismatch) it cleaves the RNA molecule. (**7**): End products of RNA-dependent mRNA amplification. Antisense RNA is truncated at its 3′ end and sense portion of the chimeric RNA is truncated at the 5′ end and acquires the cleaved-off antisense RNA fragment; its translation would result in the complete original polypeptide. *Bottom panel*: (**3′–7′**) correspond to (**3–7**) of the middle panel. The ICE element is situated within a segment of antisense RNA corresponding to the coding region of the progenitor mRNA. The chimeric RNA end product is 5′-truncated within its coding region. Potential outcome of its translation are described in the main text. Please note that this figure was shown in [10].

In the second stage of the chimeric phase of mRNA amplification (marked "2" in the middle panel of Figure 1) a helicase complex separates the sense and antisense RNA strands. It mounts the 3′-terminal poly(A) segment in double-stranded configuration with its 5″-terminal poly(U) complement and proceeds in the 5′ direction separating the strands and modifying on average every fifth nucleotide of the sense strand thus preventing the re-annealing of the strands. When separation is completed the sense strand (the progenitor mRNA molecule) can be utilized again, repeatedly, in additional amplification rounds.

The feasibility of the third stage of the chimeric phase of mRNA amplification (marked "3" in the middle panel of Figure 1) depends vitally on two requirements. The first requirement is the presence within the antisense RNA strand of two complementary segments. One is the 3′-terminal complementary element, designated TCE; it is always, by definition, 3′-terminal. The second element can be positioned anywhere within the antisense RNA strand, it is designated the internal complementary element, ICE. The complementation of the TCE and ICE does not have to be perfect, just sufficient to form a stable structure. It is important to mention that in the scenario depicted in the middle panel of Figure 1 the ICE element is located within the portion of antisense RNA corresponding to the 5′ untranslated region (5′UTR) of the mRNA progenitor molecule. The second requirement is that the TCE and ICE are topographically compatible, in other words mutually accessible within the folded configuration of the antisense RNA strand. If both requirements are satisfied, the folding of the antisense RNA strand results in an RNA molecule in self-primed conformation.

In the fourth stage of the chimeric phase of mRNA amplification (marked "4" in the middle panel of Figure 1), RdRp extends the 3′ terminus of the self-primed antisense RNA strand. During this extension, RdRp transcribes the 5′-terminal portion of the antisense RNA strand, thus generating the sense RNA strand. The extension concludes with synthesis of 3′-terminal poly(A) and results in a 5′-truncated portion of the progenitor mRNA covalently attached to the 3′ terminus of the antisense RNA strand in a hairpin-like configuration. Since this molecule consists of both sense and antisense components, it is chimeric, which explains the designation of this phase as the chimeric phase of mRNA amplification. Since this molecule is an intermediate in the amplification process, it is referred to as "the chimeric RNA intermediate" and the site of the initiation of the extension (i.e., of the transition of the antisense into sense orientation) is designated "the chimeric junction". Importantly, since in the depicted scenario the ICE element of the antisense RNA strand is positioned within its portion corresponding to the 5′UTR of the mRNA progenitor molecule, the sense orientation component of the chimeric RNA intermediate retains the entire coding region of the progenitor, including the translation-initiating AUG codon.

In the fifth stage of the chimeric phase of mRNA amplification (marked "5" in the middle panel of Figure 1), the same helicase complex that was invoked above mounts the 3′-terminal poly(A) segment of the sense-orientation component of the chimeric RNA intermediate and moves along it, separating RNA strands and making frequent nucleotide modifications thus preventing re-annealing of the strands. In the sixth stage (marked "6" in the middle panel of Figure 1) the helicase complex reaches the single-stranded portion of a hairpin structure. At this point there are two possibilities. First, the TCE/ICE structure has no mismatches. In this case, a component of the helicase complex or an associated activity cleaves at the 5′ end of the TCE. In the second possibility there is a mismatch or multiple mismatches in the TCE/ICE structure. In this case, the helicase complex cleaves at the mismatch. Here, again, there are two possibilities. If the relationship of the remaining 5′ portion of the TCE with the corresponding segment of the ICE is unstable, this is the end of the process. If, however, following the cleavage, the self-primed conformation of the folded antisense RNA remains stable, the extension could reoccur, and the process repeated. The chimeric junction in this case would shift 5′ from the initial chimeric junction, a process designated "chimeric junction shift" [58]; with multiple TCE/ICE mismatches this can take place more than once until the extension can no longer occur.

The seventh stage (marked "7" in the middle panel of Figure 1) depicts two RNA end products of the chimeric phase of mRNA amplification. One end product is the antisense RNA or rather what remains of it because it is truncated in its 3′ portion. More precisely, it has lost its entire TCE element (if the TCE and ICE had no mismatches) or a part thereof (if the cleavage occurred at the TCE/ICE mismatch). It cannot any longer form a self-primed structure. Its further processing amounts to an intracellular polymerase chain reaction and is described in Section 79 below.

Another end product is mRNA. It is distinct from the progenitor mRNA molecule in two ways. First, it is chimeric. It comprises the sense-orientation portion, which acquires at its 5′ end a portion of the antisense RNA molecule, more precisely the same portion that was cleaved off antisense RNA in the sixth stage. The second difference is that its sense-orientation portion is 5′-truncated (in comparison with the progenitor mRNA). However, because (in this scenario) the ICE is located within the portion of the antisense strand corresponding to the 5′UTR of the progenitor mRNA, the truncation removes only a terminal part of the 5′UTR. Thus, the chimeric RNA end product retains the intact coding region and can be translated into the polypeptide identical to that encoded by the progenitor mRNA.

11. Chimeric mRNA Amplification Can Occur Asymmetrically and Result in mRNA Encoding Only a CTF of the Progenitor mRNA-Encoded Protein

In the scenario presented in the preceding section the ICE element of the antisense RNA strand is positioned within its portion corresponding to the 5′UTR of the progenitor mRNA molecule. Consequently, the resulting chimeric mRNA end product retains the entire coding region of and is translated into a polypeptide identical to that encoded by the progenitor mRNA. But this is only one option of many. Whereas the position of the TCE element of the antisense RNA strand is fixed (it is always 3′-terminal by definition), the position of the ICE element is highly variable; potentially it can be localized anywhere within the antisense RNA strand. The positioning of the ICE defines the potential of the chimeric RNA end product to produce, upon translation, a polypeptide, and can lead to several interesting translational outcomes (discussed in detail in [58]).

The present section addresses the scenario where the ICE element of the antisense RNA strand is positioned within its portion corresponding to the coding region of the progenitor mRNA molecule. This scenario is illustrated in the bottom panel of Figure 1. In this scenario stages 3′ through 7′ correspond and are conceptually identical to stages 3 through 7 described in the preceding section. The final outcome, however, is substantially different. Due to the position of the ICE element, only a 3′-terminal portion of the coding region (and the rest of mRNA) would be transcribed from the antisense RNA template. As depicted in stage 7′, following separation of strands and the cleavage, one of the end products, namely the antisense RNA, would be very similar in its structure to that depicted in stage 7. It would be missing at its 3′ end only the TCE element or a part thereof. The chimeric RNA end product, however, would be very different from its counterpart shown in stage 7 in that its coding region would be 5′-truncated. The translation outcome for this chimeric RNA end product would depend on the position of the first functional translation initiation codon (marked "AUG", not to be confused with the "AUG" in the top and middle panels of Figure 1). If it is located within the truncated coding region and, if it is in-frame with the coding nucleotide sequence, translation of the chimeric RNA end product would yield a CTF of the progenitor mRNA-encoded polypeptide. Since only the 3′-terminal portion of the progenitor mRNA is amplified, which encodes only the C-terminal portion of the original protein, this process is referred to as "asymmetric" amplification.

12. Evaluating the Eligibility of an mRNA Species for the RNA-Dependent Amplification Process

Implications of the asymmetric RNA-dependent mRNA amplification and their potentially considerable relevance to Alzheimer's disease are apparent. If human AβPP mRNA can be amplified asymmetrically and if in the resulting chimeric RNA end product the first translation initiation codon happens to be the AUG encoding Met671 of AβPP, the C99 fragment would be produced, and produced at a high rate, independently of AβPP. However, the probability that human AβPP mRNA is eligible for such an asymmetric RNA-dependent amplification process is, ostensibly, rather small if not negligible. Indeed, the AUG encoding Met671 of human AβPP is located over two thousand nucleotides downstream from the 5′ terminus of AβPP mRNA. Since the segment of mRNA encoding the ICE of the antisense strand has to be in the proximity of the AUG encoding Met671, the TCE and ICE elements would also be separated by about two thousand nucleotides. First, however these elements have to be present in the AβPP antisense RNA, and this is not a given. Second, crucially, these elements (if they occur in the first place) must be mutually accessible in the folded conformation of the antisense RNA strand, a tall order considering the extent of their separation and the complexity of RNA folding. Fortunately, the occurrence of the TCE and ICE elements and their mutual accessibility in the folded antisense RNA strand can be assessed (for any mRNA species) experimentally.

In such an experiment a specific mRNA is used as the initial template, and RdRp is modeled by RNA-dependent DNA polymerase, RdDp (also referred to as reverse transcriptase, RT). Oligo(dT) is used as a primer to initiate transcription of the antisense strand (cDNA) from the 3′-terminal poly (A) segment of mRNA. It is essential for this experiment that RNAse H is included in the reaction. It models the helicase activity of the mRNA amplification process and removes mRNA strand following the completion of reverse transcription thus enabling the folding of cDNA. There are two ways to include RNase H activity into the reaction. One, the simplest, is to use a viral preparation of RdDp, which always contains RNase H activity. Another way is to use cloned RdDp, which lacks RNase H activity, and add the latter to the reaction mix. When the reverse transcription and the removal of RNA template are completed, the newly generated antisense strand (cDNA) is folded. If it contains the TCE and ICE elements and if these elements are mutually accessible, a self-primed structure will be formed, and its 3′ end will be extended by RdDp generating a 3′-terminal segment of the sense strand. Size analysis of the resulting cDNA would determine (by comparison with the size of its mRNA template) whether the extension occurred, and, if affirmative, the nucleotide sequencing of cDNA would identify the TCE and ICE elements and define the position of the ICE.

13. Validation (1): Human AβPP mRNA Is a Legitimate Template for the Asymmetric RNA-Dependent Amplification; The Chimeric RNA End Product Encodes the C99 Fragment of AβPP

The scenario presented in the preceding section describes precisely the outcome of the experiment by Mita and co-workers [62] described in Section 9 above. As the reader might remember, for the lack of an explanation, Mita et al. defined their results (a substantially 3′-extended human AβPP cDNA) as an artifact [62]. However, upon further analysis of the structure and nucleotide sequence of the observed 3′-extended AβPP cDNA it transpired that the extension seen in [62] is not random but, in fact, a 3′-terminal portion of the sense-orientation AβPP DNA strand and that it could be generated solely by the antisense self-priming mechanism described above [74]. The question arises: why this extension occurred in Mita et al. experiment [62] but not in studies of three laboratories [19–21] who initially determined the nucleotide sequence of human AβPP cDNA? The answer is trivial and technical: Mita et al. used viral preparation of RdDp, which contained RNase H activity required for the removal of the mRNA template (i.e., for strands separation). In contrast, other groups utilized more "advanced" (and expensive) cloned preparations of RdDp lacking RNase H activity; without the removal of the mRNA template no cDNA folding could occur.

As discussed above, the determination of the nucleotide sequence of the 3′-extended human AβPP cDNA identified the TCE and ICE elements of the antisense strand, defined the position of the ICE element, and established the location of the site of initiation of the extension ("the chimeric junction site" in the terminology utilized above). Importantly, it also established the identity and location of the first translation initiation codon within the extended segment: It occurs 58 nucleotides into the extension portion, and it is the AUG encoding Met671 of AβPP! Figure 2 depicts the projected pivotal stages of the chimeric pathway of asymmetric RNA-dependent amplification of human AβPP mRNA with panels (a) to (c) paralleling stages 3′ to 7′ of Figure 1. Panel (a) of Figure 2 depicts folded self-primed conformation of the antisense human AβPP RNA following its separation from the progenitor AβPP mRNA by a helicase activity. The TCE and ICE elements, separated by about two thousand nucleotides, are highlighted in yellow. Panel (b) of Figure 2 depicts the extension of the 3′ terminus of self-primed antisense RNA resulting in formation of the chimeric RNA intermediate. The extension portion is highlighted in gray. The first translation initiation codon within the extension portion is highlighted in green; it happens to be the AUG encoding Met 671 of human AβPP mRNA. Following strands separation by a helicase activity, cleavage within the chimeric RNA intermediate (denoted by red arrow) can occur either at one of the TCE/ICE nucleotide mismatches or, as shown in Panel (b), at the 5′ end of the TCE element. Panel (c) of Figure 2 shows the resulting chimeric RNA end product. Its 5′-terminal antisense portion is shown in small letters. Its sense-orientation portion is severely 5′-truncated within the coding region of the progenitor AβPP mRNA. Its first translation initiation codon is the AUG encoding Met671 of human AβPP; conventional initiation of translation from this codon would yield the C99 fragment generated independently of AβPP. This analysis was performed and published in 1996 [74], before being subsequently expanded upon in further publications [75–77].

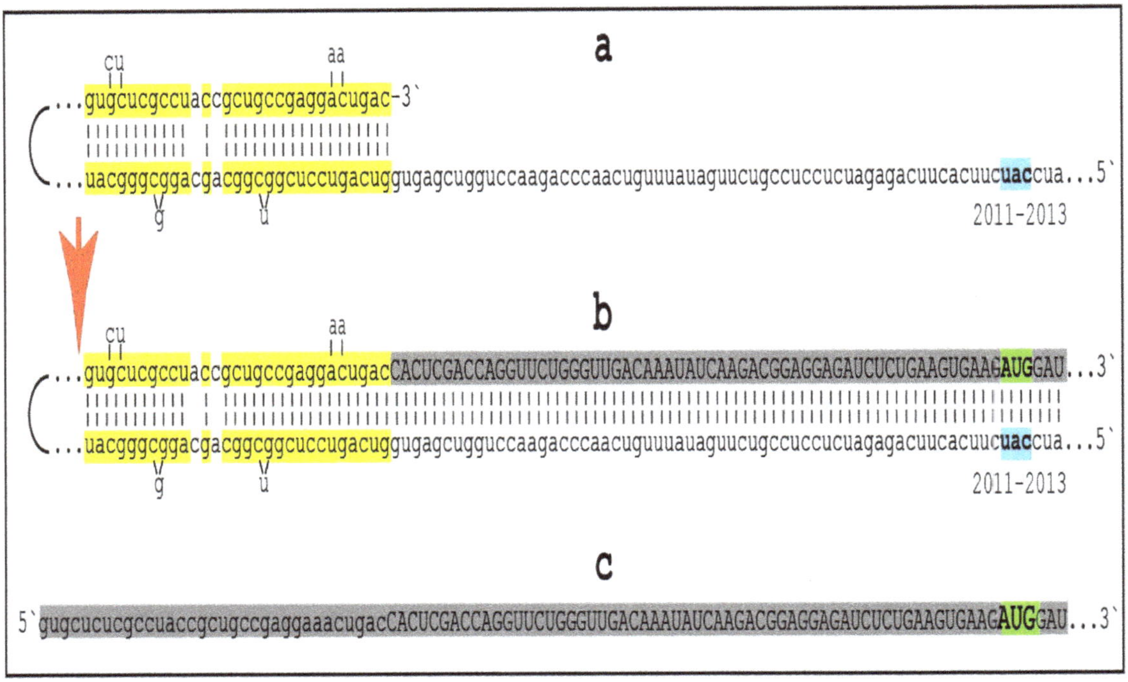

Figure 2. Human AβPP mRNA Is a Legitimate Template for the Asymmetric RNA-dependent Amplification. *Small letters*: nucleotide sequences of the relevant segments of the human antisense AβPP RNA molecule. *Large letters*: nucleotide sequence of a portion of human AβPP mRNA produced by the extension of the 3′ end of self-primed antisense RNA. *Highlighted in yellow*: the TCE and ICE components of the human antisense AβPP RNA. *2011–2013*: Positions counted in numbers of nucleotides from the 3′ terminus of the human AβPP antisense RNA. Nucleotides at these positions form the "uac" (*shown in blue*) corresponding to the "AUG" (*shown in green*) encoding Met671 of human AβPP mRNA. Segments (**a**) to (**c**) correspond to stages 3′ to 7′ of Figure 1. (**a**): Folding of human antisense AβPP RNA into self-primed structure is guided by the interaction of its TCE and ICE components. (**b**): The extension of the 3′ end of self-primed human antisense AβPP RNA produces 3′-terminal portion of AβPP RNA mRNA (*shown in gray*). Following the separation of strands by the helicase complex the hairpin-like intermediate is cleaved. (**c**) The cleavage (denoted by red arrow) produces the chimeric RNA end product (*shown in gray*). Its sense orientation portion is truncated deep within the coding region of AβPP mRNA; its translation would start from the AUG encoding Met 671 of AβPP and produce the C99 fragment independently of AβPP. Please note that this figure was shown in [10].

14. Validation (2): The Major Prediction Relevant to AβPP-Independent Production of C99 in Alzheimer's Disease Has Been Confirmed in a "Dream Experiment"

How to validate that AβPP-independent production of C99 indeed occurs in and, moreover, drives Alzheimer's disease? Through testing key predictions of this notion. The quantity of C99 produced via the amplification of AβPP mRNA in a process analogous to a massive gene amplification would greatly outbalance that of C99 generated in the AβPP proteolytic pathway. Therefore, with the AβPP mRNA amplification process operational, inhibition of beta-site cleavage by beta-secretase would be completely inconsequential and futile as a therapeutic strategy in AD. This is the major prediction. Testing it at the time when the concept was formulated, in 1996 [74], would require a major "dream experiment". A "dream" not only because the identity of beta-secretase was not yet established but also because it would require the development and testing of a drug, a financial impossibility for an academic laboratory, and experimentation on human patients (clinical trials).

Enter Big Pharma. In 1996, following the publication of [74], Merck invited one of the authors (V.V.) to make a presentation at their headquarters in King of Prussia (a town in Pennsylvania, USA). At the end of the presentation, they asked what would be a key one-line advice regarding the therapeutic strategies for AD. The answer was: do not pursue the inhibition of beta-secretase; it would be futile. Merck did not heed this advice. By 1999 the identity of beta-secretase (BACE1) was established [78–80] and the race to develop BACE1 inhibitors was on. It took over a decade and Merck won this race with the development of verubecestat, a potent, efficient, and brain (and neurons)-penetrating BACE1 inhibitor. The first promising results in animal models were obtained around 2007 [81–89], and in 2012 Merck singled on the BACE1 inhibitor designated "compound 16" [90]. Further improvements of "compound 16" resulted in verubecestat [91]. Verubecestat performed brilliantly in animal models and in early human safety trials [91], and the hopes associated with its potential use as a drug were considerable. The final test came in the form of two massive stage-III human clinical trials initiated in 2016: one for mild-to-moderate AD and another for prodromal AD [27,28]. By 2018 both trials were terminated, both prematurely, due to the complete lack of efficacy (the lack of efficacy was solely in therapeutic outcomes, mechanistically verubecestat performed in humans as effectively, if not more so, as in animal models) [27,28]. The development of verubecestat and its trials, at a combined cost of over billion dollars, constituted the "dream experiment", referred to above, and validated the notion of AβPP-independent generation of C99 in Alzheimer's disease.

15. Validation (3): Identifying the Chimeric RNA Intermediates of RNA-Dependent Human AβPP mRNA Amplification

The aim of the present section is to define precisely the nucleotide sequences to search for in order to validate the occurrence of chimeric AβPP RNA molecules in AD-affected neurons. The chimeric nature of these RNA molecules is their unique feature, and it is epitomized in the occurrence of the chimeric junction, which is a site where the antisense nucleotide sequence transits into the sense orientation. Therefore, the way to validate the occurrence of chimeric AβPP RNA molecules is to identify and sequence their regions containing both the sense and antisense components. Since the chimeric RNA end product of human AβPP mRNA amplification, containing sense/antisense junction, has to be, by definition, highly abundant, it would appear that detecting and sequencing it is simple. This, however, is not the case. When the helicase complex separates complementary segments of the chimeric RNA intermediate, as shown in Figure 1 above, it mounts the 3′-terminal poly(A) and moves along the molecule separating it from its complement. It also modifies on average every fifth nucleotide; when the newly generated chimeric RNA end product is cleaved off the chimeric RNA intermediate, it is modified along its entire length. The purpose of these modifications is, apparently, to prevent the re-annealing of separated strands and, possibly, to enable the translation of the resulting cap-less mRNA (further discussed below). Since these modifications prevent hybridization of the modified RNA with its complement, it cannot be analyzed by techniques involving nucleic acid hybridization and cDNA-based sequencing. These molecules may be ubiquitous, but they are "invisible" to current methods of detection [58]. What can be detected, however, are chimeric junction-containing regions of the chimeric RNA intermediate. This is because when the junction is generated by the extension of self-primed antisense AβPP RNA, it remains unmodified for the duration of the extension process plus the time it takes the helicase complex to traverse the length of the sense-orientation RNA strand.

The nucleotide sequence of one validation target is easy to define. It is shown in Figure 3 as a segment of its "Extension #1" portion highlighted in gray. It contains the 3′ terminus of self-primed AβPP antisense RNA extended into the sense-orientation molecule. But there could be additional sense/antisense junctions. This is because the helicase complex cleaves the chimeric RNA intermediate when it reaches its single-stranded portion, and, therefore, the cleavage can occur at a mismatch within the TCE/ICE complex. If, following such cleavage, the self-primed antisense RNA structure (i.e., the remaining portion of the TCE/ICE complex) remains stable, it can be extended again, as shown in the "Extension #2" portion of Figure 3. In such a case, a new chimeric junction is generated but it is shifted upstream from the initial one. Such a process is designated "chimeric junction shift" [58] and it can be repeated several times until, in fact, the remaining portion of the TCE/ICE complex is no longer stable. As shown in Figure 1, in case of human AβPP RNA it can occur possibly four times, every time generating new chimeric junction and defining new junction-containing regions. Nucleotide sequences of these regions (validation targets) are highlighted in gray in the "Extension #1" through "Extension #4" portions of Figure 3. These validation targets can be searched for in human neuronal cell-based AD models (discussed below).

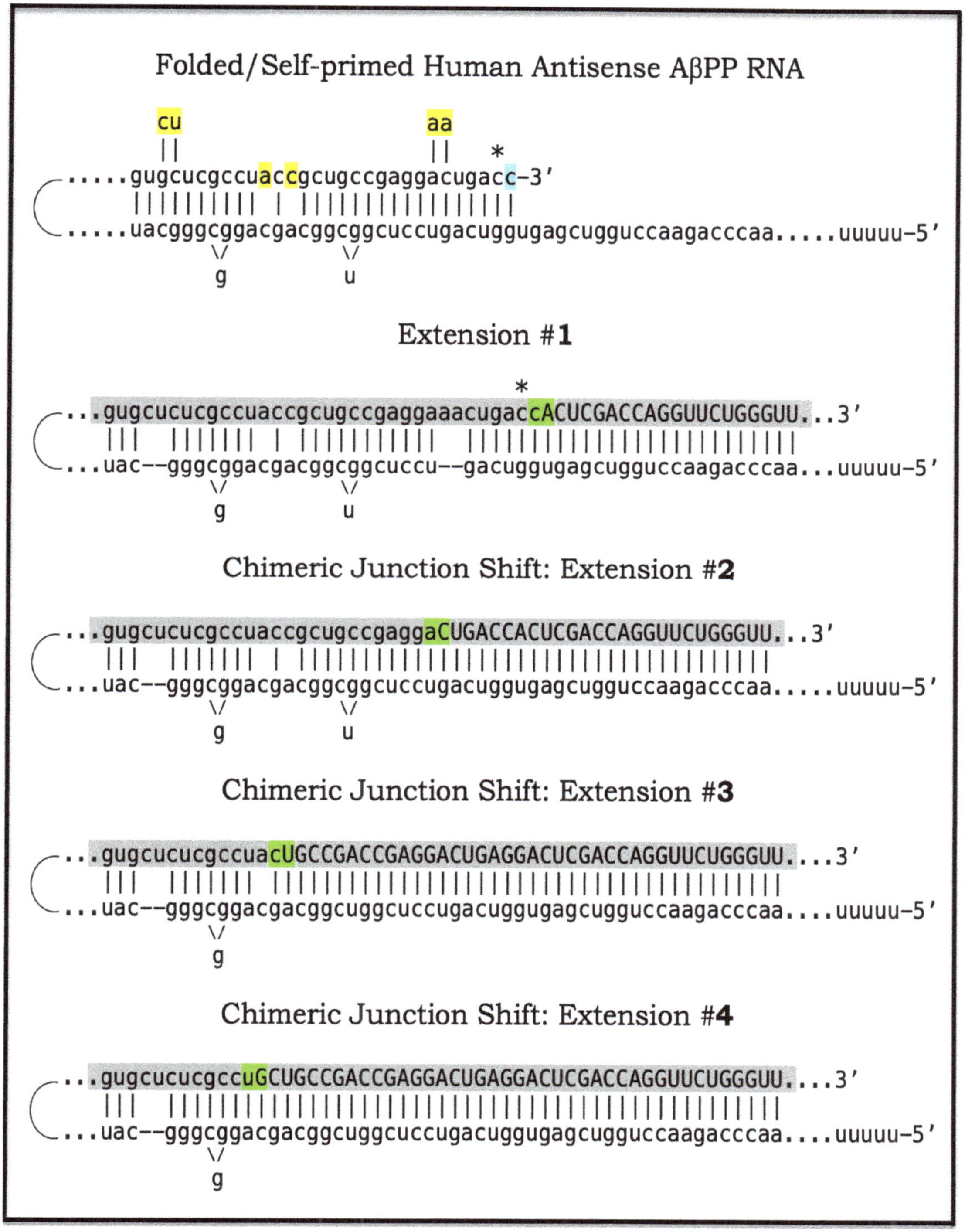

Figure 3. Anticipated nucleotide sequences of chimeric junction regions of chimeric human AβPP mRNA amplification intermediates. *Small letters*: Nucleotide sequences of the relevant portions

of self-primed antisense AβPP RNA. *Large letters*: Nucleotide sequences of sense RNA generated by the extension of self-primed human AβPP RNA. *Yellow boxes*: Mismatches within the TCE/ICE double-stranded complex. *Green boxes*: Chimeric junctions consisting of the 3′-terminal nucleotide of antisense RNA and 5′-terminal nucleotide of the sense RNA strand. *Asterisk*: The nucleotide of the antisense RNA molecule corresponding to the transcription start site positioned 149 nucleotides upstream from the translation-initiating AUG codon. *Shown in blue*: "c" transcribed from the 5′-terminal cap "G" of AβPP mRNA; it is accommodated within the self primed antisense RNA. *Shown in gray*: anticipated nucleotide sequences of the regions containing chimeric junctions. *Extension #1* of the self-primed RNA generates chimeric RNA intermediate containing the full-size TCE element of antisense RNA. *Extension #2*: Cleavage of the chimeric RNA intermediate occurs at the first TCE/ICE mismatch. Self-primed antisense RNA configuration remains stable and is extended; the chimeric junction shifts upstream from the original one. *Extension #3*: Cleavage of the chimeric RNA intermediate occurs at the second TCE/ICE mismatch. Self-primed antisense RNA configuration remains stable and is extended again; the chimeric junction shifts upstream. *Extension #4*: Cleavage of the chimeric RNA intermediate occurs at the third TCE/ICE mismatch. Self-primed antisense RNA conformation remains stable and is extended once more; the chimeric junction shifts as shown.

16. Mammalian RNA-Dependent mRNA Amplification Is Enabled and Sustained by the Integrated Stress Response

Previous investigations of mammalian RNA-dependent mRNA amplification suggested that this process is enabled and sustained by the integrated stress response (ISR) [58]. This pathway is activated in response to multiple stresses, and it is designated "integrated" because the effects of a multitude of stimuli/stressors, both physiological and pathological as well as environmental, capable of eliciting the ISR, converge on a single ("integrating") event, namely the phosphorylation of eukaryotic translation initiation factor 2 alpha (eIF2α) at its residue Ser51 [92–101]. This event is enacted by one of four constituent members of the family of eIF2α kinases: protein kinase double-stranded RNA-dependent (PKR), PKR-like endoplasmic reticulum kinase (PERK), general control non-derepressible-2 kinase (GCN2), and heme-regulated inhibitor kinase (HRI). These four kinases are extensively homological in their catalytic domains but possess distinct regulatory domains. They all operate in a similar manner involving autophosphorylation and dimerization to acquire their full activity. Each kinase, however, responds to a defined (albeit somewhat overlapping) set of stressors. Thus, PERK, which is localized at the endoplasmic reticulum (ER), is activated by ER stresses but also by ATP depletion, deprivation of glucose, and activation of various oncogenes. GCN2 responds to amino acid deprivation but also to UV and other stimuli. PKR responds to double-stranded RNA (viral infections) but also to oxidative stress, growth factor depletion, and caspase activity among other stimuli. HRI responds to heme deficiency but also to heat shock, proteasome inhibition, nitric oxide, osmotic stress, and mitochondrial dysfunction. When elicited, the ISR radically rearranges the transcriptional and translational landscapes of the cell. Global cellular protein production is severely inhibited, primarily through suppression of cap-dependent initiation of translation. Concurrently, the ISR activates cap-independent translation of a small set of selected mRNA species, including those that encode various transcription factors. It was suggested [58] that among the proteins whose production is enabled by the ISR are essential components of RdRp, not present under regular, non-ISR, conditions. This process is illustrated in Figure 4.

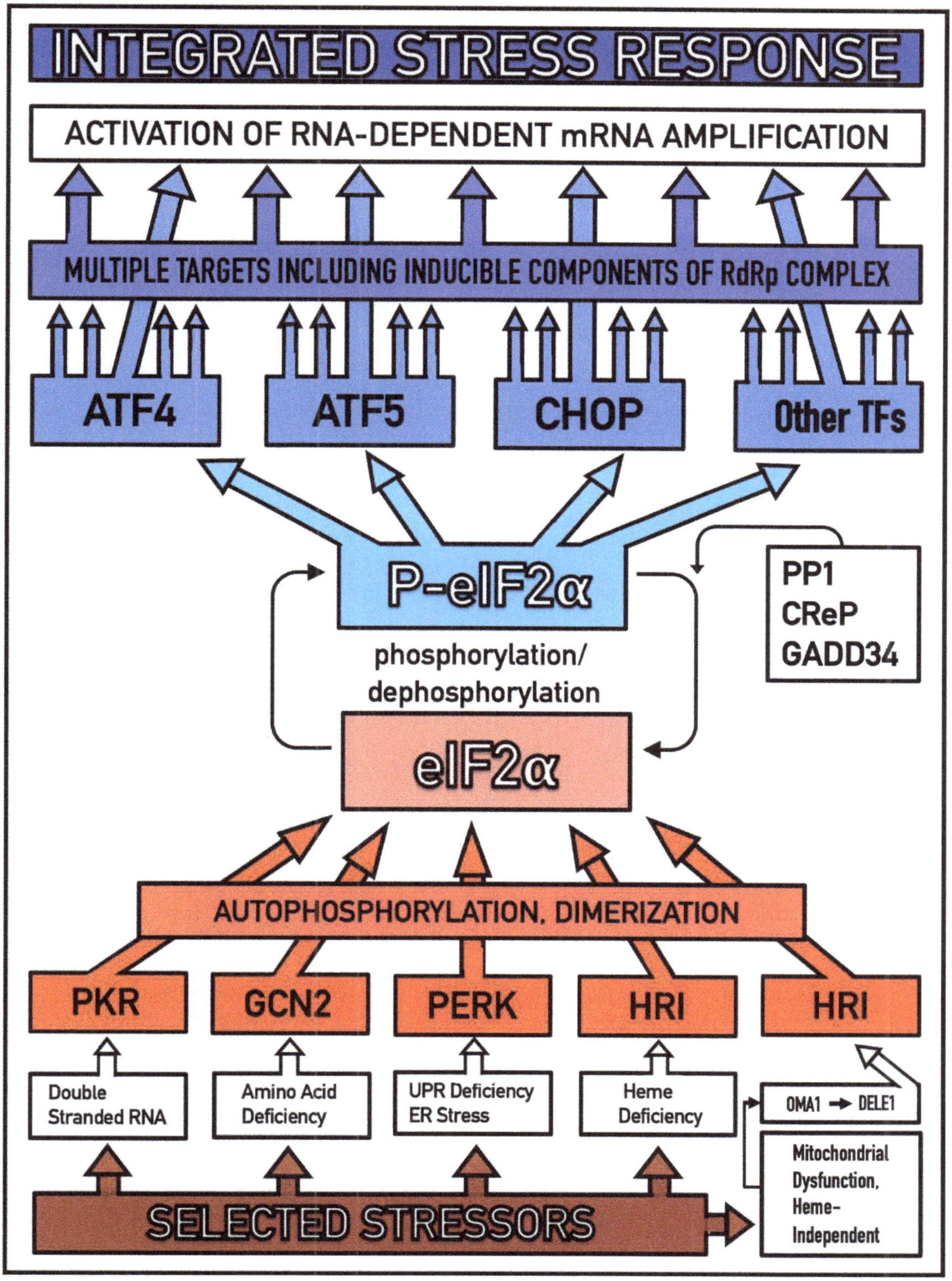

Figure 4. Mammalian RNA-dependent mRNA amplification is enabled by the integrated stress response. Distinct stressors selectively named in the Figure activate, in a stressor-specific manner, a

subset (one or more) of eIF2α kinases: PKR, GCN2, PERK, and HRI; the activation process entails autophosphorylation and dimerization of the kinases. When activated, kinases phosphorylate eIF2α at its Ser51 residue; this is the event, which integrates a multitude of signaling pathways into a single outcome; it is the essence of the integrated stress response pathway. When elicited, the ISR radically rearranges the transcriptional and translational landscapes of the cell. Global cellular protein production is severely inhibited, primarily through suppression of cap-dependent initiation of translation. Concurrently, the ISR activates cap-independent translation of a small set of selected mRNA species, including those that encode various transcription factors. Among the proteins whose production was enabled by the ISR are essential components of RdRp, not present under regular, non-ISR, conditions. Thus, ISR enables the operation of the mammalian RNA-dependent RNA amplification pathway.

In studied cases the operation of the RNA-dependent mRNA amplification pathway propagates ISR conditions and thus sustains its own activity. For example, during erythroid differentiation, the increased production of globin chains results in sequestration of heme in tetrameric hemoglobin complexes. This depletes cellular heme, activates HRI, and elicits the ISR. Under the ISR conditions RdRp is activated, and globin mRNA is amplified [66]. The resulting "turbocharged" production of globin chains further depletes heme, maintains the ISR conditions, and sustains, via the activity of RNA-dependent globin mRNA amplification pathway, its own operation. Similarly, in cells producing and depositing extracellular matrix (ECM) proteins, their elevated conventional production induces the ER stress and thus activates PERK. PERK, in turn, phosphorylates eIF2α and elicits the ISR. Under ISR conditions, RdRp is activated and mRNA species encoding ECM proteins are amplified [67]. This results in substantially increased production of ECM proteins. Consequently, PERK activity is maintained, ISR conditions propagated, and the operation of RNA-dependent amplification of mRNA encoding ECM proteins perpetualized. As described in more details in Section 26 below, in conventional Alzheimer's disease the neuronal ISR is elicited and AβPP-independent pathway of C99 production, presumably via asymmetric RNA-dependent amplification of AβPP mRNA, is activated by AβPP-derived iAβ accumulated over a certain critical threshold. This pathway, in turn, propagates ISR conditions and thus perpetuates its own operation.

17. Alzheimer's Is a Species-Specific, Possibly Uniquely Human Disease

To date, Alzheimer's disease has been detected solely in humans. It could be argued that in other species the limited lifespan does not allow AD to develop. This, apparently, is not the reason: the disease was not seen, for example, in elephants despite their longevity (over 80 years). The potential involvement of RNA-dependent amplification of AβPP mRNA explains why: in most, if not all, non-human species AβPP mRNA is ineligible for RNA-dependent amplification described above. Since mice are often used in attempts to model AD, it is prudent to analyze them in this respect as an example. As discussed above, for an mRNA species to be an eligible template for the RNA-dependent mRNA amplification process, its antisense complement must satisfy two requirements: (1) it should possess the TCE and ICE elements and (2) these elements should be mutually accessible in the folded antisense RNA molecule. However, as illustrated in Figure 5 (shown in comparison with human antisense AβPP RNA), even the first requirement is not satisfied in mouse antisense AβPP RNA. There is no better than random complementarity between segments of mouse antisense AβPP RNA corresponding to the TCE and ICE elements of its human counterpart, and the 3′ overhang would effectively prevent the priming/extension. As for the possibility that the ICE of mouse antisense AβPP RNA is positioned somewhere else in the molecule, the blast analysis of the 3′-terminal portion with the rest of the molecule showed no significant complementarity anywhere. Mouse AβPP mRNA is, therefore, not a legitimate template for RNA-dependent amplification. For the same reasons, this conclusion extends to other non-human mammalian AβPP mRNA species. It appears that Alzheimer's is species-specific if not uniquely human disease.

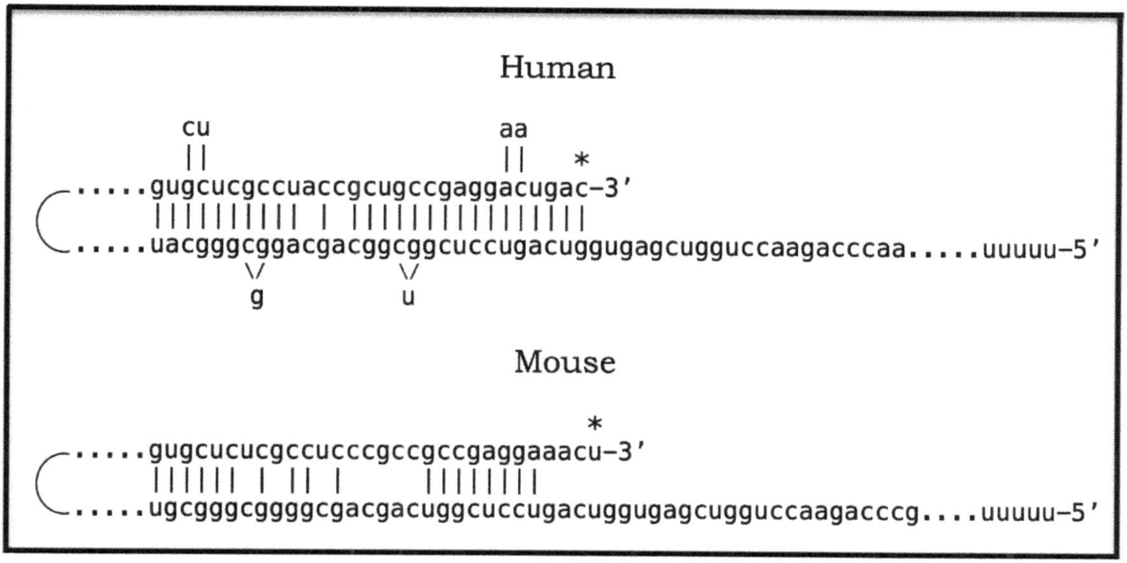

Figure 5. Mouse AβPP mRNA is not a legitimate template for RNA-dependent mRNA amplification *"Human"*: human antisense AβPP RNA folded into the self-primed configuration. *"Mouse"*: The relationship between segments of the mouse antisense AβPP RNA molecule analogous to that of the human TCE and ICE elements. *Asterisk*: The nucleotide of the antisense RNA molecule corresponding to the transcription-initiating nucleotide of either human or mouse AβPP mRNA; in both human and mouse transcription of 149 upstream from the AUG translation initiation codon. There is no better than random complementarity between segments of mouse antisense AβPP RNA corresponding to the TCE and ICE elements of its human counterpart, and the 3′ overhang would effectively prevent the priming/extension. The possibility that the ICE of mouse antisense AβPP RNA is positioned somewhere else in the molecule was excluded by the blast analysis of its 3′-terminal portion, with the rest of the molecule showing no significant complementarity anywhere. Mouse AβPP mRNA is, therefore, not a legitimate template for RNA-dependent amplification.

18. AβPP mRNA Transcribed from Human Transgenes in Animal Models Is Ineligible for RNA-Dependent Amplification

The preceding section established that in mice, as in the majority and possibly all non-human mammalian species, AβPP mRNAs cannot be amplified because their antisense counterparts lack the mutually accessible TCE and ICE elements. But human AβPP mRNA, expressed in animal models from human transgenes, is not amplified either; this is the reason why transgenic mouse models are incapable of developing the full spectrum of AD pathology. What is the reason for this failure? It could be argued that mice do not possess RNA-dependent mRNA amplification machinery, but this is not the case: they do employ a robust RNA-dependent mRNA amplification as a basic physiological tool [66,67]. This is exemplified by amplification of globin-encoding mRNAs in erythroid differentiation and of ECM proteins-encoding mRNAs during ECM deposition. Moreover, mutations, affecting (weakening) the TCE/ICE interaction in globin antisense RNA, were shown to result in various types of thalassemia [58,102]. It could also be argued that, possibly, no ISR can be elicited in mouse neurons and therefore no RNA-dependent mRNA amplification pathway can be activated. As discussed in more details in the following section, this is also not the case: iAβ, accumulated to sufficient levels, triggers the elicitation of the neuronal ISR in mice. The remaining, apparently paradoxical, answer is that somehow human AβPP mRNA is ineligible for RNA-dependent amplification in mouse neurons. This is the right answer, but it has nothing to do with mice and has everything to do with the structure of human AβPP mRNA expressed in transgenic mouse models. During the construction of human AβPP transgenes-carrying vectors the 5′ terminus of human AβPP cDNA is substantially modified for its insertion into the vector. This alters or altogether eliminates the segment encoding the TCE element of AβPP antisense RNA rendering it ineligible for RNA-dependent amplification. In this respect exogenous AβPP mRNA expressed from human transgenes is similar to endogenous mouse AβPP mRNA: both are incapable (for the same reason) of supporting their RNA-dependent amplification.

19. Transgenic Animals Overexpressing Human AβPP Model Not Alzheimer's Disease but Solely Effects of the Neuronal ISR: How to Generate an Adequate Mouse Model of AD

As mentioned above and discussed in detail below (see Section 25), in Alzheimer's disease AβPP-derived iAβ, accumulated over the critical threshold, elicits the neuronal ISR and thus activates the AβPP-independent production of C99, and, as presumed in this iteration of the ACH2.0, of iAβ that drives the disease. In transgenic mice a fraction of the vast output of human Aβ accumulates intraneuronally and triggers the elicitation of the neuronal ISR [103]. This is the integral part of the development of AD and occurs also in AD patients [104]. But in transgenic mice overexpressing human AβPP the development of AD stops at this stage because the pathway capable of generating C99 independently of AβPP is inoperative. Yet these mice display an extent of neurodegeneration and of cognitive impairment that were, until now, interpreted as symptoms of AD. It is apparent now, however, that both neurodegeneration and cognitive impairment observed in these models can be ascribed to the neuronal ISR, more specifically to the global suppression of cellular protein synthesis, which is the manifestation of the ISR. It is inevitable that the persistent neuronal ISR would, via suppression of protein production, result in cellular damage; this effect was, in fact, observed not only in neurons but in other cell types as well [105–118]. The very same neuronal ISR-caused suppression of global cellular protein production explains also cognitive impairment seen in transgenic mouse models overexpressing human AβPP. The observed impairments include defects in learning, memory formation and neuronal plasticity, i.e., in functions shown to require new neuronal protein production; it is not a surprise that the suppression of the latter causes the former [105–118]. The causal relationship between cognitive impairments in mice overexpressing human AβPP and the neuronal ISR is illustrated by observations that prevention of the neuronal ISR in these mice prevents cognitive impairment and that inhibition of the neuronal ISR (and consequent restoration of cellular protein synthesis) abrogates impairment if it already occurred [119–127]. The above considerations make it apparent that mice overexpressing human AβPP model not AD but rather solely the effects of the exogenous iAβ-elicited neuronal ISR.

Thus, the current mouse transgenic models overexpressing human AβPP are, patently, NOT AD models. The question arises, therefore, how does one generate adequate mouse AD models? This question is addressed in detail in our previous studies [7,10]. In short, the expression, at sufficient levels, of human AβPP mRNA eligible for the RNA-dependent mRNA amplification process would be both necessary and sufficient to achieve this goal (this is the core; appending FAD mutations in both AβPP and PSENs and expressing ApoE4 would significantly improve a model). The simplest approach to accomplish this is to utilize the current transgenic models overexpressing human AβPP: they are, after all, well on their way to becoming AD models in that they exhibit the neuronal ISR elicited by AβPP-derived iAβ; all that is needed is to enable operation of the AβPP-independent C99 production pathway. Modifying, via gene editing, human AβPP transgenes in existing transgenic models in such a way as to restore their 5′ termini would attain the latter. Another approach is to start from scratch by constructing new transgenic vectors expressing human AβPP mRNA identical in all respects to that produced endogenously in human neurons (or at least containing the intact 5′ terminus).

20. An Adequate Human Neuronal Cell-Based AD Model Has Been Constructed but Could Be Significantly Improved upon

Transgenic animal models of AD are essential for the advancement of the field. However, many aspects of AD can be studied and therapeutic strategies for the disease evaluated in the neuronal cell-based models. Conceivably, such a model could be the most authentic because it can utilize human neuronal cells. The advantage of employing human neuronal cells is not only in that they are from the species affected by the disease and therefore reflect more faithfully cellular AD pathology but also because they possess a mechanism capable of the AβPP-independent production of C99 (presumably the RNA-dependent AβPP mRNA amplification pathway but in unlikely case other mechanisms, discussed in Section 22 below, are involved, they are also in place). In light of considerations discussed in the preceding sections, to construct a human neuronal cell-based AD model is straightforward: Elicit sustainably the integrated stress response and the endogenous AβPP-independent C99 production pathway (presumably the RNA-dependent amplification of endogenously produced AβPP mRNA) would be activated, and cellular AD pathology would ensue. Approaches to elicit the ISR in human neuronal cells are described in detail in [7]. Briefly, the ISR could be elicited either by exogenously expressed iAβ or by any other suitable ISR-eliciting stressor. Utilization of the former would yield a cellular model of conventional AD whereas employment of the latter would result in a cellular model representing unconventional AD (for detailed discussion of unconventional AD see Section 50 below). The question is whether such a human neuronal cell-based model is feasible. The answer is affirmative: not only feasible but highly plausible. The basis for this assertion is a 2014 study by Choi and co-workers [128].

The proof of concept provided in [128] is rather inadvertent. Choi and co-workers designed their model with the aim to maximize the extent of extracellular deposition of exogenously produced Aβ in hope that this would, in accordance with the ACH, maximize the extent of cellular AD pathology. To achieve their goal, they expressed exogenously AβPP containing two FAD mutations as well as an FAD-causing mutant of PSEN1 [128]. The essential innovation of this study was that human neuronal cells were cultured in a semi-solid matrigel medium in order to prevent the diffusion of secreted Aβ. This experiment resulted in an unmitigated success, actually a triumph: for the first time an experimental model system exhibited the formation of neurofibrillary tangles. Choi and co-workers interpreted this outcome as the affirmation of the principles of the ACH (see the introductory sentence of Section 3 above) [128].

In the framework of the ACH2.0, however, the results of Choi and co-workers are interpreted very differently. Due to utilization of semi-solid matrigel medium, secreted Aβ did not diffuse but was retained in the proximity of cultured neurons; this facilitated their cellular uptake. Two mutations utilized in [128], London mutation of AβPP (V717I) and ΔE9 mutation of PSEN1 substantially increase the production and, subsequently, secretion of the Aβ42 isoform. As discussed in Section 24 below, Aβ42 is taken up by neurons twice as effectively as other isoforms of Aβ. Second AβPP mutation employed in [128] is Swedish mutation (K670N/M671L). This mutation was shown to substantially increase a fraction of C99 undergoing gamma-cleavage on the intracellular rather than on plasma membranes, a process culminating in the retention (rather than secretion) of resulting iAβ. The most important attribute of this model in the ACH2.0 perspective is that it maximizes the influx of iAβ derived from massively overproduced exogenous AβPP. When exogenous iAβ accumulates to sufficient levels (see Section 25 below), it triggers the elicitation of the ISR. This, in turn, activates the endogenous AβPP-independent production of C99. Consequently, the progression of cellular AD pathology commences and results in formation of neurofibrillary tangles. If the objective is to assure a sufficient influx of iAβ, the approaches employed in [128] are rather cumbersome; as mentioned above, they could be significantly improved upon. Nevertheless, the study by Choi and co-workers [128], interpreted in terms of the ACH2.0, serves as proof of principle for the plausibility of human neuronal cell-based models of AD.

21. Validation (4): Expression of iAβ in Human Neuronal Cells Should Trigger Cellular AD Pathology Including Formation of NFTs

In the human neuronal cell-based AD model developed by Choi and co-workers [128] and discussed in the preceding section the accumulation of iAβ occurs in an indirect and rather convoluted way. Indeed, it occurs inefficiently via the cellular uptake of secreted extracellular AβPP-derived Aβ and through the retention of iAβ resulting from gamma-cleavages occurring on intraneuronal rather than on plasma membranes. The reason for this is that this model was constructed with the purpose of maximizing the accumulation of extracellular, not intraneuronal Aβ. The goal of maximizing the accumulation of iAβ could be easily accomplished by exogenously expressing in human neuronal cells not AβPP and not C99 (because C99 contains trans-membrane domain at the C-end of its Aβ segment, and Aβ produced by gamma cleavage of C99 would be secreted) but rather Aβ, more specifically Aβ42. This can be accomplished by utilizing the AUG encoding Met671 in AβPP as translation initiation codon in the expression vector. Its primary translation product would be Met-Aβ42 (see Section 23 below); the N-terminal Met would be removed by one of multiple aminopeptidases with a broad specificity, yielding Aβ42 (discussed in Section 23 below). It can be anticipated that iAβ produced in such a manner would efficiently accumulate and rapidly reach the ISR-eliciting threshold. Consequently, the neuronal integrated stress response would be elicited (see Section 25 below), the endogenous AβPP-independent C99 generation pathway would be activated, AD cellular AD pathology would commence and progress, and neurofibrillary tangles would manifest. In a control experiment, the same procedure can be carried out in mouse neuronal cells. Since in mice the AβPP-independent C99 production pathway is inoperative (discussed above), no extensive degeneration would occur (except that caused by the exogenous iAβ-elicited neuronal ISR; see Section 19) and no NFTs would form. If the outcome obtained in human neuronal cells is as anticipated, such an experimental system would constitute an adequate cellular model system for conventional AD.

22. Alternative Modes of Initiation of Translation from the AUG Codon Encoding Met671 of AβPP

In the authors' viewpoint, in fact, a conviction, RNA-dependent amplification of human AβPP mRNA is by far the most plausible candidate for the role of a mechanism enacting the AβPP-independent production of C99. It has "built-in" human specificity observed in the field and is strongly supported by the empirical data. Its chimeric end RNA end product is cap-less. But cap-less mRNA species are translated under the ISR conditions. Moreover, its nucleotide modifications potentially ensure its translatability under the ISR (even a single m(6)A modification was shown to promote cap-independent translation [129]). Nevertheless, there are three other, albeit much less likely, possibilities of utilization of the AUG encoding Met671 of AβPP as a translation initiation codon that should be mentioned here for the reader's consideration. One such mode is the internal initiation of translation of intact AβPP mRNA from the AUG codon under discussion. This possibility was discussed in Section 7 above. It was proposed by Breimer and Denny [59] and evaluated and "ruled out" by two research groups [60,61]. However, this "ruling out", in both studies [60,61], is grossly unfounded. Breimer and Denny posited that the initiation of translation of the intact human AβPP mRNA from the AUG encoding Met671 might occur as an inducible process in neuronal cells under the AD conditions [59]. Both studies, however, utilized non-neuronal cells and were carried out certainly not under the AD conditions [60,61]. Their conclusions, therefore, should be discounted and the initiation of translation of the intact human AβPP mRNA from the AUG encoding Met671 should be considered a viable option until proven otherwise.

The remaining two possibilities of the initiation of translation of human AβPP mRNA from the AUG encoding Met671 are conceptually similar to the outcome of the RNA-dependent AβPP mRNA amplification pathway in that the translated mRNA molecules are 5′-truncated to the extent that the AUG encoding Met671 of AβPP becomes the first functional translation initiation codon. What differ are mechanisms that produce 5′-truncated AβPP mRNA. In one case, this is the internal initiation of transcription of AβPP RNA from a position within the AβPP gene proximal to but upstream of the ATG encoding Met671. This would require expression of specialized transcription factor(s) presumably induced by the neuronal ISR conditions.

In another case, the truncation is enacted by a cleavage of AβPP mRNA in proximity of but upstream from the AUG encoding Met671. Such a process would require a specialized nuclease produced under the neuronal ISR conditions. Whereas the internal initiation of transcription would produce capped 5′-truncated AβPP mRNA (unlikely to be translated under the ISR conditions), cleavage-generated 5′-truncated AβPP mRNA would be uncapped. It should be emphasized that we consider all three possibilities interesting but implausible; they do neither reflect human specificity, nor are they supported by the empirical data.

23. Validation (5): The Primary Product of Translation Initiated from the AUG Encoding Met671 of Human AβPP Is Distinguishable from Its Counterpart Generated by AβPP Proteolysis and Can Serve as a Reporter

When Breimer and Denny proposed that C99 could be generated independently of AβPP via the internal initiation of translation of AβPP mRNA at the AUG encoding Met671 [59], they presumed that it would be indistinguishable from C99 produced by the proteolytic cleavage of AβPP at the beta-site, between residues 671 and 672. The reason for this is that, at that time (1987), it was assumed that the translation-initiating methionine is always removed co-translationally by N-terminal methionine aminopeptidase (MAP). If this were the case, the primary product of translation initiated at the AUG encoding Met671 would terminate at its N-end with Asp672, and would, indeed, be indistinguishable from C99 produced by the cleavage of AβPP at the Met671/Asp672 junction. Subsequent research, however, demonstrated that this is not always the case. Interestingly, the erroneous notion that N-terminal Met is always removed co-translationally by MAP still persists in a significant fraction of the scientific community despite evidence to the contrary described below; authors are indebted to Alexander Varshavsky (Caltech) for conveying this instrumental information.

For the N-terminal methionine to be cleaved-off co-translationally, it should be accommodated, together with the following amino acid residue, within the active site of MAP. This is not always possible. More precisely, this is possible only in defined combinations, namely methionine followed by any one of the seven smallest amino acid residues. In order of increasing size, they are as follows: Gly, Ser, Cys, Thr, Pro, and Val. In cases where translation-initiating methionine is followed by any amino acid residue larger than valine (in fact, by any residue other than the above listed seven), N-terminal Met would not be removed by MAP co-translationally but would be retained in the primary translation product [130–135]. Met671 of human AβPP is followed by aspartate, which is larger than valine. Therefore, if translation were to initiate at the AUG encoding Met671 of AβPP (regardless of a mechanism enabling such an initiation of translation), the primary translation product would be not C99 but rather C100, i.e., N-terminal Met-C99. C100 would not, however, retain its N-terminal methionine for long. It was shown that in cases like this, when translation-initiating methionine is retained in the primary translation product, it is eventually removed by one of the numerous aminopeptidases with broad specificity [134,135]. Importantly, however, this would occur post-translationally; only MAP is capable of the co-translational cleavage of the N-terminal methionine. It follows that if AβPP-independent production of C100 indeed occurs in AD-affected neurons and is followed by its conversion to C99, a pool of C100 would, nevertheless, persist. This is of considerable practical significance since the occurrence of C100 would report the activity of the AβPP-independent pathway of its production. A question might be asked: why was C100 not seen in postmortem samples from AD patients? The answer is because in dead tissues protein synthesis ceases well before the activity of proteolytic enzymes stops. Consequently, without a new influx of C100, its pool would be, in its entirety, converted into C99 and thus lose its identifying N-terminal Met. On the other hand, C100 pools would exist and should be detectable in adequate models of AD, described in Sections 19 and 20 above and elsewhere [7].

24. Accumulation of Intraneuronal Aβ (iAβ) Occurs Physiologically via Two Distinct Processes

The present section picks-up the narrative from Section 5 (the intervening Sections were introduced to establish the plausibility of the AβPP-independent production of C99 in AD-affected neurons). In that section, we formulated the principles of the ACH2.0. The first of these principles is that conventional AD is triggered by iAβ accumulated over a certain critical threshold. Intraneuronal accumulation of Aβ is well established; it occurs (albeit, as argued below, at different rates) in healthy individuals, i.e., those who will not develop AD, as well as in future AD patients. Intraneuronal accumulation of Aβ takes place via two well-understood physiological mechanisms. One such mechanism is the retention of newly produced Aβ within the neuron. The immediate precursor of Aβ in the AβPP proteolytic pathway is the C99 fragment. Aβ is generated by the gamma-cleavage of C99. Usually this occurs on plasma membranes in conjunction with the secretion of newly formed Aβ. In some cases, however, the cleavage occurs on intraneuronal membranes within various organelles [136–144]. Aβ generated in this way is not secreted to the extracellular space but is retained intraneuronally as iAβ. As discussed below, certain mutations of AβPP and of presenilins facilitate the cleavage of C99 on the intraneuronal membranes; this increases the fraction of AβPP-derived Aβ retained as iAβ and elevates the rate of its influx.

Another physiologically operating mechanism resulting in accumulation of AβPP-derived iAβ is the cellular uptake of secreted Aβ [145–150]. Apparently, the oligomerization of extracellular Aβ is a prerequisite for its effective internalization [149,150]. For this reason, due to its demonstrated propensity to aggregate, the Aβ42 variant is taken up by cells at the rate twice that of other Aβ isoforms [150]. The cellular uptake of extracellular Aβ is mediated by numerous receptors [151–159]. The ligands of these receptors play an important role in and regulate the internalization of extracellular Aβ. One of such ligands is ApoE [149]. Of its several variants, ApoE4 is substantially more effective in mediating the internalization of extracellular Aβ than other ApoE variants; its occurrence, therefore, increases the rate of cellular uptake of extracellular Aβ and thus elevates the rate of influx of AβPP-derived iAβ.

25. iAβ Accumulated over the Critical Level Activates PKR and HRI Kinases and Thus Triggers the Elicitation of the Neuronal ISR

The elicitation of the integrated stress response is enacted, via phosphorylation of eIF2α, by one or more of eIF2α kinases. iAβ was shown to be capable of mediating the activation of two of those: PKR and HRI. The activation of PKR was observed, alongside with phosphorylation of eIF2α, in cell models and animal models that overexpress Aβ [160–162]. Moreover, both activated PKR and eIF2α phosphorylated at its Ser51 residue were detected in neuronal cells of AD patients [163,164]. iAβ, at sufficient cellular levels, can apparently cause phosphorylation and consequent activation of PKR in two ways. One is through PKR-activator (PACT). This conclusion is supported by the observation that PACT and activated PKR are co-localized in AD-affected neurons [165]. Another way for iAβ to cause phosphorylation and activation of PKR is via tumor necrosis factor alpha (TNFα), as was observed in transgenic mouse models that overexpress human Aβ [166].

The activation of HRI in Alzheimer's disease, on the other hand, is a corollary of the iAβ-triggered mitochondrial dysfunction in AD-affected neurons. AD-related mitochondrial distress is one of the earliest symptoms of the disease and as such was studied extensively [167–184]. The connection between iAβ-caused mitochondrial dysfunction and the activation of HRI in AD-affected neurons is as follows. One of the early events in dysfunctional mitochondria is the activation of mitochondrial protease OMA1 [185,186]. Upon its activation, OMA1 cleaves another mitochondrial protein, DELE1, in a site-specific manner [185,186]. One of the resulting fragments of DELE1 is transported across mitochondrial membranes into the cytosol [185,186]. This fragment has an affinity to HRI; upon binding, it activates the kinase. Thus, iAβ, accumulated to sufficient levels, triggers the activation of either PKR (via PACT and/or TNFα) or HRI (through OMA1 to DELE1 to HRI pathway). When PKR or HRI, or both, are activated, eIF2α is phosphorylated at its Ser51 residue and the elicitation of the neuronal ISR ensues.

26. The Neuronal ISR Enables Operation of the Self-Sustainable AβPP-Independent C99 Production Pathway; AD Commences Only When the Latter Is Activated

The rate of accumulation of AβPP-derived iAβ via its intraneuronal retention or through the cellular uptake of secreted extracellular Aβ, both occurring physiologically, could be (albeit not always; see below) sufficient to activate PKR and/or HRI and thus trigger phosphorylation of eIF2α and elicitation of the neuronal integrated stress response but it is patently insufficient to support the accretion of AβPP-derived iAβ to the AD pathology-causing levels. The AD pathology-causing levels of iAβ are attained via the activity of the AβPP-independent C99 generation pathway (discussed in the preceding sections). As described above, under the conditions of the neuronal ISR, translation of the global cellular cap-dependent mRNAs to proteins is severely suppressed. Concurrently, an array of new transcription factors is expressed, new mRNA species are transcribed, and cap-independent production of a small subset of cellular proteins is activated. Among those are essential components of the AβPP-independent C99 production pathway that are not present under the non-ISR conditions; when they are available, the pathway is activated. In the version of ACH2.0 under discussion C99 generated independently of AβPP is processed into iAβ. Its entire output is retained within neurons. Consequently, its levels rapidly increase and reach AD pathology-causing range. Not only is it toxic in its own right, it was also shown to suppress the ubiquitin-proteasome system and thus promote tau accumulation [187–190].

Thus, in this version of the ACH2.0, AD is triggered, via the elicitation of the neuronal integrated stress response, by AβPP-derived iAβ accumulated to sufficient levels and is driven by differentially derived iAβ generated by gamma-cleavage of C99 produced in the AβPP-independent pathway. Since the accumulation of AβPP-derived iAβ, regardless of its rate, is a physiological process, AD commences only with the elicitation of the neuronal ISR and consequent activation of the AβPP-independent C99 production pathway. Importantly, driving AD pathology is only one function of iAβ produced independently of AβPP. Its other major function is ascertaining the sustainability of its own production. Indeed, when the disease commences, levels of iAβ (derived from AβPP) are, by definition, already above those required for the elicitation of the neuronal ISR. Operation of the AβPP-independent C99 production pathway supplies, at the rate orders of magnitude greater than the rate of influx of AβPP-derived iAβ, additional, differentially derived iAβ. This maintains PKR and/or HRI in the activated states, sustains phosphorylation of eIF2α, and thus propagates the neuronal ISR and, consequently, perpetuates the operation of the AβPP-independent C99 (and, in this ACH2.0 version, iAβ) production pathway. This feedback loop ensures the uninterrupted production of C99 and iAβ independently of AβPP and constitutes the Engine that drives the disease. The sequence of events described above is illustrated in Figure 6.

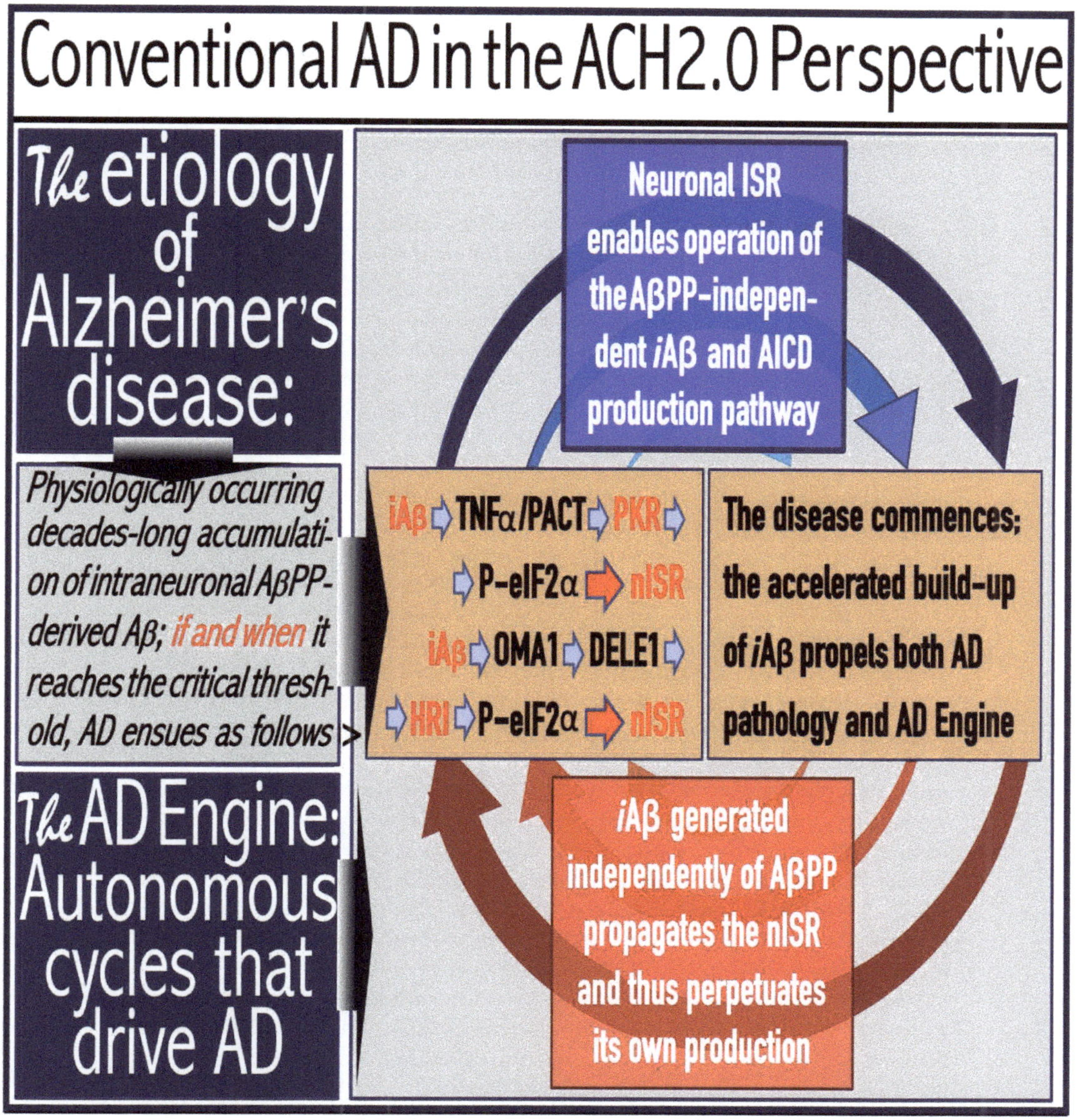

Figure 6. AD is driven by the self-sustainable AβPP-independent C99 production pathway enabled by the neuronal ISR. *eIF2α*: eukaryotic translation initiation factor 2*α*. *PKR and HRI*: kinases that phosphorylate eIF2α at its Ser51 residue. *PACT*: PKR activator. *TNFα*: tumor necrosis factor *α*. *OMA1*: mitochondrial protease activated during mitochondrial dysfunction. *DELE1*: substrate of OMA1 protease. Cleavage of DELE1 leads to the activation of HRI. Phosphorylation of eIF2α elicits the integrated stress response, which, in turn, enables the generation of components essential for the activity of the AβPP-independent pathway of C99 and, subsequently, iAβ production. iAβ: intraneuronal Aβ. AβPP-derived iAβ accumulates physiologically via the importation of extracellular Aβ and the retention of iAβ produced by gamma-cleavages on intracellular membranes in an exceedingly slow process. In most individuals it does not reach the PKR- and/or HRI-activating, the neuronal ISR-eliciting threshold within their lifetimes and no conventional AD occurs. When this threshold is

crossed, the neuronal ISR is elicited and the AβPP-independent C99 (and, subsequently, iAβ) production pathway activated. The iAβ product of the latter drives AD pathology. It also propagates the neuronal ISR and thus perpetuates its own production in the AβPP-independent pathway. It should be noted that the accumulation of iAβ produced in both the AβPP proteolytic and AβPP-independent pathways is accompanied by a proportional accumulation of AICD (AβPP Intracellular Domain). AICD was shown to be capable of interfering with numerous processes involved in AD [191–206] but its contribution to the disease, if any, remains to be elucidated.

27. The Decisive Role of the Rate of Accumulation of AβPP-Derived iAβ in the Occurrence and Timing of Alzheimer's Disease

In Figure 7 below, the T1 threshold defines levels of AβPP-derived iAβ that mediate the activation of PKR and/or HRI and thus trigger elicitation of the neuronal integrated stress response, consequent activation of the AβPP-independent C99 production pathway, and commencement of AD. The figure makes it apparent that the occurrence of AD crucially depends on the rate of accrual of AβPP-derived iAβ (the only variable in Figure 7). It is a simple logic. If no T1 crossing occurs within the lifetime of an individual, there is no AD. This, actually, describes the majority of the human population: most individuals do not develop the disease. If, however, the T1 threshold is crossed, the timing of the crossing, and, consequently, of the commencement of AD is reversely proportional to the rate of accrual of AβPP-derived iAβ: greater the latter, shorter the former. Four panels of Figure 7 depict four possibilities. First three (Panels A–C) vary only quantitatively and illustrate the proportionality invoked above. The fourth possibility (Panel D) differs qualitatively from others: the T1 threshold is not crossed, and AD does not occur. The T1 threshold appears to constitute a proverbial thin line that separates health and disease.

The concept of connectivity between the rate of accumulation of AβPP-derived iAβ and AD is validated by the analysis of mechanisms of action of all known FAD mutations, both causative and protective, which is presented in Sections 32–34 below. This connectivity, moreover, includes not only FAD mutations, but also all factors predisposing to the disease. A good example is ApoE4. As was mentioned above, all members of the ApoE family of ligands facilitate the cellular uptake of secreted extracellular Aβ; their efficiency, therefore, influence the rate of influx of AβPP-derived iAβ and, consequently, that of its accumulation. ApoE4 is substantially more efficient in facilitating the internalization of extracellular Aβ than other ApoE variants [149]. In carriers of ApoE4 allele(s), therefore, the rate of accumulation of AβPP-derived iAβ is greater than in the general population. Therefore, in these individuals the T1 threshold is crossed and, consequently, AD occurs sooner (or more often) than in the general population. They are "predisposed" to the disease and this predisposition is dose-dependent: carriers with only one ApoE4 allele are predisposed to the disease less than those carrying two ApoE4 alleles.

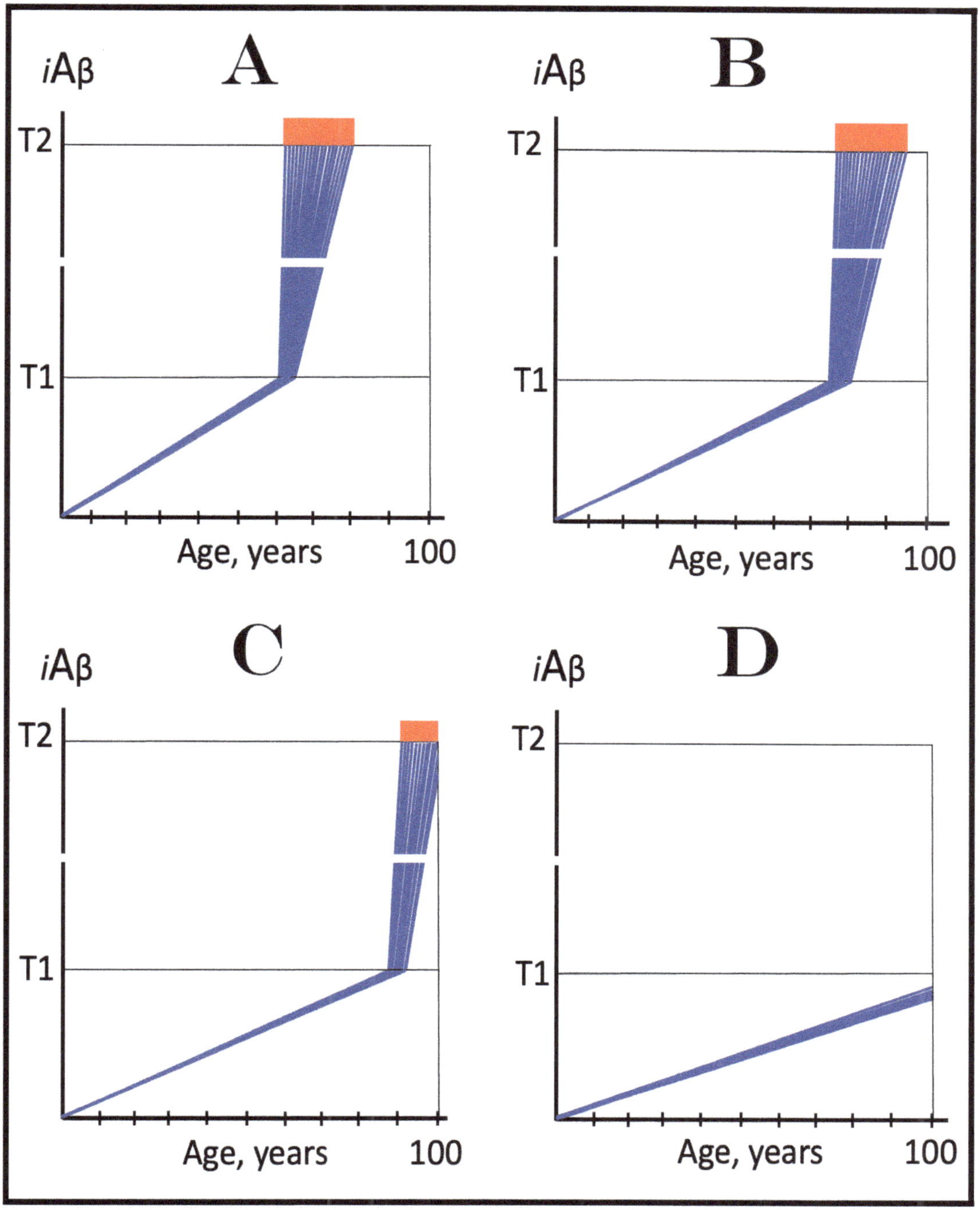

Figure 7. The occurrence and timing of conventional AD is determined by the dynamics of accumulation of AβPP-derived iAβ. *Blue lines*: Intraneuronal Aβ, iAβ. ***T1***: Threshold of cellular concentration of iAβ triggering the activation of PKR and/or HRI kinases, phosphorylation of eIF2α, the elicitation of the neuronal ISR, the initiation of the AβPP-independent C99 production pathway,

and the commencement of AD. ***T2***: Cellular concentration of iAβ triggering neuronal death via apoptosis or necroptosis [207]. *Red box*: Apoptotic zone, defined as a range of cellular concentrations of iAβ within which cells neuronal cells committed apoptosis or necroptosis or are dead. The only variable parameter in the present figure is the rate of accumulation of AβPP-derived iAβ. *Panel* (**A**): The rate of accumulation of AβPP-derived iAβ is relatively rapid. It crosses the T1 threshold at mid-sixties. PKR and/or HRI are activated, eIF2α phosphorylated, the neuronal ISR elicited, AβPP-independent C99 generation pathway initiated, and AD commences. *Panel* (**B**): The rate of accumulation of AβPP-derived iAβ is reduced. The T1 threshold is reached and crossed, and the disease commences at around the age of eighty. *Panel* (**C**): The rate of accumulation of AβPP-derived iAβ is reduced further. The T1 threshold is reached, and the disease commences only in the nineties. *Panel* (**D**): AβPP-derived iAβ does not reach the T1 threshold within the lifespan of the individual; the neuronal ISR is not elicited, the AβPP-independent C99 generation pathway is not activated, and conventional AD does not occur. Such an outcome is typical for the majority of the human population.

28. On the Conditionality of the First Stage of Conventional AD

As depicted in Figure 7 above, the kinetics of accumulation of iAβ in individuals who do not develop AD within their lifetimes is single-phased. It is, however, double-phased in those who eventually become AD patients. Phase One is the accrual of AβPP-derived iAβ until it crosses the T1 threshold. At this point the neuronal ISR is elicited, the AβPP-independent C99 (and, in this version of ACH2.0, iAβ) production pathway is activated, and the second (rapid) phase of iAβ accumulation and the disease (driven by iAβ produced in second phase) commence. Phase One of iAβ accumulation is, therefore, essential for phase Two and the disease to occur.

It is reasonable, even natural, therefore, to associate phases of iAβ accumulation with stages of AD. In these terms, phase One of iAβ accumulation is equated with stage one of AD. But this stage is a physiological process and no disease occurs at this stage. Thus, as an example, according to the above logic we must append the definition "stage one of AD" to AβPP-derived iAβ accumulation depicted in Panel D of Figure 7, which is an oxymoron: the disease not only does not occur but is not going to occur. A solution to this conundrum is simple: stage one of AD is conditional, it is rendered such only post-factum, if and when the T1 threshold was crossed and the disease actually commenced; short of this, there is no "stage one" of AD, only normal, physiologically occurring process of accumulation of AβPP-derived iAβ.

29. On the Inevitability of Conventional AD in a Sufficiently Long Lifespan

In terms of the preceding section, the incidence of AD depends on the conversion of accumulation of AβPP-derived iAβ from a physiologically occurring process into the fist stage of the disease, and this conversion is demarcated by the crossing of the T1 crossing. If the T1 threshold were not crossed within a defined lifespan, no AD would occur. But the defined lifespan is a subjective, ever changing, more precisely, ever increasing variable. In Figure 7 it is assumed to be 100 years. In Panel D, the T1 threshold is not crossed at this time limit and the disease does not occur, but in Panels A–C both the T1 crossing takes place and AD develops. As was mentioned above, at the time of Dr. Alois Alzheimer, life expectancy in Germany was only 48 years. At this "defined" lifespan the crossing of the T1 threshold would take place and AD would occur in neither of panels of Figure 7. Inversely, we can extrapolate what would happen with the increase in the "defined" lifespan. The conclusion is straightforward: provided the lifespan is long enough, the T1 threshold would be crossed, and the disease would inevitably occur. Moreover, the "long enough" is not far away. Currently, the incidence of the disease is 5% at sixty-five, 15% at seventy-five, and 35% at eighty-five. By extrapolation, it would approach 100% at about hundred-twenty

In the last hundred years, the lifetime expectancy in industrial countries has increased by 60% (from about fifty to about eighty). Given a comparable rate of increase, the longevity of 120 years and the incidence of AD approaching certainty would be reached within the next five decades. The disease is inevitable, provided the lifespan is sufficiently long, because the first stage of AD, or rather potential first stage of AD, is a physiologically occurring process; given enough time, the potential would be realized and the T1 threshold would, unavoidably, be reached. There are only two approaches to preclude the eventual certitude of developing AD. One is to limit the lifespan. Another is to develop therapies for the prevention and cure of the disease. The second approach is vastly preferable and is discussed below.

30. The Critical Role of the Extent of the T1 Threshold in the Incidence and Timing of AD and of Aging-Associated Cognitive Decline (AACD)

As was discussed above (Section 26), the incidence and the timing of conventional AD are defined by the instance when AβPP-derived iAβ crosses the neuronal ISR-eliciting T1 threshold. This crossing, in turn, depends on two factors: the rate of accumulation of AβPP-derived iAβ and the extent of the T1 threshold. The former was discussed in the preceding sections; the latter is addressed in the present section. The extent of the T1 threshold is of special importance. As discussed in the following section below, the extent of the T1 threshold in the subpopulation predisposed to non-familial conventional AD is lower than in the general population. Moreover, this parameter is decisive in the occurrence of aging-associated cognitive decline, AACD. The role of the extent of the T1 threshold in the occurrence and timing of AD is illustrated in Figure 8 (this parameter is the only variable in Figure 8). The logic of this figure is apparent: with the given rate of accumulation of AβPP-derived iAβ, the greater is the extent of the T1 threshold, the later will it be crossed and the later will AD occur (Panels A–C of Figure 8). Eventually, within a defined lifespan, the extent of the T1 threshold could be such that at the given rate of accumulation of AβPP-derived iAβ neither the T1 will be crossed, nor would AD occur (Panel D of Figure 8).

iAβ is prone to aggregate and it is reasonable to assume that above a certain level (different for different Aβ variants) its aggregates are bound to cause a degree of neuronal damage. The threshold demarcating the commencement of such neuronal damage is designated T^0 in Figure 8. If the extent of the T^0 threshold were greater than that of its T1 counterpart, this neuronal damage would be a component of AD (Panel A of Figure 8). If, however, the extent of the T^0 threshold were smaller than that of the T1, the damage would occur but would be insufficient to trigger the elicitation of the neuronal integrated stress response. In terms of the ACH2.0, AACD is defined as the neuronal damage occurring in the range of concentrations of AβPP-derived iAβ between the T^0 and T1 thresholds. It follows that for AACD to occur, the T^0 threshold must be lower than the T1. Thus, no AACD occurs in Panel A of Figure 8. In Panel B through D, the crossing of the T^0 threshold triggers AACD. At the given extent of the T^0 threshold, its duration increases with the increase in the extent of the T1 threshold. In Panels B and C of Figure 8 the T1 threshold is eventually crossed and AACD morphs into AD. In Panel D of Figure 8, due to the limited lifespan, the T1 threshold is not crossed. In this scenario, no neuronal ISR is elicited, no AβPP-independent C99 production pathway is activated, no AD occurs. AACD, however, commences with the crossing of the T^0 threshold and progresses until the end of the lifespan.

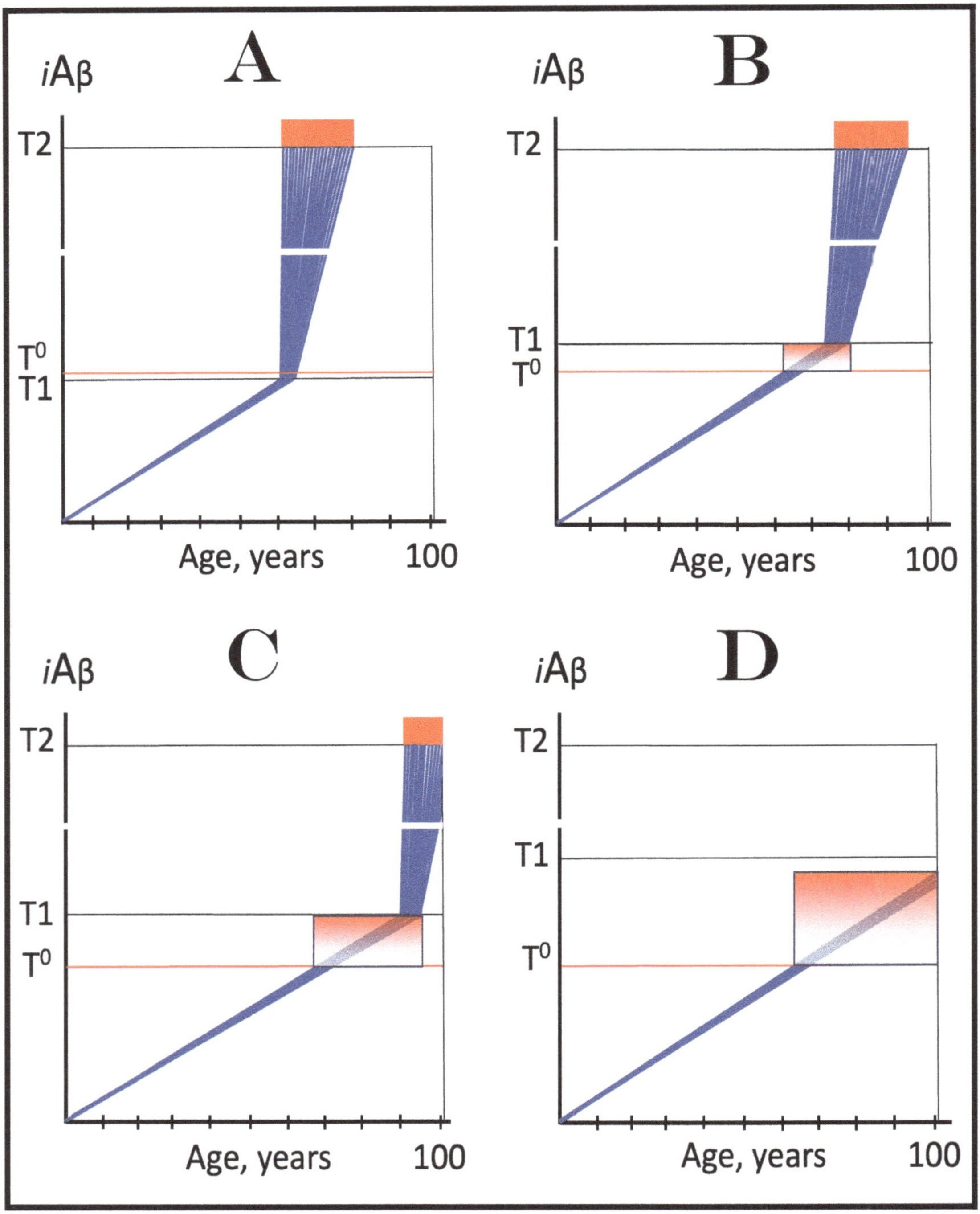

Figure 8. Role of the extent of the T1 threshold in the incidence and timing of AD and of aging-associated cognitive decline, AACD. *Blue lines*: Intraneuronal Aβ, iAβ. T^0: Threshold of cellular concentration of AβPP-derived iAβ triggering the neuronal damage, which manifests as AACD. *T1*: Threshold of cellular concentration of iAβ triggering the activation of PKR and/or HRI kinases,

phosphorylation of eIF2α, the elicitation of the neuronal ISR, the initiation of the AβPP-independent C99 production pathway, and the commencement of AD; its extent is the only variable in the present figure. ***T2***: Cellular concentration of iAβ triggering neuronal death via apoptosis or necroptosis. *Red box*: Apoptotic zone, defined as a range of cellular concentrations of iAβ within which cells neuronal cells committed apoptosis or necroptosis or are dead. *Pink boxes*: The range of cellular concentrations of AβPP-derived iAβ concentrations between the T^0 and the highest extent reached by AβPP-derived iAβ or the T1 threshold that support the progression of AACD. The condition commences with the crossing of the T^0 threshold and morphs into AD with the T1 crossing (if no T1 occurs, AACD persists for the remaining portion of the lifetime). It follows that AACD can occur only if the extent of the T^0 threshold is smaller than that of the T1. *Panel* (**A**): The T1 threshold is crossed, the neuronal ISR is elicited, the AβPP-independent C99 and iAβ production pathway is activated, and AD commences at about sixty years of age. The extent of the T^0 threshold is greater than that of the T1; consequently, no AACD occurs. *Panel* (**B**): The extent of the T1 threshold is above that of the T^0 threshold. When the T^0 is crossed by AβPP-derived iAβ, AACD commences and morphs into AD following the T1 crossing at about seventy-five years of age. *Panel* (**C**): The T1 threshold is increased further and is crossed at about ninety years of age. Since the extent of the T^0 does not change, AACD commences at the same time as the preceding panel, but its duration increases. *Panel* (**D**): The extent of the T1 threshold is such that it is not crossed by AβPP-derived iAβ within the lifetime of the individual; no conventional AD occurs. On the other hand, AACD commences with the T^0 crossing and persists for the remaining lifetime.

31. "Sporadic AD" Is Not Sporadic at All: The Disease Affects a Well-Defined, Non-Random Subpopulation; "Non-Familial AD" Is a Much Better Designation

"Sporadic" in "sporadic AD" has, uncomfortably, fatalistic connotations. It implies that, starting from a typical age of the onset of the disease, about sixty-five, the disease can strike anyone at random. This, however, is patently not the case. It follows from the preceding sections that individuals with high rate of accumulation of AβPP-derived iAβ and with low extent of the neuronal ISR-eliciting T1 threshold are selectively prone, "predisposed", to the disease. This is consistent with the findings, discussed above, that carriers of the ApoE4 allele are more likely to develop the disease, and, moreover, carriers of two ApoE4 alleles are much more likely to develop AD than carriers of other ApoE alleles. This is because ApoE4 is significantly (and in a dose-dependent manner, i.e., two versus one alleles) more efficient in facilitation of the cellular uptake of extracellular Aβ than other ApoE variants. The utilization of ApoE4 increases the influx of AβPP-derived iAβ and thus elevates both its rate of accumulation and the degree of predisposition to the disease. ApoE4 is only one example. Doubtless, numerous other factors related to the production, importation, and clearing of both extra- and intraneuronal Aβ influence the rate of AβPP-derived accumulation and, consequently, the degree of predisposition to AD.

As for the extent of the T1 threshold, it is apparently lower in the subpopulation that develops non-familial conventional AD than in the general population. This conclusion is arrived at as follows. The age of onset of AACD is greater than that of non-familial ("sporadic") AD [208,209]. It follows that in the AACD-developing subpopulation, which is significantly larger than the non-familial AD-developing subpopulation, the extent of the T^0 and, of course, T1 thresholds are greater than that of the T1 threshold in the subpopulation predisposed to non-familiar AD. The correlation of the low extent of the T1 threshold and the high rate of incidence of non-familial AD is consistent with the conclusion reached in the preceding sections that the former predisposes to the latter. Thus, the subpopulation that is likely to develop non-familial AD can be defined as a category of individuals with the rate of accumulation of AβPP-derived iAβ above the average and with the extent of the T1 threshold below the average seen in the general population. It can be stated with certainty that any factor that contributes to the increase in the rate of accumulation of AβPP-derived iAβ and/or to the lowering of the T1 threshold predisposes to AD.

32. Validation (6): ALL FAD Mutations Exert Their Effect by Accelerating the Rate of Accumulation of AβPP-Derived iAβ or Lowering the Extent of the T1 Threshold—Type One FAD Mutations

To validate the concept that conventional AD is triggered when AβPP-derived iAβ crosses the T1 threshold and that, consequently, the rate of its accumulation and the extent of the T1 threshold define the incidence and timing of the disease one does not have to venture far. The analysis of mechanisms of action of the FAD-causing mutation as well as of the mutation that protects from AD and AACD makes a powerful case. Strikingly, all such mutations, both of AβPP and of presenilins, without a single exception, exert their effect either by accelerating or decelerating the rate of accumulation of AβPP-derived iAβ or by lowering the extent of the T1 threshold. The present and the following sections analyze mutations that cause early-onset AD (FAD). They are numerous and, on the basis of their mechanisms of action, can be divided into two types. The present section considers Type One of FAD mutations. This type comprises mutations that increase the influx of AβPP-derived iAβ and thus elevate its rate of accumulation and accelerate the crossing of the T1 threshold and, consequently, the commencement of AD. It is exemplified by the Flemish and Swedish AβPP mutations, as well as by certain presenilin mutations. Flemish A692G mutation affects the physiological activity of BACE2 [210]. The main function of this enzyme is regulation of the production of Aβ and of the levels of iAβ. Its major activity is the cleavage within the iAβ or within the Aβ segment of AβPP or C99. Cleavage within iAβ directly reduces its levels whereas cleavage within the Aβ segment of AβPP or C99 does it indirectly by reducing the production and, consequently, secretion and uptake of Aβ. The Flemish mutation suppresses this activity of BACE2. As an effect, less iAβ is eliminated, more Aβ is secreted, and more extracellular Aβ is imported as iAβ. This increases the influx of AβPP-derived iAβ, elevates its rate of accumulation and accelerates the T1 crossing and commencement of the disease; hence, early-onset AD.

The Swedish AβPP mutation [211] and certain presenilin mutations [212] achieve the same outcome in a different way. They increase the fraction of C99 cleavages occurring on intracellular membranes within various cellular organelles, as described in Section 24 above. Aβ resulting from these cleavages of AβPP-derived C99 is nor secreted but is retained within neuronal cells as iAβ. As a result, the influx of AβPP-derived iAβ is increased, the rate of its accumulation is elevated, the crossing of the T1 threshold and commencement of AD are accelerated, and early-onset AD ensues. The above considerations are illustrated in Figure 9, which compares wild type (Panels A through D) with mutants (Panels A′ through D′) and considers two scenarios. In one, the extent of the T^0 threshold is greater than that of the T1 threshold; no AACD occurs. In the other scenario, the extents of the T^0 and T1 thresholds are reversed. In these cases, AD (if it occurs) is preceded by AACD.

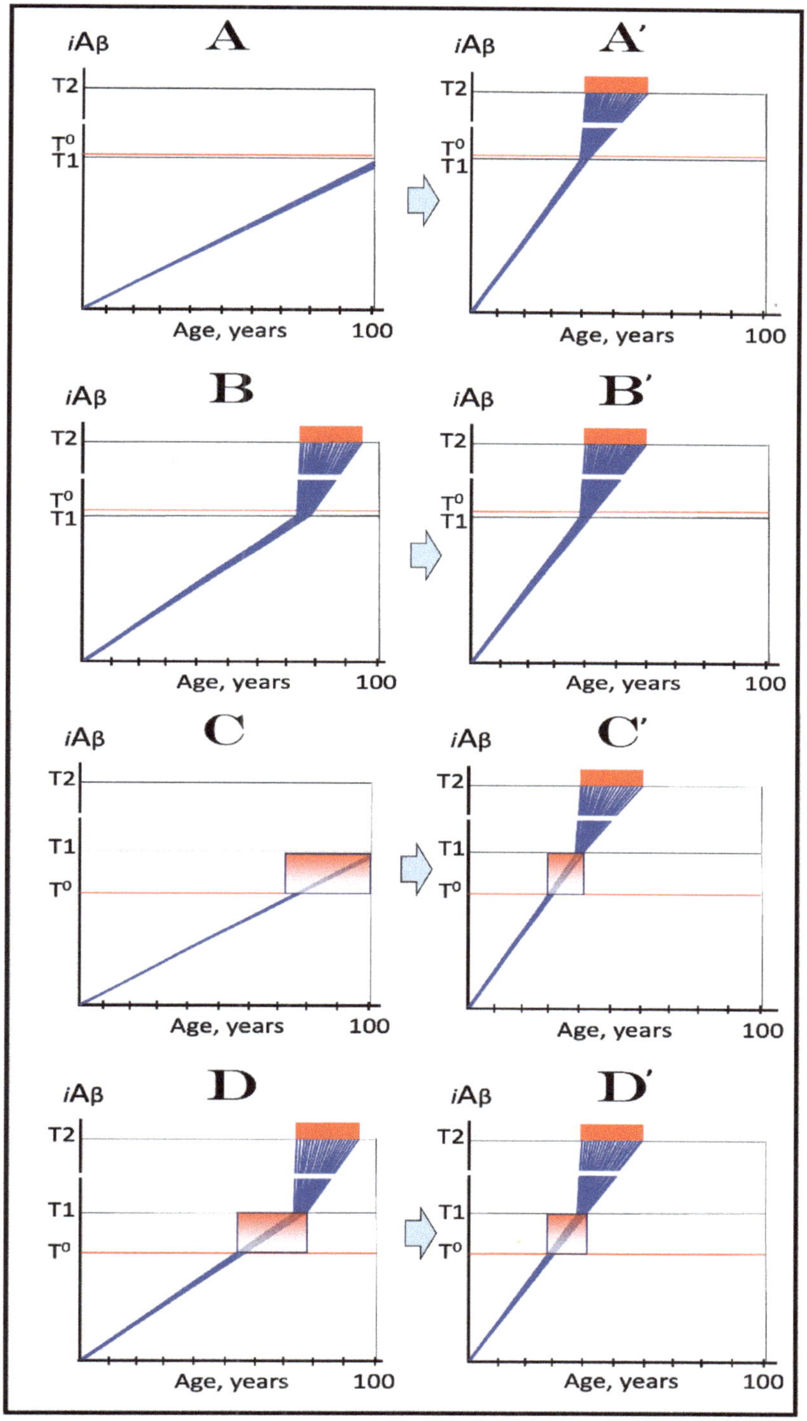

Figure 9. Familial AD mutations that exert their effect by accelerating the rate of accumulation of AβPP-derived iAβ. *Blue lines*: Intraneuronal Aβ, iAβ. T^0: Threshold of cellular concentration of

AβPP-derived iAβ triggering the neuronal damage, which manifests as AACD. *T1*: Threshold of cellular concentration of iAβ triggering the activation of PKR and/or HRI kinases, phosphorylation of eIF2α, the elicitation of the neuronal ISR, the initiation of the AβPP-independent C99 production pathway, and the commencement of AD; its extent is the only variable in the present figure. *T2*: Cellular concentration of iAβ triggering neuronal death via apoptosis or necroptosis. *Red box*: Apoptotic zone, defined as a range of cellular concentrations of iAβ within which cells neuronal cells committed apoptosis or necroptosis or are dead. *Pink boxes*: The range of cellular concentrations of AβPP-derived iAβ concentrations between the T^0 and the highest extent reached by AβPP-derived iAβ or the T1 threshold that support the progression of AACD. The condition commences with the crossing of the T^0 threshold and morphs into AD with the T1 crossing; it can occur only if the extent of the T^0 threshold is smaller than that of the T1. *Panels* (**A–D**): dynamics of iAβ accumulation in wild-type AβPP carriers. *Panels* (**A,B**): the T^0 levels are above those of T1; no AACD occurs. Panel (**A**): the T1 threshold is not crossed; no disease occurs. *Panel* (**B**): the T1 threshold is crossed; this is followed by late onset AD. *Panels* (**C,D**): the T^0 threshold is lower than T1. *Panel* (**C**): the T1 threshold is not reached; when the T^0 threshold is crossed, AACD commences and persists for the remaining lifespan. *Panel* (**D**): AACD commences when the T^0 is crossed and morphs into late onset AD upon the T1 crossing. *Panels* (**A′–D′**): Dynamics of iAβ accumulation in carriers of FAD mutations, which elevate the rate of accumulation of AβPP-derived iAβ. As a result, the T1 threshold is crossed sooner. and early-onset AD follows. *Panels* (**A′,B′**): the T^0 levels are above those of T1; no AACD occurs. *Panels* (**C′,D′**): the extent of the T^0 threshold is smaller than that of the T1; AACD commences with the T^0 crossing and morphs into the early-onset AD when the T1 threshold is reached.

33. Validation (7): ALL FAD Mutations Exert Their Effect by Accelerating the Rate of iAβ Accumulation or Lowering the Extent of the T1 Threshold—Type Two FAD Mutations

Type One FAD-causing mutations, described in the preceding section, increase the influx of AβPP-derived iAβ, elevate the rate of its accumulation, and accelerate the T1 crossing and commencement of conventional AD. Type Two FAD-causing mutations, both of AβPP and presenilins, do the same. In addition, they also lower the extent of the T1 threshold. Mechanistically, all Type Two FAD-causing mutations result in significantly increased production of Aβ42 [147]. The increase in the Aβ42 fraction of Aβ plays out both inside and outside the neuronal cells. When outside, extracellular Aβ42 is, as discussed above, taken up at the rate twice that of other Aβ variants. The reason for this is, apparently, the propensity of Aβ42 to aggregate. A prerequisite for effective cellular uptake of extracellular Aβ is its oligomerization, which is facilitated by the aggregate-forming potential of Aβ42. The increased rate of importation reproduces the effect of Type One of FAD mutation: the influx of AβPP-derived iAβ is increased, the rate of its accumulation elevated, and the crossing of the T1 threshold and commencement of AD accelerated. When inside the neuronal cells, iAβ42 lowers the extent of the T1 threshold. This is because the T1 is defined as the cellular concentration of iAβ that triggers activation of PKR and/or HRI and consequent elicitation of the neuronal ISR. iAβ42 is more toxic than other variants of iAβ. This is due to the same propensity to aggregate that increases the efficiency of its cellular uptake. Consequently, a cellular concentration of iAβ42 smaller than that of other iAβ variants is required to generate a certain level of cellular stress and/or toxicity to trigger the elicitation of the neuronal ISR, i.e., the T1 threshold for iAβ42 is lower than for other iAβ isoforms.

Effects of Type Two FAD-causing mutations are depicted schematically in Figure 10. Panels A through D show wild types whereas Panels A′ through D′ depict Type Two FAD-causing mutants. The effects of mutations are driven by two factors: the increase in the rate of accumulation of the AβPP-derived iAβ and the reduction in the extent of the T1 threshold. Cumulatively, this results in a substantial acceleration of both the T1 crossing and commencement of AD. In cases where the extent of the T^0 threshold is greater than that of the T1 threshold, no AACD occurs. When the inverse is true, AD (when it occurs) is preceded by AACD. It should be noted that due to a steeper rate of AβPP-derived iAβ accumulation the duration of AACD could be significantly shorter in mutants than in their wild-type counterparts. Also of note is that in Panels A′ through D′ the extent of the T2 thresholds would likely be also lowered due to the increased toxicity of iAβ42.

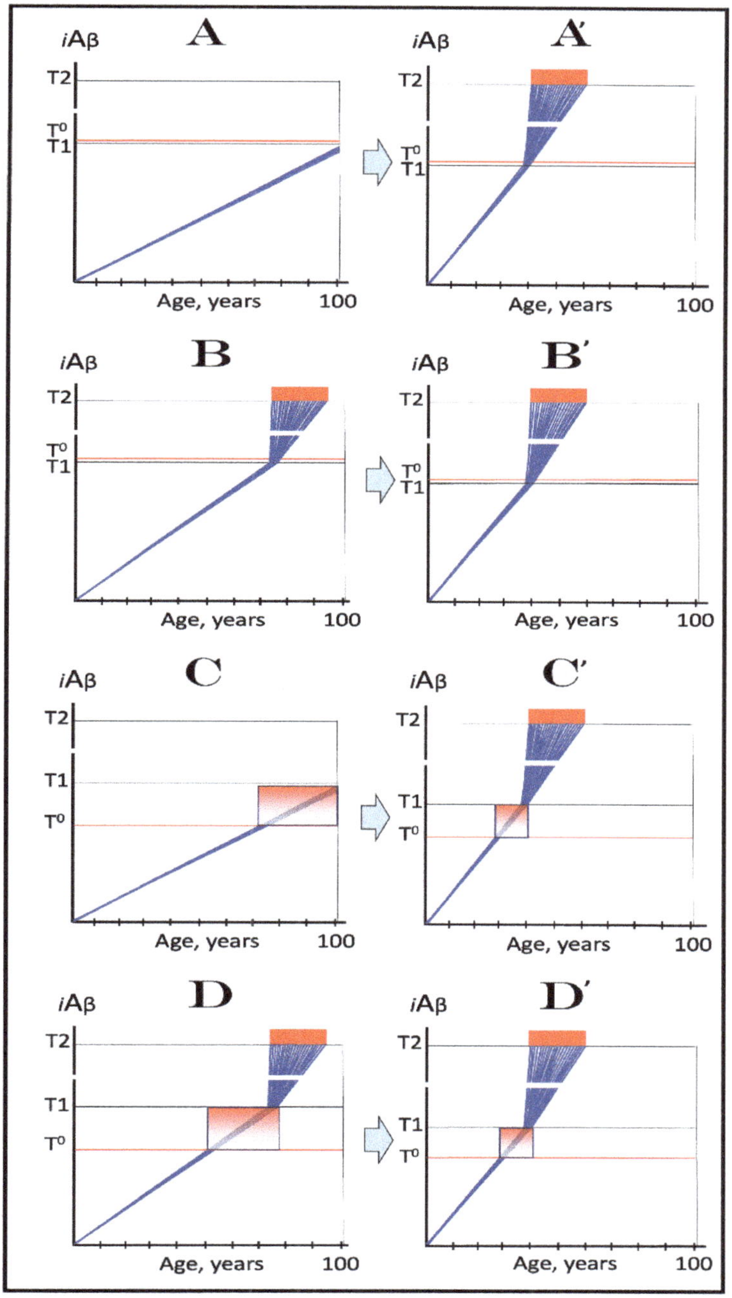

Figure 10. FAD mutations that exert their effect by accelerating the rate of iAβ Accumulation and lowering the extent of T1. *Blue lines*: Intraneuronal Aβ, iAβ. $\boldsymbol{T^0}$: Threshold of cellular concentration of

AβPP-derived iAβ triggering the neuronal damage, which manifests as AACD. ***T1***: Threshold of cellular concentration of iAβ triggering the activation of PKR and/or HRI kinases, phosphorylation of eIF2α, the elicitation of the neuronal ISR, the initiation of the AβPP-independent C99 production pathway, and the commencement of AD; its extent is the only variable in the present figure. ***T2***: Cellular concentration of iAβ triggering neuronal death via apoptosis or necroptosis. *Red box*: Apoptotic zone, defined as a range of cellular concentrations of iAβ within which cells neuronal cells committed apoptosis or necroptosis or are dead. *Pink boxes*: The range of cellular concentrations of AβPP-derived iAβ concentrations between the T^0 and the highest extent reached by AβPP-derived iAβ or the T1 threshold that support the progression of AACD. The condition commences with the crossing of the T^0 threshold and morphs into AD with the T1 crossing; it can occur only if the extent of the T^0 threshold is smaller than that of the T1. *Panels* (**A–D**): dynamics of iAβ accumulation in wild-type AβPP carriers. *Panels* (**A,B**): the T^0 levels are above those of T1; no AACD occurs. *Panel* (**A**): the T1 threshold is not crossed; no disease occurs. Panel (**B**): the T1 threshold is crossed; this is followed by late onset AD. *Panels* (**C,D**): the T^0 threshold is lower than T1. *Panel* (**C**): the T1 threshold is not reached; when the T^0 threshold is crossed, AACD commences and persists for the remaining lifespan. Panel (D): AACD commences when the T^0 is crossed and morphs into late onset AD upon the T1 crossing. *Panels* (**A′–D′**): Dynamics of iAβ accumulation in carriers of FAD mutations, which not only elevate the rate of accumulation of AβPP-derived iAβ but also reduce the extent of the T1 threshold. As a result, the T1 threshold is crossed sooner, and early-onset AD follows. Panels (**A′,B′**): the T^0 levels are above those of T1; no AACD occurs. Panels (**C′,D′**): the extent of the T^0 threshold is smaller than that of the T1; AACD commences with the T^0 crossing and morphs into the early-onset AD when the T1 is reached.

34. Validation (8): The Icelandic AβPP Mutation Protects from AD and AACD by Lowering the Influx of AβPP-Derived iAβ and Decelerating Its Rate of Accumulation

The preceding two sections validate the concept that the incidence and timing of conventional AD are determined by the crossing of the T1 threshold, which, in turn, is defined by the rate of accumulation of AβPP-derived iAβ and by the extent of the T1 threshold. The analysis described above is based on the elucidation of the mechanisms of action of the FAD-causing mutations. The present section provides a complementary validation of the same concept, arriving at it from the opposite direction. It does so by analyzing the mechanism of action of the only known AD-protective mutation. The Icelandic AβPP mutation A673T confers to its carriers protection from both AD and AACD [43,44]. This is a remarkable finding. It confirms that Aβ is involved in AACD to no lesser extent than in AD, a notion articulated by the ACH2.0 and discussed in Section 29 above. The mechanism underlying the protective effect of the Icelandic mutation has been elucidated. At its core is the increase in the rate of BACE1-mediated cleavage within iAβ and Aβ segments of AβPP and C99. As discussed above and as indicated by its designation (beta-site AβPP-cleaving enzyme), BACE1 cleaves primarily at the beta-site, i.e., between the residues 671 and 672 of AβPP; this is its major activity. Its minor activity is the cleavage at the β′-site situated ten amino acid residues downstream from the β-site. Cleavage at the β′-site both degrades iAβ and prevents its formation by the proteolysis of AβPP or C99. The Icelandic mutation increases the efficiency of cleavages at the β′-site. The increase is relatively small, only 30%, but its effects are highly consequential. Cleavage within iAβ degrades the peptide and increases the rate of its efflux. Cleavages within Aβ segments of AβPP and C99 reduce the production and secretion of Aβ and, consequently, the rate of its importation as iAβ. Thus, the rate of the influx of AβPP-derived iAβ is reduced and that of its efflux is increased. As a result, the rate of accumulation of AβPP-derived iAβ is decelerated, and it does not reach the T^0 and T1 thresholds within the lifespan of the mutation carriers or reaches them much later in life than in the wild type.

The above-described effects of the Icelandic mutation are illustrated in Figure 11. Panels A through C depict three principal scenarios of the disease in wild-type individuals. In Panel A, the extent of the T^0 threshold is greater than that of T1 and AD but no AACD occurs. In Panel B of Figure 11, AD is preceded by AACD. In Panel C the extent of the T1 threshold is such that it is not reached by AβPP-derived iAβ within the lifetime of the individual and no AD occurs. AACD, on the other hand, commences following the T^0 crossing and persists for the remaining lifetime. Panels A′ through C′ of Figure 11 show effects of the Icelandic mutation in the above-described scenarios. In all, the rate of accumulation of AβPP-derived iAβ is reduced. In none does it reach either the T^0 or T1 thresholds. Consequently, neither AD nor AACD occur in all three scenarios, hence the Icelandic mutation confers protection from both conditions.

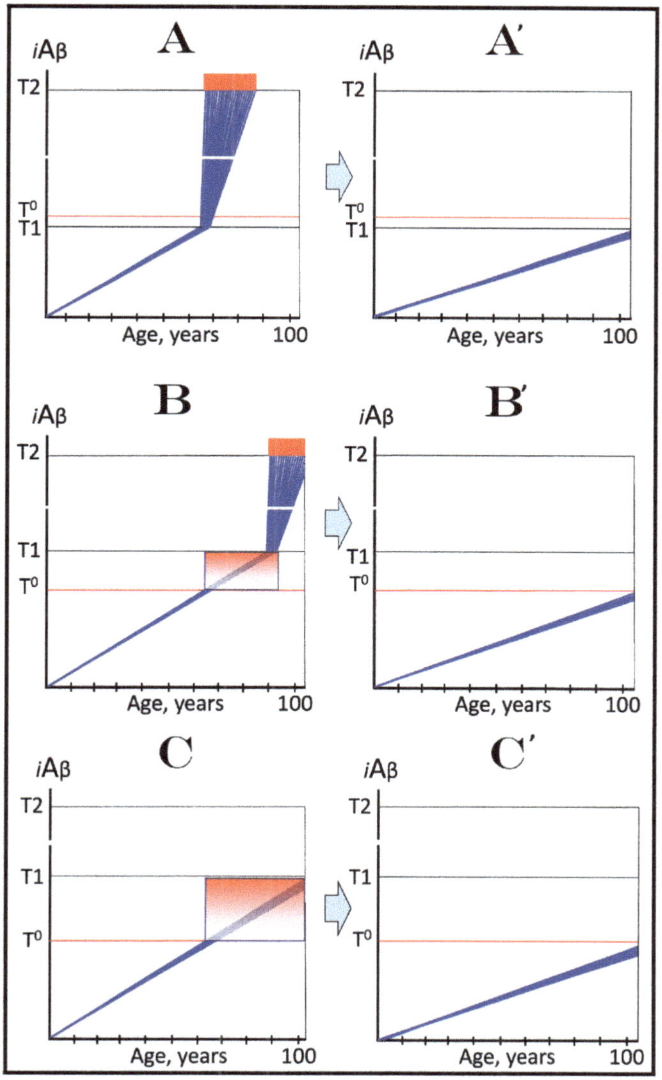

Figure 11. The Icelandic AβPP mutation protects from AD and AACD by lowering the influx of AβPP-derived iAβ and decelerating its rate of accumulation. *Blue lines*: Intraneuronal Aβ, iAβ.

T*^{0}**: Threshold of cellular concentration of AβPP-derived iAβ triggering the neuronal damage, which manifests as AACD. ***T1: Threshold of cellular concentration of iAβ triggering the activation of PKR and/or HRI kinases, phosphorylation of eIF2α, the elicitation of the neuronal ISR, the initiation of the AβPP-independent C99 production pathway, and the commencement of AD; its extent is the only variable in the present figure. ***T2***: Cellular concentration of iAβ triggering neuronal death via apoptosis or necroptosis. *Red box*: Apoptotic zone, defined as a range of cellular concentrations of iAβ within which cells neuronal cells committed apoptosis or necroptosis or are dead. *Pink boxes*: The range of cellular concentrations of AβPP-derived iAβ concentrations between the T^0 and the highest extent reached by AβPP-derived iAβ or the T1 threshold that support the progression of AACD. The condition commences with the crossing of the T^0 threshold and morphs into AD with the T1 crossing; it can occur only if the extent of the T^0 threshold is smaller than that of the T1. *Panels* (**A–C**) illustrate principal forms of conventional AD and/or AACD that occur in wild-type AβPP carriers. *Panel* (**A**): The T1 threshold is below the AACD-triggering T^0 threshold; no AACD occurs. Upon the T1 crossing the neuronal ISR is elicited, the AβPP-independent C99 and iAβ generation pathway activated, and conventional AD commences. *Panel* (**B**): The extent of the T^0 threshold is smaller than that of the T1. When AβPP-derived iAβ crosses the T^0, AACD commences and morphs into conventional AD upon the T1 crossing. *Panel* (**C**): The T1 threshold is not reached within the lifetime of the individual. When AβPP-derived iAβ crosses the T^0 threshold, AACD commences and persists for the remaining lifetime. *Panels* (**A′–C′**) illustrate the protective effect of the Icelandic AβPP. In each panel the influx of AβPP-derived iAβ is reduced and its rate of accumulation lowered. *Panel* (**A′**): AβPP-derived iAβ does not reach the T1 threshold within the lifetime of the individual; no AD occurs. *Panels* (**B′**,**C′**): AβPP-derived iAβ reaches neither the T^0 nor T1 threshold within the lifetime of the individual; neither AACD nor conventional AD occurs.

35. Icelandic AβPP Mutation as a Guide for AD and AACD Therapies

Protective effects of the Icelandic mutation provide the ultimate guidance for prevention of AD and AACD: suppress the rate of accumulation of AβPP-derived iAβ, prevent it from reaching the T^0 and T1 thresholds, and you would prevent both conditions. To emulate the Icelandic mutation, the suppression of accumulation of AβPP-derived iAβ should commence in infancy. This, however, is not entirely practical. The later in life the treatment starts, the more challenging it becomes. If at the time of commencement of the treatment levels of AβPP-derived iAβ were relatively close to the T^0 and/or T1 thresholds, the reduction in the rate of its accumulation would not be sufficient; the rate would have to be actually reversed in order to achieve the protective effect. This strategy is illustrated in Figure 12. Panel A shows the dynamics of iAβ accumulation in the wild-type individual. The extent of the T^0 threshold is greater than that of the T1, and AD commences upon the crossing of the latter. In Panel A′, the individual is treated prior to the T1 crossing. The rate of accumulation of AβPP-derived iAβ is reversed (or reduced to zero). The T1 threshold would not be crossed, and AD would not occur for the duration of the treatment. In Panel B, AD is preceded by AACD. The latter commences upon the T^0 crossing and morphs into the latter when the T1 threshold is crossed. In Panel B′, the treatment starts prior to the T^0 crossing. The rate of accumulation of AβPP-derived iAβ is reversed, the T^0 and T1 thresholds would not be crossed and AACD and AD would not occur for the duration of the treatment. In Panel C of Figure 12, the T1 threshold is not crossed within the lifespan of the individual. AACD commences with the T^0 crossing and persists for the remaining lifetime. If treatment were to begin prior to the T^0 crossing, AACD could be prevented. In Panel C′, the treatment starts after the T^0 crossing. The rate of accumulation of AβPP-derived iAβ is reversed and the progression of AACD stopped for the duration of the treatment.

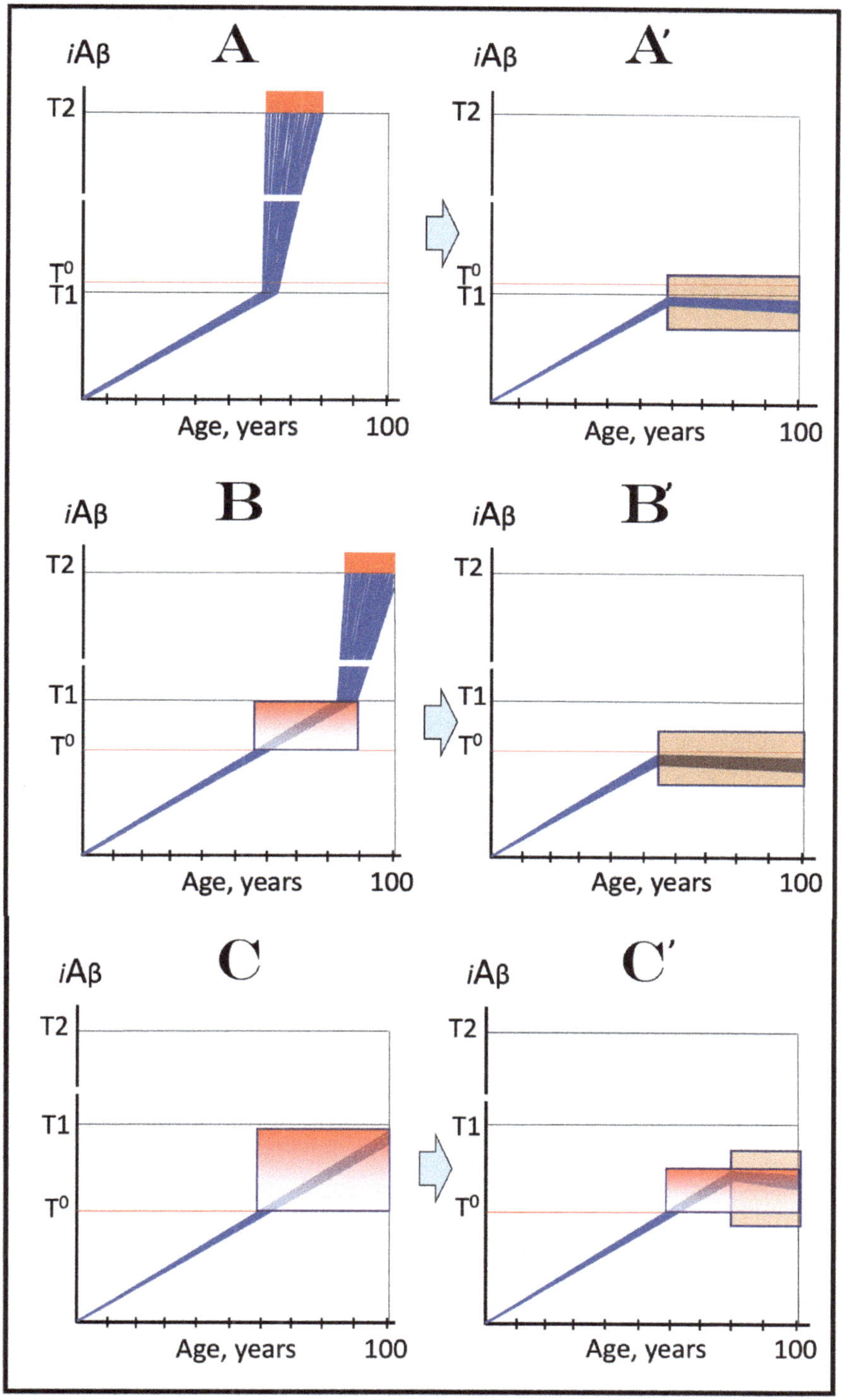

Figure 12. Therapeutic strategies for conventional AD and AACD suggested by the Icelandic AβPP mutation. *Blue lines*: Intraneuronal Aβ, iAβ. T^0: Threshold of cellular concentration of AβPP-derived

iAβ triggering the neuronal damage, which manifests as AACD. ***T1***: Threshold of cellular concentration of iAβ triggering the activation of PKR and/or HRI kinases, phosphorylation of eIF2α, the elicitation of the neuronal ISR, the initiation of the AβPP-independent C99 production pathway, and the commencement of AD; its extent is the only variable in the present figure. ***T2***: Cellular concentration of iAβ triggering neuronal death via apoptosis or necroptosis. *Red box*: Apoptotic zone, defined as a range of cellular concentrations of iAβ within which cells neuronal cells committed apoptosis or necroptosis or are dead. *Pink boxes*: The range of cellular concentrations of AβPP-derived iAβ concentrations between the T^0 and the highest extent reached by AβPP-derived iAβ or the T1 threshold that support the progression of AACD. The condition commences with the crossing of the T^0 threshold and morphs into AD with the T1 crossing; it can occur only if the extent of the T^0 threshold is smaller than that of the T1. *Orange boxes*: Duration of the treatment with the drug that reverses the rate of accumulation of AβPP-derived iAβ. *Panels* (**A–C**): Dynamics of iAβ accumulation in AD-affected neurons of the untreated AD and/or AACD patients. *Panel* (*A*): The T1 threshold is below the AACD-triggering T^0 threshold; no AACD occurs. Upon the T1 crossing, the neuronal ISR is elicited, the AβPP-independent C99 and iAβ generation pathway activated, and conventional AD commences. *Panel* (**B**): The extent of the T^0 threshold is smaller than that of the T1. When AβPP-derived iAβ crosses the T^0, AACD commences and morphs into conventional AD upon the T1 crossing. *Panel* (**C**): The T1 threshold is not reached within the lifetime of the individual When AβPP-derived iAβ crosses the T^0 threshold, AACD commences and persists for the remaining lifetime. *Panels* (**A′–C′**): Dynamics of AβPP-derived iAβ in individuals treated with a drug that suppresses its influx and reveres its rate of accumulation. In Panels (**A′**,**B′**), the drug is administered prior to the crossing of the T^0 and/or T1 thresholds. Neither threshold is reached for the duration of the treatment, nor does AD or AACD occur. In Panel (**C′**), the drug is administered after the T^0 crossing but prior to the T1 crossing. The drug reverses the accumulation of AβPP-derived iAβ and stops the progression of AACD. The T1 threshold would not be reached, and conventional AD would not occur for the duration of the treatment.

36. ACH-Based AD Drugs Could Be Effective If Administered Preventively

The preventive strategies discussed in the preceding section can potentially be enacted with ACH-based drugs. ACH-based drugs are agents designed, under the ACH guidance, to deplete extracellular Aβ, the presumed cause of AD (in the ACH). They include drugs that directly interact with extracellular Aβ, such as Aβ-specific antibodies, as well as drugs suppressing the production and, subsequently, secretion of Aβ, such as BACE1 inhibitors. Since, as discussed above, the ACH appears to be unsustainable, it could be reasoned that ACH-based AD drugs would be ineffective. This, however, is not the case. Inadvertently, ACH-based drugs should be effective in the prevention (but not the treatment; as discussed below) of conventional AD in the framework of the ACH2.0. This is because, by impacting levels of extracellular Aβ or the production of Aβ in the AβPP proteolytic pathway, they affect the rate of accumulation of AβPP-derived iAβ. The logic is apparent. The smaller the pool of extracellular Aβ, the lower the rates of its cellular uptake, of the influx of iAβ, and of the accumulation of AβPP-derived iAβ. In cases of the suppression of Aβ production in the AβPP proteolytic pathway, e.g., with BACE1 inhibitors, the impact is even greater: not only would the rate of the importation of extracellular Aβ be lowered due to the reduction of its pool, the retention of newly generated AβPP-derived iAβ would also be reduced. Therefore, if administered preventively, i.e., prior to the crossing of the T1 threshold, ACH-based drugs would either reduce or reverse the rate of accumulation of AβPP-derived iAβ. This would either delay or prevent both the T1 crossing and the commencement of AD. Likewise, if administered prior to the T^0 crossing, ACH-based drugs could delay or preclude the occurrence of AACD.

The potential preventive effects of the ACH-based AD drugs are illustrated in Figure 13. Panel A of Figure 13 depicts the dynamics of accumulation of iAβ and progression of AD in untreated individuals. In the first stage AβPP-derived iAβ accumulates and crosses the T1 threshold. PKR and/or HRI kinases are activated, eIF2α phosphorylated at its Ser51, the neuronal ISR elicited, and the AβPP-independent C99 (and iAβ in this version of the ACH2.0) production pathway initiated, and iAβ reaches the AD pathology-causing range; when it crosses the T2 threshold, neurons commit apoptosis. In Panel B of Figure 13, the ACH-based drug is administered prior to the T1 crossing. The influx of AβPP-derived iAβ is reduced but the reduction in the influx of AβPP-derived iAβ is such that its rate of accumulation is only lowered, not reversed. Its level reaches and crosses the T1 threshold; AβPP-independent C99 and iAβ production pathway is activated and AD commences. In this scenario the treatment does not prevent AD but only delays its occurrence. In Panel C of Figure 13, the administration of the ACH-based drug also commenced prior to the T1 crossing. The influx of AβPP-derived iAβ is reduced and the rate of its accumulation is reversed. The T1 threshold will not be crossed, and conventional AD will not occur for the duration of the treatment.

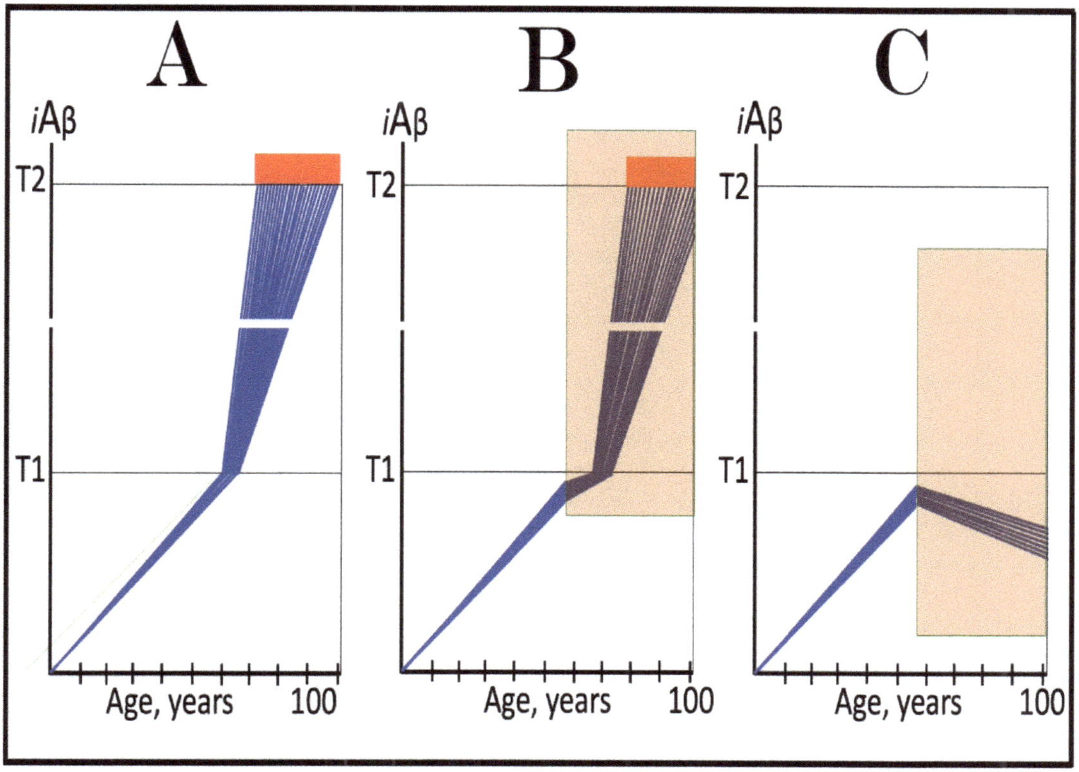

Figure 13. ACH-based AD drugs could be effective if administered preventively. *Blue lines*: Intraneuronal Aβ, iAβ. ***T1***: Threshold of cellular concentration of iAβ triggering the activation of PKR and/or HRI kinases, phosphorylation of eIF2α, the elicitation of the neuronal ISR, the initiation of the AβPP-independent C99 production pathway, and the commencement of AD. ***T2***: Cellular concentration of iAβ triggering neuronal death via apoptosis or necroptosis. *Red box*: Apoptotic zone, defined as a range of cellular concentrations of iAβ within which cells neuronal cells committed

apoptosis or necroptosis or are dead. *Orange boxes*: Duration of the treatment with the ACH-based drug that suppresses or reverses the rate of accumulation of AβPP-derived iAβ. *Panels* (**A–C**): Dynamics of iAβ accumulation in AD-affected neurons of the untreated and treated AD and/or AACD patients. *Panel* (**A**): Dynamics of the accumulation of iAβ in the untreated individual. AβPP-derived iAβ accumulates via two physiological mechanisms described in the main text. When it crosses the T1 threshold, PKR and/or HRI kinases are activated, eIF2α phosphorylated, the neuronal ISR elicited, the AβPP-independent C99 production pathway initiated, and conventional AD commences. *Panel* (**B**): The drug suppresses the influx of AβPP-derived iAβ, but its accumulation continues albeit at a reduced rate. Eventually, subject to sufficient longevity, it would cross the T1 threshold. The neuronal ISR would be elicited, the AβPP-independent C99 production pathway would be activated, and conventional AD would commence. In this scenario the drug delays the occurrence of AD. *Panel* (**C**): The drug suppresses the influx of AβPP-derived iAβ and reverses its rate of accumulation. The T1 threshold would not be reached, and conventional AD would not occur for the duration of the treatment. In this scenario the drug prevents the occurrence of conventional AD.

37. Modulators of Gamma-Secretase Could Elevate the T1 Threshold and Thus Be Potentially Effective in the Prevention of Conventional AD

Inhibitors of gamma-secretase are, potentially, ACH-based drugs for the same reason that inhibitors of beta-secretase are; both would reduce the production of Aβ in the AβPP proteolytic pathway and thus decrease its extracellular pool. This approach was indeed attempted [213–215] but it produced highly deleterious outcomes and was abandoned. Eventually, the reason for this failure became apparent. Gamma-secretase is a component of the Notch signaling pathway. As such, it has hundreds of targets and therefore cannot be interfered with without adverse effects. Gamma-secretase cleaves at multiple positions within a short segment of the C99 fragment of AβPP, and attempts have been made to develop modulators of gamma-secretase that would shift the position of the cleavage site toward shorter, more benign versions of Aβ. These efforts were met with variable success [215–218]. Like other ACH-based AD drugs, modulators of gamma-secretase would also be effective in the framework of the ACH2.0, but for different reasons. These modulators are unlikely to change either the size of the extracellular Aβ pool or, apparently, the efficiency of its cellular uptake. What they can do, however, is to elevate the extent of the T1 threshold. The logic of this assertion is the same as in the case iAβ42, which lowers the extent of the T1 because it is relatively (to other iAβ variants) more toxic and less of its needed to achieve a certain level of cellular stress. With shorter, relatively less toxic iAβ species, more of it would be needed to attain the required level of cellular stress, and the extent of the T1 threshold would increase. This is a scenario that was considered in Section 30 above. In the individual treated with such a modulator the crossing of the T1 thresholds and the commencement of conventional AD would be either delayed or prevented.

The above scenario is illustrated in Figure 14. Panel A of Figure 14 depicts the accumulation of iAβ and progression of AD in the untreated individual. Panel B of Figure 14 shows the same parameters in the individual treated with a modulator of gamma-secretase. The rate of accumulation of AβPP-derived iAβ is not affected. But because it is less toxic, more of it is needed to trigger the elicitation of the neuronal integrated stress response and the extent of the T1 threshold is elevated. Consequently, it takes longer to reach and cross it. AD commences and progresses, but it is delayed. In Panel C of Figure 14, the elevation of the extent of the T1 threshold in the individual treated with a modulator of gamma-secretase is such that AβPP-derived iAβ does not reach it within the lifetime of the treated individual. Neither the T1 crossing, nor conventional AD would occur in this scenario. It should be noted that if AD does occur in the treated individual (Panel

B of Figure 14), the extent of the T2 threshold would likely also be elevated due to the lover toxicity of iAβ produced independently of AβPP. In such a case, gamma-secretase modulators could be effective in symptomatic AD but only in the initial version of the ACH2.0 under discussion; they would be completely ineffective in symptomatic AD in the current Version Three of the ACH2.0 (see Section 66 below).

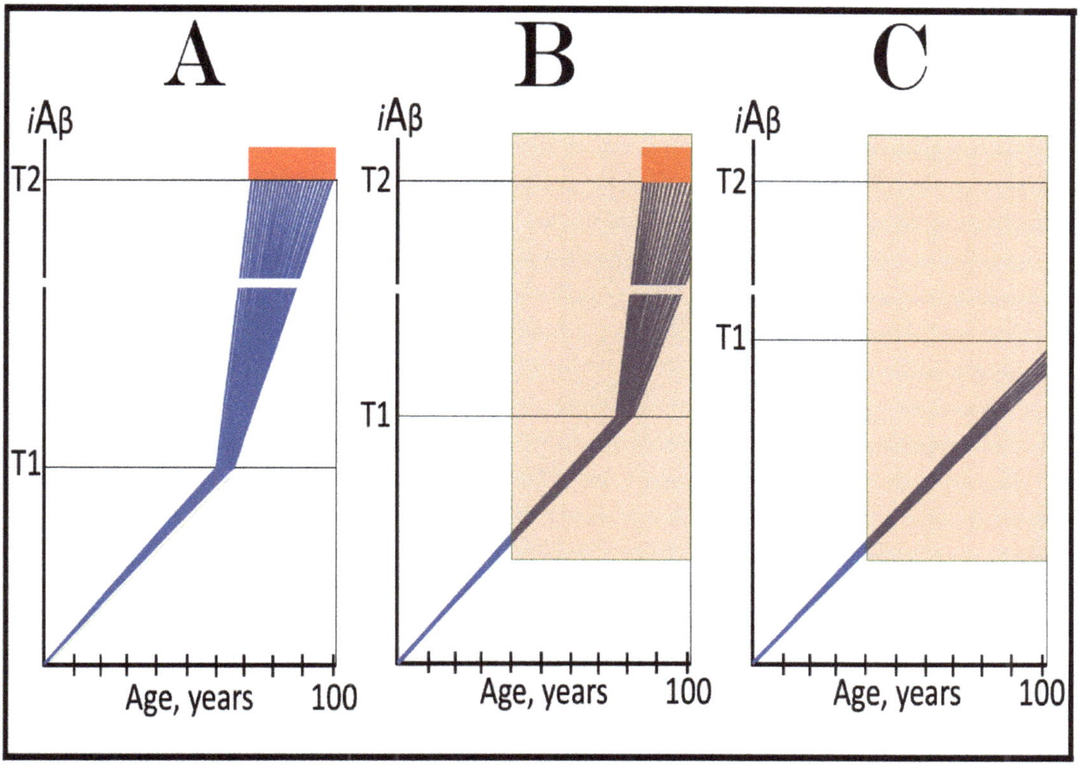

Figure 14. Modulators of gamma-secretase may potentially protect from conventional AD via the elevation of the T1 threshold. *Blue lines*: Intraneuronal Aβ, iAβ. ***T1***: Threshold of cellular concentration of iAβ triggering the activation of PKR and/or HRI kinases, phosphorylation of eIF2α, the elicitation of the neuronal ISR, the initiation of the AβPP-independent C99 production pathway, and the commencement of AD. ***T2***: Cellular concentration of iAβ triggering neuronal death via apoptosis or necroptosis. *Red box*: Apoptotic zone, defined as a range of cellular concentrations of iAβ within which cells neuronal cells committed apoptosis or necroptosis or are dead. *Orange boxes*: Duration of the treatment with the gamma-secretase modulating drug that shifts the production of Aβ toward shorter, more benign species. *Panel* (**A**): Dynamics of the accumulation of iAβ in the untreated individual. AβPP-derived iAβ accumulates via two physiological mechanisms described in the main text. When it crosses the T1 threshold, PKR and/or HRI kinases are activated, eIF2α phosphorylated, the neuronal ISR elicited, the AβPP-independent C99 production pathway initiated, and conventional AD commences. *Panel* (**B**): In the presence of the drug shorter, more benign species of Aβ are produced and accumulated as AβPP-derived iAβ. More of it is needed to attain the same level of cellular toxicity and cellular stress than with regular Aβ species. Consequently, the extent of the T1 threshold and the timing of its crossing by AβPP-derived iAβ increase. In this scenario the treatment delays the commencement of conventional AD. *Panel* (**C**): The treatment-caused increase in the extent of the T1 threshold is such that it is not reached, and conventional AD does not occur for the duration of the treatment. In this scenario, the drug prevents the occurrence of AD.

38. ACH-Based Drugs Are Completely Ineffective in Symptomatic AD

Whereas, as described in two preceding sections, ACH-based drugs could potentially be effective in the prevention of the disease, it can be stated categorically that they cannot work, in any way, form, or shape in symptomatic AD. More precisely, they cannot work following the crossing of the T1 threshold. The reason for this is the operation of the AβPP-independent C99 and (in this version of the ACH2.0) iAβ production pathway. From the instance of its activation by the neuronal integrated stress response, this pathway is completely independent from the influx of AβPP-derived iAβ (or from the absence thereof). As discussed in Section 26 above, it is fully autonomous and self-sustainable: its output propagates the neuronal ISR and thus perpetuates its own activity. Moreover, the output of the AβPP-independent C99 and iAβ production pathway is orders of magnitude greater than that of the AβPP proteolytic pathway. Therefore, even if the influx of AβPP-derived iAβ completely ceases, this would affect neither the continuous operation of the AβPP-independent C99 and iAβ production pathway nor the progression of AD.

The above considerations are illustrated in Figure 15. In this figure, the mode of the production of iAβ is color-coded. AβPP-derived iAβ is presented as blue lines whereas iAβ generated independently of AβPP (in the same neuronal cells) is depicted in red. In Panel A of Figure 15, the administration of the ACH-based drug starts when all affected neurons have already crossed the T1 threshold and AD commenced. The influx of AβPP-derived iAβ is reduced and its rate of accumulation is lowered. This, however, has no effect on the continuous operation of the AβPP-independent C99 and iAβ production pathway and on the progression of AD. The only difference in Panel B of Figure 15 is that the ACH-based drug-mediated reduction in the influx of AβPP-derived iAβ is such that the rate of its accumulation is reversed, and its levels are decreasing. This however does not change the outcome: the AβPP-independent C99 and iAβ production pathway remains fully operative and the progression of AD is unimpeded.

Two conceptually different scenarios are presented in Panels C and D of Figure 15. In both panels, the administration of the ACH-based drug commences at the time when either symptoms of AD have manifested or its biomarker(s) occurred, but a fraction of the affected neurons have not yet crossed the T1 threshold. In both panels, in the neurons that crossed the T1 threshold the activity of the AβPP-independent C99 and iAβ production pathway has commenced and remains unaffected and unimpeded by the drug. The rate of accumulation of AβPP-derived iAβ, however, is affected. In Panel C of Figure 15, it is reduced but accumulation continues. Eventually, levels of AβPP-derived iAβ in the initially under-T1 neurons would reach the T1 threshold. At this point the AβPP-independent C99/iAβ production pathway would be activated and the progression of cellular AD pathology would commence. The drug would be effective in delaying the onset of AD pathology in the initially under-T1 neuronal fraction, but its effect would be only marginal because this neuronal fraction is marginal.

In Panel D of Figure 15, the ACH-based drug-mediated reduction in the influx of AβPP-derived iAβ is such that the rate of its accumulation is reversed. Its levels in the initially under-T1 neurons are decreasing and would not reach the T1 threshold for the duration of the treatment. The drug would be effective in preventing the onset of AD pathology in the initially under-T1 neurons. Its effect, however, would be marginal because the fraction of the initially under-T1 neurons is.

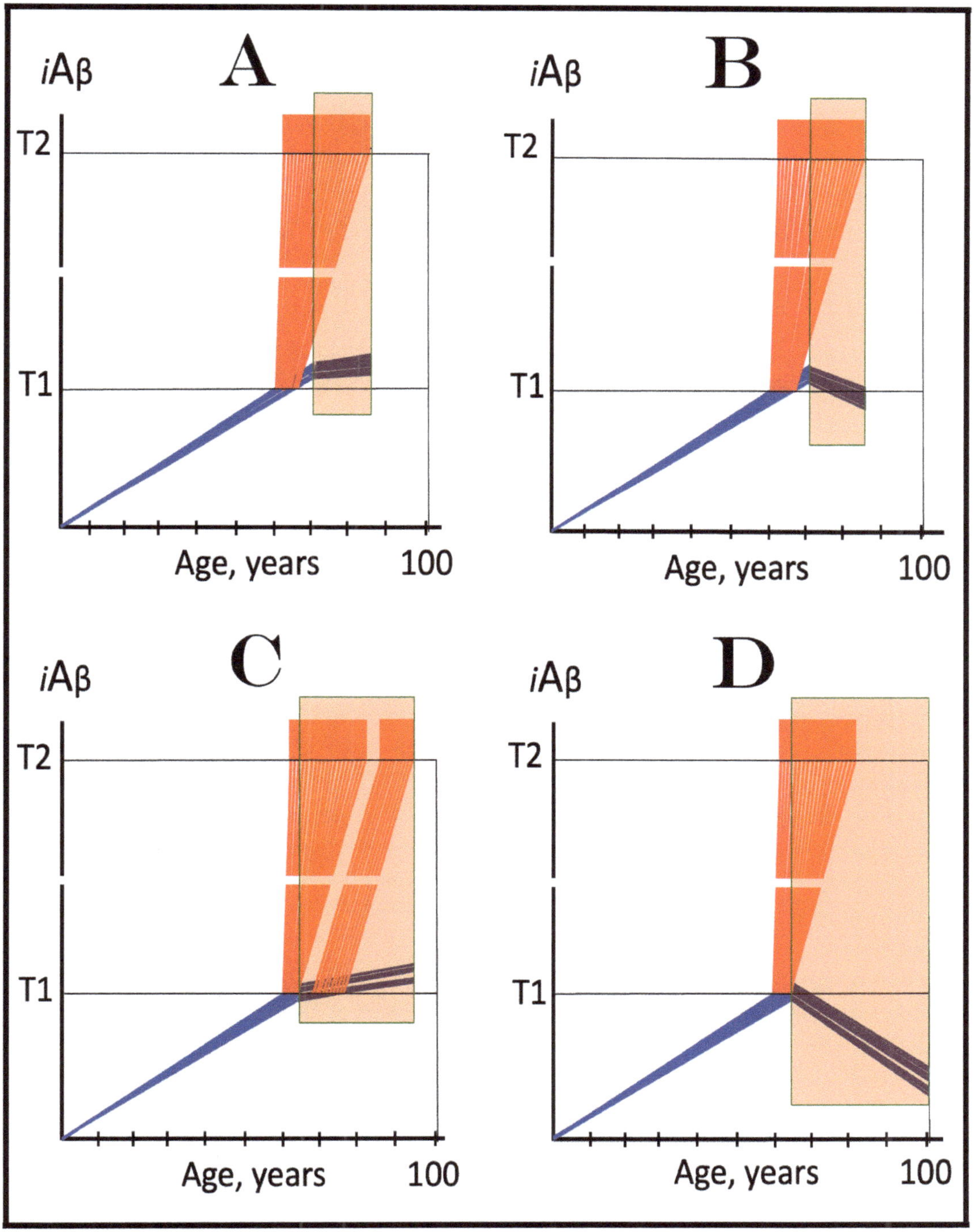

Figure 15. ACH-based drugs are ineffective when the AβPP-independent C99/iAβ production pathway is operational. *Blue lines*: AβPP-derived iAβ. *Red lines*: iAβ generated independently of AβPP.

T1: Threshold of cellular concentration of iAβ triggering the activation of PKR and/or HRI kinases, phosphorylation of eIF2α, the elicitation of the neuronal ISR, the initiation of the AβPP-independent C99 production pathway, and the commencement of AD. ***T2***: Cellular concentration of iAβ triggering neuronal death via apoptosis or necroptosis. *Red box*: Apoptotic zone, defined as a range of cellular concentrations of iAβ within which cells neuronal cells committed apoptosis or necroptosis or are dead. *Orange boxes*: Duration of the treatment with the ACH-based drug that suppresses or reverses the rate of accumulation of AβPP-derived iAβ. *Panel* (**A**): AβPP-derived iAβ accumulates via two physiological mechanisms described in the main text. When it crosses the T1 threshold PKR and/or HRI kinases are activated, eIF2α phosphorylated, the neuronal ISR elicited, the AβPP-independent C99 production pathway initiated, and conventional AD commences. ACH-based drugs administered at this point reduce significantly the rate of accumulation of AβPP-derived iAβ. They have, however, little, if any, effect on both the production of iAβ in the AβPP-independent pathway and the progression of the disease. *Panel* (**B**): As in the preceding panel, the ACH-based drug is administered when all affected neurons have already crossed the T1 threshold, and the AβPP-independent C99 production pathway has already been activated. The reduction in the influx of AβPP-derived iAβ is such that the rate of its accumulation is reversed. Since the operation of the self-sustainable AβPP-independent C99 production pathway is not affected, this reversal is inconsequential for the progression of the disease. Panels (**C**,**D**): The ACH-based drug is administered when a fraction of the affected neurons has not yet crossed the T1 threshold. In over-T1 neurons, the AβPP-independent C99/iAβ production pathway has been activated, and the drug does not affect its operation. In *Panel* (**C**), the drug reduces the influx of AβPP-derived iAβ, but its accumulation continues. The initially under-T1 neurons eventually cross the T1 threshold, the AβPP-independent C99/iAβ production pathway is activated, and cellular AD pathology commences. In *Panel* (**D**), the drug reverses the rate of accumulation of AβPP-derived iAβ. The initially under-T1 neurons do not cross the T1 threshold and remain AD pathology-free for the duration of the treatment.

39. Validation (9): Effects of Lecanemab and Donanemab in Early AD Are Due to the Timing of Their Administration: They Affect Preventively Only a Marginal Fraction of Neurons That Has Not Yet Crossed the T1 Threshold

The outcomes predicted in scenarios depicted in Panels C and D of Figure 15 in the preceding section were, in fact, verified in clinical trials of two ACH-based drugs, lecanemab and donanemab [29–33]. Both drugs are monoclonal antibodies designed to reduce levels of extracellular Aβ. Both trials produced positive, however marginal, results: cognitive decline of treated patients continued but at a decreased rate. These results were interpreted as an affirmation of the notion that the reduction in levels of extracellular Aβ reduced its adverse cytotoxic effect. However, this interpretation is inconsistent with the outcomes of earlier trials of ACH-based drugs (conducted at more advanced stages of AD) where levels of extracellular Aβ were reduced to no lesser extent without any positive outcomes whatsoever. The interpretation of the results of lecanemab and donanemab clinical trials in terms of the ACH2.0 resolves this controversy.

The selection of subjects in both trials was based on the appearance of a serum biomarker unavailable in earlier clinical trials of ACH-based AD drugs. This biomarker (phosphorylated tau) is currently one of the earliest if not the earliest biomarker of AD. Accordingly, subjects of these trials were at very early stages of the disease, substantially earlier than in any previous clinical AD trial. As argued above, ACH-based drugs are utterly ineffective following the T1 crossing. It follows from their marginal effect that at the time of the commencement of drugs' administration a fraction of the affected neurons has not yet crossed the T1 threshold and that the observed outcomes resulted from the impact of the drugs on this neuronal fraction (referred to as "initially under-T1"). Moreover, it follows that both drugs acted not curatively (i.e., not on neurons undergoing AD pathology) but only preventively (i.e., preventing the onset of AD pathology).

The above interpretation is illustrated in Figure 16. Panel A of Figure 16 depicts the initial (i.e., at the time of the commencement of drug's administration) state of the levels of iAβ in affected neurons. In most neurons the T1 threshold has been crossed, the AβPP-independent C99/iAβ production pathway activated, and the progression of cellular AD pathology commenced. A fraction of the neurons is, however, still under-T1 at this stage. Panel B of Figure 16 shows how the initial state would evolve in the untreated patient. The initially under-T1 neurons would cross the T1 threshold; cellular AD pathology would progress in all affected neurons, and the disease would reach its end stage. In Panel C of Figure 16, the ACH-based drug is ineffective in the over-T1 neurons where the AβPP-independent C99/iAβ production pathway is operative. In the initially under-T1 neuronal fraction, it reduces the influx of AβPP-derived iAβ and lowers the rate of its accumulation. Eventually, however, this neuronal fraction would cross the T1 threshold, and the progression of AD pathology would commence. In Panel D of Figure 16, the drug is also affecting only the initially under-T1 neuronal fraction. The reduction in the influx of AβPP-derived iAβ in these neurons is such that the rate of its accumulation is reversed, and its levels are declining. The T1 threshold would not be crossed, and AD pathology would not commence in these neurons for the duration of the treatment. In the scenarios depicted in both Panels C and D, the effect would be marginal because the drug affects only a marginal neuronal fraction.

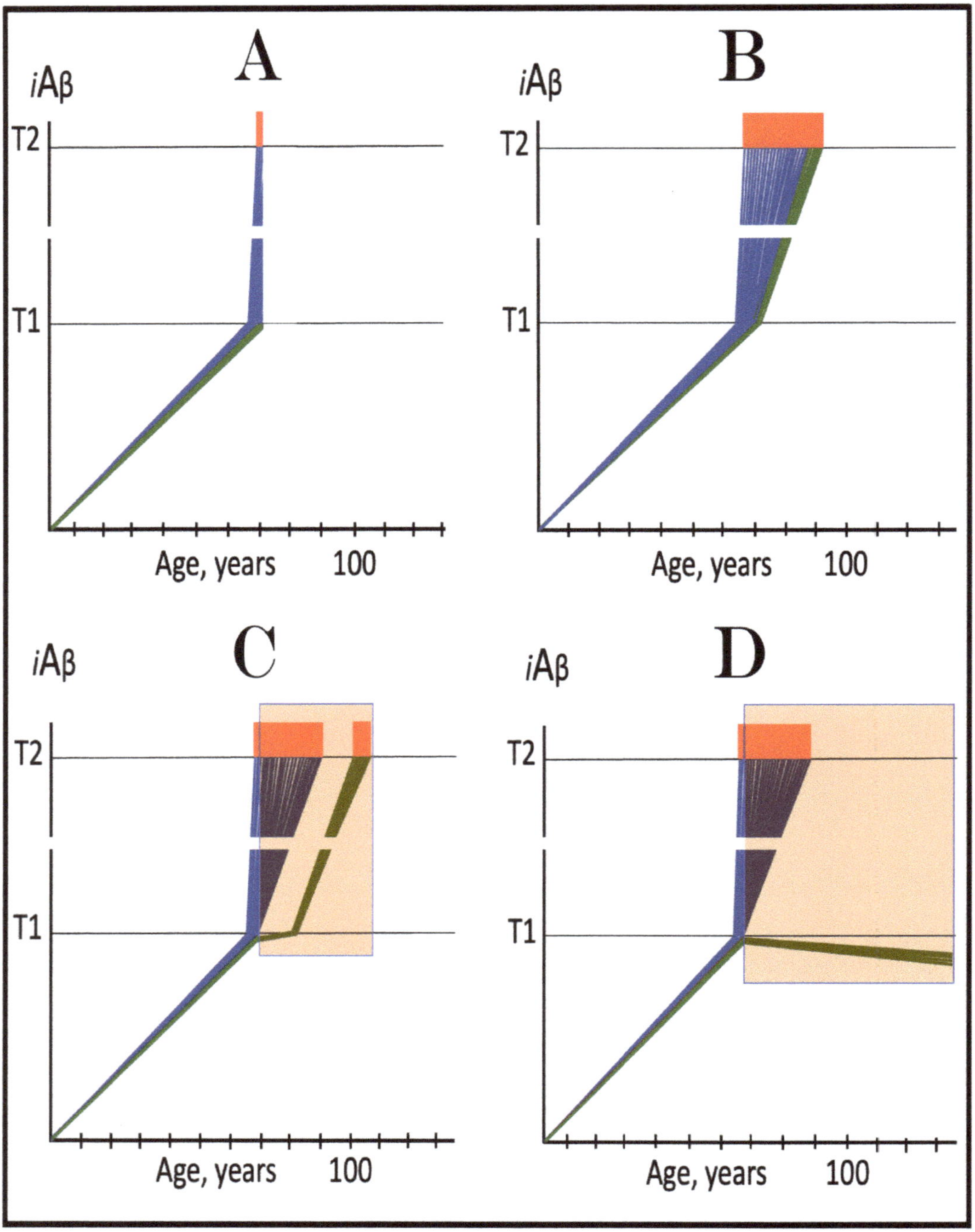

Figure 16. Effect of Lecanemab and Donanemab in early AD: only a marginal fraction of neurons that have not yet crossed the T1 threshold is affected. *Blue lines*: Intraneuronal Aβ, iAβ. *Green lines*: Levels of iAβ in the neurons that have not yet crossed the T1 threshold at the commencement of the treatment. ***T1***: Threshold of cellular concentration of iAβ triggering the activation of PKR and/or HRI kinases, phosphorylation of eIF2α, the elicitation of the neuronal ISR, the initiation of the AβPP-independent

C99 production pathway, and the commencement of AD. ***T2***: Cellular concentration of iAβ triggering neuronal death via apoptosis or necroptosis. *Red box*: Apoptotic zone, defined as a range of cellular concentrations of iAβ within which cells neuronal cells committed apoptosis or necroptosis or are dead. *Orange boxes*: Duration of the treatment with lecanemab or donanemab. *Panel* (**A**): The initial state of levels of iAβ at the commencement of the drug's administration. The bulk of the affected neurons have crossed the T1 threshold. The neuronal ISR has been elicited and the AβPP-independent C99/iAβ production pathway has been activated. Some neurons have reached the T2 threshold and AD symptoms have manifested. *Panel* (**B**): The evolution of the initial state in the untreated AD patient. All affected neurons cross the T1 threshold. The disease progresses and reaches the end stage. *Panels* (**C,D**): The evolution of the initial state in treated patients. The drug has no effect whatsoever on the operation of the self-sustainable AβPP-independent C99/iAβ production pathway; it affects only the rate of accumulation of AβPP-derived iAβ. *Panel* (**C**): The drug suppresses the influx of AβPP-derived iAβ, but its accumulation continues, albeit at a reduced rate. Eventually, subject to sufficient longevity, it would cross the T1 threshold in the initially under-T1 neurons. The neuronal ISR would be elicited, the AβPP-independent C99 production pathway would be activated, and cellular AD pathology would commence. *Panel* (**D**): The drug suppresses the influx of AβPP-derived iAβ and reverses its rate of accumulation. The T1 threshold would not be reached, and AD pathology would not occur in the initially under-T1 neurons for the duration of the treatment. In both scenarios, however, the effect is only marginal because the fraction of the drug-affected neurons is.

40. ANY ACH-Based Drug Has a Proverbial Snowball's in Hell Chance of Being Effective in Advanced AD

It is important to emphasize that lecanemab and donanemab, discussed in the preceding section, do not act conceptually differently or more efficiently than other ACH-based drugs. They exhibited their marginal effect (and therefore were FDA-approved as AD drugs) only because their makers were fortunate to be able to conduct trials at the early enough stages of AD (guided by phosphorylated tau serum biomarker not available previously) when a marginal neuronal subpopulation did not yet cross the T1 threshold and could be treated preventively. It can be predicted with considerable certainty that most, if not all, ACH-based drugs would yield similar results if tested at similarly early stages of AD.

The bottom line is that ACH-based drugs may have an effect only in situations where the AβPP-independent C99/iAβ production pathway is inoperative. This is why they performed brilliantly in transgenic mouse models overexpressing human AβPP. In these models, for the reasons explained above, the AβPP-independent C99/iAβ generation pathway is inoperative, and they do not develop AD. Neurodegeneration and cognitive impairment exhibited by these models are the consequence of the neuronal ISR elicited (via PKR and/or HRI activation) by AβPP-derived iAβ accumulated over the T1 threshold. ACH-based drugs reduce the levels of iAβ to those below the T1 threshold and thus reverse the neuronal ISR, hence the effect.

This is why ACH-based drugs are legitimate AD and AACD preventive agents. Moreover, they may be effective in the treatment of AACD (where by definition, levels of iAβ are below the T1 threshold). What they principally cannot do, regardless of their effectiveness in depleting the extracellular Aβ pool or suppressing the production of Aβ in the AβPP proteolytic pathway, is be effective in settings where the AD Engine, i.e., the self-sustainable AβPP-independent C99/iAβ generation pathway is active. This is an unadulterated impossibility in terms of the ACH2.0; hence there is a "snowball in hell" chance of the ACH drugs being effective in symptomatic AD.

41. ACH2.0-Based AD Drugs: A Definition

ACH2.0-base AD drugs are the agents capable of the prevention and treatment of AD (and AACD) in terms of the ACH2.0 theory of the disease. Mechanistically, they are defined by their targets and on this basis can be divided into three categories. One comprises agents capable of direct interference with the operation of the AβPP-independent C99/iAβ production pathway. This pathway is the active core of the disease, its engine, and, therefore, constitutes the major therapeutic target. Presuming that the AβPP-independent C99/iAβ production pathway is underpinned by RNA-dependent amplification of AβPP mRNA, in authors' opinion by far the most plausible scenario, targeting it is challenging. As described above (see Section 10), RNA-dependent mRNA amplification is a major physiological process crucial, for example, in erythropoiesis and in the deposition of extracellular matrix proteins, and therefore, it cannot be interfered with systemically without deleterious consequences. The only way to interfere with it is specifically, i.e., to target its specific substrate, in this case AβPP mRNA or the antisense AβPP RNA. This is feasible, and promising possibilities are discussed in detail in Section 79 below.

Another category includes drugs targeting the enabler and sustainer of the AβPP-independent C99/iAβ production pathway, namely the neuronal integrated stress response. The latter provides components essential for the operation of the former. Inhibition of the neuronal ISR would cease the production of these components and disable the AD-driving pathway. ISR inhibitors are available and their effect in the prevention and treatment of conventional AD are analyzed in the two following sections. The ISR is a major physiological survival tool and its systemic long-term inhibition is bound to be detrimental. On the other hand, transient administration of ISR inhibitors in concert with other therapies could be crucially important in successful treatment of AD, as described in Section 48 below.

The two categories above discuss approaches that would affect AD but not AACD. The third category of ACH2.0-based drugs would affect both. It includes the agents that deplete the trigger and the actual driver of the disease, iAβ (in this version of the ACH2.0, but the concept holds up in other ACH2.0 versions as well, as discussed below). Preventively, the depletion of iAβ (in this case AβPP-derived iAβ) would preclude the crossing of the T^0 and T1 threshold and the occurrence of AACD and AD. With the AβPP-independent C99/iAβ production pathway operational, the depletion of iAβ (in this case mostly iAβ produced independently of AβPP) would deprive the disease of its driver and stop its progression. Importantly, the sufficient, below the T1 threshold, depletion of iAβ would also cease the propagation of the neuronal ISR and cause its reversal. This, in turn, would deprive the AβPP-independent C99/iAβ production pathway of its essential components and thus disable it. This opens up a possibility that is indeed remarkable. Following the transient depletion, and provided it is deep enough, the accumulation of iAβ would resume from a low baseline, supported only by the AβPP proteolytic pathway, and it may not reach the T1 threshold within the remaining lifetime of the treated individual. In such a case, a single, one-time-only iAβ depletion treatment may prevent the occurrence of AD (and AACD) in healthy individuals and stop the progression of the disease in AD patients. This possibility is further discussed in Sections 44–48 below.

42. Suppression of the Neuronal Integrated Stress Response in the Prevention of Conventional Alzheimer's Disease

Suppression of the neuronal ISR in order to prevent AD is an intuitive approach. Indeed, the AβPP-independent C99/iAβ production pathway cannot be activated and therefore AD cannot occur in the absence of the neuronal ISR. It follows that the prevention of the elicitation of the neuronal ISR would prevent the disease. This approach is illustrated in Figure 17. Panel A of Figure 17 shows the initial state of the levels of AβPP-derived iAβ in the neurons of the healthy individual who would develop AD if left untreated. Panel B of Figure 17 depicts the evolution of the initial state in the untreated individual. AβPP-derived iAβ accumulates unimpeded and reaches the T1 threshold. This triggers activation of PKR and/or HRI, phosphorylation of eIF2α, elicitation of the neuronal ISR, and initiation of the AβPP-independent C99/iAβ production pathway. AD commences and progresses until it reaches its end stage. Panel C of Figure 17 presents the evolution of the initial state in the individual treated with the ISR-inhibiting drug. AβPP-derived iAβ reaches and crosses the T1 threshold but the neuronal ISR cannot be elicited. The influx of iAβ is supported solely by the AβPP proteolysis and it continues to accumulate at a slow pre-T1 crossing rate. Its level would not reach the AD pathology-causing range for the duration of the treatment.

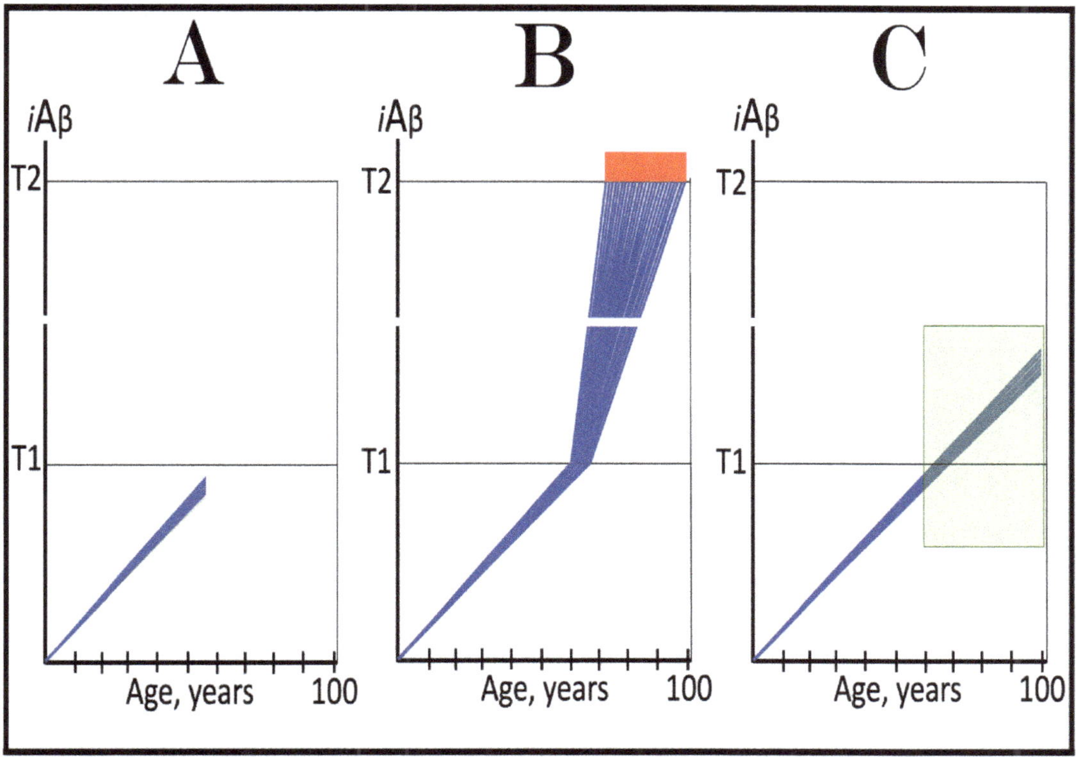

Figure 17. Suppression of the neuronal integrated stress response in the prevention of conventional AD. *Blue lines*: Intraneuronal Aβ, iAβ. ***T1***: Threshold of cellular concentration of iAβ triggering the activation of PKR and/or HRI kinases, phosphorylation of eIF2α, the elicitation of the neuronal ISR, the initiation of the AβPP-independent C99 production pathway, and the commencement of AD. ***T2***: Cellular concentration of iAβ triggering neuronal death via apoptosis or necroptosis. *Red box*: Apoptotic zone, defined as a range of cellular concentrations of iAβ within which cells neuronal cells committed apoptosis or necroptosis or are dead. *Green box*: Duration of the treatment with the ISR-inhibiting drug. *Panel* (**A**): The initial state of the levels of AβPP-derived iAβ in the neurons of the healthy individual who would develop AD if untreated. AβPP-derived iAβ has been accumulating but its levels are below the T1 threshold *Panel* (**B**): The evolution of the initial state in the untreated individual. AβPP-derived iAβ accumulates unimpeded and crosses the T1 threshold. This triggers activation of PKR and/or HRI, phosphorylation of eIF2α, elicitation of the neuronal ISR, and initiation of the AβPP-independent C99/iAβ production pathway. AD commences and progresses until it reaches its end stage. *Panel* (**C**): The evolution of the initial state in the individual treated with the ISR-inhibiting drug. AβPP-derived iAβ reaches and crosses the T1 threshold but the neuronal ISR cannot be elicited. The influx of iAβ is supported solely by the AβPP proteolysis and it continues to accumulate at a slow pre-T1 crossing rate. Its level would not reach the AD pathology-causing range for the duration of the treatment.

43. Suppression of the Neuronal Integrated Stress Response in the Treatment of Conventional Alzheimer's Disease

In the treatment of conventional AD, the administration of the ISR-inhibiting drug commences when the AβPP-independent C99/iAβ production pathway is operational and the disease progresses. This approach is illustrated in Figure 18. Panel A of Figure 18 depicts the initial state of the levels of iAβ in the affected neurons of the AD patient. At this point AβPP-derived iAβ has crossed the T1 threshold in all affected neurons, the neuronal ISR has been elicited, AβPP-independent C99/iAβ production pathway activated, and AD symptoms have manifested. Left untreated, the disease would progress until it reaches the end stage, as shown in Panel B of Figure 18. Panel C of Figure 18 shows the evolution of the initial state in the presence of the ISR-inhibiting drug. The neuronal ISR is suppressed and no longer supplies the components required for the operation of the AβPP-independent C99/iAβ production pathway. The operation of the pathway ceases. However, the accumulation of iAβ, supported now only by the AβPP proteolysis, continues, albeit at a slow pre-T1 crossing rate. The progression of AD does not stop; it continues but at a significantly lower rate for the duration of the treatment.

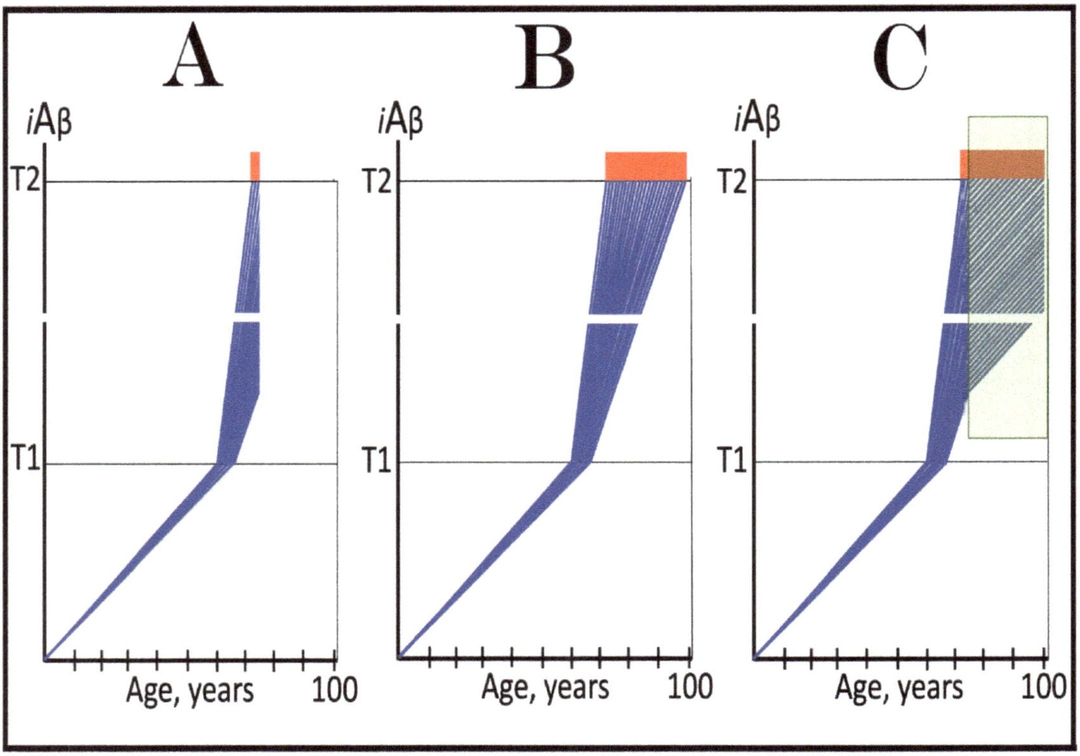

Figure 18. Suppression of the neuronal integrated stress response in the treatment of conventional AD. *Blue lines*: Intraneuronal Aβ, iAβ. ***T1***: Threshold of cellular concentration of iAβ triggering the activation of PKR and/or HRI kinases, phosphorylation of eIF2α, the elicitation of the neuronal ISR, the initiation of the AβPP-independent C99 production pathway, and the commencement of AD. ***T2***: Cellular concentration of iAβ triggering neuronal death via apoptosis or necroptosis. *Red box*: Apoptotic zone, defined as a range of cellular concentrations of iAβ within which cells neuronal cells

committed apoptosis or necroptosis or are dead. *Green box*: Duration of the treatment with the ISR-inhibiting drug. *Panel* (**A**): The initial state of the levels of AβPP-derived iAβ in the neurons of the AD patient. AβPP-derived iAβ has crossed the T1 threshold in all affected neurons, the neuronal ISR has been elicited, the AβPP-independent C99/iAβ production pathway activated, the T2 threshold crossed in a neuronal fraction, and AD symptoms have manifested. *Panel* (**B**): The disease progresses and reaches its end stage. *Panel* (**C**): The evolution of the initial state in the individual treated with the ISR-inhibiting drug. The supply of components essential for the activity of the AβPP-independent C99/iAβ production pathway stops and it operation ceases. The accumulation of iAβ, however, continues but at a slow pre-T1 crossing rate. The rate of the progression of the disease is reduced for the duration of the treatment.

Importantly, in both the present and the preceding sections the beneficial effect of the suppression of the neuronal ISR requires a continuous long-term administration of the drug. However, the long-term systemic inhibition of the ISR is, apparently, not feasible. The ISR is a major physiological cellular survival tool, and its systemic inhibition is likely to cause severe adverse effects. Such an outcome is strongly indicated by studies of eIF2α Ser51Ala KI mice [219]. Ser51 of aIF2α is the amino acid residue, which is phosphorylated in the integrated stress response. If the phosphorylation in this position were prevented, there would be no ISR. This is precisely what occurs when Ser51 is replaced by alanine. Ala51 of eIF2α cannot be phosphorylated and, consequently, the integrated stress response cannot be elicited in the carriers of this mutation. It is a predictive equivalent of the long-term systemic inhibition of the ISR, and the outcome is highly deleterious: mice with a homozygous mutation eIF2α Ser51Ala survived only up to eighteen hours after their birth [219]. These findings suggest that other types of drugs and alternative approaches are required for the prevention and treatment of AD.

44. BACE1 and BACE2 Activators Are Potential ACH2.0-Based iAβ Depletion Drugs

Any agent capable of specific degradation of iAβ is a potential ACH2.0-based drug. For example, directly acting iAβ-specific degradation agents such as appropriately designed proteolysis-targeting chimeras (PROTACs) and molecular-glue degraders (MGDs) would constitute ACH2.0-based AD drugs. Fortunately, we do not have to venture far afield to find two outstanding candidates capable of the targeted degradation of iAβ, they are front and center: BACE1 and BACE2, or, more precisely, intra-iAβ cleaving activities of BACE1 and BACE2. Both enzymes are capable of cleaving at the beta-site of AβPP, as indicated by their designations. In BACE1 cleavage at the β-site is its major activity. In addition, BACE1 also cleaves at the β′-site, ten amino acids downstream from the β-site [43,44] and between amino acid residues 34 and 35 of Aβ, the latter generating an intermediate in the iAβ clearance process [220–223]. Potential beneficial outcomes of these cleavages were demonstrated in transgenic mouse model and in neuronal cells overexpressing BACE1. This substantially increased both the efficiency and the proportion of cleavages at the β′ and Aβ34/35 sites [220–227].

In BACE2 cleavages within iAβ, or within the Aβ segment of AβPP and C99, are the major activity of the enzyme. These cleavages, occurring physiologically, take place at positions Aβ19 and Aβ20 [228,229]. The therapeutic potential of these cleavages is underscored by the observation that when BACE2 activity was suppressed in transgenic models, the production of Aβ as well as the extent of its extracellular deposition significantly increased. The notion that the increased activity of BACE2 confers benefits and its reduced activity is detrimental is supported by the observations made in human pluripotent cells-derived brain organoids, where increased activity of BACE2 protected from the iAβ-mediated toxicity and neuronal death whereas the reduction in BACE2 activity correlated with the increased neurodegeneration [230]. Thus, the activation of intra-iAβ cleaving potential of BACE1 and, especially, BACE2 (because intra-iAβ cleavages are its major activity) is, apparently, a plausible therapeutic strategy. Moreover, simultaneous activation of both BACE1 and BACE2 can yield synergetic benefits:

these enzymes not only target different positions within iAβ and the Aβ segment of AβPP and C99 but are also localized in different intracellular compartments [231].

45. Validation (10): Augmentation of the Efficiency of the BACE1-Mediated Intra-iAβ Cleavage Protects from AD and AACD Whereas Suppression of the BACE2-Mediated Intra-iAβ Cleavage Causes AD

How to test the proposition that the activation of BACE1 and BACE2 is a valid therapeutic strategy? The answer is simple: develop suitable and efficient activators of both enzymes (more precisely, of their intra-iAβ cleaving activities) and test them in human clinical trials. If the proposition is valid, activated BACE enzymes would protect from AD and AACD. In a decisive complementary trial, unfeasible for the ethical reasons, BACE intra-iAβ cleaving activities would be suppressed. If the proposition were valid, this would cause AD. Both trials, in fact, were already carried out by nature (which is, apparently, not concerned by the ethical considerations), in the form of Aβ mutations; the results of these "natural" trials soundly validate the proposition.

One of the mutations referred to above is the Icelandic Aβ mutation. It is in the proximity of the β′-site and, consequently, affects, more specifically, increases the efficiency of cleavage at this position (in iAβ as well as in Aβ segments of AβPP and C99) by BACE1. This increase has a two-pronged effect. First, it reduces (by cleavages within Aβ segments of AβPP and C99) the rate of production of Aβ. This decreases the rate of secretion of Aβ, reduces the extracellular Aβ pool, lowers the rate of the uptake of extracellular Aβ, and thus diminishes the rate of influx of AβPP-derived iAβ. Second, it increases, by the cleavage within iAβ, the rate of its efflux. Consequently, with the decreased influx and increased efflux, the rate of accumulation of AβPP-derived iAβ decreases and the timing of the T1 crossing increases. The extent of the Icelandic mutation-mediated increase in the efficiency of cleavage at the β′-site is relatively small, only 30%. It is sufficient, however, to confer on its carriers protection from both AD and AACD [43,44].

Another mutation under discussion in the present section is the Flemish Aβ mutation. It substitutes the amino acid residue at the position Aβ21, a position in the contiguous proximity to BACE2 intra-iAβ cleavage sites (Aβ19 and Aβ29). This affects, more specifically, decreases to below the physiological the efficiency of cleavages at the positions Aβ19 and Aβ20 in iAβ as well as in Aβ segments of AβPP and C99. Since fewer cleavages occur within Aβ segments of AβPP and C99, the rates of production and secretion of Aβ increase. This results in an increase in the extracellular Aβ pool and elevated rates of its uptake, leading to an influx of AβPP-derived iAβ accordingly. Concurrently, the reduction in the rate of cleavages within iAβ lowers the rate of its efflux. With the increased influx and decreased efflux, the rate of accumulation of AβPP-derived iAβ increases. The crossing of the T1 threshold and, consequently, AD would occur sooner than in the wild type. Indeed, in the carriers of the Flemish mutation the onset of AD begins as early as thirty-five years of age [232–238].

The above results leave little if any doubt that effective activators of intra-iAβ cleaving capabilities of BACE1 and BACE2 would constitute efficient AD and AACD drugs. The following three sections analyze their anticipated effects in the prevention and treatment of conventional Alzheimer's disease.

46. Activators of BACE1 and/or BACE2 in the Prevention of Conventional AD and the Prevention and Treatment of AACD: A Single Transient Therapy Can Potentially Protect for Life

As discussed in the preceding sections, activators of the intra-iAβ cleaving capabilities of BACE1 and BACE2 (referred to henceforth as "activators of BACE1 and BACE2" and their effect designated as "activation of BACE1 and BACE2") are potential AD and AACD drugs. The present section considers their anticipated effect in the prevention of conventional AD and the prevention and treatment of AACD. In all three scenarios the treatment is administered prior to the crossing of the T1 threshold by AβPP-derived iAβ, i.e., prior to the activation of the AβPP-independent C99/iAβ production pathway. In such a setting, it can be anticipated that activated BACE1 and/or BACE2 would be very effective in depleting iAβ. Indeed, only a 30% increase in the efficiency of cleavage at only the β′-site (the Icelandic mutation) prevents the accumulation of AβPP-derived iAβ to the T1 level. Adequate activators can potentially increase the efficiency of cleavages not only at the β′-site but also at the position Aβ34 (by BACE1) and at the positions Aβ19 and Aβ20 (by BACE2) to the extent much greater than 30%. The required duration of the treatment would be defined by its efficiency. The aim is to deplete iAβ close to its initial baseline. With the anticipated activity of sufficiently activated BACE1 and/or BACE2, the desired depletion can be accomplished in a relatively short time. The treatment, in such a case, would be transient. When the drug is withdrawn, the de novo accumulation of AβPP-derived iAβ, supported only by the AβPP proteolysis, would commence from a low baseline at the pre-treatment rate. The accumulation of AβPP-derived iAβ is a very slow process. It does not reach the T1 threshold within the lifetime of most individuals, and in those who develop non-familial AD, it takes over six decades for the T1 crossing to occur. This implies that if the transient iAβ depletion treatment via the activation of BACE1 and/or BACE2 is administered at midlife, around fifties, and if the depletion is sufficiently deep, iAβ levels would reach neither the T^0 nor T1 thresholds within the lifetime of the treated individual, and neither AACD nor AD would occur. Thus, a single treatment, timely administered, can potentially protect from both AD and AACD for life.

The above strategy is illustrated in Figure 19. Panel A of Figure 19 depicts the accumulation of iAβ and the progression of AD in the untreated individual. In the first stage the accumulation of iAβ is supported solely by the AβPP proteolysis. The extent of the T^0 threshold is greater than that of the T1 threshold; no AACD occurs in this scenario. When the T1 threshold is crossed, PKR and/or HRI are activated, eIF2α phosphorylated, the neuronal integrated stress response elicited and the AβPP-independent production of C99/iAβ initiated. Levels of the latter rapidly increase and reach the AD pathology-causing range. When they reach the T2 threshold, neuronal loss ensues. Panel B of Figure 19 shows the outcome of the transient treatment with activators of BACE1 and/or BACE2. The treatment is administered prior to the T1 crossing, and AβPP-derived iAβ is substantially depleted. Following the withdrawal of the drug, the de novo accumulation commences from low baseline and proceeds at the pre-treatment rate. Neither the T1 threshold would be crossed, nor AD would occur within the lifetime of the treated individual.

Panel C of Figure 19 presents the dynamics of accumulation of iAβ and the progression of AACD and AD in a scenario where the extent of the T^0 threshold is smaller than that of the T1 threshold. When AβPP-derived iAβ reaches the T^0 threshold, AACD commences and persists until the T1 crossing and the activation of the AβPP-independent C99/iAβ production pathway. At this point AACD morphs into AD. If activators of BACE1 and/or BACE2 were administered transiently prior to the T^0 crossing, and if the depletion of iAβ were deep enough, iAβ, accumulating de novo following the treatment, would not reach the T^0 threshold within the remaining lifetime of the treated individual and no AACD (and neither AD) would occur. If, as shown in Panel D of Figure 19, the treatment was administered post-T^0 crossing, but prior to the T1 crossing, iAβ would be substantially depleted well below the T^0 threshold. At this point AACD would be cured. The accumulation of AβPP-derived iAβ would start de novo from a low baseline and proceed at the pre-treatment rate. It would not reach the T^0 threshold within the lifetime cf the treated individual and neither AACD would recur nor AD would occur.

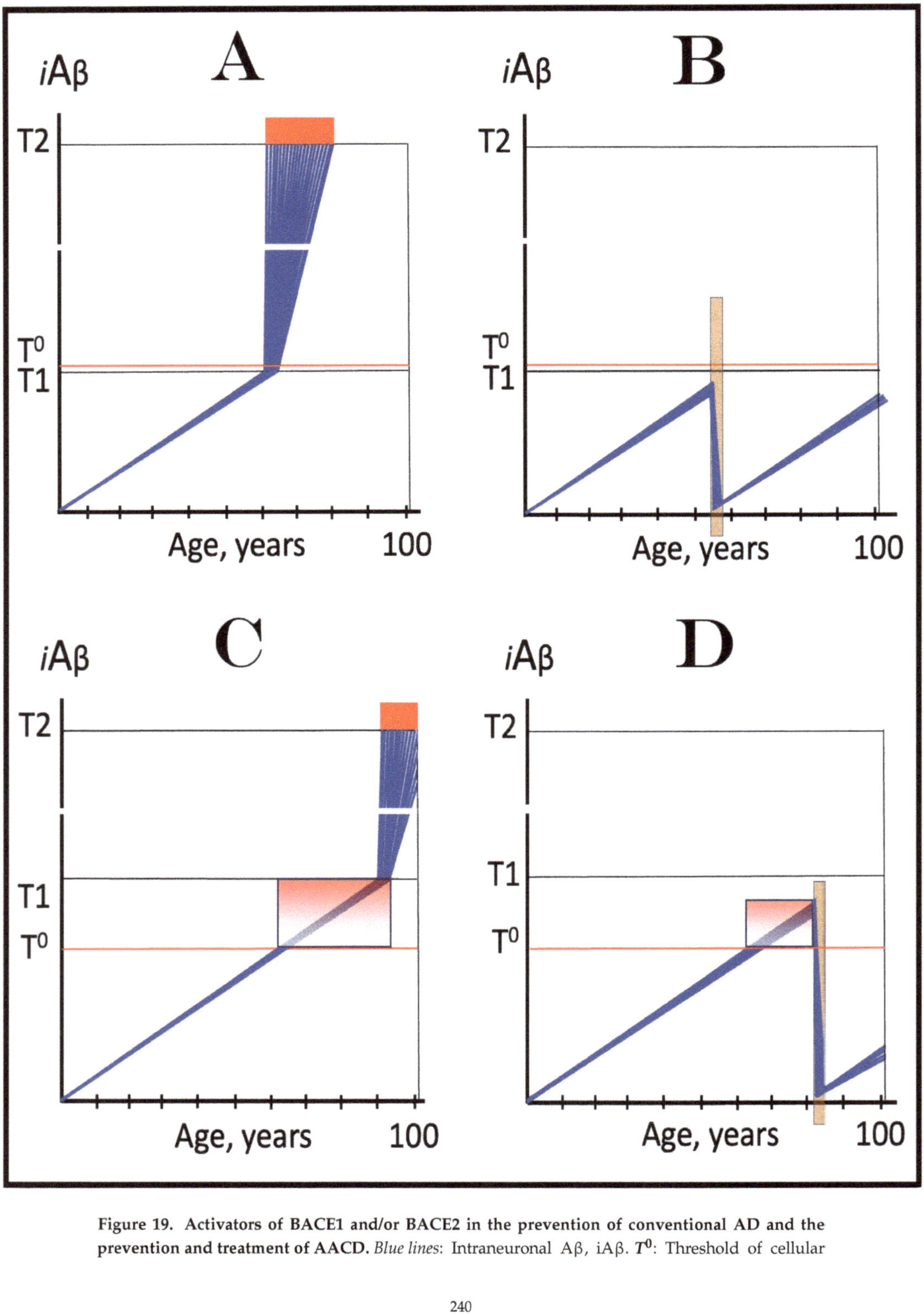

Figure 19. Activators of BACE1 and/or BACE2 in the prevention of conventional AD and the prevention and treatment of AACD. *Blue lines*: Intraneuronal Aβ, iAβ. $\boldsymbol{T^0}$: Threshold of cellular

concentration of AβPP-derived iAβ triggering the neuronal damage, which manifests as AACD. ***T1***: Threshold of cellular concentration of iAβ triggering the activation of PKR and/or HRI kinases, phosphorylation of eIF2α, the elicitation of the neuronal ISR, the initiation of the AβPP-independent C99 production pathway, and the commencement of AD; its extent is the only variable in the present figure. ***T2***: Cellular concentration of iAβ triggering neuronal death via apoptosis or necroptosis. *Red box*: Apoptotic zone, defined as a range of cellular concentrations of iAβ within which cells neuronal cells committed apoptosis or necroptosis or are dead. *Pink boxes*: The range of cellular concentrations of AβPP-derived iAβ concentrations between the T^0 and the highest extent reached by AβPP-derived iAβ or the T1 threshold that support the progression of AACD. The condition commences with the crossing of the T^0 threshold and morphs into AD with the T1 crossing; it can occur only if the extent of the T^0 threshold is smaller than that of the T1. *Orange boxes*: Duration of the treatment with the BACE1- and/or BAEC2-activating drug. *Panel* (**A**): Dynamics of the accumulation of iAβ in the untreated AD patient. AβPP-derived iAβ accumulates via two physiological mechanisms described in the main text. When it crosses the T1 threshold PKR and/or HRI kinases are activated, eIF2α phosphorylated, the neuronal ISR elicited, the AβPP-independent C99 production pathway initiated, and conventional AD commences. The T^0 threshold is above the T1; there is no AACD. *Panel* (**B**): Activators of BACE1 and/or BACE2 are administered transiently prior to the T1 crossing. As a result, AβPP-derived iAβ is substantially depleted. Following the withdrawal of the drug, the de novo accumulation commences from low baseline and proceeds at the pre-treatment rate. Neither the T1 threshold would be crossed, nor AD would occur within the lifetime of the treated individual. *Panel* (**C**): Dynamics of the accumulation of iAβ in the untreated AACD/AD patient. When AβPP-derived iAβ reaches the T^0 threshold, AACD commences and persists until the T1 crossing, the activation of the AβPP-independent C99/iAβ production pathway, and the commencement of AD. At this point AACD morphs into AD. *Panel* (**D**): Activators of BACE1 and/or BACE2 are administered transiently following the T^0 crossing and the commencement of AACD but prior to the T1 crossing. As a result, iAβ is substantially depleted well below the T^0 threshold. At this point AACD is cured. The accumulation of AβPP-derived iAβ starts de novo from low baseline and proceeds at the pre-treatment rate. It would not reach the T^0 threshold within the lifetime of the treated individual and neither AACD would recur nor AD would occur.

47. Problem with the Utilization of Activators of BACE1 and/or BACE2 in the Treatment of Conventional AD: With the AβPP-Independent C99 Generation Pathway Operational the Rate of Degradation of iAβ Is Unlikely to Match or Exceed That of Its Influx

The depletion of iAβ to a low baseline would be as effective therapeutically in the treatment of symptomatic conventional AD as in the prevention of the disease provided it could be accomplished. However, the problem with the activators of BACE1 and/or BACE2 is that, on their own, they are, apparently, incapable of such a feat. The source of this problem is the operation of the AβPP-independent C99/iAβ production pathway. As described above, the efficiency of this pathway is apparently comparable to that of a massive gene amplification, with the rate of iAβ generation potentially orders of magnitude greater than the rate of iAβ production in the AβPP proteolytic pathway. In such a case, it is inconceivable that the degradation of iAβ by activated BACE1 and/or BACE2, however efficient, could match the influx of iAβ generated independently of AβPP. In such a case, even if employed long-term, the activation of BACE1 and/or BACE2 would only reduce the rate of iAβ accumulation and, at best, slow down the progression of AD.

The anticipated outcome of the utilization of activators of BACE1 and BACE2 in the treatment of conventional AD in the version of the ACH2.0 under discussion is illustrated in Figure 20. Panel A of Figure 20 shows the initial state of the levels of iAβ in the affected neurons of the AD patient at the commencement of the treatment. AβPP-derived iAβ has crossed the T1 threshold in all affected neurons, the neuronal integrated stress response has been elicited and the AβPP-independent C99/iAβ production pathway activated; a fraction of the neurons has crossed the T2 threshold and AD symptoms have manifested. Panel B depicts the evolution of the initial state in the untreated AD patient. The accumulation of iAβ, mainly in the AβPP-independent C99/iAβ production pathway, continues and AD pathology progresses. When a sufficient neuronal fraction reaches the T2 threshold and commits apoptosis, the disease reaches its end stage. Panel C presents the evolution of the initial state in the presence of the activators of BACE1 and/or BACE2. The rate of degradation of iAβ via its internal cleavages increases but it cannot match, not to say exceed, the rate of the influx of iAβ produced independently of AβPP. The rate of the accumulation of iAβ decreases and the progression of AD slows down for the duration of the treatment, but the disease persists. This is better than nothing but not good enough. Fortunately, the solution to this problem exists and is presented in the following section.

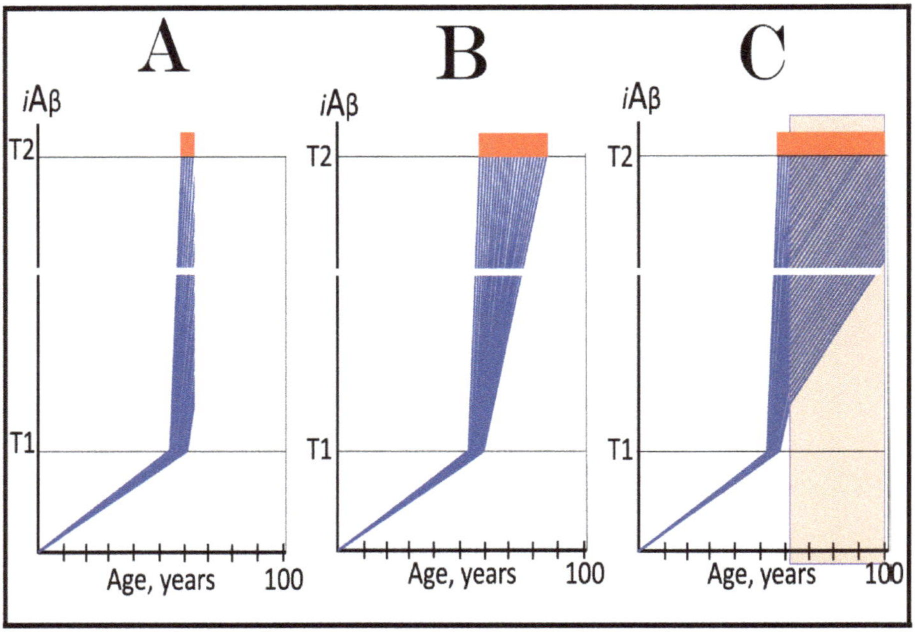

Figure 20. With the AβPP-independent C99/iAβ production pathway operational, the rate of degradation of iAβ by activated BACE1 and/or BACE2 cannot match or exceed that of its influx. *Blue lines*: Intraneuronal Aβ, iAβ. ***T1***: Threshold of cellular concentration of iAβ triggering the activation of PKR and/or HRI kinases, phosphorylation of eIF2α, the elicitation of the neuronal ISR, the initiation of the AβPP-independent C99 production pathway, and the commencement of AD. ***T2***: Cellular concentration of iAβ triggering neuronal death via apoptosis or necroptosis. *Red box*: Apoptotic zone, defined as a range of cellular concentrations of iAβ within which cells neuronal cells committed apoptosis or necroptosis or are dead. *Orange box*: Duration of the treatment with the BACE1- and/or BACE2-activating drug. *Panel* (**A**): The initial state of the levels of AβPP-derived iAβ in the neurons of the AD patient. AβPP-derived iAβ has crossed the T1 threshold in all affected neurons, the neuronal ISR has been elicited, the AβPP-independent C99/iAβ production pathway activated, the T2 threshold crossed in a neuronal fraction, and AD symptoms have manifested. *Panel* (**B**): The disease progresses and reaches its end stage. *Panel* (**C**): The evolution of the initial state in the individual treated with the BACE1- and/or BACE2-activating drug. The rate of degradation of iAβ via its internal cleavages increases but it cannot match the rate of the influx of iAβ produced independently of AβPP. The rate of the accumulation of iAβ decreases and the progression of AD slows down for the duration of the treatment, but the disease persists.

48. Solution: Cessation of the Influx of iAβ Produced Independently of AβPP by the Transient Inhibition of the Neuronal ISR in the Presence of Activators of BACE1 and/or BACE2

As discussed in the preceding section, the problem with the employment of the activators of BACE1 and BACE2 in the treatment of conventional AD is that the rate of the influx of iAβ produced independently of AβPP is too great to be matched by the rate of degradation of iAβ by activated BACE1 and/or BACE2. The solution is simple: suppress the influx of iAβ produced independently of AβPP for the duration of the depletion treatment or even better, stop it altogether. If this was accomplished, the influx of iAβ would be supported only by the AβPP proteolysis, and the rate of its depletion by activated BACE1 and/or BACE2 would be comparable to that anticipated in the prevention of the disease. This is doable and can be accomplished by transient inhibition of the neuronal integrated stress response executed concurrently with the treatment by the activators of BACE1 and/or BACE2.

It was argued above (see Section 43) that the long-term systemic inhibition of the integrated stress response is, apparently, not feasible. This is correct. But in the application under discussion the transient suppression of the ISR would suffice (and, apparently, would not be deleterious; indeed, transient inhibition of ISR did not cause adverse effects in mouse models [119–127]). The reversal of the neuronal ISR would deprive the AβPP-independent C99/iAβ generation pathway of its essential components; its operation would stop, and the influx of its product would cease. The situation in such a case would be similar to that encountered in the prevention of AD: the influx of iAβ would be supported solely by the AβPP proteolysis. Activators of BACE1 and/or BACE2, administered concurrently with ISR inhibitors, would rapidly and efficiently deplete iAβ. The administration of ISR inhibitors in this application is transient. Their presence is essential only for the depletion of iAβ to levels below the T1 thresholds. This would ensure that, when they are withdrawn, the neuronal ISR would not be re-elicited and the AβPP-independent C99/iAβ production pathway reactivated. There are no obvious limitations for the duration of the presence of activators of BACE1 and BACE2, but they also are needed only transiently, until the levels of iAβ are reduced to sufficiently low extents. Following the withdrawal of BACE1 and/or BACE2 activators, the accumulation of AβPP-derived iAβ would resume. However, its levels would not reach the T1 thresholds and AD would not recur.

The strategy under discussion is illustrated in Figure 21. Panels A and B of Figure 21 are similar, in fact, identical to the corresponding panels of Figure 20. Panel C of Figure 21 is drastically different from its counterpart in Figure 20. It depicts a composite therapy. Inhibitor(s) of the ISR are administered transiently and concurrently with activator(s) of BACE1 and/or BACE2. Suppression of the neuronal ISR disables the operation of the AβPP-independent C99/iAβ production pathway and the influx of its product stops. This allows activated BACE1 and/or BACE2 to efficiently deplete iAβ. With the AβPP-independent C99/iAβ production pathway inoperative, the progression of AD ceases. Following the withdrawal of both ISR inhibitors and BACE activators, de novo accumulation of iAβ commences from a low baseline; it is supported solely by the AβPP proteolysis. It would not reach the T1 threshold, the AβPP-independent C99/iAβ production pathway would not be reactivated, and the disease would not recur within the remaining lifetime of the treated AD patient.

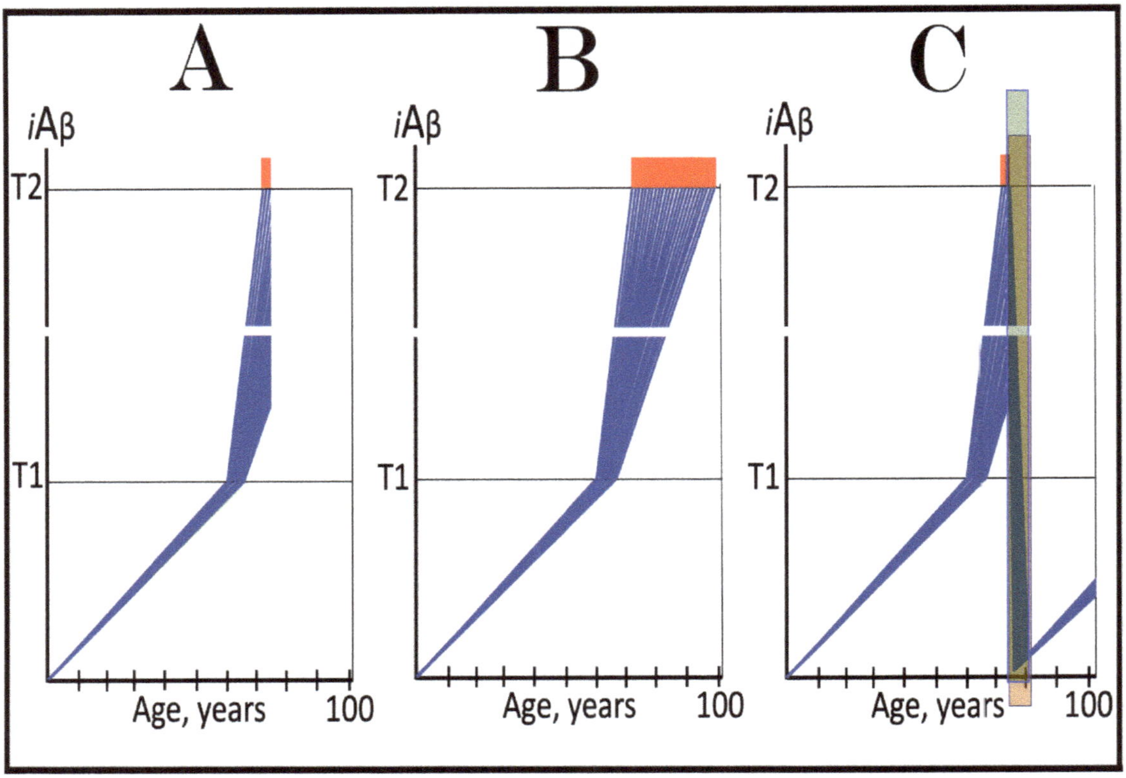

Figure 21. Concurrent transient BACE1 and/or BACE2 activation and ISR inhibition in the treatment of conventional AD. *Blue lines*: Intraneuronal Aβ, iAβ. ***T1***: Threshold of cellular concentration of iAβ triggering the activation of PKR and/or HRI kinases, phosphorylation of eIF2α, the elicitation of the neuronal ISR, the initiation of the AβPP-independent C99 production pathway, and the commencement of AD. ***T2***: Cellular concentration of iAβ triggering neuronal death via apoptosis or necroptosis. *Red box*: Apoptotic zone, defined as a range of cellular concentrations of iAβ within which cells neuronal cells committed apoptosis or necroptosis or are dead. *Green box*: Duration of the treatment with the ISR-inhibiting drug. *Orange box*: Duration of the treatment with the BACE1- and/or BACE2-activating drug. *Panel* (**A**): The initial state of the levels of AβPP-derived iAβ in the neurons of the AD patient. AβPP-derived iAβ has crossed the T1 threshold in all affected neurons, the neuronal ISR has been elicited, the AβPP-independent C99/iAβ production pathway activated, the T2 threshold crossed in a neuronal fraction, and AD symptoms have manifested. *Panel* (**B**): The disease progresses and reaches its end stage. *Panel* (**C**): The evolution of the initial state in the treated patient. The ISR-inhibiting drug is administered transiently and concurrently with the activator of BACE1 and/or BACE2. Suppression of the neuronal ISR disables the operation of the AβPP-independent C99/iAβ production pathway and the influx of its product stops. This allows activated BACE1 and/or BACE2 to efficiently deplete iAβ. Following the withdrawal of both ISR inhibitors and BACE activators, the AβPP-independent C99/iAβ production pathway remains inoperative, and the progression of AD ceases. De novo accumulation of iAβ commences from a low baseline supported solely by the AβPP proteolysis. It would not reach the T1 threshold, the AβPP-independent C99/iAβ production pathway would not be reactivated, and the disease would not recur within the remaining lifetime of the treated AD patient.

49. Progression of AD Is Defined by the Rate of the Production of iAβ in the AβPP-Independent Pathway and by the Extent of the T2 Threshold: Sequential Occurrence of AD Pathology in Defined Brain Compartments

AβPP-derived iAβ reaches and crosses the T1 threshold in the affected neurons within a narrow temporal window [1,4], and this crossing denotes the commencement of conventional AD. Therefore, the rate of accumulation of AβPP-derived iAβ and the extent of the T1 threshold decide the occurrence of conventional AD (or the lack thereof) and determine its timing. Once the disase occurs, its progression is also defined by two major parameters. One is the dynamics of accumulation of iAβ produced independently of AβPP (in this variant of the ACH2.0), which, in turn, is defined by the efficiency of the AβPP-independent C99/iAβ generation pathway. Another is the extent of the T2 threshold. The greater the rate of accumulation of iAβ produced independently of AβPP and the smaller the extent of the T2 threshold, the faster would be the progression of the disease and the shorter its duration. Conversely, the smaller the rate of accumulation of iAβ produced independently of AβPP and the greater the extent of the T2 threshold, the slower the progression of AD and the longer its duration.

Variations in these two parameters, the rate of accumulation of iAβ produced independently of AβPP and the extent of the T2 threshold, can explain one of the principal features of Alzheimer's disease, namely the sequential occurrence of AD pathology in defined compartments of the brain [239–241]. Indeed, it appears that the AD-associated neurodegeneration commences within the defined layer of the entorhinal cortex. This is followed, sequentially, by its occurrence in the hippocampus, temporal cortex, frontoparietal cortex, and subcortical nuclei [239]. The sequential development of AD pathology is observed even within defined areas of a specific brain region. Thus, in the hippocampus, AD pathology spreads sequentially from the CA1 area to CA2, CA3, and, finally, DG areas [240]. One explanation of such a temporal order is that, at the given rate of accumulation of iAβ produced independently of AβPP, the extent of the T2 threshold differs in different brain compartments and even sub-compartments. The greater it is, the later the neurodegeneration manifests. Another explanation is that, at the given extent of the T2 threshold, the rate of accumulation of iAβ produced independently of AβPP varies in distinct brain compartments and sub-compartments. The greater it is, the sooner the neurodegeneration manifests. Finally, of course, these two variables can modulate concurrently.

The variability discussed above is illustrated in Figure 22 where different brain compartments are schematically denoted by different colors. Panel A of Figure 22 depicts the effects of the variable extent of the T2 threshold at the background of the given rate of accumulation of iAβ produced independently of AβPP. The greater the former, the later the neurodegeneration manifests. Panel B of Figure 22 shows the combined effect of modulation of both the extent of the T1 threshold and the rate of accumulation of iAβ produced independently of AβPP.

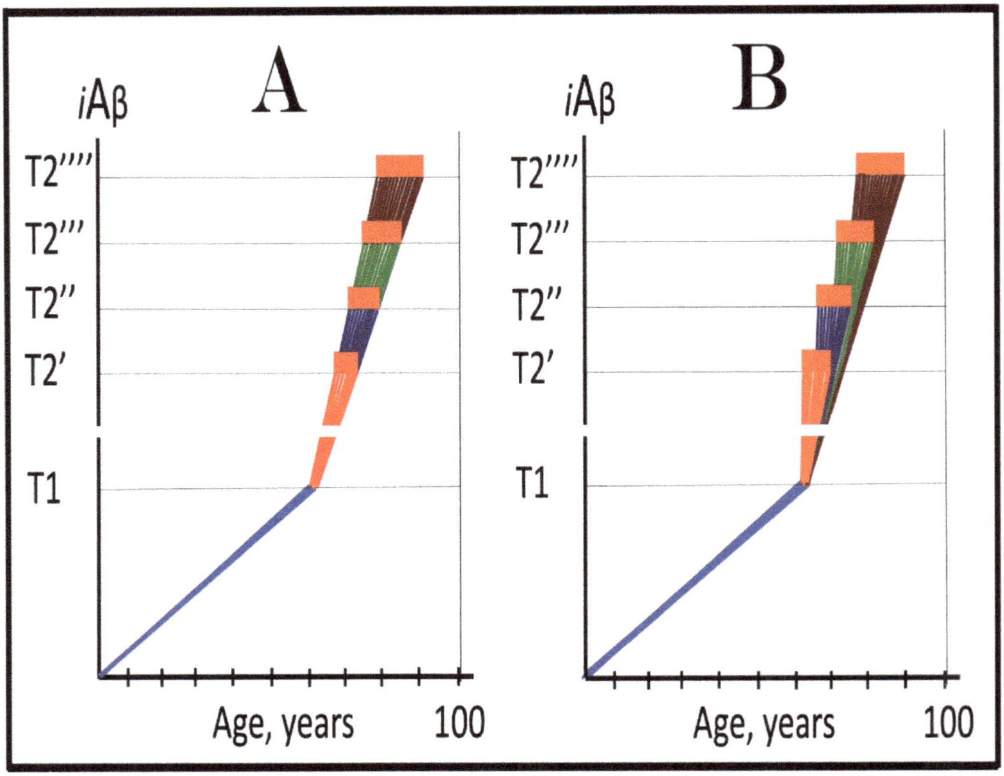

Figure 22. Sequential occurrence of AD pathology in the defined brain compartments as the consequence of the variability in the rates of AβPP-independent iAβ accumulation and the extents of the T2 threshold. *Blue lines*: Intraneuronal Aβ, iAβ. ***T1***: Threshold of cellular concentration of iAβ triggering the activation of PKR and/or HRI kinases, phosphorylation of eIF2α, the elicitation of the neuronal ISR, the initiation of the AβPP-independent C99 production pathway, and the commencement of AD. ***T2***: Cellular concentration of iAβ triggering neuronal death via apoptosis or necroptosis. ***T2′, T2″, T2‴, T⁗***: Variable T2 thresholds in the defined regions of the brain. *Red, blue, green, and brown lines above the T1 threshold*: iAβ accumulation in the AD-affected neurons in various regions of the brain. *Red box*: Apoptotic zone, defined as a range of cellular concentrations of iAβ within which cells neuronal cells committed apoptosis or necroptosis or are dead. *Panel* (**A**): Effects of the variable extent of the T2 threshold at the background of the given rate of accumulation of iAβ produced independently of AβPP on the sequential occurrence of AD pathology in the defined regions of the brain. The greater the former, the later the neurodegeneration manifests. *Panel* (**B**): the combined effect of modulation of both the extent of the T1 threshold and the rate of accumulation of iAβ produced independently of AβPP on the sequential occurrence of AD pathology in the defined brain compartment.

Importantly, the above interpretation of the sequential occurrence of AD pathology in the defined compartments of the brain has consequential therapeutic implications and underscores the significance of the earliest possible intervention in AD. Indeed, if the composite therapy described in Section 48 above is successful, the progression of the disease in the selectively affected brain compartments will cease and not recur. The depletion of iAβ to low baselines would take place also in brain compartments not yet affected (or affected insignificantly) by the disease. Therefore, in the treated patients, these compartments would be protected by the treatment and would stay AD pathology-free for the remaining lifespan.

50. The ACH2.0, Version Two: Alzheimer's Is a Disease of the Neuronal ISR; Unconventional AD Is Triggered by Stressors Distinct from AβPP-Derived iAβ and Driven by iAβ Generated Independently of AβPP

The initial version of the ACH2.0, discussed above, defines conventional AD. In this version, by far the most important player is iAβ generated differentially in two distinct pathways. iAβ derived in the AβPP proteolytic pathway and accumulated over the T1 threshold triggers the activation of the AβPP-independent C99/iAβ generation pathway, and it is iAβ produced in the latter, which drives the disease. The importance of the neuron-specific component, iAβ, produced in two distinct ways, in the neuron-specific disease, AD, overshadowed the role of a non-neuron-specific intermediate component in transition from AβPP-derived iAβ to iAβ generated independently of AβPP, namely the integrated stress response. But it is the neuronal integrated stress response, which supplies essential constituents of the AβPP-independent production of C99/iAβ. Moreover, it not just "switches on" the process: the neuronal ISR maintains the activity of the AβPP-independent C99/iAβ production pathway, thus enabling it; if the ISR conditions are reversed, the operation of the AβPP-independent C99/iAβ generation pathway ceases. AD was defined, in terms of the ACH, as the disease of Aβ, and the notion remained ingrained and, therefore, hard to overcome. But it is, in fact, the neuronal ISR, which triggers the disease by delivering its essential components, and it is not only necessary but also sufficient for the disease to occur [8,10]. AD is, therefore, not a disease of Aβ but a disorder of the neuronal integrated stress response: elicit, IN ANY WAY, the sustained neuronal ISR and Alzheimer's disease will ensue.

It follows that the sustained elicitation of the neuronal ISR, even by stressors other than AβPP-derived iAβ and even at the levels of AβPP-derived iAβ far below the T1 threshold, would trigger the activation of the AβPP-independent C99/iAβ production pathway and, consequently, the commencement of AD pathology. The category of Alzheimer's disease caused by sustained neuronal ISR elicited by stressors distinct from AβPP-derived iAβ has been designated "unconventional AD" [8–10] to distinguish it from "conventional AD", which is defined as the disease caused by the neuronal ISR elicited solely by AβPP-derived iAβ. This separation into two categories seems ostentatiously asymmetrical: the only difference appears to be one specific ISR-eliciting stressor (AβPP-derived iAβ) versus a multitude of potential ISR-eliciting stressors. This, however, is not the case. First, conventional AD is the prevalent category of the disease, accounting for about 60% of all cases of dementia. Second, all unconventional AD cases can be grouped together (and separately from conventional cases) on the basis of the following criteria. Whereas in conventional AD the AβPP-independent C99/iAβ production pathway is always self-sustainable from the instance of its activation, in unconventional AD it takes some time for it to become self-sustainable: if the initial ISR-eliciting unconventional stressor were withdrawn too early, the activity of the AβPP-independent C99/iAβ generation pathway would be only transient. This is why, to trigger AD, the unconventionally elicited neuronal ISR state should be sufficiently prolonged in order to ensure that the AβPP-independent C99/iAβ production pathway becomes self-sustainable. This aspect of unconventional AD is further discussed in Section 52 below.

Thus, beside AβPP-derived iAβ (designated "conventional stressor") that triggers conventional AD, any "unconventional" (i.e., distinct from AβPP-derived iAβ) stressor capable of sustainably activating one or more of the four kinases of eIF2α can potentially initiate unconventional AD. Such stressors are abundant and, therefore, so are their sources, i.e., conditions potentially resulting in unconventional AD. They include brain trauma (traumatic brain injury, TBI) [242–244] and chronic traumatic encephalopathy, CTE [245].

These conditions can stem from viral as well as from bacterial infections [246–254]. Thus, an infection of viral encephalitis elevates more than thirtyfold the probability of developing unconventional AD [246–250] whereas various bacterial infections make the incidence of AD more than tenfold likely [251–254]. Notable conditions that have strong association with AD encompass neuroinflammation, and chronic systemic and even localized inflammations including, for example, rheumatoid- and osteoarthritis [255–266]. Cases of dementia related to the above conditions are often categorized as AD-like dementia, ADLD or AD-related dementia, ADRD. In the framework of the ACH2.0 version under discussion, the majority, if not all, cases of ADRD and ADLD probably belong to the category of unconventional AD.

Recently, it has been shown that distincly different sources of unconventional stressors can interact with and even stimulate one another. Thus, herpes simplex virus in general and HSV-1 in particular were shown in multiple studies to be associated with Alzheimer's disease [267–270]. HSV-1 is known to be neurotropic and capable of establishing dormant state in nerve tissues, e.g., in the sensory root ganglia of the trigeminal nerve [268]. In a recent study employing human brain organoid tissue model harboring dormant HSV-1 it was demonstrated that repetitive mechanical injury (i.e., a simulation of CTE) activates dormant HSV-1 and induces phenotypes associated with AD [271].

Specific signaling pathways connecting the above discussed conditions with the activation of one or more of eIF2α kinases remain to be elucidated. However, there are apparently at least two features that are common to all conditions listed above. One such feature is compromised blood–brain barrier, BBB [272]. BBB defects, in turn, allow the neuronal penetration of stressors potentially capable of eliciting the unconventional neuronal ISR. Another common feature is the suppression of the cerebral blood flow, CBE [273–279]. There are, potentially, numerous ways for CBE to trigger the elicitation of the neuronal ISR but one stands out because of its involvement in conventional AD. CBE is known to trigger mitochondrial dysfunction in neuronal cells [280,281]. This is sufficient to activate, via the OMA1 to DELE1 to HRI signaling pathway, the HRI [185,186], and thus to trigger the elicitation of the neuronal ISR and, consequently, unconventional AD [8]. The bottom line of the above considerations is that conventional and unconventional forms of AD differ only in the manner of elicitation of the neuronal ISR. Once the AβPP-independent C99/iAβ production pathway, activated by the neuronal ISR, becomes self-sustainable, conventional and unconventional forms of AD are mechanistically indistinguishable. This relationship between two forms of the disease is illustrated in Figure 23.

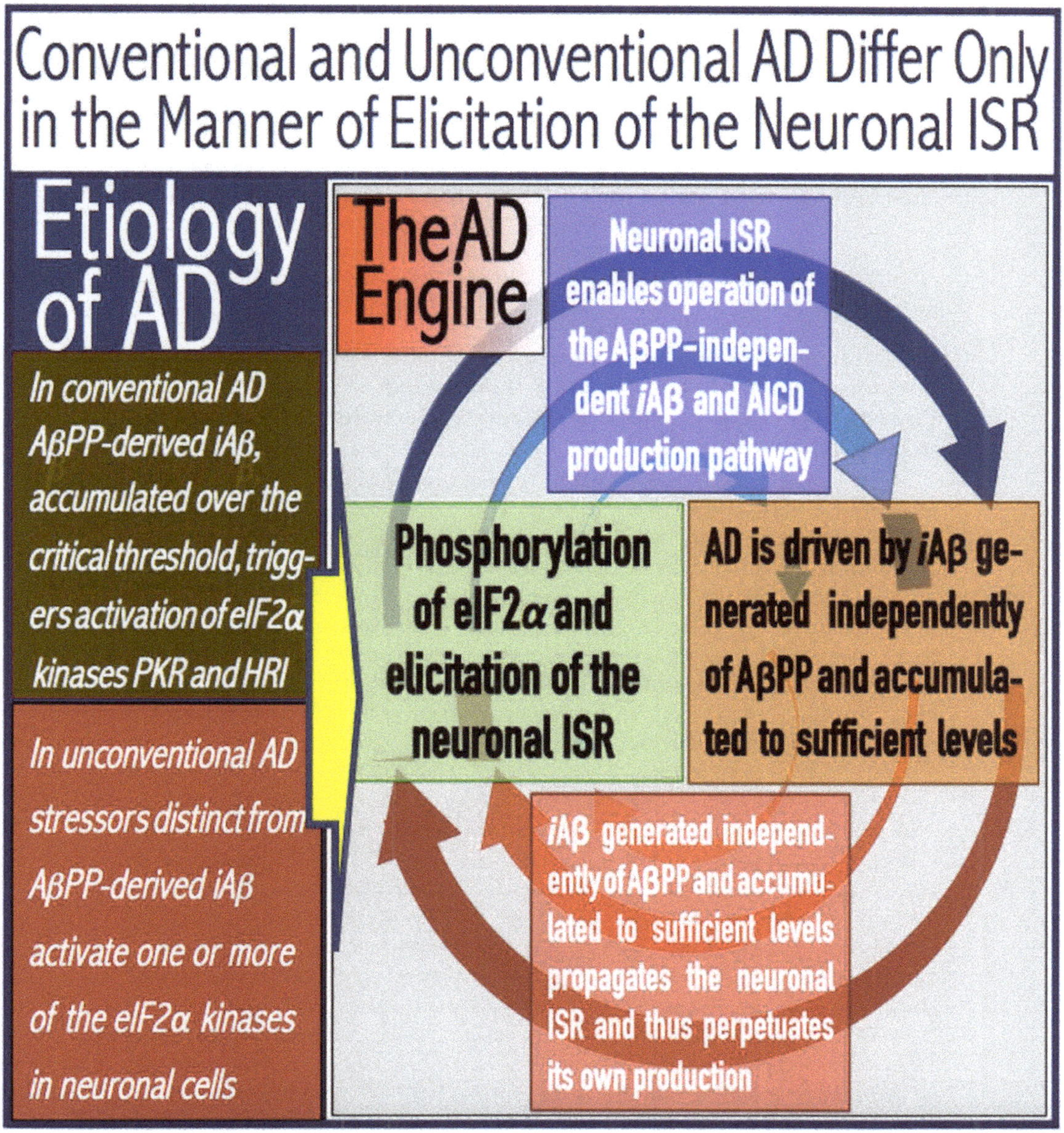

Figure 23. Alzheimer's is a multiform disease of the neuronal ISR: Conventional AD is triggered by AβPP-derived iAβ whereas neuronal ISR-eliciting stressors distinct from AβPP-derived iAβ initiate unconventional AD. *eIF2α*: eukaryotic translation initiation factor 2α. *PKR and HRI*: kinases that phosphorylate eIF2α at its Ser51 residue. *PACT*: PKR activator. *TNFα*: tumor necrosis factor α. *OMA1*: mitochondrial protease activated during mitochondrial dysfunction. *DELE1*: substrate of OMA1 protease. Cleavage of DELE1 leads to the activation of HRI. Phosphorylation of eIF2α elicits the integrated stress response, which, in turn, enables the generation of components essential for the activity of the AβPP-independent pathway of C99 and iAβ production. iAβ: intraneuronal Aβ. The accumulation of iAβ produced in both the AβPP proteolytic and AβPP-independent pathways is accompanied by a proportional accumulation of AICD, AβPP Intracellular Domain. AICD was shown to be capable of interfering with numerous processes involved in AD but its contribution to the disease

remains to be elucidated. AβPP-derived iAβ accumulates physiologically via the importation of extracellular Aβ and the retention of iAβ produced by gamma-cleavages on intracellular membranes in an exceedingly slow process. In most individuals it does not reach the PKR- and/or HRI-activating, the neuronal ISR-eliciting threshold within their lifetimes and no conventional AD occurs. When this threshold is crossed, the neuronal ISR is elicited and the AβPP-independent C99 (and, subsequently, iAβ) production pathway activated, and conventional AD ensues. In unconventional AD stressors that are distinct from AβPP-derived iAβ and capable of activating one or more of eIF2α kinases trigger the elicitation of the neuronal ISR. In both conventional and unconventional forms of AD the iAβ product of the AβPP-independent pathway of C99 and iAβ production drives AD pathology. It also propagates the neuronal ISR and thus perpetuates its own production in the AβPP-independent pathway. Thus, conventional and unconventional AD differ only in the manner of the elicitation of the neuronal integrated stress response.

51. Validation (11): Sustained Elicitation of the ISR by Unconventional Stressors Should Activate the AβPP-Independent C99 Production Pathway and Trigger Cellular AD Pathology, Including NFT Formation, in Human Neuronal Cells

The concept that sustained elicitation of the neuronal integrated stress response is necessary and sufficient to cause AD, more precisely unconventional AD, can be validated with human neuronal cells. Above, in Section 20, we argued that, as experiments by Choi and co-workers demonstrated [128], human neuronal cells constitute an adequate model of AD. They possess functional machinery underlying the operation of the AβPP-independent C99/iAβ production pathway. In the approach described in [128] this pathway is activated by the neuronal ISR elicited by exogenously expressed human iAβ accumulated over the T1 threshold. Its operation drives cellular AD pathology, which manifests in the formation of neurofibrillary tangles [128]. If sustained neuronal ISR is all it takes (i.e., is necessary and sufficient) to activate the AβPP-independent C99/iAβ generation pathway, the same outcome can be attained via the utilization of unconventional ISR-eliciting stressors (i.e., stressors distinct from iAβ) such as, for example, those causing mitochondrial dysfunction (and, consequently activating, via the OMA1 to DELE1 to HRI signaling pathway, the HRI kinase). If the concept of unconventional AD were valid, cellular AD pathology would develop and manifest as NFTs. Another anticipated reporter is N-terminal Met-C99 (C100); it would, as discussed in Section 23 above, report the activity of the AβPP-independent C99/iAβ production pathway. As a control, several unrelated stressors could be employed; this would rule out that the outcome is a side effect of a stressor rather than of the neuronal ISR. The same confirmation can be obtained by performing the experiment in the presence of an ISR inhibitor; this should abrogate the outcome (i.e., the occurrence of NFTs and of C100). Another control option is to perform the same experiment in mouse neuronal cells. As discussed above (Section 17), the AβPP-independent C99/iAβ production pathway is inoperative in mice. Therefore, neither NFTs nor C100 should be observed in mouse neuronal cells under the sustained neuronal ISR conditions. In Section 21 above we described the utilization of human neuronal cells as a model system for conventional AD. Positive outcomes of the experiment proposed in the present section would not only validate the concept formulated above but also establish human neuronal cells as an adequate model system for unconventional AD.

52. Unconventionally Activated AβPP-Independent C99/iAβ Production Pathway Becomes Self-Sustainable Only After a Lag Period

Two principal features differentiate unconventional AD from its conventional counterpart. One such feature was discussed above: in unconventional AD the elicitation of the neuronal ISR is triggered not by AβPP-derived iAβ (as in conventional AD) but rather by a variety of stressors that are explicitly not iAβ. Another distinguishing feature is that whereas in conventional AD the activation of the AβPP-independent C99/iAβ production pathway equates with the commencement of the disease, the unconventional activation, in fact, even multiple unconventional activations of this pathway could be transient and do not necessarily result in AD. This is because, in contrast to its conventionally activated counterpart, the unconventionally activated AβPP-independent C99/iAβ production pathway is initially not self-sustainable.

In conventional AD, at the instance of the activation of the AβPP-independent C99/iAβ production pathway levels of AβPP-derived iAβ have crossed the ISR-eliciting T1 threshold rendering the pathway self-sustainable. Indeed, iAβ produced independently of AβPP significantly elevates the total iAβ levels thus ensuring that it remains well over-T1. This maintains the activity of PKR and/or HRI kinase, propagates the neuronal ISR conditions and thus sustains the activity of the AβPP-independent C99/iAβ production pathway. The pathway is, therefore, self-sustainable from the instance of its activation. The withdrawal of the initial ISR-eliciting stressor (in this case AβPP-derived iAβ) would have no effect whatsoever on the operation of the pathway (this is the reason why ACH-based AD drugs are inapplicable in symptomatic AD).

On the other hand, the unconventional activation of the AβPP-independent C99/iAβ production pathway occurs always at the levels of AβPP-derived iAβ below the T1 threshold (otherwise the activation would occur conventionally). iAβ produced independently of AβPP would rapidly accumulate and eventually reach and cross the T1 threshold. At this point the AβPP-independent C99/Aβ production pathway would be rendered self-sustainable, i.e., sustained by its own product, which propagates the neuronal ISR state and thus maintains the activity of the pathway, and unconventional AD would commence. The withdrawal of the initial (unconventional) ISR-eliciting stress at this stage would have no effect on the continuous operation of the AβPP-independent C99/iAβ production pathway. If, however, the initial unconventional stressor would be withdrawn before the T1 threshold is reached, the ISR conditions would be reversed and the activity of the AβPP-independent C99/iAβ production pathway would cease; its operation would be only transient, and the disease would not commence. This aspect of the unconventional activation of the AβPP-independent C99/iAβ production pathway is further discussed in the following sections below.

53. Unconventional Activation of the AβPP-Independent C99/iAβ Production Pathway: Effect of the Long-Term Presence of Unconventional ISR-Eliciting Stressors

Figure 24 illustrates the effect of the long-term presence of an unconventional stressor capable of eliciting the integrated stress response in neuronal cells. Panel A of Figure 24 depicts the kinetics of accumulation of AβPP-derived iAβ in the healthy individual. The rate of the accumulation of iAβ is such that it does not reach the T1 threshold. eIF2α kinases are not activated, eIF2α is not phosphorylated, the neuronal ISR is not elicited, the AβPP-independent C99/iAβ production pathway is not initiated, and AD does not occur. In Panel B of Figure 24, the unconventional (i.e., distinct from AβPP-derived iAβ) stressor occurs when the levels of AβPP-derived iAβ are well below the T1 threshold and persists for the remaining portion of the lifetime. The neuronal ISR is unconventionally elicited, and the AβPP-independent C99/iAβ production pathway is unconventionally activated. iAβ generated independently of AβPP rapidly accumulates. When its levels cross the T1 threshold, the AβPP-independent C99/iAβ production pathway is rendered self-sustainable and unconventional AD commences.

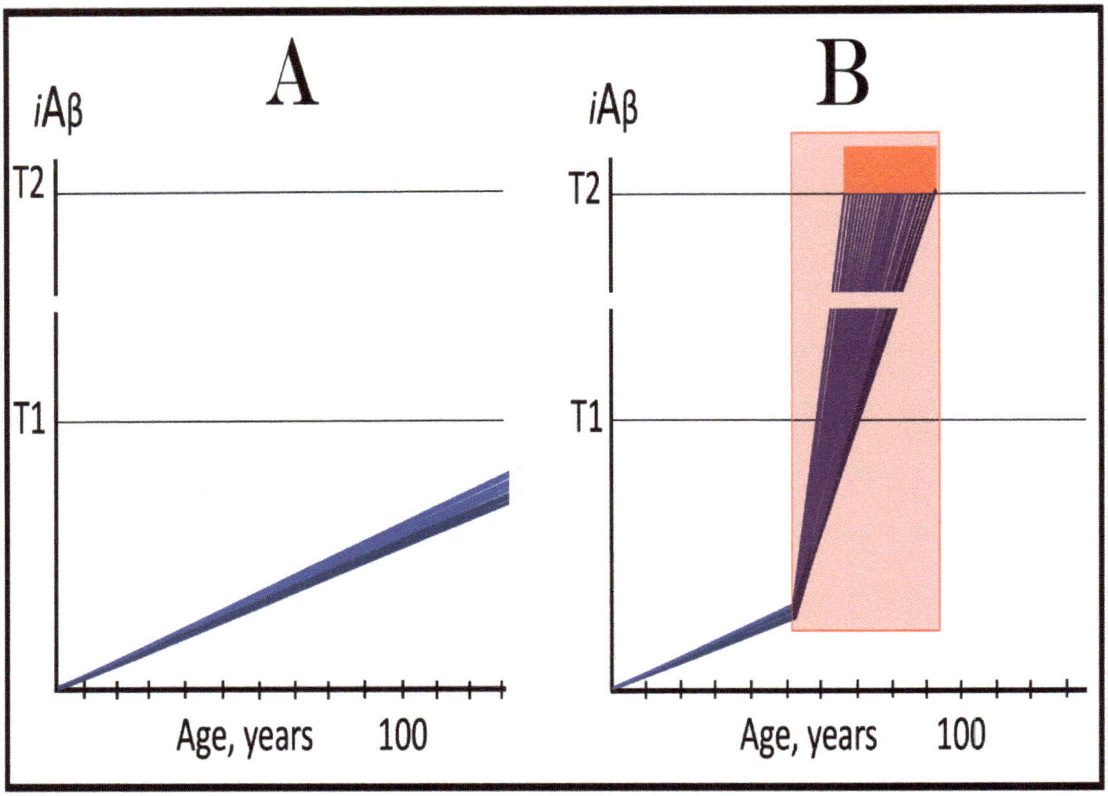

Figure 24. Unconventional activation of the AβPP-independent C99/iAβ production pathway: Effect of the long-term presence of unconventional ISR-eliciting stressors. *Blue lines*: Intraneuronal Aβ, iAβ. ***T1***: Threshold of cellular concentration of iAβ triggering the activation of PKR and/or HRI kinases, phosphorylation of eIF2α, the elicitation of the neuronal ISR, the initiation of the AβPP-independent C99 production pathway, and the commencement of AD. ***T2***: Cellular concentration of iAβ triggering neuronal death via apoptosis or necroptosis. *Red box*: Apoptotic zone, defined as a range of cellular concentrations of iAβ within which cells neuronal cells committed apoptosis or necroptosis or are dead. *Pink box*: Duration of the occurrence of the unconventional stressor capable of activating one or more of eIF2α kinases and triggering the elicitation of the neuronal ISR. *Panel* (**A**): Kinetics of the accumulation of AβPP-derived iAβ in the healthy individual. The rate of accumulation of iAβ is such that it does not reach the T1 threshold. eIF2α kinases are not activated, eIF2α is not phosphorylated, the neuronal ISR is not elicited, the AβPP-independent C99/iAβ production pathway is not initiated, and AD does not occur. *Panel* (**B**): Kinetics of the iAβ accumulation in the presence of the unconventional stressor. The unconventional (i.e., distinct from AβPP-derived iAβ) stressor occurs when the levels of AβPP-derived iAβ are well below the T1 threshold and persists for the remaining portion of the lifetime. The neuronal ISR is unconventionally elicited, and the AβPP-independent C99/iAβ production pathway is unconventionally activated. iAβ generated independently of AβPP rapidly accumulates. When its levels cross the T1 threshold, the AβPP-independent C99/iAβ production pathway is rendered self-sustainable and unconventional AD commences.

54. Unconventional Activation of the AβPP-Independent C99/iAβ Production Pathway: Effects of the Transient Presence of Unconventional ISR-Eliciting Stressors

Figure 25 considers the effects of the transient presence of the unconventional stressor capable of eliciting the neuronal ISR. The outcomes of the transient occurrence of unconventional stressors are multiple and defined by whether or not the levels of iAβ (produced mainly independently of AβPP) have crossed the T1 threshold at the time of withdrawal of the unconventional stressor. In Panel A of Figure 25, the unconventional stressor occurs when the levels of AβPP-derived iAβ are well below the T1 threshold. The rate of accumulation of the latter is such that it would not reach the T1 threshold within the individual's lifetime, and no AD would develop (projected accumulation of AβPP-derived iAβ in the absence of the unconventional stressor is shown as broken lines). Following the occurrence of the unconventional stressor, the neuronal ISR is unconventionally elicited and the AβPP-independent C99/iAβ production pathway unconventionally activated. The levels of iAβ produced independently of AβPP rapidly increase. But before they reach the T1 threshold, the unconventional stressor is withdrawn. The ISR conditions are reversed, and without the supply of its essential components the operation of the AβPP-independent C99/iAβ production pathway ceases. The accumulation of AβPP-derived iAβ resumes from an elevated baseline but proceeds at a low rate, supported only by the AβPP proteolysis. It does not reach the T1 threshold within the lifetime of the individual and no AD occurs.

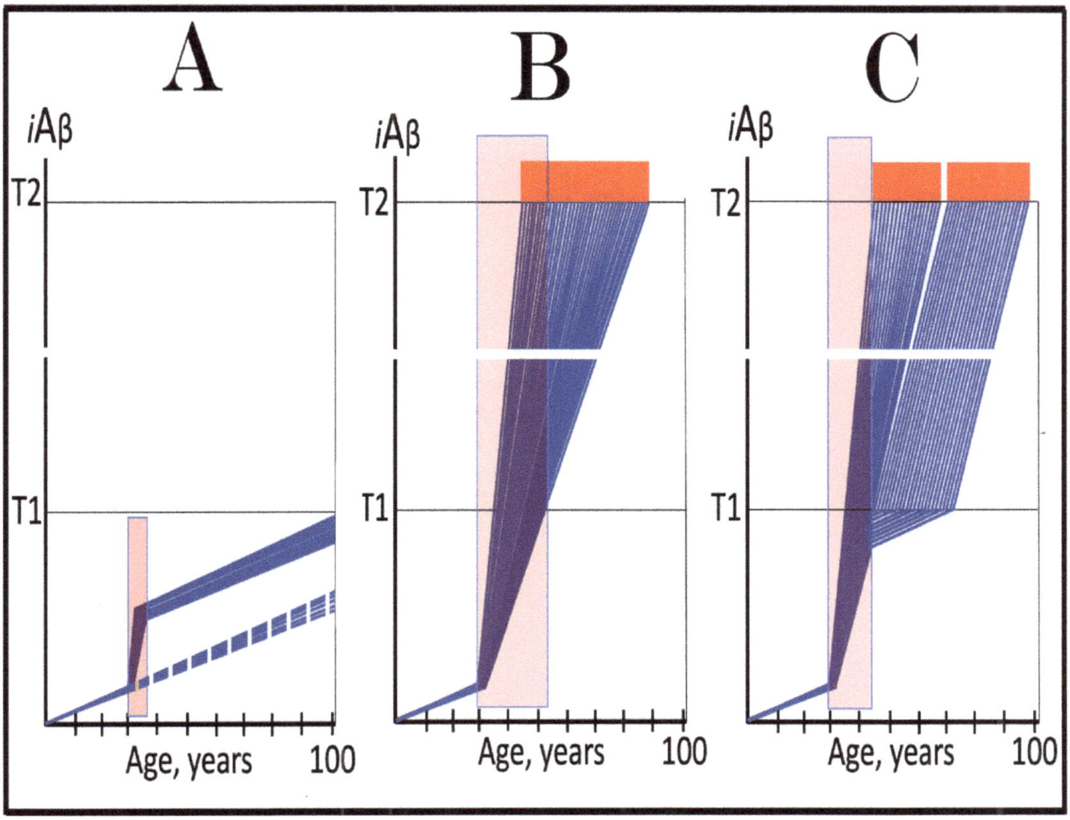

Figure 25. Unconventional activation of the AβPP-independent C99/iAβ production pathway: effects of the transient presence of unconventional ISR-eliciting stressors. *Blue lines*: Intraneuronal

Aβ, iAβ. ***T1***: Threshold of cellular concentration of iAβ triggering the activation of PKR and/or HRI kinases, phosphorylation of eIF2α, the elicitation of the neuronal ISR, the initiation of the AβPP-independent C99 production pathway, and the commencement of AD. ***T2***: Cellular concentration of iAβ triggering neuronal death via apoptosis or necroptosis. *Red box*: Apoptotic zone, defined as a range of cellular concentrations of iAβ within which cells neuronal cells committed apoptosis or necroptosis or are dead. *Pink boxes*: Duration of the occurrence of the unconventional stressor capable of activating one or more of eIF2α kinases and triggering the elicitation of the neuronal ISR. *Broken lines*: The projected dynamics of the accumulation of AβPP-derived iAβ in the absence of the episodes of unconventional activity of the AβPP-independent C99 production pathway. *Panel* (**A**): The unconventional stressor occurs when AβPP-derived iAβ is well below the T1 threshold and accumulates at such a rate that it would not reach the T1 threshold within the individual's lifetime (projected accumulation of AβPP-derived iAβ in the absence of the unconventional stressor is shown as broken lines). Following the occurrence of the unconventional stressor, the neuronal ISR is unconventionally elicited and the AβPP-independent C99/iAβ production pathway unconventionally activated. The levels of iAβ produced independently of AβPP rapidly increase. But before they reach the T1 threshold, the unconventional stressor is withdrawn. The ISR conditions are reversed, and the operation of the AβPP-independent C99/iAβ production pathway ceases. The accumulation of AβPP-derived iAβ resumes from an elevated baseline but proceeds at a low rate, supported only by the AβPP proteolysis. It does not reach the T1 threshold within the lifetime of the individual and no AD occurs. *Panel* (**B**): The unconventional stressor persists for a considerable duration. The neuronal ISR is unconventionally elicited, the AβPP-independent C99/iAβ production pathway is unconventionally activated, and iAβ produced independently of AβPP rapidly accumulates. When the unconventional stressor is withdrawn, iAβ has crossed the T1 threshold in all affected neurons. The AβPP-independent C99/iAβ production pathway has attained self-sustainability and the withdrawal of the initial unconventional stressor has no effect whatsoever on its continuous operation. *Panel* (**C**): The unconventional stressor is withdrawn when iAβ produced independently of AβPP has crossed the T1 threshold in only a fraction of affected neurons. In these neurons the AβPP-independent C99/iAβ production pathway has attained self-sustainability and its continuous operation is not affected by the withdrawal of the initial unconventional stressor. In the neurons that have not yet crossed the T1 threshold, the withdrawal of the unconventional stressor terminates the operation of the AβPP-independent C99/iAβ production pathway. The accumulation of AβPP-derived iAβ resumes and when it crosses the T1 threshold the neuronal ISR is elicited, the AβPP-independent C99/iAβ production pathway is activated, and the progression of AD pathology in these neurons commences.

In Panel B of Figure 25, the unconventional stressor occurs at the low levels of AβPP-derived iAβ and persists for a considerable duration. Following its occurrence, the neuronal ISR is unconventionally elicited, the AβPP-independent C99/iAβ production pathway is unconventionally activated, and iAβ produced independently of AβPP rapidly accumulates. When the unconventional stressor is withdrawn, the levels of iAβ have crossed the T1 threshold in all affected neurons. The AβPP-independent C99/iAβ production pathway has attained self-sustainability and the withdrawal of the initial unconventional stressor has no effect whatsoever on its continuous operation.

In Panel C of Figure 25, the unconventional stressor is withdrawn when iAβ produced independently of AβPP has crossed the T1 threshold in only a fraction of affected neurons. In these neurons the AβPP-independent C99/iAβ production pathway has attained self-sustainability and its continuous operation is not affected by the withdrawal of the initial unconventional stressor. In the neurons that have not yet crossed the T1 threshold, the withdrawal of the unconventional stressor terminates the operation of the AβPP-independent C99/iAβ production pathway. The accumulation of AβPP-derived iAβ resumes and when it crosses the T1 threshold the neuronal ISR is elicited, the AβPP-independent C99/iAβ production pathway is activated, and the progression of AD pathology in these neurons commences.

55. Numerous Transient Microbursts of Unconventional Activity of the AβPP-Independent C99/iAβ Production Pathway Potentially Contribute to the "Normal" Under-T1 Accumulation of iAβ; Conventional AD Always Contains an Unconventional Component

The present section develops further the concept illustrated in Panel A of Figure 25, namely that a burst of the unconventional activity of the AβPP-independent C99/iAβ production pathway occurring at AβPP-derived iAβ levels below the T1 threshold may have benign consequences and does not necessarily result in AD. Conceivably, such bursts, more appropriately "microbursts", of unconventional activity of the AβPP-independent C99/iAβ production pathway may occur dozens of times during the lifetime of an individual, reflecting numerous episodes of viral or bacterial infection and of transient inflammation events, occurring without causing AD (because levels of iAβ remain below the T1 threshold and the AβPP-independent C99/iAβ production pathway does not attain self-sustainability) and thus remaining undetected and unconsequential. Indeed, many, if not most, of the transient viral and bacterial infections may be accompanied by the transient microbursts of unconventional activity of the AβPP-independent C99/iAβ production pathway. When infection ends, the unconventional neuronal ISR-eliciting stressor is withdrawn and the activity of the AβPP-independent C99/iAβ production pathway ceases. It follows that the "normal" kinetics of accumulation of iAβ in the healthy individual are not uniformly linear (as depicted in Panel A of Figure 24) but composed of the linear portions of AβPP-derived iAβ accumulation interspersed by dozens of micro-steps, each resulting from the transient unconventional activation ("microburst") of the AβPP-independent C99/iAβ production pathway; as long as the levels of iAβ do not cross the T1 threshold, no AD occurs.

Above, in Section 24, it was stated that prior to reaching the T1 threshold iAβ accumulates physiologically in two ways. One is the retention of Aβ generated by gamma-cleavages occurring on intraneuronal membranes within various cellular organelles. Another way is the importation of secreted extracellular AβPP-derived Aβ. In view of the above considerations, we must add yet another way: the production of iAβ in microbursts of the activity of the AβPP-independent C99/iAβ generation pathway. This is not a physiological mechanism, but it is "normal" in that it potentially occurs routinely and repeatedly in every individual; it either remains uncosequential (if the T1 threshold is not crossed) or contributes to the occurrence of AD (if the T1 threshold is crossed).

The above notion is illustrated in Figure 26. Panel A of Figure 26 shows the under-T1 accumulation of iAβ in neurons of the healthy individual. Blue lines depict iAβ derived from AβPP whereas red lines show iAβ produced in microbursts independently of AβPP. Every time the unconventional ISR-eliciting stressor appears, and the burst occurs, the levels of iAβ are rapidly increased. Each burst concludes with the withdrawal of the unconventional stressor and the resumption of the slow accumulation of AβPP-derived iAβ from the elevated baseline. Despite numerous microbursts, the levels of iAβ do not reach the T1 threshold within the lifetime of the individual and no AD occurs.

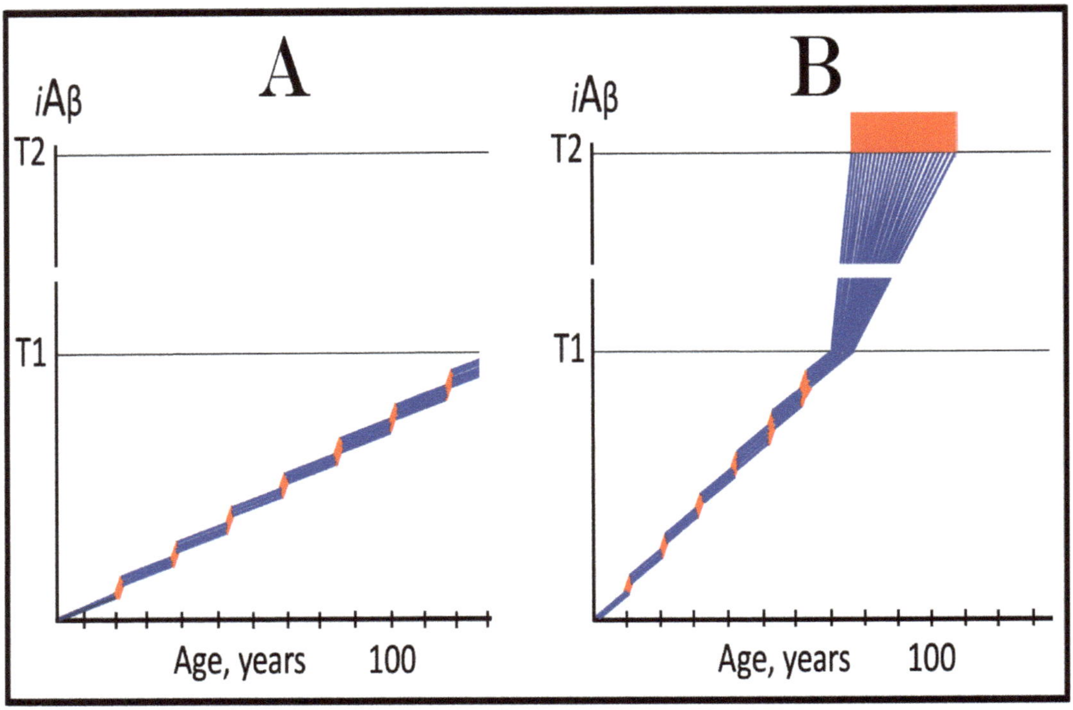

Figure 26. Microbursts of unconventional activity of the AβPP-independent C99/iAβ production pathway contribute to the normal Under-T1 accumulation of iAβ. *Blue lines*: Intraneuronal Aβ, iAβ. iAβ produced in the microbursts of the unconventional activity of the AβPP-independent C99/iAβ generation pathway. ***T1***: Threshold of cellular concentration of iAβ triggering the activation of PKR and/or HRI kinases, phosphorylation of eIF2α, the elicitation of the neuronal ISR, the initiation of the AβPP-independent C99 production pathway, and the commencement of AD. ***T2*** Cellular concentration of iAβ triggering neuronal death via apoptosis or necroptosis. *Red box* Apoptotic zone, defined as a range of cellular concentrations of iAβ within which cells neuronal cells committed apoptosis or necroptosis or are dead. *Panel* (**A**): The accumulation of iAβ in neurons of the healthy individual. Unconventional neuronal ISR-eliciting stressors appear for only a short duration but on numerous occasions. Every time their occurrence causes a microburst of the activity of the AβPP-independent C99/iAβ generation pathway and the levels of iAβ rapidly increase. Each burst concludes with the withdrawal of the unconventional stressor and the resumption of the slow accumulation of AβPP-derived iAβ from the elevated baseline. Despite numerous microbursts, the levels of iAβ do not reach the T1 threshold within the lifetime of the individual and no AD occurs. *Panel* (**B**): The rate of accumulation of AβPP-derived iAβ combined with the extents and number of microbursts generating iAβ independently of AβPP are such that the T1 threshold is reached and crossed. The neuronal ISR is sustainably elicited, the self-sustainable AβPP-independent C99/iAβ generation pathway is activated, and conventional AD commences and progresses. Thus, numerous microbursts of the unconventional activity of the AβPP-independent C99/iAβ production pathway potentially occur "normally", without adverse effects, in healthy individuals. On the other hand, conventional AD potentially always contains an unconventional component in the form of the microbursts of the unconventional activity of the AβPP-independent C99/iAβ generation pathway.

In Panel B of Figure 26, the rate of accumulation of AβPP-derived iAβ combined with the extents and number of microbursts generating iAβ independently of AβPP is such that the T1 threshold is reached and crossed. The neuronal ISR is sustainably elicited, the self-sustainable AβPP-independent C99/iAβ generation pathway is activated, and conventional AD commences and progresses. Thus, numerous microbursts of the unconventional activity of the AβPP-independent C99/iAβ production pathway potentially occur "normally", without adverse effects, in healthy individuals. On the other hand, conventional AD potentially always contains an unconventional component in the form of the contribution of iAβ generated in numerous microbursts of the unconventional activity of the AβPP-independent C99/iAβ generation pathway to reaching and crossing the T1 threshold.

56. Traumatic Brain Injury (TBI) and Chronic Traumatic Encephalopathy (CTE) as Causes of Unconventional AD

In Panel A of Figure 25, shown above, the duration of the operation of the unconventionally activated AβPP-independent C99/iAβ generation pathway is substantial yet AβPP-derived iAβ, accumulating from an elevated baseline following the withdrawal of the unconventional stressor, does not reach the T1 threshold within the lifetime of the individual, and no AD occurs. However, with a sufficient increase in the duration of the operation of the unconventionally activated AβPP-independent C99/iAβ generation pathway and consequent elevation of the baseline of the resumed accumulation of AβPP-derived iAβ (to levels still below the T1) (following the withdrawal of an unconventional stressor), the latter would inevitably cross the T1 threshold and AD would occur. In such a case the timing of the T1 crossing by AβPP-derived iAβ and of the commencement of AD would be inversely proportional to the duration of the operation of the unconventionally activated AβPP-independent C99/iAβ generation pathway.

The above scenario is illustrated in Panel A of Figure 27. The duration of the operation of the unconventionally activated AβPP-independent C99/iAβ generation pathway is such that upon its termination the levels of iAβ produced independently of AβPP are relatively close to but still below the T1 threshold. The accumulation of AβPP-derived iAβ resumes from this elevated baseline and relatively soon reaches and crosses the T1 threshold. PKR and/or HRI are activated, eIF2α is phosphorylated, the neuronal ISR is elicited, the self-sustainable AβPP-independent C99/iAβ generation pathway is initiated, and AD commences. In terms of the ACH2.0 this scenario describes how traumatic brain injury causes AD. TBI provides the unconventional neuronal ISR-eliciting stressor (probably via the suppression of CBF) and thus enables the unconventional activation and transient operation of the AβPP-independent C99/iAβ generation pathway. Upon withdrawal of the unconventional stressor, AβPP-derived iAβ resumes its accumulation from the elevated (but still under-T1) baseline and when the T1 is crossed, the disease commences. The time period between the TBI event and the commencement of AD is defined by the severity of TBI and, consequently, by the duration of the presence of the unconventional neuronal ISR-eliciting stressor.

Panel B of Figure 27 illustrates how chronic traumatic encephalopathy leads to AD. In CTE, the severity of trauma is substantially smaller than in TBI, but it does provide an unconventional neuronal ISR-eliciting stressor (probably by the same mechanism as TBI, namely the suppression of CBF). Consequently, the duration of the transient operation of the unconventionally activated AβPP-independent C99/iAβ generation pathway is shorter and the extent of the elevation of the baseline of the resumed accumulation of AβPP-derived iAβ is smaller. But in CTE traumatic events occur repeatedly (e.g., during boxing, football, hockey matches, or any other contact events), and after each event the baseline of the resumed accumulation of AβPP-derived iAβ is elevated further until, eventually and inevitably, the T1 threshold is crossed, the self-sustainable AβPP-independent C99/iAβ generation pathway is activated, and AD commences. Importantly, in both panels the disease is due solely to TBI and CTE and consequent episodes of the unconventional activity of the AβPP-independent C99 generation pathway; in their absence AβPP-derived iAβ would not reach (broken blue lines) the T1 threshold within the lifetime of the individuals and no AD would occur.

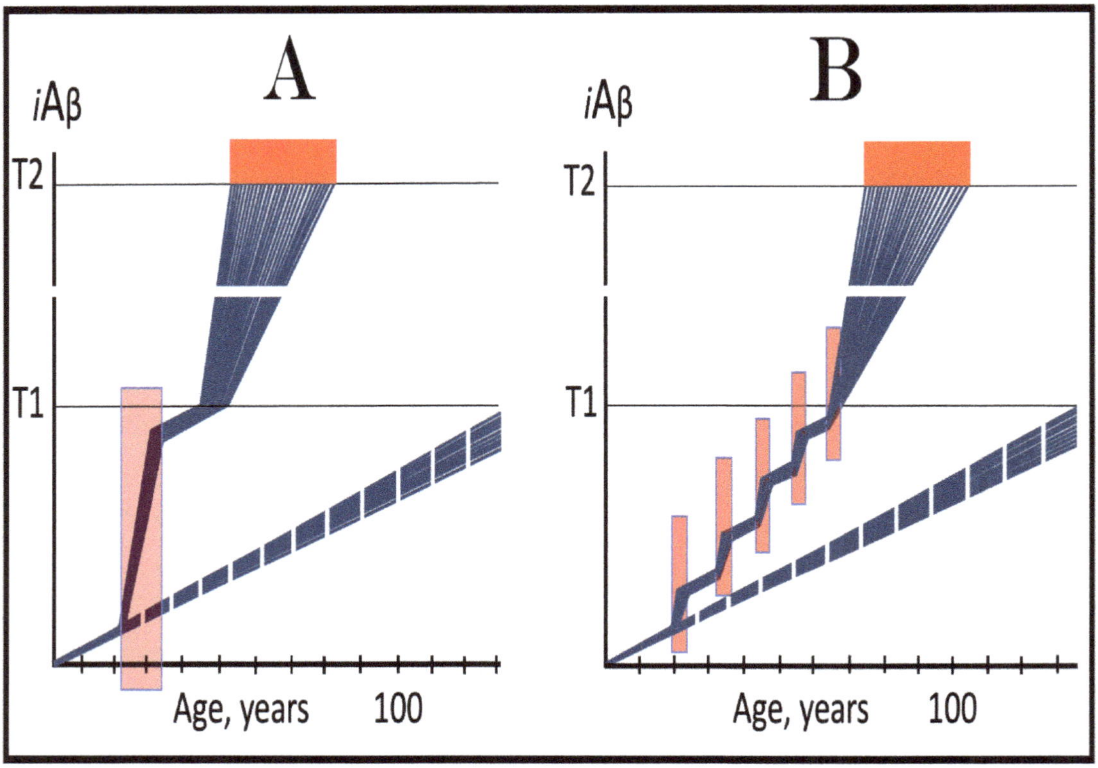

Figure 27. TBI and CTE cause unconventional AD via the unconventional transient activation of the AβPP-independent C99/iAβ generation pathway. *Blue lines*: Intraneuronal Aβ, iAβ. ***T1***: Threshold of cellular concentration of iAβ triggering the activation of PKR and/or HRI kinases, phosphorylation of eIF2α, the elicitation of the neuronal ISR, the initiation of the AβPP-independent C99 production pathway, and the commencement of AD. ***T2***: Cellular concentration of iAβ triggering neuronal death via apoptosis or necroptosis. *Red box*: Apoptotic zone, defined as a range of cellular concentrations of iAβ within which cells neuronal cells committed apoptosis or necroptosis or are dead. *Pink boxes*: Duration of the occurrence of the unconventional stressor capable of activating one or more of eIF2α kinases and triggering the elicitation of the neuronal ISR. *Broken lines*: The projected dynamics of the accumulation of AβPP-derived iAβ in the absence of the episodes of unconventional activity of the AβPP-independent C99 production pathway; no T1 would be crossed within the lifetime of the individual, no AD would occur. *Panel* (**A**): TBI provides, probably via the suppression of CBF, the unconventional neuronal ISR-eliciting stressor and thus enables the unconventional activation and transient operation of the AβPP-independent C99/iAβ generation pathway. Upon withdrawal of the unconventional stressor, AβPP-derived iAβ resumes its accumulation from the elevated (but still under-T1) baseline. When the T1 is crossed, PKR and/or HRI are activated, eIF2α is phosphorylated, the neuronal ISR is elicited, the self-sustainable AβPP-independent C99/iAβ generation pathway is initiated, and AD commences. The time period between the TBI event and the commencement of AD is defined by the severity of TBI and, consequently, by the duration of the presence of the unconventional neuronal ISR-eliciting stressor. *Panel* (**B**): In CTE, the severity of trauma is substantially smaller

than in TBI, but it does provide, potentially through the suppression of CBF, an unconventional neuronal ISR-eliciting stressor. The duration of the transient operation of the unconventionally activated AβPP-independent C99/iAβ generation pathway is shorter and the extent of the elevation of the baseline of the resumed accumulation of AβPP-derived iAβ is smaller in CTE than in TBI. But in the former traumatic events occur repeatedly, and after each event the baseline of the resumed accumulation of AβPP-derived iAβ is elevated further until, eventually, the T1 threshold is crossed, the self-sustainable AβPP-independent C99/iAβ generation pathway is activated, and AD commences.

57. ACH-Based Drugs Would Have No Effect Whatsoever in Unconventional AD

As discussed above (see Sections 36 and 37), in terms of the ACH2.0, the ACH-based AD drugs exert their potentially preventive effect (on conventional AD) by reducing the rate of accumulation of AβPP-derived iAβ or elevating the extent of the T1 threshold. This translates into a delay or prevention of the crossing of the T1 threshold by AβPP-derived iAβ. Consequently, no ISR is elicited, the AβPP-independent C99/iAβ production pathway is not activated, and no AD occurs. ACH-based drugs, however, have no effect on the AβPP-independent C99/iAβ production pathway and thus are completely ineffective when this pathway is operational. It follows that ACH-based drugs can have no effect whatsoever in unconventional AD. Indeed, the occurrence of unconventional ISR-eliciting stressors does not depend in any way on and can take place at any level of AβPP-derived iAβ. Modulating the latter would have no conceivable effect on the former. Moreover, when the AβPP-independent C99/iAβ production pathway is unconventionally activated, ACH-based drugs cannot interfere in any way with its operation. On the other hand, ACH2.0-based drugs can be effective in unconventional AD. Their anticipated effects in both the prevention and treatment of the disease are discussed in the following sections below.

58. ISR Inhibitors in the Prevention and Treatment of Unconventional AD

The present and following sections are concerned with the therapeutic strategies for unconventional AD. In these considerations, it is assumed that unconventional stressors capable of the elicitation of the neuronal ISR, once they appear, persist for the remaining lifetime (their withdrawal prior to the T1 crossing would render the disease, provided it eventually occurs, conventional, and their removal following the T1 crossing is inconsequential since the AβPP-independent C99/iAβ production pathway is self-sustainable at this point). Unconventional AD is driven by the self-sustainable AβPP-independent C99/iAβ production pathway, which is activated by the sustained neuronal ISR. The suppression of the neuronal integrated stress response is, therefore, an intuitive therapeutic strategy in both the prevention and treatment of the disease. Such therapeutic strategies are presented in Figure 28.

Panel A of Figure 28 depicts schematically the effects of inhibition of the neuronal integrated stress response in the prevention of unconventional AD. In this panel, the unconventional stressor has occurred at the levels of AβPP-derived iAβ well below the T1 threshold. The neuronal ISR has been unconventionally elicited and the AβPP-independent C99/iAβ production pathway unconventionally activated. iAβ levels have rapidly increased but have not yet reached the T1 threshold. The administration of ISR inhibitors at this point would reverse the ISR conditions, deprive the AβPP-independent C99/iAβ production pathway of its essential components and render it inoperative. The accumulation of iAβ would continue at a low rate, supported only by the AβPP proteolysis. The crossing of the T1 threshold would be inconsequential since no ISR can be elicited in the presence of the drug. The levels of iAβ would not reach the AD pathology-causing range and AD would not occur in treated individuals.

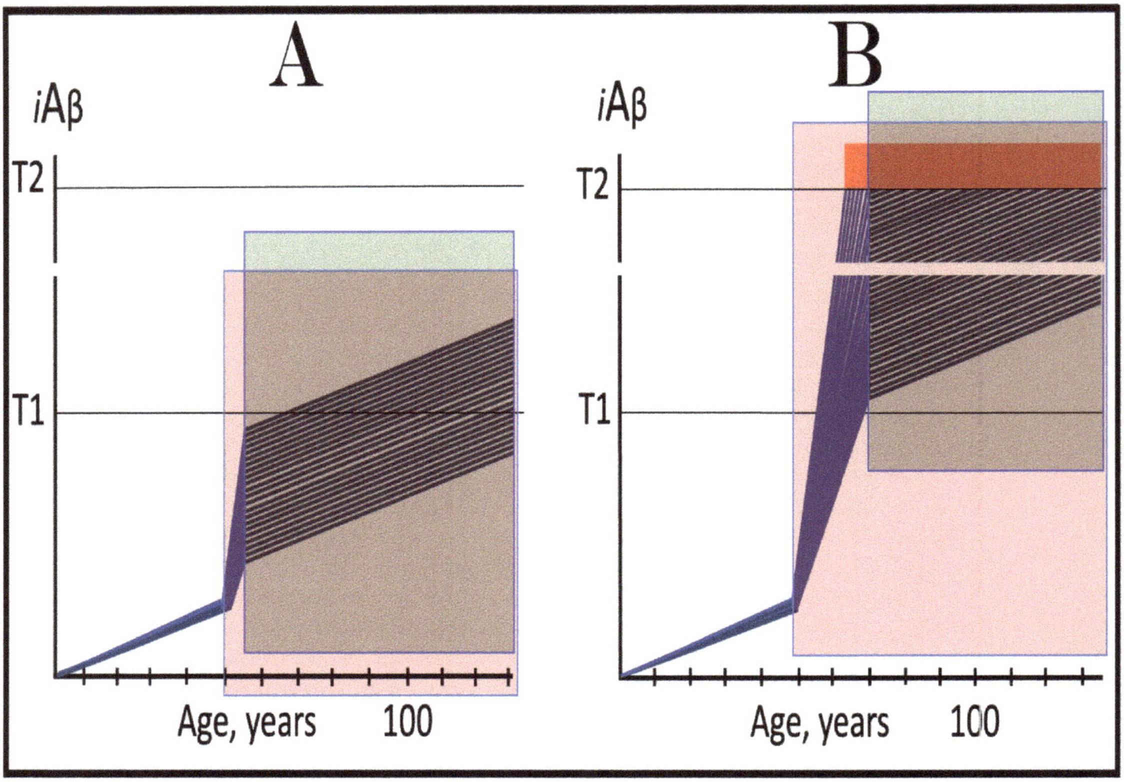

Figure 28. Inhibitors of the neuronal integrated stress response in the prevention and treatment of unconventional AD. *Blue lines*: Intraneuronal Aβ, iAβ. ***T1***: Threshold of cellular concentration of iAβ triggering the activation of PKR and/or HRI kinases, phosphorylation of eIF2α, the elicitation of the neuronal ISR, the initiation of the AβPP-independent C99 production pathway, and the commencement of AD. ***T2***: Cellular concentration of iAβ triggering neuronal death via apoptosis or necroptosis. *Red box*: Apoptotic zone, defined as a range of cellular concentrations of iAβ within which cells neuronal cells committed apoptosis or necroptosis or are dead. *Pink boxes*: The duration of the presence of the unconventional stressor capable of activating one or more of eIF2α kinases and triggering the elicitation of the neuronal ISR. *Green box*: The duration of the treatment with the ISR-inhibiting drug. *Panel* (**A**): ISR inhibitors in the prevention of unconventional AD. Following the occurrence of the unconventional stressor, the neuronal ISR has been unconventionally elicited and the AβPP-independent C99/iAβ production pathway unconventionally activated. iAβ levels have rapidly increased but have not yet reached the T1 threshold. The administration of ISR inhibitors at this point reverse the ISR conditions, deprive the AβPP-independent C99/iAβ production pathway of its essential components and render it inoperative. The accumulation of iAβ continues, but at a low rate, supported only by the AβPP proteolysis. The crossing of the T1 threshold would be inconsequential since no ISR can be elicited in the presence of the drug. The levels of iAβ would not reach the AD pathology-causing range and AD would not occur for the duration of the treatment. *Panel* (**B**): ISR inhibitors in the treatment of unconventional AD. Inhibition of the neuronal ISR deprives the AβPP-independent C99/iAβ production pathway of its essential components and its operation ceases. The rate of accumulation of iAβ, produced at this point only by the AβPP proteolysis, significantly decreases. The progression of the disease continues but at a much slower rate for the duration of the treatment.

In Panel B of Figure 28, the administration of the ISR suppressing drug commences when the T1 threshold has been crossed in all affected neurons; a fraction of the neurons has also crossed the T2 threshold, and AD symptoms have manifested. The administration of the ISR-inhibiting drug deprives the AβPP-independent C99/iAβ production pathway of its essential components, and its operation ceases. The rate of accumulation of iAβ, produced at this point only by the AβPP proteolysis, significantly decreases. Accordingly, the progression of the disease continues but at a much slower rate. Thus, both strategies, preventive and curative, would be beneficial. However, for the reasons discussed above (Section 43) they are, apparently, unfeasible due to anticipated adverse effects of systemically long-term administered ISR inhibitors. A solution would be either the development of neuron-specific ISR inhibitors or selective delivery of the drug to the neurons alone.

59. Activation of BACE1 and/or BACE2 in the Prevention of Unconventional Alzheimer's Disease

As discussed above (Section 46), it can be anticipated that transient activation of BACE1 and/or BACE2 would prevent conventional AD and AACD for the remaining portion of the lifetime of the treated individual. This is because, in the absence of the activity of the AβPP-independent C99/iAβ production pathway, iAβ is substantially depleted by activated BACE1 and/or BACE2. Following the withdrawal of the drug, the accumulation of AβPP-derived iAβ resumes from a low baseline and would reach neither T^0 nor T1 thresholds within the lifetime of the treated person. The situation in unconventional AD is fundamentally different in that the AβPP-independent C99/iAβ production pathway is operational even prior to the T1 crossing. Consequently, the rate of degradation of iAβ by activated BACE1 and/or BACE2 cannot match the rate of influx of iAβ produced independently of AβPP. The best possible outcome, therefore, is the reduction in the rate of accumulation of iAβ generated in the AβPP-independent C99/iAβ production pathway.

The above considerations are illustrated in Figure 29. Panel A of Figure 29 depicts the initial state of the levels of iAβ in individual neurons at the time of the commencement of the drug's administration. In these neurons, the occurrence of the unconventional stressor elicited the neuronal integrated stress response, which, in turn, provided components essential for the operation of the AβPP-independent C99/iAβ production pathway and thus activated the pathway. iAβ generated independently of AβPP rapidly accumulated but is still below the T1 threshold. Panel B of Figure 29 shows the evolution of the initial state in the untreated individual. The accumulation of iAβ produced independently of AβPP continues unhindered. It crosses the T1 threshold and the AβPP-independent C99/iAβ production pathway is rendered self-sustainable. AD commences and progresses until it reaches its end stage. Panel C of Figure 29 presents the evolution of the initial state in the individual treated with the BACE1 and/or BACE2-activating drug. In the presence of the drug the rate of the efflux of iAβ increases but it cannot match that of the influx of iAβ produced independently of AβPP. However, the rate of the accumulation of iAβ is reduced and the T1 crossing is delayed. Accordingly, the commencement of AD is also delayed, and its progression is slowed down for the duration of the treatment.

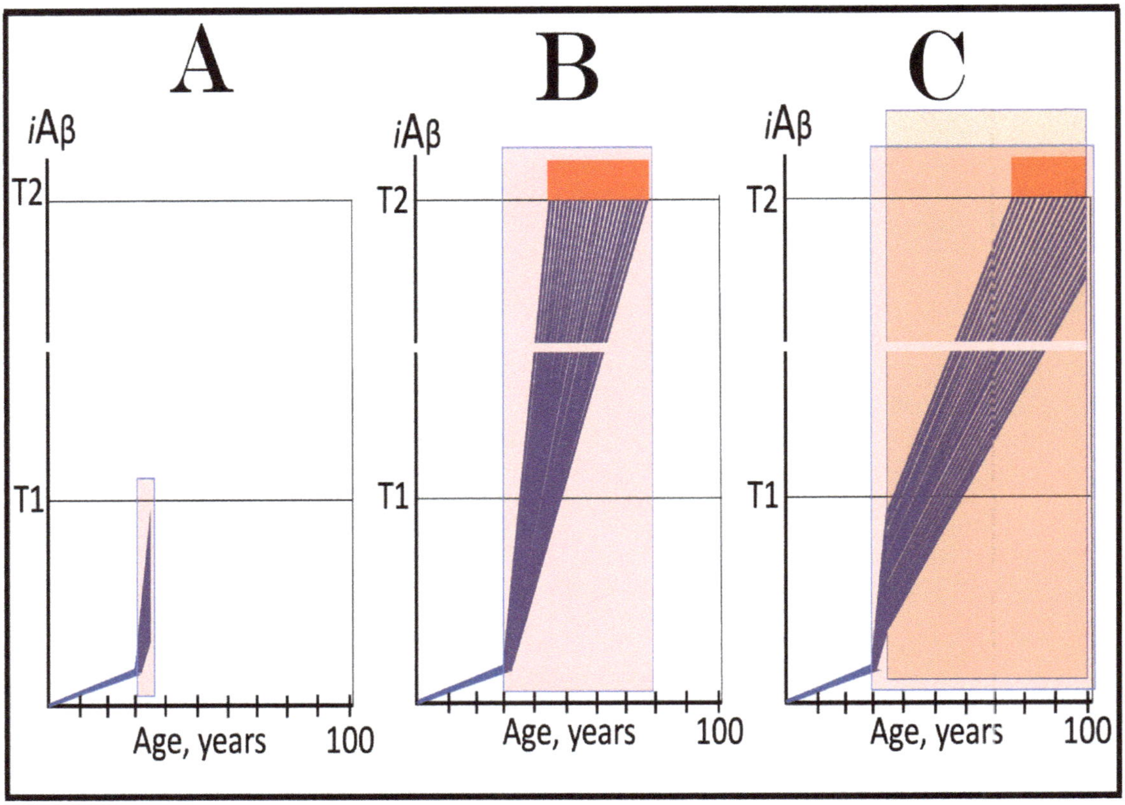

Figure 29. Activation of BACE1 and/or BACE2 in the prevention of unconventional AD. *Blue lines*: Intraneuronal Aβ, iAβ. ***T1***: Threshold of cellular concentration of iAβ triggering the activation of PKR and/or HRI kinases, phosphorylation of eIF2α, the elicitation of the neuronal ISR, the initiation of the AβPP-independent C99 production pathway, and the commencement of AD. ***T2***: Cellular concentration of iAβ triggering neuronal death via apoptosis or necroptosis. *Red box*: Apoptotic zone, defined as a range of cellular concentrations of iAβ within which cells neuronal cells committed apoptosis or necroptosis or are dead. *Pink boxes*: The duration of the presence of the unconventional stressor capable of activating one or more of eIF2α kinases and triggering the elicitation of the neuronal ISR. *Orange box*: The duration of the treatment with the BACE1 and/or BACE2-activating drug. *Panel* (**A**): The initial state of the levels of iAβ in individual neurons at the time of the commencement of the drug's administration. In these neurons, the occurrence of the unconventional stressor elicited the neuronal ISR, which, in turn, provided components essential for the operation of the AβPP-independent C99/iAβ production pathway and thus activated it. iAβ generated independently of AβPP rapidly accumulated but is still below the T1 threshold. *Panel* (**B**): The evolution of the initial state in the untreated individual. The accumulation of iAβ produced independently of AβPP continues unchecked. It crosses the T1 threshold and the AβPP-independent C99/iAβ production pathway is rendered self-sustainable. AD commences and progresses until it reaches its end stage. *Panel* (**C**): The evolution of the initial state in the individual treated with the BACE1 and/or BACE2-activating drug. In the presence of the drug, the rate of the efflux of iAβ increases but it cannot match that of the influx of iAβ produced independently of AβPP. However, the rate of the accumulation of iAβ is reduced and the T1 crossing is delayed. The commencement of AD is also delayed, and its progression is slowed down for the duration of the treatment.

60. Activation of BACE1 and/or BACE2 in the Treatment of Unconventional Alzheimer's Disease

Whereas the outcomes of the preventive application of activators of BACE1 and/or BACE2 differ radically in conventional and unconventional AD, they are conceptually similar in the treatment of both forms of the disease. This is because in both, at the time of treatment, the AβPP-independent C99/iAβ production pathway is operational (in the preventive application of BACE1 and/or BACE2-activating drugs this pathway is operational in unconventional AD but inoperative in the conventional form of the disease). Effects of BACE1 and/or BACE2-activating drugs in symptomatic unconventional AD are considered in Figure 30. Panel A of Figure 30 shows the initial state of the levels of iAβ at the commencement of drug's administration. At this stage, the neuronal integrated stress response has been unconventionally elicited and the AβPP-independent C99/iAβ production pathway unconventionally activated at the levels of AβPP-derived iAβ well below the T1 threshold. iAβ produced independently of AβPP has rapidly accumulated, crossed the T1 threshold in all affected neurons, and the AβPP-independent C99/iAβ production pathway was rendered self-sustainable. AD commenced and progressed; a fraction of the neurons has crossed the T2 threshold, and AD symptoms manifested. Panel B of Figure 30 depicts the evolution of the initial state in the untreated individual. iAβ produced independently of AβPP continues to accumulate unhindered, further neurons cross the T2 threshold, and the disease enters its end stage. Panel C of Figure 30 presents the evolution of the initial state in the presence of the BACE1 and/or BACE2-activating drug. iAβ is being degraded by the intra-iAβ cleaving activities of BACE1 and/or BACE2 but the rate of its efflux is significantly smaller than that of its influx. Its accumulation continues and the disease progresses, although at a decreased rate.

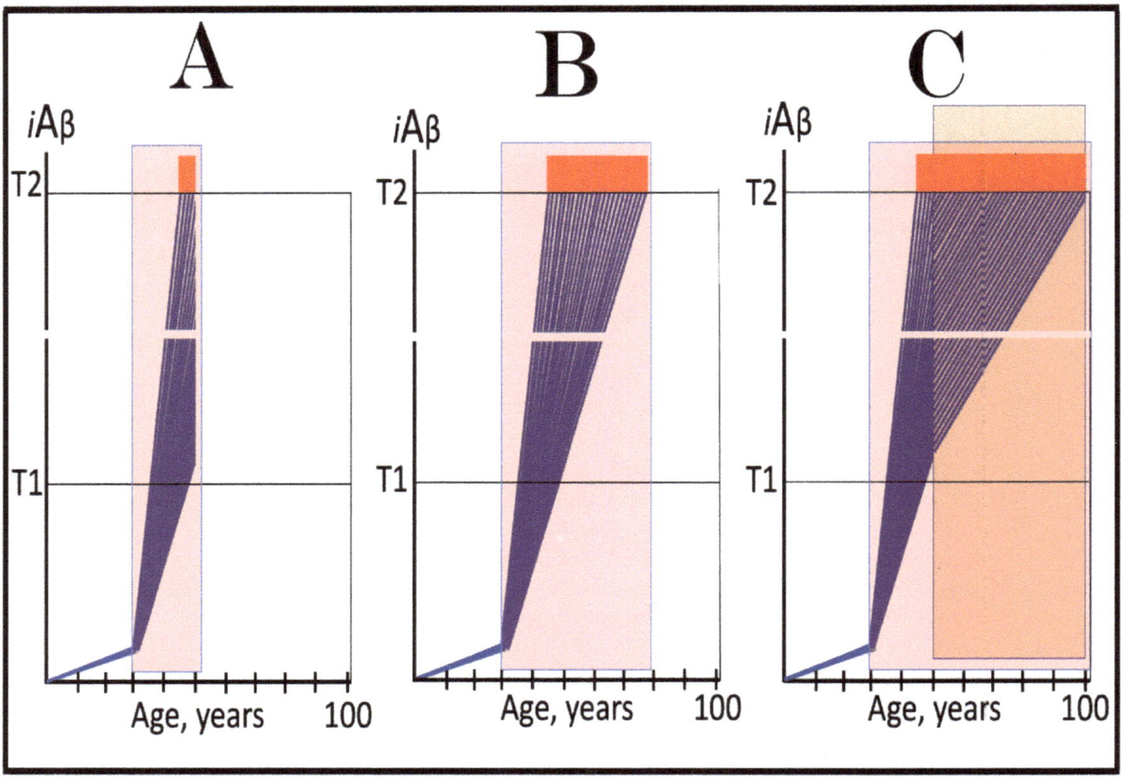

Figure 30. Activation of BACE1 and/or BACE2 in the treatment of unconventional AD. *Blue lines*: Intraneuronal Aβ, iAβ. ***T1***: Threshold of cellular concentration of iAβ triggering the activation of PKR and/or HRI kinases, phosphorylation of eIF2α, the elicitation of the neuronal ISR, the initiation of the AβPP-independent C99 production pathway, and the commencement of AD. ***T2***: Cellular concentration of iAβ triggering neuronal death via apoptosis or necroptosis. *Red box*: Apoptotic zone, defined as a range of cellular concentrations of iAβ within which cells neuronal cells committed apoptosis or necroptosis or are dead. *Pink boxes*: The duration of the presence of the unconventional stressor capable of activating one or more of eIF2α kinases and triggering the elicitation of the neuronal ISR. *Orange box*: The duration of the treatment with the BACE1 and/or BACE2-activating drug. *Panel* (**A**): The initial state of the levels of iAβ in individual neurons at the time of the commencement of the drug's administration. The neuronal ISR has been unconventionally elicited and the AβPP-independent C99/iAβ production pathway unconventionally activated at the levels of AβPP-derived iAβ well below the T1 threshold. iAβ produced independently of AβPP rapidly accumulated and crossed the T1 threshold in all affected neurons, and the AβPP-independent C99/iAβ production pathway was rendered self-sustainable. AD commenced and progressed; a fraction of the neurons crossed the T2 threshold, and AD symptoms manifested. *Panel* (**B**): The evolution of the initial state in the untreated individual. iAβ produced independently of AβPP continues to accumulate unhindered, additional neurons cross the T2 threshold, and the disease enters its end stage. *Panel* (**C**): The evolution of the initial state in the individual treated with the BACE1 and/or BACE2-activating drug. iAβ is being degraded by the intra-iAβ cleaving activities of BACE1 and/or BACE2 but the rate of its efflux is significantly smaller than that of its influx. Its accumulation continues and the disease progresses, although at a decreased rate, for the duration of the treatment.

61. Composite Transient ISR Inhibition/Long-Term BACE Activation Therapy in the Prevention and Treatment of Unconventional AD

The BACE1 and/or BACE2 activation therapy in the prevention and treatment of unconventional AD, considered in two preceding sections, is beneficial. However, it neither prevents the disease nor does it stop the progression of AD. On the other hand, this therapy can be significantly improved upon. One of the main reasons for its inefficiency is that it commences at the high rates of iAβ production (in the AβPP-independent pathway) and no depletion of iAβ takes place during the therapy. A solution for this is the composite therapy described above for the treatment of conventional AD (see Section 48). In this approach, inhibitors of the ISR are administered transiently and concurrently with activators of BACE1 and/or BACE2, also administered transiently. The former disable the AβPP-independent C99/iAβ production pathway and interrupt the influx of iAβ generated independently of AβPP. Consequently, the latter are capable of depleting iAβ substantially below the T1 threshold. When drugs are withdrawn, the de novo accumulation of iAβ commences from a low baseline and (since it is conventional AD) is supported only by the AβPP proteolysis. Its levels do not reach the T1 threshold, and the disease does not recur within the lifetime of the treated patient.

The same composite therapy can be applied in the prevention and treatment of unconventional AD. In this case, however, following the withdrawal of the ISR inhibitor (which can be applied only transiently to avoid adverse effects), the neuronal ISR would be re-elicited and the AβPP-independent C99/iAβ production pathway reactivated. Therefore, to reduce the rate of accumulation of iAβ generated independently of AβPP, it is advantageous to continue the activation of BACE1 and/or BACE2 long-term. Effects of such a strategy in the prevention and treatment of unconventional AD are illustrated in Figure 31. In Panel A of Figure 31, the neuronal ISR is unconventionally elicited and the AβPP-independent C99/iAβ production pathway unconventionally activated at the levels of AβPP-derived iAβ well below the T1 threshold. The levels of iAβ generated independently of AβPP rapidly increase but at the time of the commencement of the composite therapy do not yet reach the T1 threshold. The ISR inhibitor disables the AβPP-independent C99/iAβ production pathway and stops the influx of iAβ generated independently of AβPP. The concurrent activation of BACE1 and/or BACE2 substantially depletes iAβ. Following the withdrawal of the ISR inhibitor, the neuronal ISR is re-elicited and the AβPP-independent C99/iAβ production pathway reactivated. The continuous presence of the BACE activator reduces the rate of accumulation of iAβ produced independently of AβPP and its levels increase relatively slowly. As shown, they do not reach the AD pathology-causing range and AD does not occur for the duration of the treatment.

In Panel B of Figure 31, iAβ produced independently of AβPP has crossed the T1 threshold in all affected neurons. The AβPP-independent C99/iAβ production pathway has been rendered self-sustainable and AD has commenced and progressed. At the time of the commencement of the composite therapy a fraction of the neurons had crossed the T2 threshold and AD symptoms manifested. The presence of the ISR inhibitor abrogates the influx of iAβ produced independently of AβPP and allows the activator of BACE1 and/or BACE2 to substantially reduce iAβ levels. Following the withdrawal of the ISR inhibitor, the neuronal ISR is unconventionally re-elicited and the AβPP-independent C99/iAβ production pathway unconventionally reactivated. However, the continuous presence of the BACE1 and/or BACE2 activator maintains a relatively low rate of accumulation of iAβ. As depicted in Panel B, it does not reach the AD pathology-causing levels and AD does not recur for the duration of the treatment.

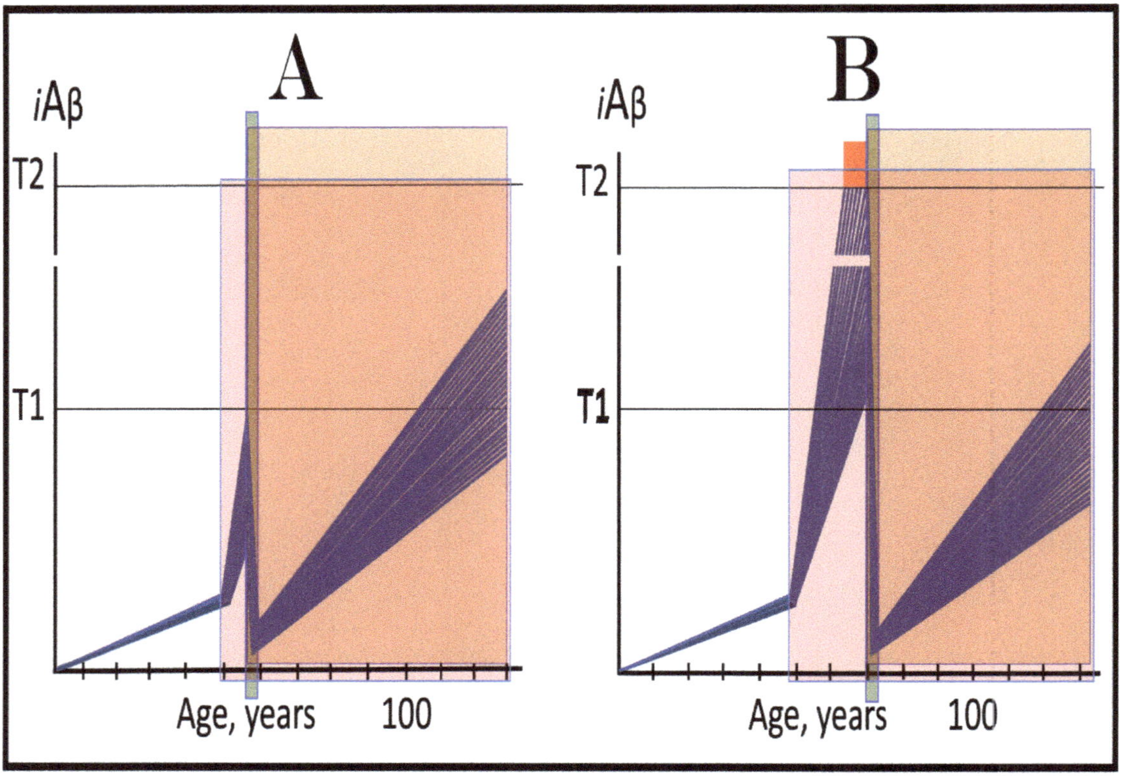

Figure 31. Composite transient ISR inhibition/long-term BACE activation therapy in the prevention and treatment of unconventional AD. *Blue lines*: Intraneuronal Aβ, iAβ. *T1*: Threshold of cellular concentration of iAβ triggering the activation of PKR and/or HRI kinases, phosphorylation of eIF2α, the elicitation of the neuronal ISR, the initiation of the AβPP-independent C99 production pathway, and the commencement of AD. *T2*: Cellular concentration of iAβ triggering neuronal death via apoptosis or necroptosis. *Red box*: Apoptotic zone, defined as a range of cellular concentrations of iAβ within which cells neuronal cells committed apoptosis or necroptosis or are dead. *Pink boxes*: The duration of the presence of the unconventional stressor capable of activating one or more of eIF2α kinases and triggering the elicitation of the neuronal ISR. *Green box*: The duration of the treatment with the ISR-inhibiting drug. *Orange box*: The duration of the treatment with the BACE1 and/or BACE2-activating drug. *Panel* (**A**): The neuronal ISR is unconventionally elicited and the AβPP-independent C99/iAβ production pathway unconventionally activated at the levels of AβPP-derived iAβ well below the T1 threshold. The levels of iAβ generated independently of AβPP rapidly increase but at the time of the commencement of the composite therapy do not yet reach the T1 threshold. The ISR inhibitor disables the AβPP-independent C99/iAβ production pathway and stops the influx of iAβ generated independently of AβPP. The concurrent activation of BACE1 and/or BACE2 substantially depletes iAβ. Following the withdrawal of the ISR inhibitor, the neuronal ISR is re-elicited and the AβPP-independent C99/iAβ production pathway reactivated. The continuous presence of the BACE activator reduces the rate of accumulation of iAβ produced independently of AβPP and its levels increase relatively slowly. They do not reach the AD pathology-causing range and AD does not occur for the duration of the treatment. *Panel* (**B**): iAβ produced independently of AβPP has crossed the T1 threshold in all affected neurons and the AβPP-independent C99/iAβ production pathway has been rendered self-sustainable. At the time of the commencement of the composite therapy a fraction of the neurons had crossed the T2 threshold and AD symptoms manifested. The presence of the ISR inhibitor abrogates the influx of iAβ produced independently of AβPP and allows

the activator of BACE1 and/or BACE2 to substantially reduce iAβ levels. Following the withdrawal of the ISR inhibitor, the neuronal ISR is unconventionally re-elicited and the AβPP-independent C99/iAβ production pathway unconventionally reactivated. However, the continuous presence of the BACE1 and/or BACE2 activator maintains a relatively low rate of accumulation of iAβ. It does not reach the AD pathology-causing levels and AD does not recur for the duration of the treatment.

62. Transient Composite ISR Inhibition/BACE Activation Therapy in the Prevention and Treatment of Unconventional AD

Life-long activation of the intra-iAβ cleavage at the β′-site confers only benefits and no adverse effects in the carriers of the Icelandic AβPP mutations. This indicates that the long-term activation of BACE1 and/or BACE2 would likely not be harmful. On the other hand, a relatively long-term inhibition of BACE1 in AD patients resulted in substantial adverse effects [27,28]. However, if the long-term utilization of BACE1 and/or BACE2 activators turns out to be problematic, their transient application as a part of the composite therapy would also be beneficial and would offer a viable therapeutic option. Therefore, the present and following sections consider effects of the composite transient ISR inhibition/BACE activation therapy in the prevention and treatment of unconventional AD.

These effects are illustrated schematically in Figure 32. Panel A of Figure 32 depicts the anticipated effect of the concurrent transient administration of both, the ISR inhibitor and BACE1 and/or BACE2 activator in the prevention of unconventional AD. In this panel, the neuronal integrated stress response has been unconventionally elicited and the AβPP-independent C99/iAβ production pathway unconventionally activated at the levels of AβPP-derived iAβ well below the T1 threshold. iAβ generated independently of AβPP has rapidly accumulated but at the time of the implementation of the composite therapy is still below the T1 threshold. The treatment with the ISR inhibitor disables the AβPP-independent C99/iAβ production pathway and terminates the influx of its iAβ product. This empowers a substantial depletion of iAβ by concurrently activated BACE1 and/or BACE2. When both drugs, the ISR inhibitor and BACE activator, are withdrawn, the neuronal IST is unconventionally re-elicited and the AβPP-independent C99/iAβ production pathway unconventionally reactivated. The accumulation of iAβ resumes from a low baseline but is unimpeded and proceeds rapidly. When it crosses the T1 threshold the AβPP-independent C99/iAβ production pathway is rendered self-sustainable, AD commences and progresses, eventually reaching its end stage. The composite therapy does not prevent the disease but provides a reprieve of several AD-free years.

Panel B of Figure 32 depicts the anticipated effect of the concurrent transient administration of the ISR inhibitor and BACE1 and/or BACE2 activator in the treatment of unconventional AD. In this panel, the neuronal integrated stress response has been unconventionally elicited and the AβPP-independent C99/iAβ production pathway unconventionally activated at the low levels of AβPP-derived iAβ. iAβ generated independently of AβPP has rapidly accumulated and crossed the T1 threshold in all affected neurons. The AβPP-independent C99/iAβ production pathway has been rendered self-sustainable and AD commenced. At the time of the implementation of the composite therapy a fraction of the neurons has crossed the T2 threshold and AD symptoms manifested. Inhibition of the neuronal ISR deprives the AβPP-independent C99/iAβ production pathway of its essential components, disables it, and stops the influx of iAβ generated independently of AβPP. The concurrent activation of BACE1 and/or BACE2 substantially depletes iAβ. Upon the completion of the treatment and the withdrawal of the drugs the neuronal ISR is re-elicited and the AβPP-independent C99/iAβ production pathway reactivated. De novo accumulation of iAβ commences from a low baseline but proceeds unimpeded and rapidly. When it reaches the T1 threshold, the disease recurs. Thus, the transient composite therapy does not prevent the recurrence of AD; it buys, however, a few disease-free years.

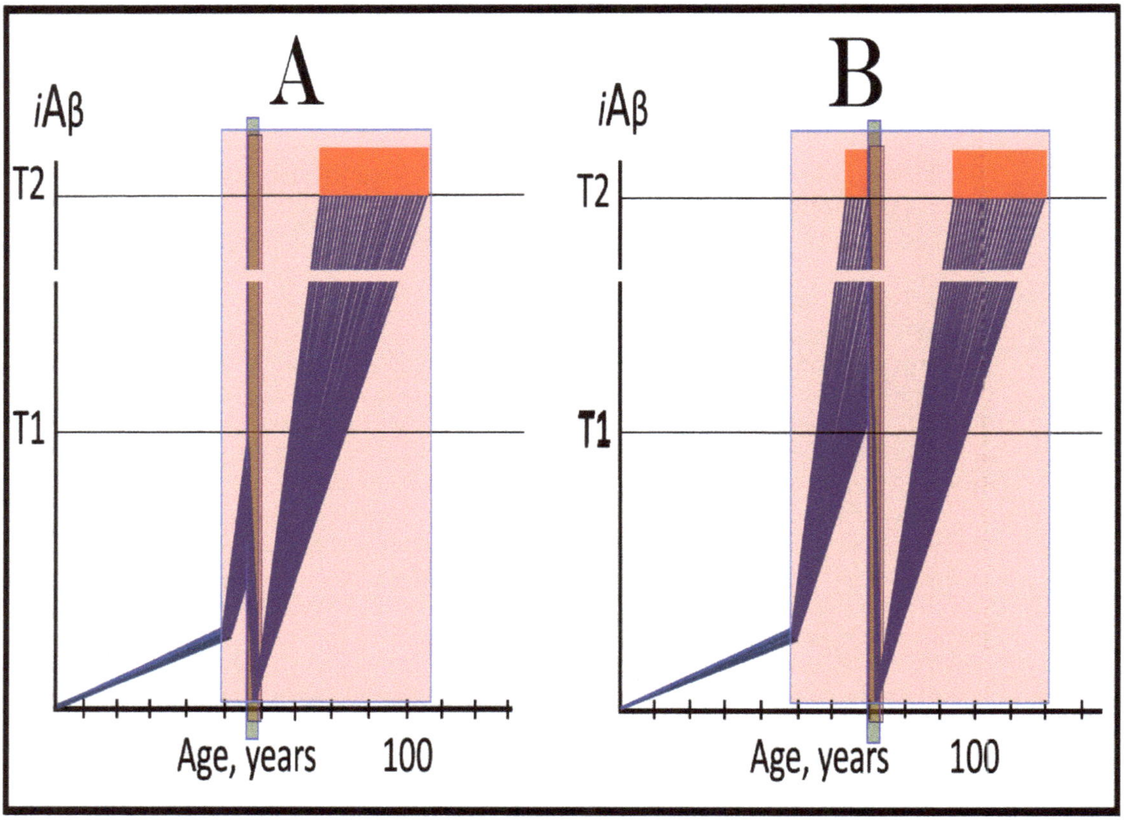

Figure 32. Transient composite ISR inhibition/BACE activation therapy in the prevention and treatment of unconventional AD. *Blue lines*: Intraneuronal Aβ, iAβ. ***T1***: Threshold of cellular concentration of iAβ triggering the activation of PKR and/or HRI kinases, phosphorylation of eIF2α, the elicitation of the neuronal ISR, the initiation of the AβPP-independent C99 production pathway, and the commencement of AD. ***T2***: Cellular concentration of iAβ triggering neuronal death via apoptosis or necroptosis. *Red box*: Apoptotic zone, defined as a range of cellular concentrations of iAβ within which cells neuronal cells committed apoptosis or necroptosis or are dead. *Pink boxes*: The duration of the presence of the unconventional stressor capable of activating one or more of eIF2α kinases and triggering the elicitation of the neuronal ISR. *Green box*: The duration of the treatment with the ISR-inhibiting drug. *Orange box*: The duration of the treatment with the BACE1 and/or BACE2-activating drug. *Panel* (**A**): The effect of the transient concurrent ISR inhibition and BACE1 and/or BACE2 activation in the prevention of unconventional AD. The neuronal ISR has been unconventionally elicited and the AβPP-independent C99/iAβ production pathway unconventionally activated at the levels of AβPP-derived iAβ well below the T1 threshold. iAβ generated independently of AβPP has rapidly accumulated but at the time of the implementation of the composite therapy is still below the T1 threshold. The treatment with the ISR inhibitor disables the AβPP-independent C99/iAβ production pathway and terminates the influx of its iAβ product. This enables a substantial depletion of iAβ by concurrently activated BACE. When both drugs are withdrawn, the neuronal IST is unconventionally re-elicited and the AβPP-independent C99/iAβ production pathway unconventionally reactivated. The accumulation of iAβ resumes from a low baseline but proceeds rapidly. When it crosses the T1 threshold the AβPP-independent C99/iAβ production pathway is rendered self-sustainable and AD commences. The composite therapy does not prevent the disease but provides a short reprieve. *Panel* (**B**): The effect of the transient concurrent ISR inhibition and BACE1 and/or BACE2 activation in

the treatment of unconventional AD. At the time of the implementation of the composite therapy all affected neurons have crossed the T1 threshold, a fraction of the neurons has crossed the T2 threshold, and AD symptoms manifested. Inhibition of the neuronal ISR deprives the AβPP-independent C99/iAβ production pathway of its essential components, disables it, and stops the influx of iAβ generated independently of AβPP. The concurrent activation of BACE1 and/or BACE2 substantially depletes iAβ. Upon the completion of the treatment and the withdrawal of the drugs the neuronal ISR is re-elicited and the AβPP-independent C99/iAβ production pathway reactivated. De novo accumulation of iAβ commences from a low baseline but proceeds rapidly. When it reaches the T1 threshold, the disease recurs.

63. Recurrent Transient Composite ISR Inhibition/BACE Activation Therapy in the Prevention and Treatment of Unconventional AD

As described in the preceding section, the transient composite therapy consisting of the concurrent administration of the inhibitors of the integrated stress response and activators of BACE1 and/or BACE2 can neither prevent the occurrence of unconventional AD nor preclude its recurrence following the treatment. In both cases, however, the implementation of this therapy provides a temporary reprieve of several disease-free years. It follows that if, at strategically timed intervals defined by suitable biomarkers, the therapy would be repeated, the duration of the reprieve would be multiplied proportionally to the number of repeats. Potentially, the reprieve could be extended for the remaining portion of the lifetime of the treated individual.

The above strategy is illustrated in Figure 33. Panel A of Figure 33 depicts the implementation of the recurrent transient composite ISR inhibition/BACE activation therapy in the prevention of unconventional AD. In this panel, the neuronal integrated stress response has been unconventionally elicited and the AβPP-independent C99/iAβ production pathway unconventionally activated at the low levels of AβPP-derived iAβ. iAβ generated independently of AβPP has rapidly accumulated but at the time of the implementation of the therapy has not yet reached the T1 threshold. The concurrent administration of ISR inhibitors and BACE activators disables the AβPP-independent C99/iAβ production pathway and substantially depletes iAβ. Upon the completion of the treatment the neuronal ISR is unconventionally re-elicited and the AβPP-independent C99/iAβ production pathway unconventionally reactivated. The accumulation of iAβ produced independently of AβPP resumes but from a low baseline. Before it reaches the T1 threshold, at a point indicated by suitable biomarkers, the composite therapy is repeated recurrently. The T1 threshold is not crossed, and the disease does not occur for the duration of the treatments.

Panel B of Figure 33 depicts the implementation of the recurrent transient composite ISR inhibition/BACE activation therapy in the treatment of unconventional AD. The first round of the transient composite therapy is administered when iAβ produced independently of AβPP have already crossed the T1 threshold in all affected neurons, the AβPP-independent C99/iAβ production pathway has been rendered self-sustainable, some neurons have crossed the T2 threshold, and AD symptoms have already manifested. The concurrent transient inhibition of the neuronal ISR and activation of BACE1 and/or BACE2 disable the AβPP-independent C99/iAβ production pathway, stop the influx of iAβ generated independently of AβPP, and substantially deplete iAβ. When drugs are withdrawn, the neuronal ISR is unconventionally re-elicited and the AβPP-independent C99/iAβ production pathway unconventionally reactivated. The accumulation of iAβ resumes from a low baseline. Before it reaches the AD pathology-causing levels, at a point defined by the appropriate biomarkers, the composite therapy is re-administered and repeated recurrently as needed. The AD pathology-causing range of iAβ concentrations is not reached and the disease does not recur for the duration of the treatments. On the alternative approach, effectively converting unconventional AD into the conventional form of the disease, see Section 78 below.

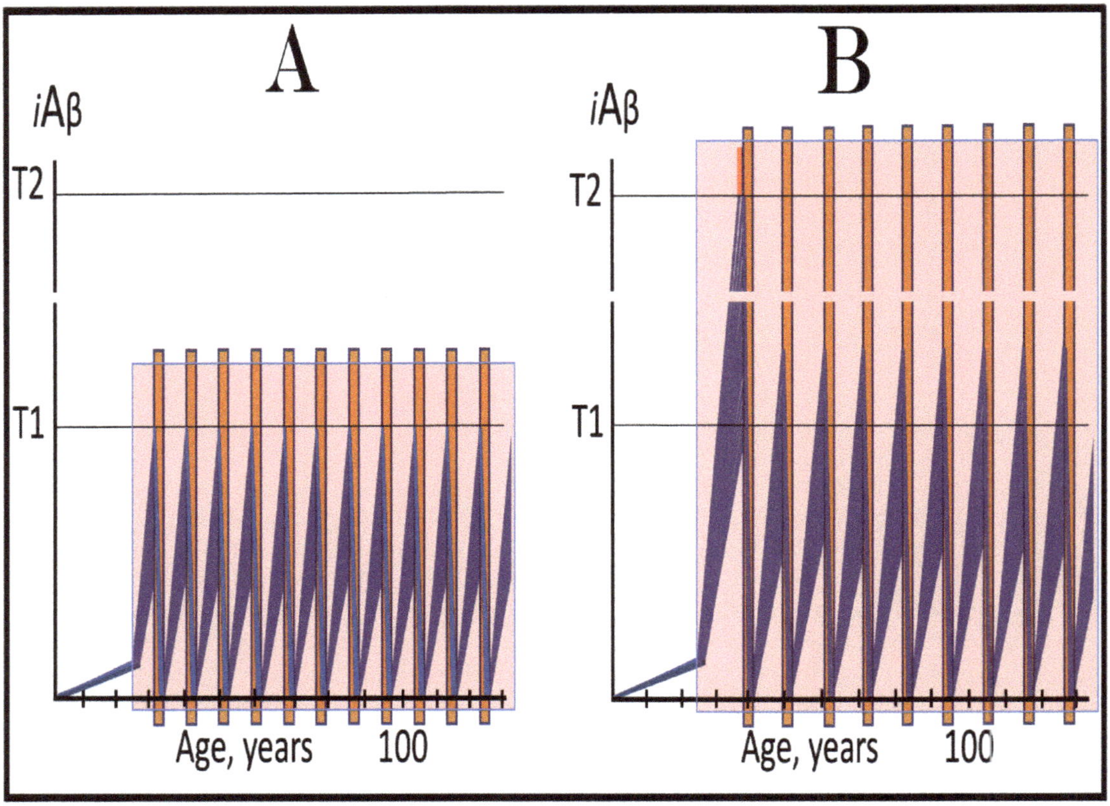

Figure 33. Recurrent transient composite ISR inhibition/BACE activation therapy in the prevention and treatment of unconventional AD. *Blue lines*: Intraneuronal Aβ, iAβ. ***T1***: Threshold of cellular concentration of iAβ triggering the activation of PKR and/or HRI kinases, phosphorylation of eIF2α, the elicitation of the neuronal ISR, the initiation of the AβPP-independent C99 production pathway, and the commencement of AD. ***T2***: Cellular concentration of iAβ triggering neuronal death via apoptosis or necroptosis. *Red box*: Apoptotic zone, defined as a range of cellular concentrations of iAβ within which cells neuronal cells committed apoptosis or necroptosis or are dead. *Pink boxes*: The duration of the presence of the unconventional stressor capable of activating one or more of eIF2α kinases and triggering the elicitation of the neuronal ISR. *Brown boxes*: The duration of the composite treatment with concurrently administered ISR inhibitors and BACE activators. *Panel* (**A**): Recurrent administration of the composite therapy in the prevention of unconventional AD. At the time of the implementation of the therapy the neuronal ISR has been unconventionally elicited and the AβPP-independent C99/iAβ production pathway unconventionally activated at the low levels of AβPP-derived iAβ. iAβ generated independently of AβPP has rapidly accumulated but has not yet reached the T1 threshold. The concurrent administration of ISR inhibitors and BACE activators disables the AβPP-independent C99/iAβ production pathway and substantially depletes iAβ. Upon the completion of the treatment, the neuronal ISR is unconventionally re-elicited and the AβPP-independent C99/iAβ production pathway unconventionally reactivated. The accumulation of iAβ produced independently of AβPP resumes but from a low baseline. Before it reaches the T1 threshold, at a point indicated by suitable biomarkers, the composite therapy is repeated recurrently. The T1 threshold is not crossed, and the disease does not occur for the duration of the treatments. *Panel* (**B**): Recurrent administration of the composite therapy in the prevention of unconventional AD. The first round of the transient composite therapy is administered when iAβ produced independently of AβPP have already crossed the T1 threshold in all affected neurons, the AβPP-independent C99/iAβ production pathway has been rendered self-sustainable, some neurons have crossed the T2 threshold, and AD symptoms have

already manifested. The concurrent transient inhibition of the neuronal ISR and activation of BACE1 and/or BACE2 disable the AβPP-independent C99/iAβ production pathway, stop the influx of iAβ generated independently of AβPP, and substantially deplete iAβ. When drugs are withdrawn, the neuronal ISR is unconventionally re-elicited and the AβPP-independent C99/iAβ production pathway unconventionally reactivated. The accumulation of iAβ resumes from a low baseline. Before it reaches the AD pathology-causing levels, at a point defined by the appropriate biomarkers, the composite therapy is re-administered and repeated recurrently as needed. The AD pathology-causing range of iAβ concentrations is not reached and the disease does not recur for the duration of the treatment.

64. On the Difficulties of Overcoming a Dogma: Ingrained Notions Are "Set in Stone"; Quantum Leaps as Breaks Through the Inertia of Preexisting Assumptions

The following sections outline the third version of the ACH2.0 theory of AD. The transition from the initial version of ACH2.0 to its second version was conceptually fundamental but logically trivial. Yet it took considerable time to make the leap from one to another. Initially, the ACH2.0 posited that AβPP-derived iAβ triggers AD by eliciting the neuronal integrated stress response. The ISR, in turn, enables the AβPP-independent C99/iAβ production pathway, and it is iAβ generated independently of AβPP, which drives AD pathology. Thus, the neuronal ISR initiates AD. It follows trivially that ANY stressor (other than AβPP-derived iAβ) capable of eliciting the neuronal ISR would trigger the disease (i.e., unconventional AD). The difficulty in formulating this concept was the ingrained notion that AD is triggered and driven by Aβ. An ingrained notion forms, apparently, a durable neural circuit, an equivalent of being "set in stone", and the inertia of preexisting assumptions is very difficult to overcome; it involves the necessity to disassemble and replace a durable neural circuit. This is why quantum leaps, at least some of them, can be regarded as breaking through the inertia of the ingrained notion. This is also why newcomers to a field are disproportionately successful in executing quantum jumps: they are blessed by ignorance.

The breakthrough from Version Two to Version Three of the ACH2.0, which we are about to discuss in the following sections, is neither conceptually less fundamental nor less trivial than the shift between the first two versions. Moreover, the nature of the challenge of formulating it is the same as encountered in the first shift: the presumed role of Aβ in AD. The second shift stems from our understanding of effects of the integrated stress response. The ISR fundamentally restructures the translational (and transcriptional) landscape of the cell. This entails the global suppression of the cellular protein synthesis. Translation of the decisive bulk of cellular protein species is inhibited and there is no reason to assume that the constituents of the AβPP proteolytic pathway, i.e., AβPP, BACE enzymes, components of the gamma-secretase complex, and, consequently, Aβ are exempted from this process. Moreover, as discussed in the following sections below, this notion is supported by the empirical data. The difficulty in making this leap and the reason that it (again) took a considerable time to implement it is that it (again) downsizes the role of Aβ in AD, in fact, eliminates Aβ, which epitomizes AD in the previously ingrained notion, as the driver of the disease. However, both shifts, from the first version of the ACH2.0 to the second and from the second to the third are dictated by simple logic, and it is the logic, which we must follow because in the end it is always right.

65. Lessons from Transgenic Mouse Models: Production of Aβ Is Suppressed Following the Elicitation of the Neuronal ISR

As discussed above, mouse transgenic models overexpressing human AβPP develop the neuronal integrated stress response. It is elicited by AβPP-derived iAβ accumulated over the T1 threshold. The accumulation of iAβ in mice occurs via the same mechanisms that operate in humans, namely, the retention of intraneuronal Aβ produced by gamma cleavages on the intracellular membranes and the uptake of secreted extracellular Aβ. In transgenic mouse models both mechanisms are very efficient for the following reasons. First, these models express human AβPP mRNA (which is, however, ineligible for RNA-dependent amplification due to its modified 5′ terminus) from numerous, in some cases over a hundred, transgenes. Consequently, the rate of the exogenous production of Aβ is very high. Accordingly, the intraneuronally retained fraction of Aβ (produced by gamma cleavages on the intracellular membranes) translates into a greatly increased cellular concentration of iAβ. The increased rate of exogenous production of Aβ also elevates its extracellular pool; this significantly increases the rate of its cellular uptake. The rate of accumulation of exogenous AβPP-derived iAβ in mouse transgenic models is further augmented by the utilization of the FAD-causing mutations and AD-predisposing factors. Thus, human AβPP genes used in these models usually contain a combination of FAD-causing mutations. Transgenes encoding PSENs harboring FAD mutations and those encoding the ApoE4 isoform are also utilized in mouse models.

Cumulatively, these mutations significantly augment the rate of accumulation of exogenously produced AβPP-derived iAβ. Indeed, the mutations employed shift the production of Aβ toward its Aβ42 isoform, which, following its secretion is taken up twice as efficiently as other Aβ isoforms. Moreover, exogenously expressed ApoE4 is much more efficient in facilitating the cellular uptake of Aβ than other ApoE variants. In addition, some AβPP mutations usually utilized in human transgenes (such as Swedish AβPP mutation) as well as certain PSEN mutations substantially increase the fraction of gamma-cleavages occurring on the intraneuronal membranes; this increases the rate of the retention of exogenous AβPP-derived iAβ. The overall rate of accumulation of exogenous AβPP-derived iAβ is, therefore, considerable. Moreover, the utilization of exogenous Aβ42 reduces the extent of the T1 threshold, as discussed above. Consequently, iAβ reaches the T1 threshold and elicits the neuronal integrated stress response early in life. This is reflected in the occurrence of eIF2α phosphorylated at its Ser51, the elicitation of the neuronal ISR, and in the neuronal ISR-mediated mild neurodegeneration and cognitive impairments (caused by the global suppression of cellular protein synthesis) observed in these models. If exogenously produced iAβ were to continue accumulating at the same rate following the T1 crossing and elicitation of the neuronal ISR, a massive neurodegeneration and neuronal loss, caused by sheer volume of iAβ and its unavoidable aggregation, would be inevitable. This, however, does not happen. The most plausible explanation is that, following the elicitation of the neuronal ISR, the AβPP proteolytic pathway is suppressed as a part of the global suppression of cellular protein synthesis [10]. Since iAβ cannot be produced independently of AβPP in these models, its rate of accumulation is substantially reduced or even reversed, and the extents of both neurodegeneration and cognitive impairment do not increase significantly with time.

The above interpretation is illustrated in Figure 34. In Panel A of Figure 34, exogenously produced AβPP-derived iAβ accumulates at a considerable rate and reaches the T1 threshold relatively early in life. PKR and/or HRI are activated, eIF2α is phosphorylated at its Ser51 residue, and the neuronal integrated stress response is elicited. This causes mild neurodegeneration and cognitive impairment (both due to the ISR-mediated suppression of cellular protein synthesis). Were the accumulation of exogenous AβPP-derived iAβ to continue at the pre-T1 crossing rate (shown by broken blue lines), it would inevitably cause massive neurodegeneration, increasingly severe cognitive impairment and, upon reaching the T2 threshold, massive neuronal loss. Instead, the production of constituents of the AβPP proteolytic pathway is suppressed in the framework of the neuronal ISR together with that of the decisive bulk of cellular proteins. Consequently, the rate of accumulation of AβPP-derived iAβ is substantially reduced and the degree of the neuronal damage and of cognitive impairment does not change significantly with time.

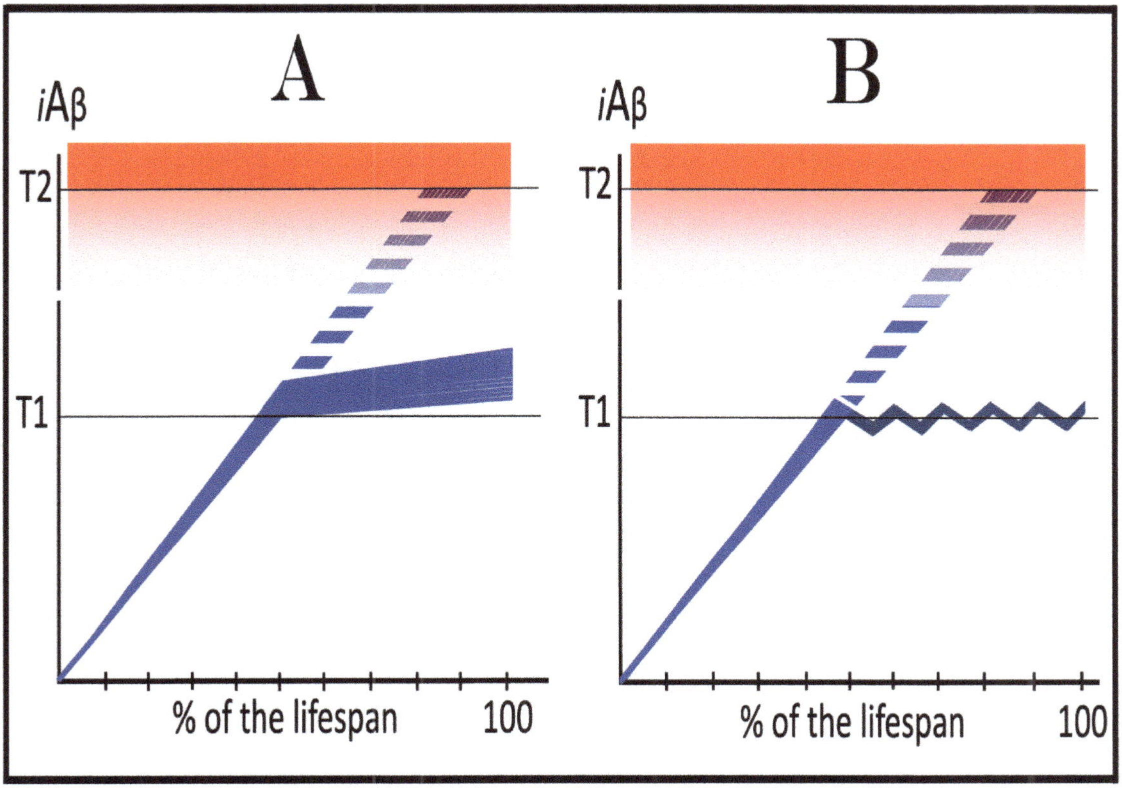

Figure 34. The neuronal ISR inhibits the production of Aβ and suppresses the rate of iAβ accumulation: lessons from transgenic mouse models. *Blue lines*: Intraneuronal Aβ, iAβ. ***T1***: Threshold of cellular concentration of iAβ triggering the activation of PKR and/or HRI kinases, phosphorylation of eIF2α, the elicitation of the neuronal ISR, the initiation of the AβPP-independent C99 production pathway, and the commencement of AD. ***T2***: Cellular concentration of iAβ triggering neuronal death via apoptosis or necroptosis. *Red box*: Apoptotic zone, defined as a range of cellular concentrations of iAβ within which cells neuronal cells committed apoptosis or necroptosis or are dead. *Pink box*: Range of AD pathology-causing concentrations of iAβ. *Broken blue lines*: the projected dynamics of accumulation of AβPP-derived iAβ occurring at the pre-T1 crossing rate; it would cause massive neurodegeneration and neuronal loss. *Panel* (**A**): Exogenously produced AβPP-derived iAβ rapidly accumulates and crosses the T1 threshold. PKR and/or HRI are activated, eIF2α is phosphorylated at its Ser51 residue, and the neuronal ISR is elicited. ISR-mediated suppression of cellular protein synthesis causes mild neurodegeneration and cognitive impairment. Were the accumulation of exogenous AβPP-derived iAβ to continue at the pre-T1 crossing rate (shown by broken blue lines), it would inevitably cause massive neuronal loss. Instead, the production of constituents of the AβPP proteolytic pathway is suppressed in the framework of the neuronal ISR together with that of the decisive bulk of cellular proteins. Consequently, the rate of accumulation of AβPP-derived iAβ is substantially reduced and the degree of the neuronal damage and of cognitive impairment does not change significantly with time. *Panel* (**B**): The rate of accumulation of AβPP-derived iAβ is reversed under ISR conditions, and its levels are decreasing. When iAβ levels reverse-cross the T1 threshold the ISR state is also reversed. This restores cellular protein synthesis and enables the production of the components of the AβPP proteolytic pathway and, consequently, the generation of iAβ. With its rate of accumulation restored, levels of iAβ increase. When they cross the T1 threshold, the neuronal ISR is re-elicited, the AβPP proteolytic pathway and the production of iAβ are suppressed, and the rate of accumulation of the latter is again reversed. The cycle repeats, and the levels of iAβ oscillate around the T1 threshold.

In Panel B of Figure 34, following the T1 crossing and the elicitation of the neuronal integrated stress response, the rate of accumulation of AβPP-derived iAβ is reversed, and its levels are decreasing. This creates an interesting situation. When iAβ levels reverse-cross the T1 threshold the ISR state is abrogated and is no longer in effect. This restores a "normal" cellular protein synthesis and enables the production of the components of the AβPP proteolytic pathway and, consequently, the generation of iAβ. With its rate of accumulation restored, levels of iAβ increase. When they cross the T1 threshold, the neuronal integrated stress response is re-elicited, the AβPP proteolytic pathway and the production of iAβ are suppressed, and the rate of accumulation of the latter is again reversed. The cycle repeats, and the levels of iAβ oscillate around the T1 threshold.

66. The ACH2.0, Version Three: The Neuronal ISR Suppresses AβPP Proteolytic Pathway but Tau Is Translated from IRES; AD, in Its Both Conventional and Unconventional Forms, Is Driven by C99 Generated Independently of AβPP

Version Three of the ACH2.0 posits that in both, conventional and unconventional forms of Alzheimer's disease the neuronal integrated stress response-caused suppression of the global cellular protein synthesis includes all constituents of the AβPP proteolytic pathway, namely AβPP, BACE enzymes, components of the gamma-secretase complex, and, consequently, Aβ and iAβ. The picture that emerges in this paradigm is as follows. The neuronal integrated stress response is elicited either by AβPP-derived iAβ accumulated over the T1 threshold (conventional AD) or by a stressor distinct from iAβ at the levels of AβPP-derived iAβ well below the T1 threshold (unconventional AD). The neuronal ISR provides components essential for the operation of the AβPP-independent C99 generation pathway and this pathway is activated. Concurrently, under the neuronal ISR conditions the global cellular protein synthesis, including that of all components of the AβPP proteolytic pathway, is suppressed. The accumulation of AβPP-derived iAβ is either suppressed or reversed. The production of C99 in the AβPP-independent pathway, on the other hand, proceeds unimpeded. However, since the production of the components of the gamma-secretase complex is suppressed under the neuronal ISR conditions, C99 produced independently of AβPP is processed no further. It rapidly accumulates, and it is C99 generated independently of AβPP, which drives AD pathology. Accordingly, in this version of the ACH2.0 the pathway underlying AD pathology is designated as AβPP-independent C99 (not C99/iAβ) production pathway. The notion that C99 generated independently of AβPP drives AD pathology is strongly supported by the observation that in AD patients C99 accumulates in the disease-affected neurons but not in the regions of the brain not affected by the disease, and that the extent of its accumulation correlates with the severity of cognitive impairment [282,283]. The qualifications of C99 generated independently of AβPP to serve as the driver of AD pathology are discussed in the following section below. The present Version Three of the ACH 2.0 is illustrated schematically in Figure 35. It should be emphasized that in Version Three of the ACH2.0 the neuronal integrated stress response is sustained in the disease by C99 (generated independently of AβPP) at the levels above the ISR-eliciting T1 threshold. In the preceding versions of the ACH2.0 AβPP-derived iAβ was the only "conventional" neuronal ISR-eliciting stressor; in the present Version Three C99 is also the conventional ISR-eliciting stressor; not only does it drive AD pathology, it also propagates the neuronal ISR state.

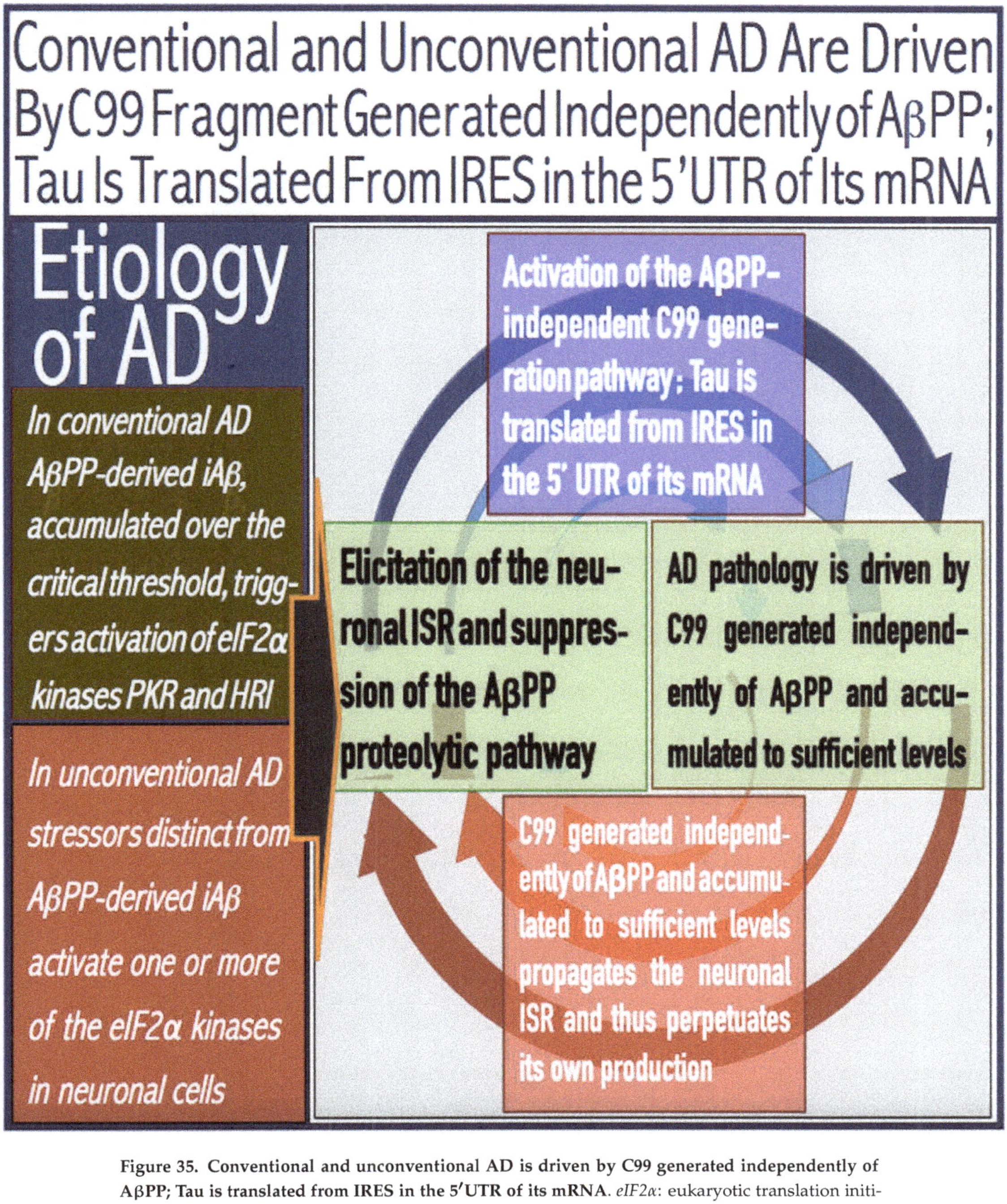

Figure 35. Conventional and unconventional AD is driven by C99 generated independently of AβPP; Tau is translated from IRES in the 5′UTR of its mRNA. *eIF2α*: eukaryotic translation initiation factor 2α. *PKR and HRI*: kinases that phosphorylate eIF2α at its Ser51 residue. *PACT*: PKR activator. *TNFα*: tumor necrosis factor α. *OMA1*: mitochondrial protease activated during mitochondrial dysfunction. *DELE1*: substrate of OMA1 protease. Cleavage of DELE1 leads to the activation of HRI. Phosphorylation of eIF2α elicits the integrated stress response, which, in turn, enables the generation of components essential for the activity of the AβPP-independent pathway of C99 and iAβ production. iAβ: intraneuronal Aβ. AβPP-derived iAβ accumulates physiologically via the

importation of extracellular Aβ and the retention of iAβ produced by gamma-cleavages on intracellular membranes in an exceedingly slow process. In most individuals it does not reach the PKR- and/or HRI-activating, the neuronal ISR-eliciting threshold within their lifetimes and no conventional AD occurs. When this threshold is crossed, the neuronal ISR is elicited, and the AβPP-independent C99 production pathway activated, and conventional AD ensues. In unconventional AD stressors that are distinct from AβPP-derived iAβ and capable of activating one or more of eIF2α kinases trigger the elicitation of the neuronal ISR. In both cases the neuronal ISR provides components essential for the operation of the AβPP-independent C99 production pathway. Concurrently, under the neuronal ISR conditions the global cellular protein synthesis, including that of all components of the AβPP proteolytic pathway, is suppressed. Consequently, the accumulation of AβPP-derived iAβ is also suppressed. The production of C99 in the AβPP-independent pathway, on the other hand, proceeds unimpeded; due to the deficiency of gamma secretase it is processed no further. It rapidly accumulates, and it is C99 generated independently of AβPP, which drives AD pathology. Translation of tau protein, on the other hand, persists under ISR conditions because its mRNA contains an internal ribosomal entry site in the 5′UTR.

Thus, in AD-affected neurons the driver of AD pathology, C99, is produced in a neuronal ISR-compatible process. But what happens with the other major constituent of AD pathology, namely tau protein? It is, evidently, produced and accumulates in the disease. This implies that its production is not suppressed by the neuronal ISR and, therefore, must occur in an ISR-compatible manner. This is indeed the case. Whereas the production of Aβ (and of the bulk of cellular proteins) is suppressed in AD-affected neurons under the ISR conditions, the production of tau protein continues because the human tau mRNA contains an internal ribosomal entry site (IRES) in its 5′ untranslated region [284]. Translation from this site is unaffected by the ISR and results in the complete polypeptide.

67. C99 Generated Independently of AβPP Meets Qualification Criteria for the Driver of AD: C99 as "Substance X" Predicted in [2]

A possibility that an agent other than Aβ or iAβ, dubbed "substance X", can drive AD pathology was entertained in our previous study [2]. The study [2] considered it a given that conventional AD is triggered by iAβ accumulated over the PKR and/or HRI activating threshold. Consequently, eIF2α is phosphorylated and the neuronal integrated stress response elicited. A small subset of cellular proteins, newly expressed under the ISR conditions, contains components essential for the operation of the substance X-generating pathway. The pathway is activated, and its product, substance X, accumulates and drives the disease. This study [2] concluded that in order to qualify as the driver of AD, such an agent (substance X) must meet two requirements. First, it has to be sufficiently toxic to be capable of promoting AD pathology. Second, it has to be capable of maintaining the neuronal integrated response state and, consequently, of sustaining the operation of the pathway of its own production. By positing that the disease is driven by C99 generated independently of AβPP, the present Perspective equates the C99 fragment with substance X. But does C99 meet the stated above qualification criteria for the driver of AD?

The answer to the above question is, apparently, affirmative. Indeed, the C99 fragment is capable of triggering the neuronal integrated stress response in multiple ways, including, for example, the ER stress [285] or causing mitochondrial distress [286]. The former entails the activation of the PERK kinase and the latter results, via the OMA1 to DELE1 to HRI signaling pathway, in the activation of HRI kinase. In both cases, eIF2α is phosphorylated at its Ser51 residue and the neuronal ISR is elicited. With regard to its toxicity, C99 was shown to interfere with lipid metabolism and to cause the loss of lipid homeostasis. This occurs through the interaction of C99 with MAM (mitochondria-associated ER membranes) [287]. Intraneuronal accumulation of C99 was also shown to trigger endosomal/autophagic/lysosomal (EAL) dysfunction, which is a recognized neuropathological feature of AD [288]. In this case, the deleterious effect of the C99 fragment apparently results from its aggregation within EAL-vesicle membranes; this leads to disrupted lysosomal proteolysis and to autophagic impairment. Both of these conditions, if sustained, are capable of initiating and maintaining cascades of cellular pathology. Thus, the C99 fragment apparently possesses the toxicity required to drive AD pathology and is also capable of propagating the neuronal ISR state, consequently perpetuating the AβPP-independent pathway of its own production. Thus, it meets qualification criteria for the driver of AD.

68. Validation (12): Inhibition of Gamma-Secretase in Human Neuronal Cell-Based AD Model Would Not Prevent or Abrogate Cellular AD Pathology

The notion that AD is driven by C99 generated independently from AβPP can potentially be validated in a human neuronal cell-based AD model. In this application, the sustained neuronal integrated stress response can be elicited unconventionally, by a variety of stressors other than iAβ (e.g., mitochondrial distress). The utilization of unconventional, rather than conventional (iAβ-triggered) AD model is important in order to exclude the possibility that cellular AD pathology, if observed, is caused by exogenously produced iAβ. It can be anticipated that the neuronal ISR would activate the AβPP-independent C99 generation pathway and that, with a sufficient duration of the presence of the unconventional stressor, this pathway would be rendered self-sustainable. It is expected that, as described above, this would lead to the development of cellular AD pathology and manifest in the formation of NFTs. If cellular AD pathology in such an experiment is indeed driven by C99, the inclusion of gamma-secretase inhibitors would not prevent AD pathology, and the addition of inhibitors when NFTs have already formed would not abrogate it. This is in contrast to a possibility that C99 produced independently of AβPP is processed into iAβ, which actually drives AD pathology. In such a case, the inclusion of gamma-secretase inhibitors would prevent AD pathology (and the formation of NFTs) or, at least, modulate it.

69. Dynamics of Conventional and Unconventional AD in the ACH2.0 Version Three Perspective

From the perspective of the ACH2.0, Version Three, the dynamics of AD in its conventional form is defined by two sets of factors. One is AβPP-derived iAβ, more precisely, the rate of its accumulation, and the extent of the T1 threshold. Its role in the version under discussion is, in fact, identical to that in the preceding versions of ACH2.0. The occurrence of AD is defined by AβPP-derived iAβ reaching and crossing the T1 threshold; if the latter is not crossed within the lifetime of an individual, the former does not occur. If the disease does occur, the timing of its commencement is directly proportional to the extent of the T1 threshold and inversely proportional to the rate of accumulation of AβPP-derived iAβ. The relationships between the rate of accumulation of AβPP-derived iAβ, the extent of the T1 threshold, and the occurrence and timing of conventional AD are discussed in-depth in Sections 27–31 above. All conclusions reached in those sections apply fully in Version Three of the ACH2.0.

Another set of factors determining the dynamics of conventional AD is the rate of accumulation of C99 produced independently of AβPP and the extent of the T2 threshold. In this respect, in Version Three of the ACH2.0, C99 generated in the AβPP-independent pathway plays the same role as iAβ, presumably produced independently of AβPP, does in the preceding versions of ACH2.0. The rate of the progression of AD is directly proportional to the rate of accumulation of C99 produced independently of AβPP, whereas the duration of the disease is inversely proportional to the rate of accumulation of C99 generated in the

AβPP-independent pathway and directly proportional to the extent of the T2 threshold. This dynamic is examined (for iAβ produced in the AβPP-independent pathway as the driver of AD) in Section 49 above. The conclusions arrived at in that section are conceptually fully applicable in Version Three of the ACH2.0 as well.

The dynamics of conventional AD in terms of the accumulation of AβPP-derived iAβ and of C99 generated independently of AβPP is illustrated concisely (it shows neither the effects of the variable rates of accumulation of both nor the outcomes of the variable extents of the T1 and T2 thresholds) in Panel A of Figure 36. In the first phase of AD AβPP-derived iAβ accumulates via the retention of a fraction of Aβ resulting from gamma-cleavages occurring on the intracellular membranes within various cellular organelles and the cellular uptake of a fraction of secreted extracellular Aβ (discussed in Section 24 above). When its levels reach and cross the T1 threshold, PKR and/or HRI are activated, eIF2α is phosphorylated at its Ser51 residue, and the neuronal integrated stress response is elicited. This results in two concurrently occurring developments. First, global cellular protein synthesis is severely suppressed. This includes all constituents of the AβPP proteolytic pathway: AβPP, BACE enzymes, components of the gamma-secretase complex, and, consequently, Aβ. The influx of AβPP-derived iAβ is inhibited at both sources (suppressed production of Aβ, suppressed retention; diminishing extracellular pool of Aβ, diminishing uptake), and the rate of its accumulation is reduced. In the second, and concurrent, development, the neuronal ISR enables the production of components essential for the operation of the AβPP-independent C99 generation pathway. The pathway is activated and is self-sustainable from the instance of its activation because the ISR state is initially sustained by over-T1 AβPP-derived iAβ and eventually by C99 after it crosses the T1 threshold. C99 produced independently of AβPP drives AD pathology; when it crosses the T2 threshold the disease enters its end stage. Thus, in Version Three of the ACH2.0 the role of Aβ, more precisely of iAβ, is limited to triggering the neuronal integrated stress response, and even this limited function is restricted to the conventional form of AD; it plays little or no role in the progression of AD. It also retains its causative role in AACD. If the extent of the T^0 threshold were below that of the T1 threshold, AACD would commence upon the crossing of the former and morph into AD when the latter is crossed.

The dynamics of unconventional AD is different. It is less complex as far as iAβ is concerned. The disease is triggered at the levels of AβPP-derived iAβ below the T1 threshold, and therefore iAβ plays no substantive role in unconventional AD. The dynamics is, however, more complex due to the potential variability of the timing and, especially, the duration of the unconventionally elicited neuronal integrated stress response. The former (timing) is defined by the occurrence of an unconventional stressor capable of eliciting the neuronal ISR (any stressor distinct from iAβ and C99). The latter, on the other hand, is determined by the duration of the presence of the unconventional stressor. In unconventional AD the neuronal ISR is unconventionally elicited and, accordingly, the AβPP-independent C99 generation pathway is unconventionally activated. C99, produced independently of AβPP, rapidly accumulates. If the duration of the activity of the AβPP-independent C99 generation pathway were insufficient for the T1 crossing by C99, the former would be only transient and not necessarily result in AD (further discussed below). If and when C99 crosses the T1 threshold, the pathway becomes self-sustainable, and AD commences and progresses; the T2 crossing denotes the entrance of the disease into the end stage. The rate of its progression is directly proportional to the rate of accumulation of C99 produced independently of AβPP, and its duration is directly proportional to the extent of the T2 threshold and inversely proportional to the rate of accumulation of C99 generated in the AβPP-independent pathway.

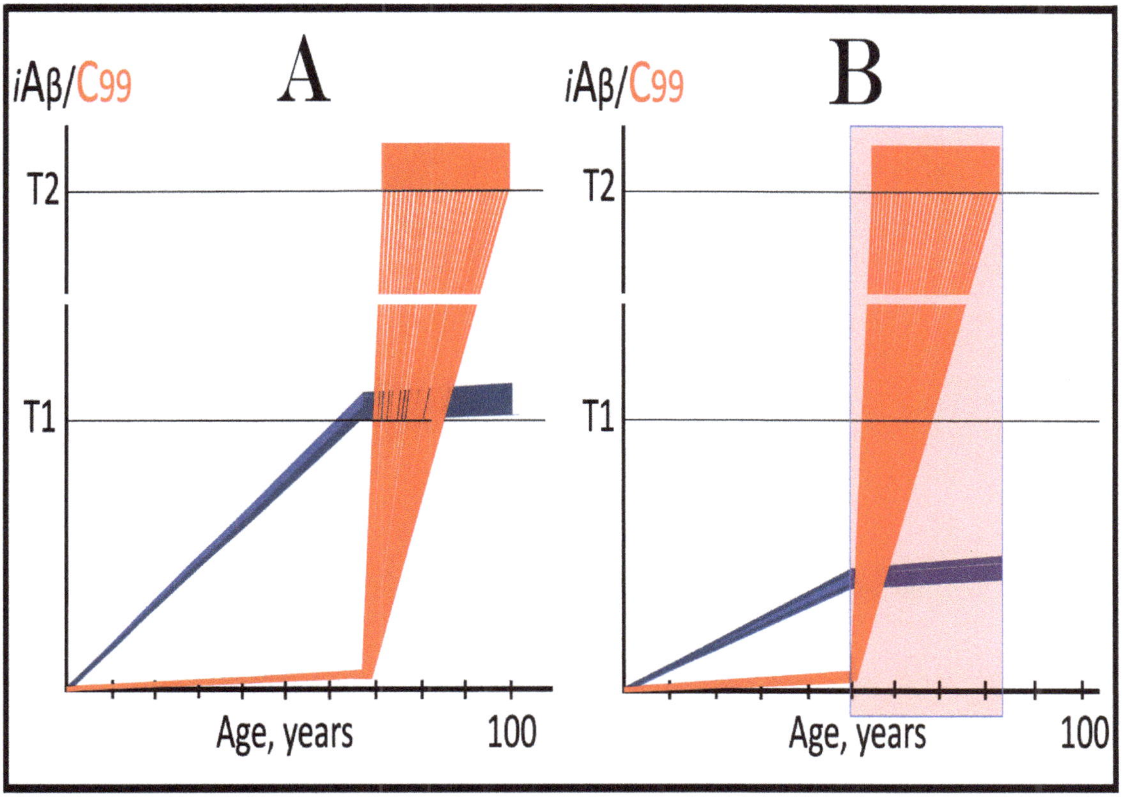

Figure 36. Dynamics of conventional and unconventional AD in the ACH2.0 Version Three perspective. *Blue lines*: Intraneuronal Aβ, iAβ. *Red lines*: The C99 fragment of AβPP. ***T1***: Threshold of cellular concentration of iAβ triggering the activation of PKR and/or HRI kinases, phosphorylation of eIF2α, the elicitation of the neuronal ISR, the initiation of the AβPP-independent C99 production pathway, and the commencement of AD. ***T2***: Cellular concentration of iAβ triggering neuronal death via apoptosis or necroptosis. *Red box*: Apoptotic zone, defined as a range of cellular concentrations of iAβ within which cells neuronal cells committed apoptosis or necroptosis or are dead. *Pink box*: The duration of the presence of the unconventional stressor capable of activating one or more of eIF2α kinases and triggering the elicitation of the neuronal ISR. *Panel* (**A**): Dynamics of conventional AD. AβPP-derived iAβ accumulates physiologically via mechanisms discussed in the main text. When it crosses the T1 threshold, PKR and/or HRI are activated, eIF2α is phosphorylated at its Ser51 residue, and the neuronal ISR is elicited. Global cellular protein synthesis, including all constituents of the AβPP proteolytic pathway: AβPP, BACE enzymes, components of the gamma-secretase complex, and, consequently Aβ, is severely suppressed. The influx of AβPP-derived iAβ is inhibited, and the rate of its accumulation is reduced. Concurrently, the neuronal ISR enables the production of components essential for the operation of the AβPP-independent C99 generation pathway. The pathway is self-sustainable from the instance of its activation because the ISR state is initially sustained by over-T1 AβPP-derived iAβ and eventually by C99 after it crosses the T1 threshold. C99 produced independently of AβPP drives AD pathology; when it crosses the T2 threshold the disease enters its end stage. *Panel* (**B**): Dynamics of unconventional AD. The neuronal ISR is unconventionally elicited, and the AβPP-independent C99 generation pathway is unconventionally activated at the levels of AβPP-derived iAβ well below the T1 threshold. Under the neuronal ISR conditions the AβPP proteolytic pathway is suppressed and the rate of accumulation of AβPP-derived iAβ is substantially reduced. On the other hand, C99 generated independently of AβPP rapidly accumulates and crosses the T1 threshold (provided the unconventional stressor persists for a sufficient duration). Following

the T1 crossing, the AβPP-independent C99 production pathway is rendered self-sustainable, and AD commences and progresses. When the T2 threshold is reached in a sufficient neuronal fraction, the disease enters its end stage.

As stated above, the above-invoked complexity is due in part to the possibility that the presence of the unconventional stressor can be transient. Its withdrawal after the T1 crossing by C99 would be inconsequential: the neuronal ISR state would be sustained by the product of the AβPP-independent C99 production pathway, i.e., the pathway would be self-sustainable, and the disease would proceed unimpeded. On the other hand, if the unconventional stressor is withdrawn prior to the crossing of the T1 threshold by C99, the ISR state would be reversed, and the AβPP-independent C99 generation pathway would be rendered inoperative. Thus, transient unconventional activation of the AβPP-independent C99 production pathway could be benign. If, however, it occurs recurrently, every time the accumulation of C99 would resume from an elevated threshold until, eventually it crosses the T1 threshold, the AβPP-independent C99 generation pathway becomes self-sustainable, and the disease commences. The possible outcomes of transient unconventional activation of the AβPP-independent C99 production pathway were discussed above (Sections 52 and 53) with an assumption that AD is driven by iAβ generated independently of AβPP. Nevertheless, conceptually, the conclusions reached above apply fully in Version Three of the ACH2.0 as well.

The dynamics of unconventional AD in terms of the accumulation of C99 generated independently of AβPP is presented in Panel B of Figure 36. In this panel, the neuronal integrated stress response is unconventionally elicited, and the AβPP-independent C99 generation pathway is unconventionally activated at the levels of AβPP-derived iAβ well below the T1 threshold. Under the neuronal ISR conditions the AβPP proteolytic pathway is suppressed and the rate of accumulation of AβPP-derived iAβ is substantially reduced. On the other hand, C99 generated independently of AβPP rapidly accumulates and crosses the T1 threshold (it is assumed in this panel that the duration of the presence of the unconventional stressor is sufficient for the T1 crossing to occur in all affected neurons) Following the T1 crossing, the AβPP-independent C99 production pathway is rendered self-sustainable, and AD commences and progresses. When the T2 threshold is reached in a sufficient neuronal fraction, the disease enters its end stage. Thus, in unconventional AD, Aβ and iAβ play little or no role at all in both triggering the disease and driving AD pathology.

70. ACH-Based AD Drugs from the Perspective of the ACH2.0, Version Three

ACH-based AD drugs are agents designed under the ACH guidance with the purpose of reducing the levels of extracellular Aβ. This group of drugs includes Aβ-specific antibodies designed to sequester extracellular Aβ and BACE1 inhibitors intended to suppress the production of Aβ and, consequently, its secretion and extracellular concentration. Initially, gamma-secretase inhibitors were also part of this group of drugs, but their utilization was ruled out due to severe adverse effects. Instead, the effort switched to the development of gamma-secretase modulators, agents that shift the position of gamma-cleavage within the C99 fragment toward shorter, more benign Aβ species. Above (Sections 36 and 37), we reasoned that, in the initial version of the ACH2.0, ACH-based drugs are legitimate (albeit not very efficient) conventional AD-preventive agents. This is because by reducing the extracellular pool of Aβ they reduce the rate of its cellular uptake, and by suppressing its production (with BACE1 inhibitors) they reduce both the cellular uptake of Aβ (due to its reduced extracellular pool) and its retention (less produced, less retained). As a result, the rate of accumulation of AβPP-derived iAβ is decreased, the T1 crossing is either delayed or prevented, and the occurrence of AD is either delayed or precluded. As for gamma-secretase modulators, the more benign iAβ species would elevate the T1 threshold (by the same logic that more toxic Aβ42 would lower it; see Section 33 above), and the T1 threshold crossing also be either delayed or prevented.

The above conclusions remain fully in effect in Version Three of the ACH2.0 but only for the conventional form of AD and only for its prevention but not the treatment. In conventional AD the neuronal integrated stress response can be elicited, and the disease triggered only by AβPP-derived iAβ accumulated above the T1 threshold. If no T1 crossing occurs, neither does the disease; if the T1 crossing is delayed, the occurrence of AD is also delayed. The preventive prospect of ACH-based drugs in conventional AD, however, is their only potential utility in Version Three of the ACH2.0. Indeed, in unconventional AD these drugs are inapplicable preventively because stressors other than AβPP-derived iAβ trigger the disease. As for the treatment of AD, ACH-based drugs are inapplicable in both, conventional and unconventional forms of the disease because in both AD pathology is driven by self-sustainable AβPP-independent C99 production pathway. Aβ and iAβ play no role at this stage and applying ACH-based drugs would amount to attacking an imaginary target. This applies also to modulators of gamma-secretase. In the initial version of the ACH2.0, such agents could potentially confer benefits in the treatment of AD by elevating the T2 threshold and thus slowing down the progression of the disease. This is principally not the case in Version Three of ACH2.0, where the AβPP proteolytic pathway is suppressed in symptomatic AD and there is little or no gamma-secretase to modulate. On the other hand, as discussed in the following sections below, ACH2.0-based drugs retain their therapeutic potential in both the prevention and treatment of both conventional and unconventional forms of the disease in terms of the ACH2.0, Version Three.

71. ISR Inhibitors in the Prevention of Conventional AD: The ACH2.0, Version Three Perspective

The elicitation of the neuronal integrated stress response remains the AD-triggering event in Version Three of the ACH2.0. It follows that the prevention of the neuronal ISR by a therapeutic intervention would potentially prevent the disease for the duration of the treatment. This notion, in its application to conventional AD, is considered in Figure 37. Panel A of Figure 37 depicts the initial state of the levels of AβPP-derived iAβ at the time of the commencement of the treatment. It has been accumulating via both the cellular uptake of a fraction of secreted extracellular Aβ and the retention of iAβ produced by gamma-cleavages of C99 on the intraneuronal membranes within various cellular organelles, but its levels have not yet reached the T1 threshold. Panel B of Figure 37 illustrates the evolution of the initial state in the absence of the ISR-inhibiting treatment. The accumulation of AβPP-derived iAβ continues unhindered and it crosses the T1 threshold. PKR and/or HRI are activated, eIF2α is phosphorylated at its Ser51 residue, and the neuronal integrated stress response is elicited. Under the ISR conditions the global cellular protein synthesis, including that of the constituents of the AβPP proteolytic pathway, is suppressed and the rate of accumulation of AβPP-derived iAβ is substantially reduced. Concurrently, the neuronal ISR enables the production of components essential for the operation of the AβPP-independent C99 production pathway, and the pathway is activated. C99 generated independently of AβPP rapidly accumulates and drives the progression of AD pathology. Panel C of Figure 37 shows the evolution of the initial state in the individual treated with the ISR-inhibiting drug. AβPP-derived iAβ continues to accumulate and crosses the T1 threshold. The drug precludes the elicitation of the neuronal integrated stress response, and the AβPP-independent C99 production pathway remains inoperative. The accumulation of AβPP-derived iAβ continues at the pre-T1 crossing rate; it does not, however, reach the

AD pathology-causing levels. Thus, ISR inhibitors administered prior to the T1 crossing by AβPP-derived iAβ can prevent the occurrence of conventional AD for the duration of the treatment. On the other hand, if AβPP-derived iAβ crosses the T^0 threshold, even if it is above the T1 threshold, as shown in Figure 37, AACD would commence and persist for the remaining portion of the lifetime; it would be unaffected by the treatment. Importantly, the treatment has to be administered for the lifetime. If the drug were withdrawn, the neuronal ISR would be elicited, and conventional AD would commence.

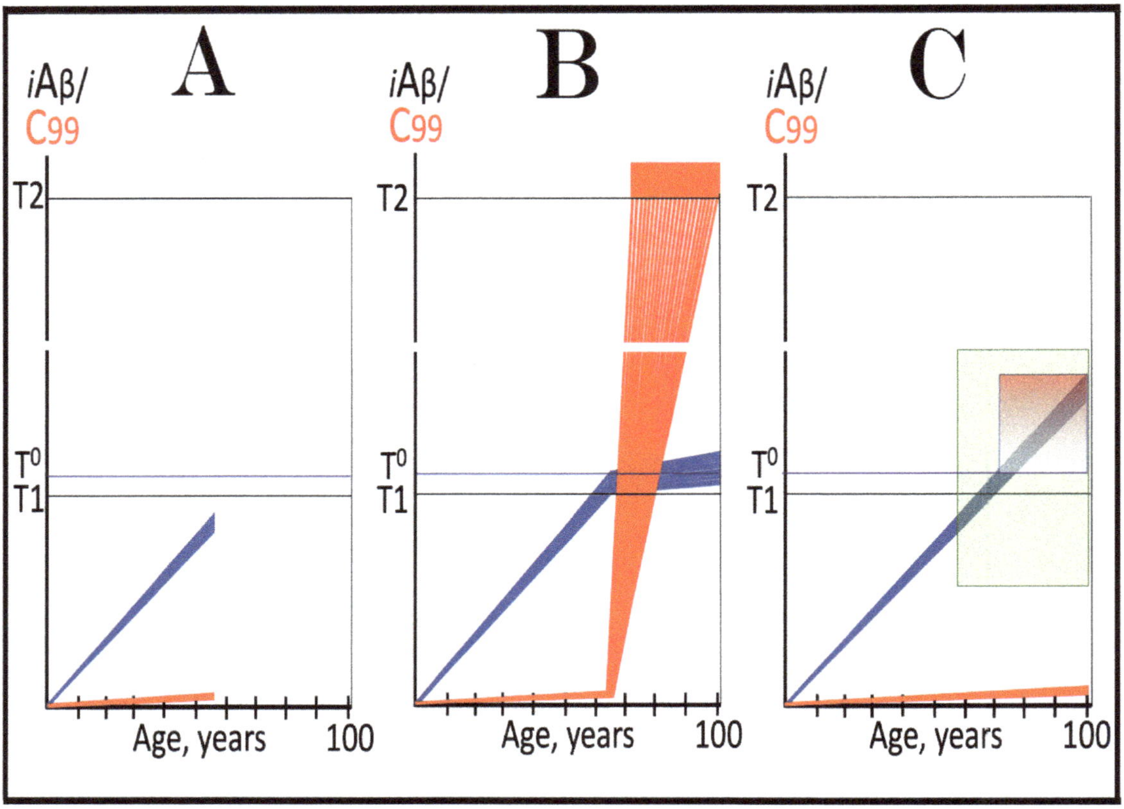

Figure 37. ISR inhibitors in the prevention of conventional AD: the ACH2.0, Version Three perspective. *Blue lines*: Intraneuronal Aβ, iAβ. *Red lines*: The C99 fragment of AβPP. ***T^0***: Threshold of cellular concentration of AβPP-derived iAβ triggering the neuronal damage, which manifests as AACD. ***T1***: Threshold of cellular concentration of iAβ triggering the activation of PKR and/or HRI kinases, phosphorylation of eIF2α, the elicitation of the neuronal ISR, the initiation of the AβPP-independent C99 production pathway, and the commencement of AD; its extent is the only variable in the present figure. ***T2***: Cellular concentration of iAβ triggering neuronal death via apoptosis or necroptosis. *Red box*: Apoptotic zone, defined as a range of cellular concentrations of iAβ within which cells neuronal cells committed apoptosis or necroptosis or are dead. *Pink box*: The range of cellular concentrations of AβPP-derived iAβ concentrations between the T^0 and the highest extent reached by AβPP-derived iAβ or the T1 threshold that support the progression of AACD. The condition commences with the crossing of the T^0 threshold and morphs into AD with the T1 crossing and elicitation of the ISR; usually it can occur only if the extent of the T^0 threshold is smaller than that of the T1. *Green box*: Duration of the treatment with the ISR-inhibiting drug. *Panel* (**A**): The initial state of the levels of AβPP-derived iAβ in the neurons of the healthy individual who would develop AD if untreated. AβPP-derived iAβ has been accumulating but its levels are below the T1 threshold.

Panel (**B**): AβPP-derived iAβ crosses the T1 threshold, PKR and/or HRI are activated, eIF2α is phosphorylated at its Ser51 residue, and the neuronal ISR is elicited. Under the ISR conditions the global cellular protein synthesis, including that of the constituents of the AβPP proteolytic pathway, is suppressed and the rate of accumulation of AβPP-derived iAβ is substantially reduced. Concurrently, the neuronal ISR enables the production of components essential for the operation of the AβPP-independent C99 production pathway, and the pathway is activated. C99 generated independently of AβPP rapidly accumulates and drives the progression of AD pathology. *Panel* (**C**): The evolution of the initial state in the individual treated with the ISR-inhibiting drug. AβPP-derived iAβ accumulates and crosses the T1 threshold. The drug precludes the elicitation of the neuronal ISR, and the AβPP-independent C99 production pathway remains inoperative. The accumulation of AβPP-derived iAβ continues at the pre-T1 crossing rate; it does not reach the AD pathology-causing levels for the duration of the treatment. On the other hand, since the ISR is suppressed, if AβPP-derived iAβ crosses the T^0 threshold, even if it is above the T1 threshold, AACD would commence and persist for the remaining portion of the lifetime; it would be unaffected by the treatment.

72. ISR Inhibitors in the Treatment of Conventional AD: The ACH2.0, Version Three Perspective

In the ACH2.0, Version Three, AD is driven by C99 generated in the AβPP-independent pathway. The operation of this pathway is contingent on the sustained occurrence of the neuronal ISR state. The inhibition of the latter would deprive the former of its essential components; its operation would cease, and the progression of AD potentially arrested. This scenario is considered in Figure 38. Panel A of Figure 38 shows the initial state of the levels of AβPP-derived iAβ and C99 generated independently of AβPP at the commencement of the treatment. In this panel, the levels of AβPP-derived iAβ have crossed the T1 threshold in all affected neurons of the AD patient, and the neuronal integrated stress response has been elicited. The AβPP proteolytic pathway has been suppressed and the rate of accumulation of AβPP-derived iAβ substantially reduced. Concurrently, the AβPP-independent C99 generation pathway has been activated. C99 accumulated, reached the T2 threshold in a fraction of the neurons, and AD symptoms manifested. Panel B of Figure 38 depicts the evolution of the initial state in the untreated patient. C99 produced independently of AβPP continues to accumulate, reaches the T2 threshold in a sufficient fraction of the neurons, and the disease enters its end stage. Panel C of Figure 38 presents the evolution of the initial state in the AD patient treated with the ISR-inhibiting drug. The ISR state is reversed. The activity of the AβPP proteolytic pathway is restored, and the accumulation of AβPP-derived iAβ resumes at the pre-T1 crossing rate. It would not, however, reach the AD pathology-causing levels. On the other hand, the operation of the AβPP-independent C99 generation pathway ceases. Levels of C99, processed by now available gamma-secretase decline, and the progression of AD is arrested for the duration of the treatment. The processing of C99 generated in the AβPP-independent pathway by gamma-secretase would produce a temporary peak of Aβ, but without the influx of C99 produced independently of AβPP this peak would dissipate. Thus, the long-term treatment with ISR inhibitors can stop the progression of conventional AD.

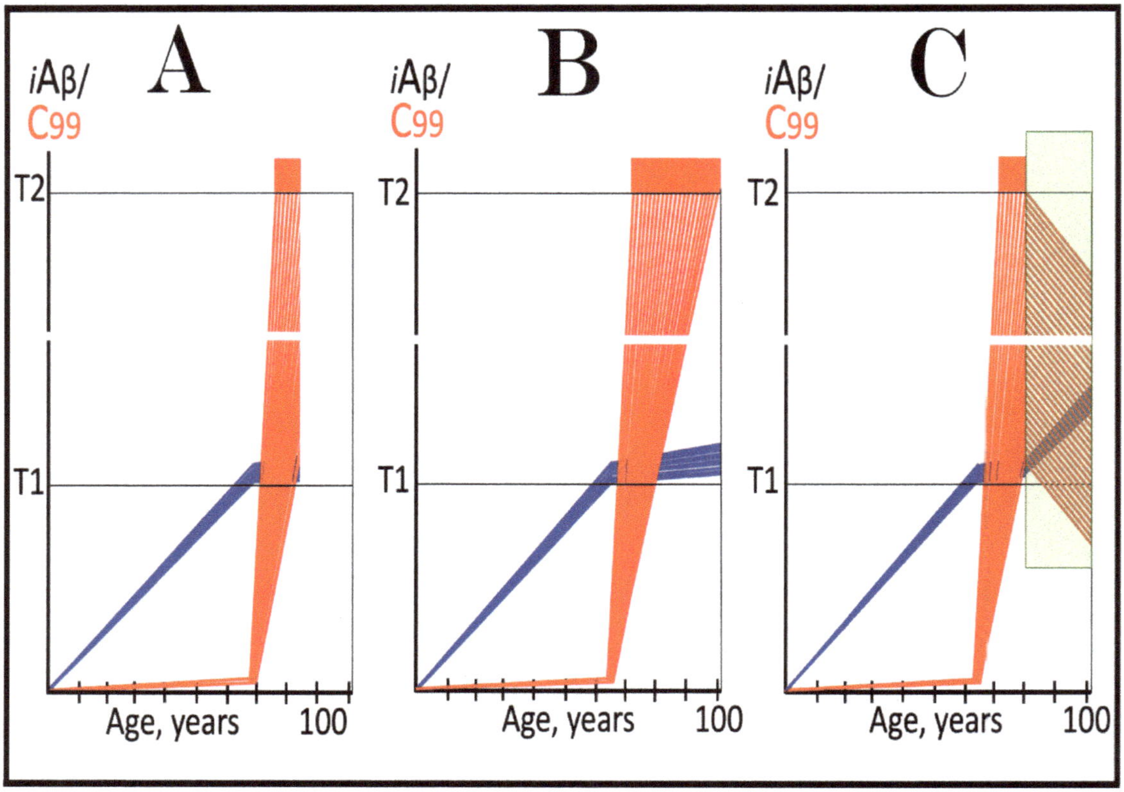

Figure 38. ISR inhibitors in the treatment of conventional AD: the ACH2.0, Version Three perspective. *Blue lines*: Intraneuronal Aβ, iAβ. *Red lines*: The C99 fragment of AβPP. ***T1***: Threshold of cellular concentration of iAβ triggering the activation of PKR and/or HRI kinases, phosphorylation of eIF2α, the elicitation of the neuronal ISR, the initiation of the AβPP-independent C99 production pathway, and the commencement of AD; its extent is the only variable in the present figure. ***T2***: Cellular concentration of iAβ triggering neuronal death via apoptosis or necroptosis. *Red box*: Apoptotic zone, defined as a range of cellular concentrations of iAβ within which cells neuronal cells committed apoptosis or necroptosis or are dead. *Green box*: Duration of the treatment with the ISR-inhibiting drug. *Panel* (**A**): The initial state of the levels of AβPP-derived iAβ and C99 in the neurons of the AD patient. AβPP-derived iAβ has crossed the T1 threshold in all affected neurons, and the neuronal ISR has been elicited. The AβPP proteolytic pathway has been suppressed and the rate of accumulation of AβPP-derived iAβ substantially reduced. Concurrently, the AβPP-independent C99 generation pathway has been activated. C99 accumulated and reached the T2 threshold in a fraction of the neurons, and AD symptoms manifested. *Panel* (**B**): The evolution of the initial state in the untreated patient. C99 produced independently of AβPP continues to accumulate and reaches the T2 threshold in a sufficient fraction of the neurons, and the disease enters its end stage. *Panel* (**C**): The evolution of the initial state in the AD patient treated with the ISR-inhibiting drug. The ISR state is reversed. The activity of the AβPP proteolytic pathway is restored, and the accumulation of AβPP-derived iAβ resumes at the pre-T1 crossing rate. It would not, however, reach the AD pathology-causing levels. On the other hand, the operation of the AβPP-independent C99 generation pathway ceases. Levels of C99, processed by now available gamma-secretase decline, and the progression of AD is arrested for the duration of the treatment.

73. ISR Inhibitors in the Prevention and Treatment of Unconventional AD: The ACH2.0, Version Three Perspective

In unconventional AD the neuronal integrated stress response is elicited by a variety of stressors distinct from iAβ or C99. This occurs at the levels of AβPP-derived iAβ below the T1 threshold (the T1 crossing triggers conventionally the neuronal ISR, and the appearance of unconventional ISR-eliciting stressors at this point would be redundant and inconsequential). The unconventionally elicited neuronal ISR suppresses the AβPP proteolytic pathway and substantially reduces the rate of accumulation of AβPP-derived iAβ. At the same time, it activates the AβPP-independent C99 generation pathway. C99 produced independently of AβPP rapidly accumulates, and AD commences when it crosses the T1 threshold. This is because prior to the T1 crossing, the AβPP-independent C99 production pathway is not self-sustainable, i.e., if the unconventional stressor were withdrawn, the operation of the pathway would cease. Thus, if the neuronal ISR is inhibited prior to the T1 crossing by C99 produced independently of AβPP, the AβPP-independent C99 generation pathway would be rendered inoperative, C99 produced independently of AβPP would not cross the T1 threshold, and AD would be prevented.

This scenario is illustrated in Panel A of Figure 39. In this panel, the neuronal integrated stress response has been unconventionally elicited and the AβPP-independent C99 generation pathway unconventionally activated at the levels of AβPP-derived iAβ well below the T1 threshold. The neuronal ISR suppresses the AβPP proteolytic pathway and reduces the rate of accumulation of AβPP-derived iAβ. At the same time C99 produced independently of AβPP rapidly accumulates but at the time of the commencement of treatment with an ISR-inhibiting drug it is still below the T1 threshold. The drug reverses the ISR state, restores the activity of the AβPP proteolytic pathway, and the accumulation of the AβPP-derived iAβ resumes at its initial (pre-ISR) rate. It also deprives the AβPP-independent C99 generation pathway of its essential components, and the production of C99 in the AβPP-independent manner ceases. Its levels decline and would not reach the T1 threshold and AD would be prevented for the duration of the treatment. AβPP-derived iAβ may cross the T1 threshold, but in the presence of the ISR inhibitor this would be inconsequential, as described in the preceding sections.

Panel B of Figure 39 illustrates another scenario where the unconventional elicitation of the neuronal integrated stress response suppressed the AβPP proteolytic pathway, reduced the rate of accumulation of AβPP-derived iAβ, and activated the AβPP-independent C99 production pathway. In this scenario C99 generated in the unconventionally activated AβPP-independent pathway accumulates over the T1 threshold. The pathway is rendered self-sustainable, and AD commences. At the time of the commencement of treatment with the ISR inhibitor, C99 produced independently of AβPP has crossed the T2 threshold in a fraction of the neurons, and AD symptoms have manifested. The reversal of the neuronal ISR state enables the activity of the AβPP proteolytic pathway, and the accumulation of AβPP-derived iAβ resumes at the pre-ISR rate. It also disables the AβPP-independent C99 production pathway. The levels of C99 generated independently of AβPP are declining and the progression of the disease is arrested for the duration of the treatment. AβPP-derived iAβ may potentially cross the T1 threshold but no AD would be triggered in the presence of the drug. Thus, the long-term inhibition of the neuronal ISR may stop the progression of unconventional AD.

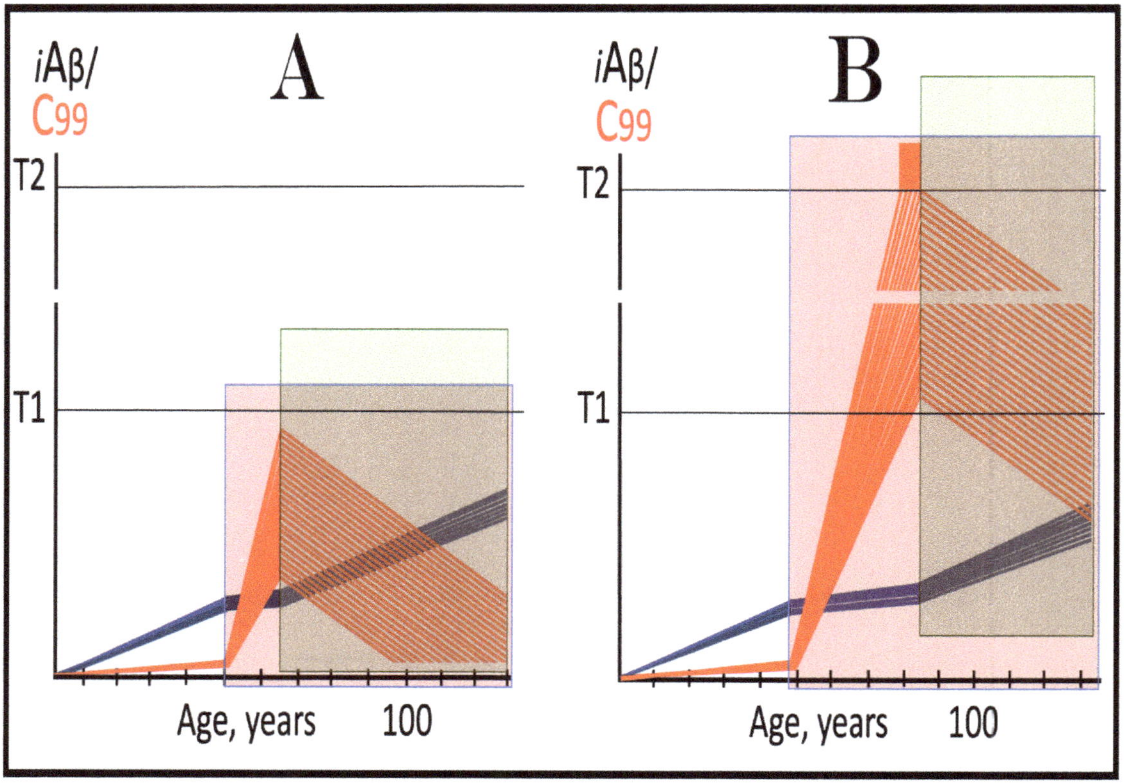

Figure 39. ISR inhibitors in the prevention and treatment of unconventional AD: the ACH2.0, Version Three perspective. *Blue lines*: Intraneuronal Aβ, iAβ. *Red lines*: The C99 fragment of AβPP. ***T1***: Threshold of cellular concentration of iAβ triggering the activation of PKR and/or HRI kinases, phosphorylation of eIF2α, the elicitation of the neuronal ISR, the initiation of the AβPP-independent C99 production pathway, and the commencement of AD; its extent is the only variable in the present figure. ***T2***: Cellular concentration of iAβ triggering neuronal death via apoptosis or necroptosis. *Red box*: Apoptotic zone, defined as a range of cellular concentrations of iAβ within which cells neuronal cells committed apoptosis or necroptosis or are dead. *Pink boxes*: The duration of the presence of the unconventional stressor capable of activating one or more of eIF2α kinases and triggering the elicitation of the neuronal ISR. *Green box*: Duration of the treatment with the ISR-inhibiting drug. *Panel* (**A**): ISR inhibitors in the prevention of unconventional AD. Following the occurrence of the unconventional stressor, the neuronal ISR has been unconventionally elicited and the AβPP-independent C99 production pathway unconventionally activated. The neuronal ISR suppresses the AβPP proteolytic pathway and reduces the rate of accumulation of AβPP-derived iAβ. Concurrently, C99 produced independently of AβPP rapidly accumulates but is still below the T1 threshold. The drug reverses the ISR state, restores the activity of the AβPP proteolytic pathway, and the accumulation of the AβPP-derived iAβ resumes at its pre-ISR rate. It also disables the production of C99 in the AβPP-independent pathway, and its levels decline for the duration of the treatment. *Panel* (**B**): ISR inhibitors in the treatment of unconventional AD. At the time of the drug's administration, C99 produced independently of AβPP has crossed the T2 threshold in a fraction of the neurons, and AD symptoms have manifested. The drug reverses the neuronal ISR state, enables the activity of the AβPP proteolytic pathway, and the accumulation of AβPP-derived iAβ resumes at the pre-ISR rate. It also disables the AβPP-independent C99 production pathway. The levels of C99 generated independently of AβPP are declining and the progression of the disease is stopped for the duration of the treatment.

74. BACE1 and/or BACE2 Activators in the Prevention of Conventional AD and in the Prevention and Treatment of AACD: The Perspective of the ACH2.0, Version Three

The present section considers a scenario where the levels of AβPP-derived iAβ in the neurons of an individual are either below or above the AACD-triggering T^0 threshold, but in both cases below the T1 threshold. In this scenario, since the T1 threshold has not been crossed, the neuronal integrated stress response has not yet been elicited, and the AβPP-independent C99 production pathway remains inoperative. Above (Section 46) and elsewhere [9,10] it has been established that with the influx of iAβ supported solely by the AβPP proteolytic pathway, the transient activation of BACE1 and/or BACE2 would apparently be highly effective in substantially depleting AβPP-derived iAβ. The de novo accumulation of iAβ would commence from a low baseline and proceed at the pre-treatment rate. If the treatment is implemented in mid-life of an individual, accumulating AβPP-derived iAβ would reach neither the T^0 nor T1 thresholds within the remaining portion of the lifetime. Therefore, if the treatment were carried out prior to the T^0 crossing (in this scenario the extent of the T^0 threshold is smaller than that of the T1), it would prevent both conventional AD and AACD (which commences upon the T^0 crossing). On the other hand, if the treatment were implemented at the levels of AβPP-derived iAβ above the T^0 but below the T1 thresholds, it would cure AACD and prevent for life the occurrence of conventional AD and recurrence of AACD.

The above considerations are illustrated in Figure 40. Panel A of Figure 40 depicts the initial state of the levels of AβPP-derived iAβ in the neurons of an individual. At the time of the administration of the transient iAβ depletion treatment they are below the T^0 threshold. Panel B of Figure 40 shows the evolution of the initial state in the untreated individual. AβPP-derived iAβ continues to accumulate. When it crosses the T^0 threshold, AACD commences and progresses. When AβPP-derived iAβ crosses the T1 threshold, PKR and/or HRI kinases are activated, eIF2α is phosphorylated at its Ser51 residue, and the neuronal integrated stress response is elicited. Under the ISR conditions the AβPP proteolytic pathway is suppressed and the rate of accumulation of AβPP-derived iAβ is substantially reduced. Concurrently, the neuronal ISR enables the production of components essential for the operation of the AβPP-independent C99 generation pathway. As a result, the pathway is activated, and C99 produced independently of AβPP rapidly accumulates and drives the progression of AD pathology. When the T2 threshold is crossed in a sufficient fraction of the neurons, the disease reaches its end stage.

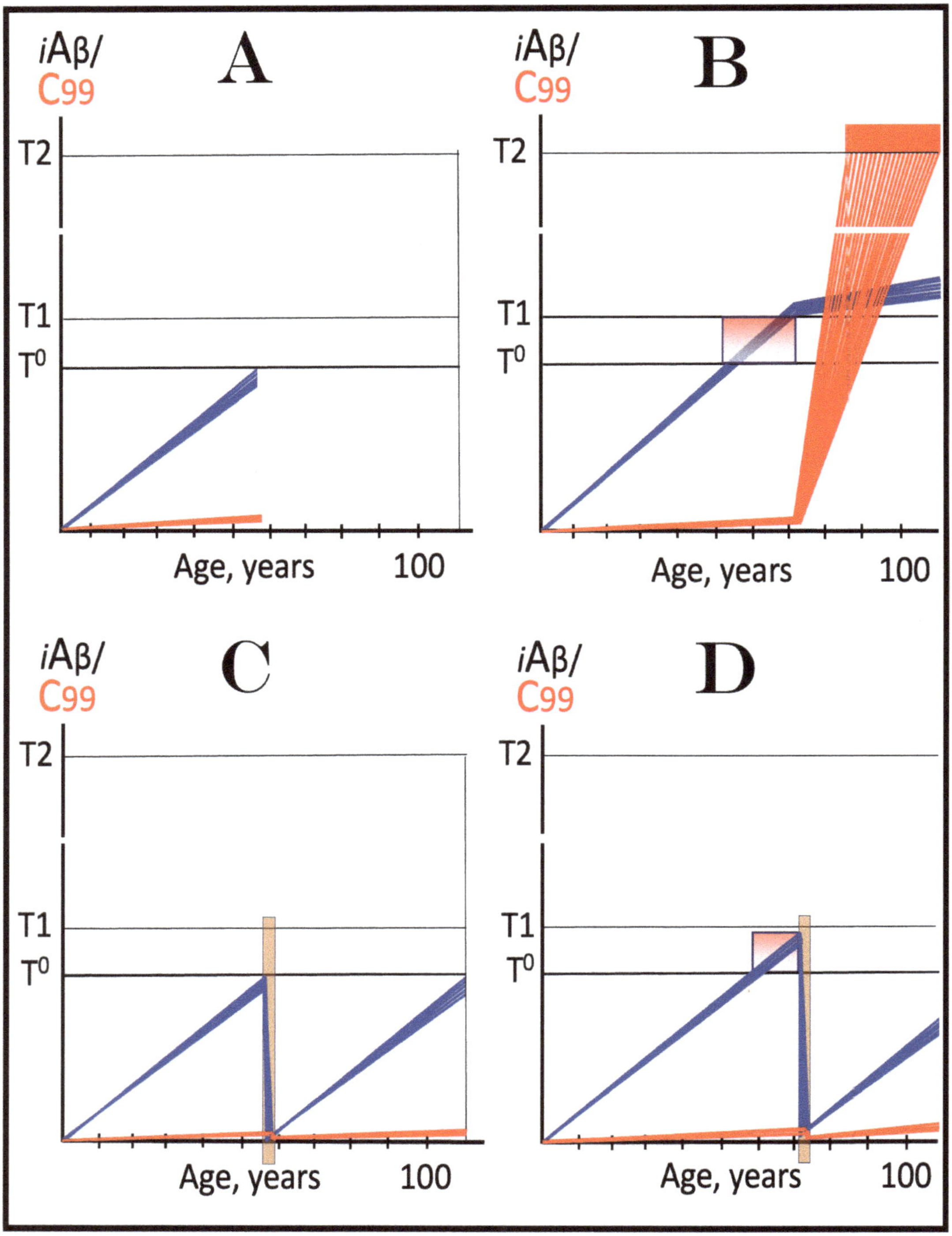

Figure 40. BACE1 and/or BACE2 activators in the prevention of conventional AD and in the prevention and treatment of AACD: the ACH2.0, Version Three perspective. *Blue lines*: Intraneuronal

Aβ, iAβ. *Red lines*: The C99 fragment of AβPP. ***T*⁰**: Threshold of cellular concentration of AβPP-derived iAβ triggering the neuronal damage, which manifests as AACD. ***T1***: Threshold of cellular concentration of iAβ triggering the activation of PKR and/or HRI kinases, phosphorylation of eIF2α, the elicitation of the neuronal ISR, the initiation of the AβPP-independent C99 production pathway, and the commencement of AD. ***T2***: Cellular concentration of iAβ triggering neuronal death via apoptosis or necroptosis. *Red box*: Apoptotic zone, defined as a range of cellular concentrations of iAβ within which cells neuronal cells committed apoptosis or necroptosis or are dead. *Pink box*: The range of cellular concentrations of AβPP-derived iAβ concentrations between the T^0 and the highest extent reached by AβPP-derived iAβ or the T1 threshold that support the progression of AACD. The condition commences with the crossing of the T^0 threshold and morphs into AD with the T1 crossing and elicitation of the ISR; it can occur only if the extent of the T^0 threshold is smaller than that of the T1. *Orange box*: Duration of the treatment with the BACE1 and/or BACE2-activating drug. *Panel* (**A**): The initial state of the levels of AβPP-derived iAβ and C99 in the neurons of an individual. At the time of the drug's administration these levels are below the T^0 threshold. *Panel* (**B**): The evolution of the initial state in the untreated individual. AβPP-derived iAβ continues to accumulate. When it crosses the T^0 threshold, AACD commences and morphs into AD upon the T1 crossing. At this point PKR and/or HRI kinases are activated, eIF2α is phosphorylated at its Ser51 residue, and the neuronal integrated stress response is elicited. Under the ISR conditions the AβPP proteolytic pathway is suppressed and the rate of accumulation of AβPP-derived iAβ is substantially reduced. Concurrently, the neuronal ISR enables the operation of the AβPP-independent C99 generation pathway. C99 produced independently of AβPP rapidly accumulates and drives the progression of AD pathology. When the T2 threshold is crossed in a sufficient fraction of the neurons, the disease reaches its end stage. *Panel* (**C**): The evolution of the initial state following the transient administration of the BACE1 and/or BACE2-activating drug. The treatment substantially depletes AβPP-derived iAβ and its accumulation resumes from low baseline and proceeds at the pre-treatment rate. Neither the T^0 or T1 thresholds are reached nor AACD or conventional AD occurs within the remaining lifetime of the treated individual. *Panel* (**D**): The outcome of the transient treatment with the BACE1 and/or BACE2 activating drug administered to the AACD patient. At the time of the implementation of the treatment AβPP-derived iAβ has already crossed the T^0 threshold and triggered the commencement of AACD. The transient administration of BACE1 and/or BACE2 activators depletes AβPP-derived iAβ well below the T^0 threshold. When the T1 is reverse-crossed by AβPP-derived iAβ, AACD is cured. The de novo accumulation of AβPP-derived iAβ commences from a low baseline. It reaches neither T^0 nor T1 thresholds within the remaining lifetime of the treated individual. Thus, a single transient iAβ depletion treatment via the activation of BACE1 and/or BACE2 prevents both the occurrence of conventional AD and recurrence of AACD.

Panel C of Figure 40 illustrates the evolution of the initial state following the transient iAβ depletion treatment via the activation of BACE1 and/or BACE2. The treatment substantially depletes AβPP-derived iAβ and its accumulation resumes from low baseline and proceeds at the pre-treatment rate. Neither the T^0 or T1 thresholds are reached, nor do AACD or conventional AD occur within the remaining lifetime of the treated individual. Panel D of Figure 40 presents the outcome of the transient treatment with the BACE1 and/or BACE2 activating drug administered to the AACD patient. At the time of the implementation of the treatment AβPP-derived iAβ has already crossed the T^0 threshold and triggered the commencement of AACD. The transient treatment with activators of BACE1 and/or BACE2 depletes AβPP-derived iAβ well below the T^0 threshold. When this threshold is reverse-crossed by AβPP-derived iAβ, AACD is cured. The de novo accumulation of AβPP-derived iAβ commences from a low baseline. It reaches neither T^0 nor T1 thresholds within the remaining lifetime of the treated individual. Thus, a single transient iAβ depletion treatment via the activation of BACE1 and/or BACE2 prevents both the occurrence of conventional AD and recurrence of AACD. It should be emphasized that treatment with directly acting iAβ-specific degradation agents such as appropriately designed proteolysis-targeting chimeras (PROTACs) and molecular-glue degraders (MGDs) would also achieve the outcomes described in the present section above.

75. BACE1 and/or BACE2 Activators Can Be Effective in the Treatment of Conventional AD Only if Administered Concurrently with ISR Inhibitors; A Single Transient Composite Therapy Could Prevent for Life the Recurrence of the Disease: The ACH2.0 Version Three Perspective

In the preceding two versions of the ACH2.0, the purpose of the activation of BACE1 and/or BACE2 as a therapeutic strategy in the treatment of conventional AD and the prevention and treatment of unconventional AD was to deplete iAβ (produced overwhelmingly in the AβPP-independent manner). The rationale for such a strategy is simple. By depleting iAβ we remove the driver of AD pathology (a presumption of versions One and Two of the ACH2.0). By depleting iAβ we presumably (in the first two versions of the ACH2.0) also remove the propagator of the neuronal integrated stress response and the sustainer of the activity of the AβPP-independent C99 (and iAβ in versions One and Two) production pathway. Without its driver, the progression of AD pathology ceases, and it takes time (in case of conventional AD, potentially in excess of the remaining lifetime) for iAβ to accumulate to the neuronal ISR-eliciting concentrations.

Whereas the efficiencies of intra-iAβ cleavages by activated BACE1 and, especially, BACE2 are, apparently, sufficient to effectively deplete iAβ produced solely by the AβPP proteolysis, they cannot match the influx of C99 (and iAβ in versions One and Two) generated independently of AβPP. Therefore, to enable the depletion of iAβ (rather than only a reduction in the rate of its accumulation), its production in the AβPP-independent pathway must be stopped for the duration of the administration of BACE1 and/or BACE2 activators. In the composite strategy described above for versions One and Two of the ACH2.0 this is achieved by the concurrent administration of ISR inhibitors. The suppression of the neuronal ISR deprives the AβPP-independent C99 production pathway of its essential components and terminates its operation. With the influx of iAβ produced independently of AβPP abrogated, activated BACE1 and BACE2 can effectively deplete total iAβ.

In Version Three of the ACH2.0 the driver of AD pathology is the C99 fragment generated independently of AβPP; it assumes the role of iAβ produced in the AβPP-independent pathway formulated in the preceding versions of the ACH2.0 in both driving the disease and propagating the neuronal integrated stress response and thus sustaining the activity of the AβPP-independent C99 generation pathway (thus rendering it self-sustainable). Therefore, to attain the objectives of the therapeutic strategy described above, C99 has to be depleted. This is in addition to the depletion of the accumulated AβPP-derived iAβ, so that its de novo accumulation would commence from a low baseline. Fortunately, the degradation and depletion of C99 can be accomplished in precisely the same way as those of iAβ, namely via the activation of intra-iAβ cleaving capabilities of BACE1 and/or BACE2. Indeed, C99 contains the Aβ segment as its integral part. The toxicity of C99 apparently emanates from that of its Aβ segment. Thus, when the Aβ segment is cleaved off by gamma-secretase, the remaining AICD loses the toxicity of C99. Moreover, when only sixteen N-terminal residues are removed from C99, the remaining C83 fragment is rendered non-toxic. It can be firmly anticipated, therefore, that intra-iAβ cleavages by activated BACE1and/or BACE2 would degrade C99 and detoxify its resulting fragments. However, there are two problems. One, the production of BACE enzymes has been suppressed under the neuronal ISR conditions. Another, just as with iAβ in the preceding versions of the ACH2.0, in its version Three the influx of C99 generated independently of AβPP is so great that to deplete it via its degradation by activated BACE1 and/or BACE2, its production in the AβPP-independent pathway must be stopped. Both problems are solved by the transient inhibition of the neuronal ISR concurrently with the activation of BACE1 and/or BACE2.

It should be emphasized that there is a principal distinction of the purpose of the inclusion of ISR inhibitors in the composite C99/iAβ depletion therapy in Version Three of the ACH2.0 versus two preceding ACH2.0 versions. In the preceding versions of the ACH2.0 the sole purpose of ISR inhibitors was to disable the AβPP-independent C99 production pathway and thus to abrogate the influx of iAβ generated independently of AβPP. This purpose remains in effect in Version Three as well: ISR inhibitors would stop the influx of C99 generated independently of AβPP. But in this version of the ACH2.0 there is a second, no less, and arguably more important purpose: to enable the production of BACE1 and BACE2. Indeed, in the ACH2.0, Version Three the neuronal ISR state suppresses all components of the AβPP proteolytic pathway, including the production of BACE enzymes. The reversal of the neuronal ISR state not only disables the AβPP-independent C99 production pathway but also ensures the availability of BACE enzymes. Since, for the reasons described above (Section 43), systemic ISR inhibitors can be administered only transiently, so BACE1 and BACE2 activators can (because BACE enzymes are both unavailable and would be inefficient under the neuronal ISR conditions). The transient composite therapy, nevertheless, would be sufficient to prevent the recurrence of conventional AD for the remaining lifetime of the treated patient.

Thus, the transient composite BACE1 and or BACE2 activation administered concurrently with the ISR inhibition is fully valid as a therapeutic strategy in ACH2.0, Version Three. It would result in the depletion of not only C99, but also of the accumulated AβPP-derived iAβ. Following the completion of the treatment of the AD patient, the accumulation of AβPP-derived iAβ would resume from a low baseline at the pre-T1 crossing rate. It would not reach the T1 threshold, and conventional AD would not recur for the duration of the treatment.

The considerations described above are illustrated in Figure 41. Panel A of Figure 41 depicts the initial state of the levels of AβPP-derived iAβ and C99, produced mainly in the AβPP-independent pathway, at the time of the administration of the composite BACE activation/ISR inhibition therapy. In this panel, AβPP-derived iAβ, accumulating via the cellular uptake of secreted extracellular Aβ and the retention of a fraction of Aβ resulting from gamma-cleavages on the intraneuronal membranes, has crossed the T1 threshold in all affected neurons. PKR and/or HRI have been activated, eIF2α phosphorylated at its Ser51 residue, and the neuronal ISR elicited. As a result, the AβPP proteolytic pathway has been suppressed and the rate of accumulation of AβPP-derived iAβ substantially reduced. Concurrently, the AβPP-independent C99 generation pathway has been activated, and AD commenced. C99 produced independently of AβPP has rapidly accumulated, crossed the T2 threshold in a fraction of the neurons, and AD symptoms manifested. Panel B of Figure 41 shows the evolution of the initial state in the untreated patient. The accumulation of C99 generated independently of AβPP continues unimpeded, more neurons cross the T2 threshold, and the disease enters its end stage. Panel C of Figure 41 presents the evolution of the initial state following the composite BACE activation/ISR inhibition therapy. The reversal of the neuronal ISR disables the AβPP-independent C99 generation pathway and ceases the influx of C99 produced independently of AβPP. It also restores the production of BACE enzymes and ensures their availability. With AβPP-independent C99 production pathway inoperative, activated BACE1 and/or BACE2 efficiently and substantially deplete both C99 and iAβ. When drugs are withdrawn, the progression of AD is arrested; the accumulation of AβPP-derived iAβ resumes from low baseline and proceeds at a slow pre-T1 crossing rate. It would not reach the T1 threshold and conventional AD would not recur in the treated individual. Thus, a single composite BACE activation/ISR inhibition treatment can prevent for life the recurrence of conventional AD. It should be emphasized that in this strategy BACE1 and/or BACE2-activating drugs can be replaced, with the same effect, with directly acting iAβ-specific degradation agents such as appropriately designed proteolysis-targeting chimeras (PROTACs) and molecular-glue degraders (MGDs).

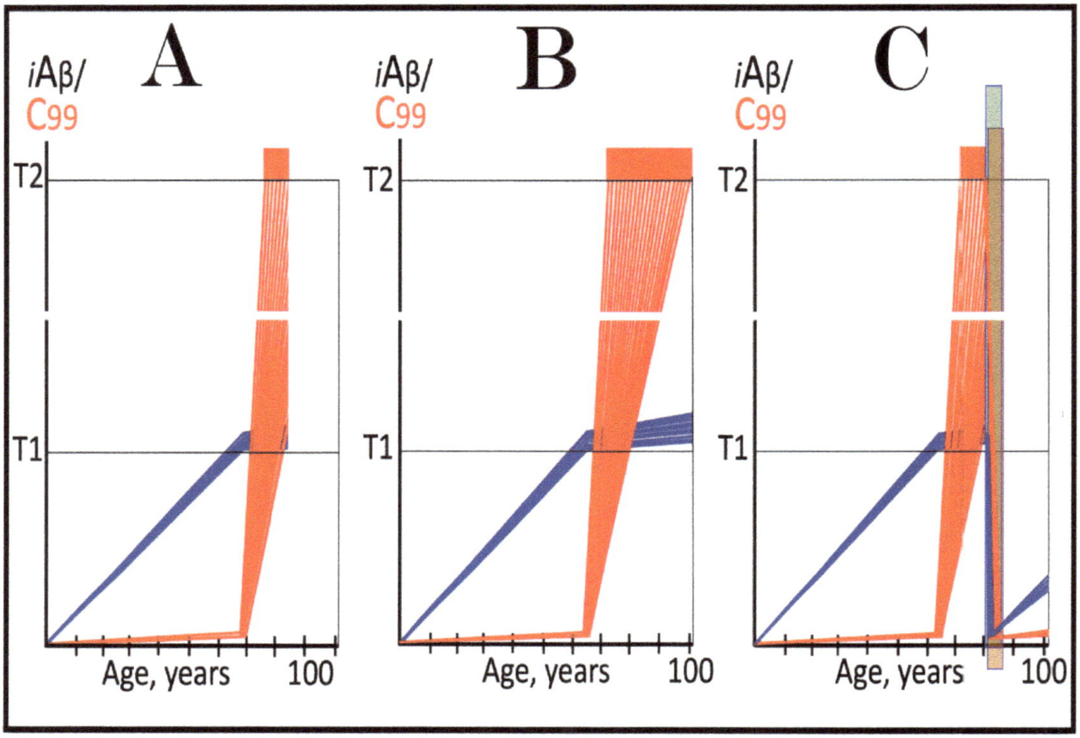

Figure 41. A single transient composite ISR inhibition/BACE activation therapy could prevent for life the recurrence of conventional AD: the ACH2.0 Version Three perspective. *Blue lines*: Intraneuronal Aβ, iAβ. *Red lines*: The C99 fragment of AβPP. ***T1***: Threshold of cellular concentration of iAβ triggering the activation of PKR and/or HRI kinases, phosphorylation of eIF2α, the elicitation of the neuronal ISR, the initiation of the AβPP-independent C99 production pathway, and the commencement of AD; its extent is the only variable in the present figure. ***T2***: Cellular concentration of iAβ triggering neuronal death via apoptosis or necroptosis. *Red box*: Apoptotic zone, defined as a range of cellular concentrations of iAβ within which cells neuronal cells committed apoptosis or necroptosis or are dead. *Green box*: Duration of the treatment with the ISR-inhibiting drug. *Orange box*: Duration of the treatment with the BACE1 and/or BACE2-activating drug. *Panel* (**A**): The initial state of the levels of AβPP-derived iAβ and C99, produced mainly in the AβPP-independent pathway, at the time of the administration of the composite BACE activation/ISR inhibition therapy. AβPP-derived iAβ has crossed the T1 threshold in all affected neurons. PKR and/or HRI have been activated, eIF2α phosphorylated at its Ser51 residue, and the neuronal ISR elicited. As a result, the AβPP proteolytic pathway has been suppressed and the rate of accumulation of AβPP-derived iAβ substantially reduced. Concurrently, the production of C99 in the AβPP-independent pathway has been activated, and AD commenced. It rapidly accumulated, crossed the T2 threshold in a fraction of the neurons, and AD symptoms manifested. *Panel* (**B**): The evolution of the initial state in the untreated patient. The accumulation of C99 generated independently of AβPP continues, more neurons cross the T2 threshold, and the disease enters its end stage. *Panel* (**C**): The evolution of the

initial state following the composite BACE activation/ISR inhibition therapy. The reversal of the neuronal ISR disables the AβPP-independent C99 generation pathway and ceases the influx of its C99 product. It also restores the production of BACE enzymes and ensures their availability. Activated BACE1 and/or BACE2 efficiently and substantially deplete both C99 and iAβ. When drugs are withdrawn, the progression of AD is arrested; the accumulation of AβPP-derived iAβ resumes from low baseline and proceeds at a slow pre-T1 crossing rate. It would not reach the T1 threshold and conventional AD would not recur in the treated individual. Thus, a single composite BACE activation/ISR inhibition treatment can prevent for life the recurrence of conventional AD.

76. Transient Composite BACE Activation/ISR Inhibition Therapy in the Prevention and Treatment of Unconventional AD: The ACH2.0 Version Three Perspective

The present section considers the potential outcomes of the transient composite BACE activation/ISR inhibition therapy in the prevention and treatment of unconventional AD. In this form of the disease AβPP-derived iAβ has little, if any, relevance. The disease is triggered by the neuronal integrated stress response elicited unconventionally, i.e., by a stressor distinct from AβPP-derived iAβ, when the levels of the latter are below the T1 threshold. The neuronal ISR activates, in turn, the AβPP-independent C99 generation pathway, and C99 produced independently of AβPP rapidly accumulates. If the duration of the occurrence of unconventional stressors and, consequently, of the ISR state is sufficient, C99 levels would cross the T1 threshold, the AβPP-independent C99 generation pathway would become self-sustainable, and the disease would commence. In the present and the following sections, we consider the scenarios where once the unconventional stressor capable of eliciting the neuronal integrated stress response appeared, it persists for the remaining lifetime.

In unconventional AD, as in its conventional counterpart, the objective of the therapeutic strategy is to substantially deplete the driver of AD pathology, in this case C99 generated independently of AβPP. Inhibiting the neuronal ISR and, concurrently activating BACE1 and/or BACE2 can accomplish this. The inhibition of the ISR enables the production of BACE enzymes. It also disables the AβPP-independent C99 generation pathway, stops the influx of C99 produced independently of AβPP, and empowers the efficient depletion of C99. BACE activators can be used only in the absence of the ISR state. Therefore, since ISR inhibitors can be employed only transiently, BACE activators would be effective for no longer duration. It is expected that at the completion of the transient composite BACE activation/ISR inhibition treatment C99 would be substantially depleted. However, unlike in conventional AD, in the unconventional form of the disease the depletion of C99 does not result in the cessation of the activity of the AβPP-independent C99 generation pathway. This is because, as soon as drugs are withdrawn, the neuronal ISR would be unconventionally re-elicited (due to the persistent presence of unconventional stressors), and AβPP-independent C99 production pathway unconventionally activated. The accumulation of C99 produced independently of AβPP would resume from a low baseline but as soon as it crosses the T1 threshold, AD would commence (or re-commence). The delay in the occurrence of the disease (in preventive cases) or in its recurrence (in the treatment scenario) would be relatively short, but still probably measured in years.

The above considerations are illustrated in Figure 42. Panel A of Figure 42 depicts the anticipated outcome of the implementation of the transient composite BACE activation/ISR inhibition therapy in the prevention of unconventional AD. In this panel, at the time of the administration of the therapy the neuronal integrated stress response has been unconventionally elicited at the levels of AβPP-derived iAβ below the T1 thresholds. Consequently, the AβPP proteolytic pathway has been suppressed and the AβPP-independent C99 generation pathway unconventionally activated. C99 produced independently of AβPP has rapidly accumulated but at the commencement of the composite therapy is still below the T1 threshold; the disease, therefore, has not yet commenced. The ISR inhibition enables the production of BACE enzymes, disables the AβPP-independent C99 generation pathway, and stops the influx of C99 produced independently of AβPP. In this setting, activated BACE1 and/or BACE2 efficiently deplete C99. When drugs are withdrawn, the neuronal ISR is unconventionally re-elicited, and the AβPP-independent C99 generation pathway unconventionally reactivated. C99 produced independently of AβPP rapidly accumulates. When it crosses the T1 threshold the disease commences and progresses.

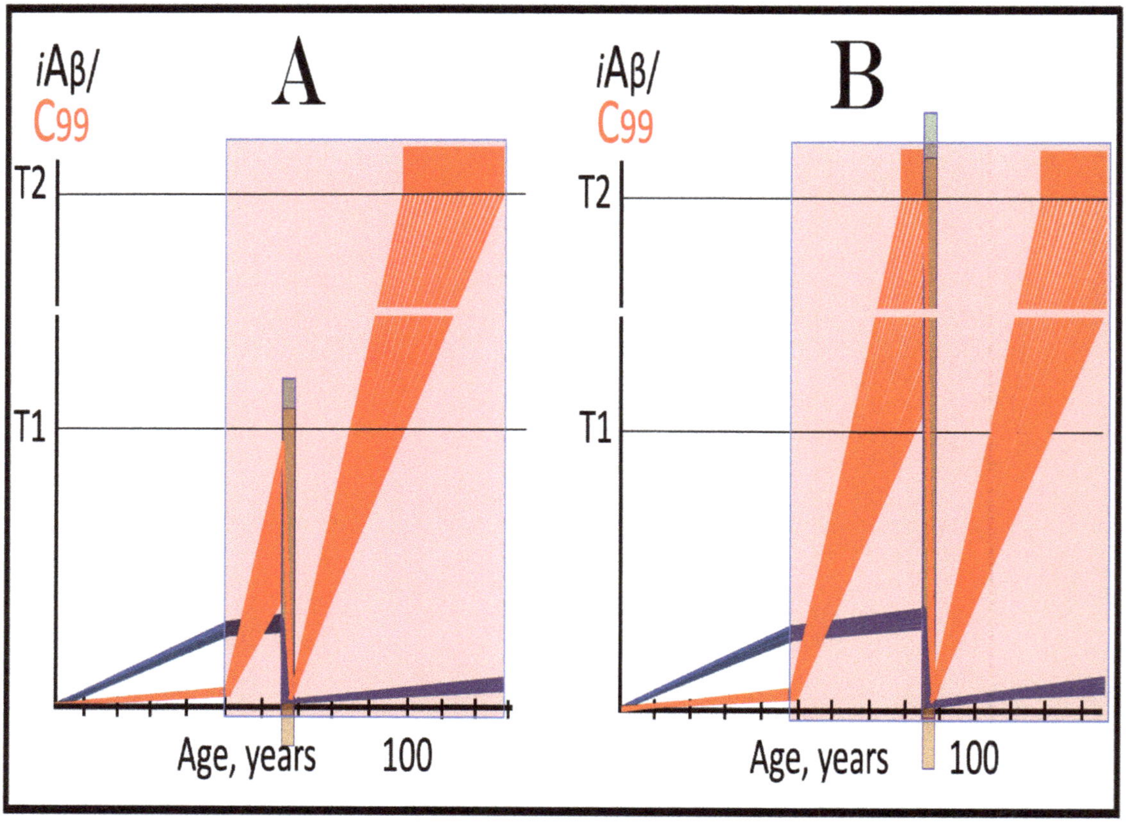

Figure 42. Transient composite ISR inhibition/BACE activation therapy in the prevention and treatment of unconventional AD: The ACH2.0 Version Three perspective. *Blue lines*: Intraneuronal Aβ, iAβ. *Red lines*: The C99 fragment of AβPP. ***T1***: Threshold of cellular concentration of iAβ triggering the activation of PKR and/or HRI kinases, phosphorylation of eIF2α, the elicitation of the neuronal ISR, the initiation of the AβPP-independent C99 production pathway, and the commencement of AD; its extent is the only variable in the present figure. ***T2***: Cellular concentration of iAβ triggering neuronal death via apoptosis or necroptosis. *Red box*: Apoptotic zone, defined as a range of cellular concentrations of iAβ within which cells neuronal cells committed apoptosis or necroptosis or are dead. *Green box*: Duration of the treatment with the ISR-inhibiting drug. *Orange box*: Duration of the treatment with the BACE1 and/or BACE2-activating drug. *Pink boxes*: The duration of the presence of the unconventional stressor capable of activating one or more of eIF2α kinases and triggering the elicitation of the neuronal ISR. *Panel* (**A**): The anticipated outcome of the transient composite BACE activation/ISR inhibition therapy in the prevention of unconventional AD. At the time of the administration of the therapy the neuronal ISR has been unconventionally elicited at the

levels of AβPP-derived iAβ below the T1 thresholds. Consequently, the AβPP proteolytic pathway and the accumulation of AβPP-derived iAβ have been suppressed, and the AβPP-independent C99 generation pathway unconventionally activated. C99 produced independently of AβPP has rapidly accumulated but is still below the T1 threshold; the disease does not commence. The ISR inhibition enables the production of BACE enzymes, disables the production of C99 in the AβPP-independent C99 pathway, and stops its influx. In this setting, activated BACE1 and/or BACE2 efficiently deplete C99 and iAβ. When drugs are withdrawn, the neuronal ISR is unconventionally re-elicited, and the AβPP-independent C99 generation pathway unconventionally reactivated. C99 produced independently of AβPP rapidly accumulates. When it crosses the T1 threshold the disease commences. *Panel* (**B**): Anticipated effects of the transient composite BACE activation/ISR inhibition therapy in the treatment of unconventional AD. At the time of the treatment the neuronal ISR has been unconventionally elicited at levels of AβPP-derived iAβ below the T1 threshold. The AβPP proteolytic pathway has been suppressed and the AβPP-independent C99 production pathway unconventionally activated. C99 generated independently of AβPP has rapidly accumulated and crossed the T1 threshold. The AβPP-independent C99 production pathway was rendered self-sustainable, and AD commenced. C99 has continued to accumulate, crossed the T2 threshold in a fraction of the neurons, and AD symptoms manifested. The composite treatment resulted in a substantial depletion of C99 and iAβ. However, upon the withdrawal of the drugs, the neuronal ISR is unconventionally re-elicited, the AβPP-independent C99 production pathway unconventionally reactivated; C99 rapidly accumulates, re-crosses the T1 threshold, and the disease recurs.

Panel B of Figure 42 shows anticipated effects of the concurrent administration of BACE activators and ISR inhibitors in the treatment of unconventional AD. In this panel, the neuronal ISR has been unconventionally elicited at levels of AβPP-derived iAβ below the T1 threshold. The AβPP proteolytic pathway has been suppressed and the AβPP-independent C99 production pathway unconventionally activated. C99 generated independently of AβPP has rapidly accumulated and crossed the T1 threshold. The AβPP-independent C99 production pathway was rendered self-sustainable, and AD commenced. C99 has continued to accumulate, crossed the T2 threshold in a fraction of the neurons, and AD symptoms manifested. The transient concurrent administration of ISR inhibitors and BACE activators led to substantial depletion of C99 in exactly the same way as described in the preceding paragraph. However, upon the withdrawal of the drugs, the neuronal ISR is unconventionally re-elicited, the AβPP-independent C99 production pathway unconventionally reactivated; C99 rapidly accumulates, re-crosses the T1 threshold, and the disease recurs. Thus, in both scenarios, preventive and curative, the transient administration of the composite BACE activation/ISR inhibition therapy is anticipated to provide a temporary reprieve from either the occurrence or recurrence of unconventional AD; the duration of such a reprieve is probably measured in years.

If, in the above-described strategy, BACE-activating drugs were substituted by directly acting iAβ-specific degradation agents such as appropriately designed proteolysis-targeting chimeras (PROTACs) and molecular-glue degraders (MGDs), similar outcomes would be attained. Importantly, utilization of directly acting iAβ-specific degradation agents instead of BACE activators in the above strategy provides an additional option of a long-term administration because, unlike BACE enzymes, once introduced, their cellular presence is not affected by the ISR state; the anticipated outcomes of this approach are analogous to those discussed (for BACE activators) in Section 61 and shown in Figure 31.

77. Recurrent Transient Composite BACE1 and/or BACE2 Activation/ISR Inhibition Therapy in the Prevention and Treatment of Unconventional AD: The ACH2.0 Version Three Perspective

As discussed in the preceding section, the activated BACE-mediated degradation of C99/iAβ is feasible only in the presence of ISR inhibitors, and since the latter can be used only transiently, so the former can. However, the transient concurrent administration of BACE activators and ISR inhibitors provides only a temporary reprieve, both preventive and curative: following the treatment, either the occurrence or the recurrence of unconventional AD is only delayed. The benefit, therefore, is transient, lasting, apparently, a few years. On the other hand, this limitation, both in the prevention and treatment of unconventional AD can be circumvented. Indeed, the duration of benefits conferred by both preventive and curative applications of the transient composite BACE activation/ISR inhibition therapy could be multiplied by repeating the treatment at strategically timed intervals. Conceptually, the treatments could continue recurrently for the remaining lifetime, ensuring that unconventional AD neither occurs in the preventively treated individual nor recurs in the treated patient.

The above strategy is illustrated schematically In Figure 43. Panel A of Figure 43 depicts the effects of the recurrent transient composite BACE activation/ISR inhibition therapy in the prevention of unconventional AD. In this panel, the neuronal integrated stress response has been unconventionally elicited at the levels of AβPP-derived iAβ below the T1 threshold. Consequently, the AβPP proteolytic pathway has been suppressed and the rate of accumulation of AβPP-derived iAβ substantially reduced. Concurrently, the neuronal ISR has activated the AβPP-independent C99 generation pathway. C99 produced independently of AβPP has rapidly accumulated but at the time of the initial administration of the transient composite BACE activation/ISR inhibition therapy has not yet reached the T1 threshold. The inhibition of the neuronal ISR enables the production of BACE enzymes, disables the AβPP-independent C99 generation pathway, stops the influx of its C99 product, and empowers the efficient depletion of the latter by activated BACE1 and/or BACE2. When drugs are withdrawn, the neuronal ISR is unconventionally re-elicited, the AβPP-independent C99 generation pathway unconventionally reactivated, and C99 produced independently of AβPP resumes its accumulation from a low baseline. Before it reaches the T1 threshold, at the time points determined by appropriate biomarkers, the treatment is repeated recurrently for the remaining portion of the lifetime; neither the T1 threshold would be crossed, nor AD would occur in the treated individual.

Panel B of Figure 43 shows the effects of the recurrent transient composite BACE activation/ISR inhibition therapy in the treatment of unconventional AD. In this panel, at the time of the initial administration of the transient composite therapy, the neuronal ISR has been unconventionally elicited, the AβPP proteolytic pathway suppressed, AβPP-independent C99 generation pathway unconventionally activated, the C99 product of the latter has crossed the T1 threshold, the pathway was rendered self-sustainable, and the disease commenced. In a fraction of the neurons C99 has crossed the T2 threshold, and AD symptoms have manifested. The concurrent administration of ISR-inhibiting and BACE-activating drugs enables the production of BACE enzymes, ceases the operation of the AβPP-independent C99 generation pathway, and blocks the influx of its C99 product. In this setting, activated BACE1 and/or BACE2 efficiently deplete C99 (and iAβ). Upon the completion of the treatment, the neuronal ISR is unconventionally re-elicited, AβPP-independent C99 production pathway unconventionally reactivated, and the accumulation of its C99 product resumes from a low baseline. It crosses the T1 threshold but before it reaches AD-pathology-causing levels, at time points defined by suitable biomarkers, the transient composite therapy is re-implemented recurrently for the remaining lifetime. The AD-pathology-causing range of C99 concentrations would not be reached, and the disease would not recur in the treated patient.

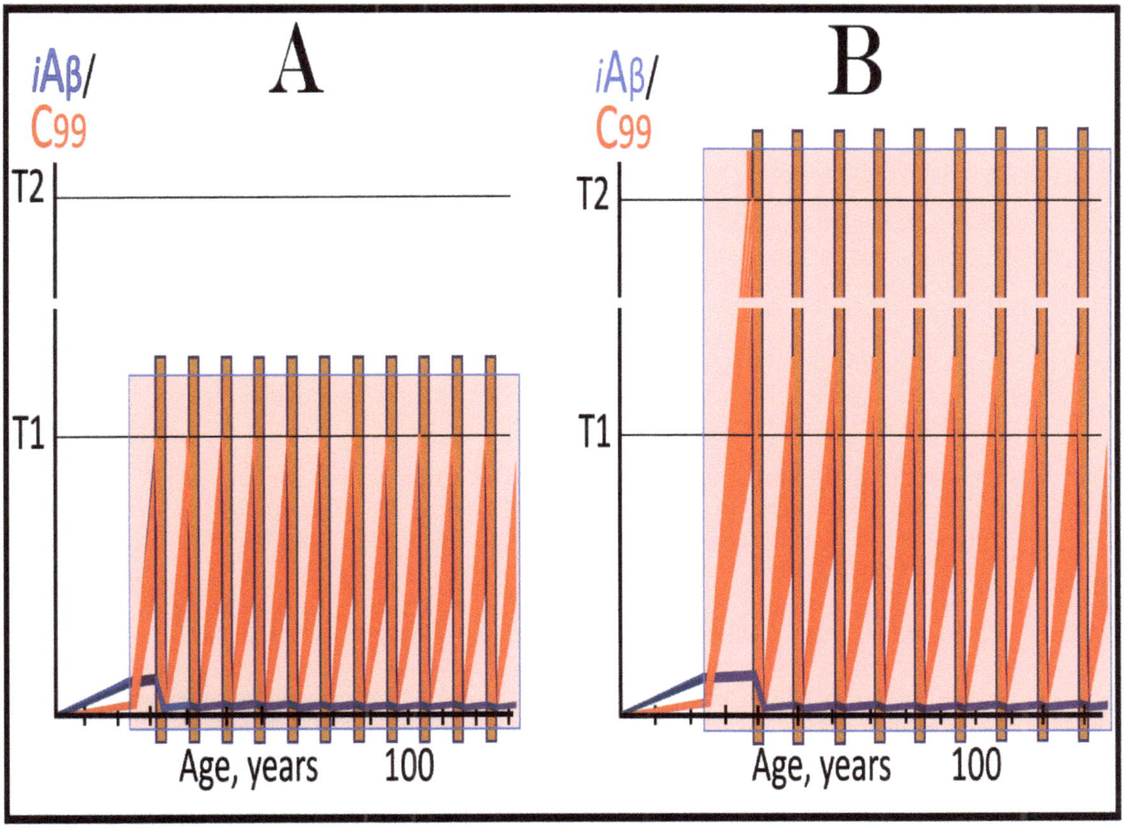

Figure 43. Recurrent transient composite BACE1 and/or BACE2 activation/ISR inhibition therapy in the prevention and treatment of unconventional AD: The ACH2.0 Version Three perspective. *Blue lines*: Intraneuronal Aβ, iAβ. *Red lines*: The C99 fragment of AβPP. ***T1***: Threshold of cellular concentration of iAβ triggering the activation of PKR and/or HRI kinases, phosphorylation of eIF2α, the elicitation of the neuronal ISR, the initiation of the AβPP-independent C99 production pathway, and the commencement of AD; its extent is the only variable in the present figure. ***T2***: Cellular concentration of iAβ triggering neuronal death via apoptosis or necroptosis. *Red box*: Apoptotic zone, defined as a range of cellular concentrations of iAβ within which cells neuronal cells committed apoptosis or necroptosis or are dead. *Pink boxes*: The duration of the presence of the unconventional stressor capable of activating one or more of eIF2α kinases and triggering the elicitation of the neuronal ISR. *Brown boxes*: The duration of the composite treatment with concurrently administered ISR inhibitors and BACE activators. *Panel* (**A**): Effects of the recurrent transient composite BACE activation/ISR inhibition therapy in the prevention of unconventional AD. At the time of the initial treatment the neuronal ISR has been unconventionally elicited. Consequently, the AβPP proteolytic pathway has been suppressed and the rate of accumulation of AβPP-derived iAβ substantially reduced. Concurrently, the production of C99 in the AβPP-independent pathway has been activated. It has rapidly accumulated but has not yet reached the T1 threshold. The inhibition of the neuronal ISR enables the production of BACE enzymes, disables the AβPP-independent C99 generation pathway, stops the influx of its C99 product, and empowers the efficient depletion of the latter by activated BACE1 and/or BACE2. When drugs are withdrawn, the neuronal ISR is unconventionally re-elicited, the AβPP-independent C99 generation pathway unconventionally reactivated, and C99 resumes its accumulation from a low baseline. Before it reaches the T1 threshold, at the time points determined by appropriate biomarkers, the treatment is repeated recurrently for the remaining portion of the lifetime;

neither the T1 threshold would be crossed, nor AD would occur in the treated individual. *Panel* (**B**): Effects of the recurrent transient composite BACE activation/ISR inhibition therapy in the treatment of unconventional AD. At the time of the initial treatment, the neuronal ISR has been unconventionally elicited, the AβPP proteolytic pathway suppressed, AβPP-independent C99 generation pathway unconventionally activated, the C99 product of the latter has crossed the T1 threshold, the pathway was rendered self-sustainable, and the disease commenced. In a fraction of the neurons C99 has crossed the T2 threshold, and AD symptoms have manifested. The concurrent administration of ISR-inhibiting and BACE-activating drugs enable the production of BACE enzymes, cease the operation of the AβPP-independent C99 generation pathway, and block the influx of its C99 product. In this setting, activated BACE1 and/or BACE2 efficiently deplete C99 and iAβ. Upon the completion of the treatment, the neuronal ISR is unconventionally re-elicited, AβPP-independent C99 production pathway unconventionally reactivated, and the accumulation of its C99 product resumes from a low baseline. It crosses the T1 threshold, but before it reaches AD-pathology-causing levels, at time points defined by suitable biomarkers, the transient composite therapy is re-implemented recurrently for the remaining lifetime. The AD-pathology-causing range of C99 concentrations would not be reached, and the disease would not recur in the treated patient.

78. Sustained Reduction/Removal of Unconventional Stressors Would Convert Unconventional AD into Conventional AD, Preventable and Treatable by a Single Transient Administration of the Composite BACE Activation/ISR Inhibition Therapy

The preceding sections demonstrate the feasibility of preventing and treating unconventional AD by recurrent applications of transient composite BACE activation/ISR inhibition therapy administered every few years for the remaining lifetime. There is, however, a better way to do it. Indeed, it is of the utmost importance to emphasize that the best way to approach the prevention and treatment of unconventional AD is to perform the transient composite BACE activation/ISR inhibition therapy (or utilize directly acting iAβ degradation agents instead of BACE activators in this approach) in conjunction with the reduction or removal of unconventional stressors capable of eliciting the neuronal integrated stress response via the elimination of their source. The unconventional stressors form the foundation of unconventional AD. By eliciting the neuronal integrated stress response, they enable the operation of the AβPP-independent C99 generation pathway and sustain it until C99 produced independently of AβPP crosses the T1 threshold and the pathway is rendered self-sustainable. If the unconventional stressors were removed (or their levels reduced below those required for the elicitation of the neuronal ISR), the outcomes of the transient composite BACE activation/ISR inhibition therapy would be as follows. The transient composite BACE activation/ISR inhibition treatment would substantially deplete C99 and AβPP-derived iAβ. When drugs are withdrawn both C99 and AβPP-derived iAβ are well below the neuronal ISR-eliciting T1 threshold. Unconventional stressors are either sustainably removed or their levels are below those needed for the elicitation of the neuronal ISR. Therefore, the neuronal ISR is not re-elicited, and the AβPP-independent C99 generation pathway remains inoperative. Under these conditions the de novo accumulation of AβPP-derived iAβ would commence from a low threshold and proceed at a slow rate. It would not cross the T1 threshold, and no AD would occur or recur within the remaining lifetime of the treated individual. The removal of unconventional neuronal ISR-eliciting stressor effectively converts unconventional AD (or perspective unconventional AD) into conventional one. One-time-only C99/iAβ depletion treatment, therefore, would be sufficient to either prevent the disease or preclude its recurrence for life.

The above considerations are illustrated in Figure 44. In Panel A of Figure 44 the neuronal ISR has been unconventionally elicited at levels of AβPP-derived iAβ below the T1 threshold. The AβPP proteolytic pathway has been suppressed and the rate of accumulation of AβPP-derived iAβ reduced. Concurrently, the AβPP-independent C99 production pathway has been unconventionally activated; its C99 product has accumulated but is still below the T1 threshold. The transient application of the composite BACE activation/ISR inhibition therapy is carried out in conjunction with the sustained removal of unconventional stressors or the reduction in their levels below those required for the elicitation of the neuronal ISR. As detailed above, when the transient BACE activation/ISR inhibition treatment is completed, both C99 and AβPP-derived iAβ are substantially depleted. Importantly, there are no (or there are insufficient) unconventional stressors, and upon the withdrawal of BACE activators and ISR inhibitors the neuronal ISR is not re-elicited, and the AβPP-independent C99 production pathway remains inoperative. In such a setting, the accumulation of AβPP-derived iAβ would resume from a low baseline and proceed at a slow rate. It would not reach the T1 threshold and AD would not occur within the lifetime of the treated individual.

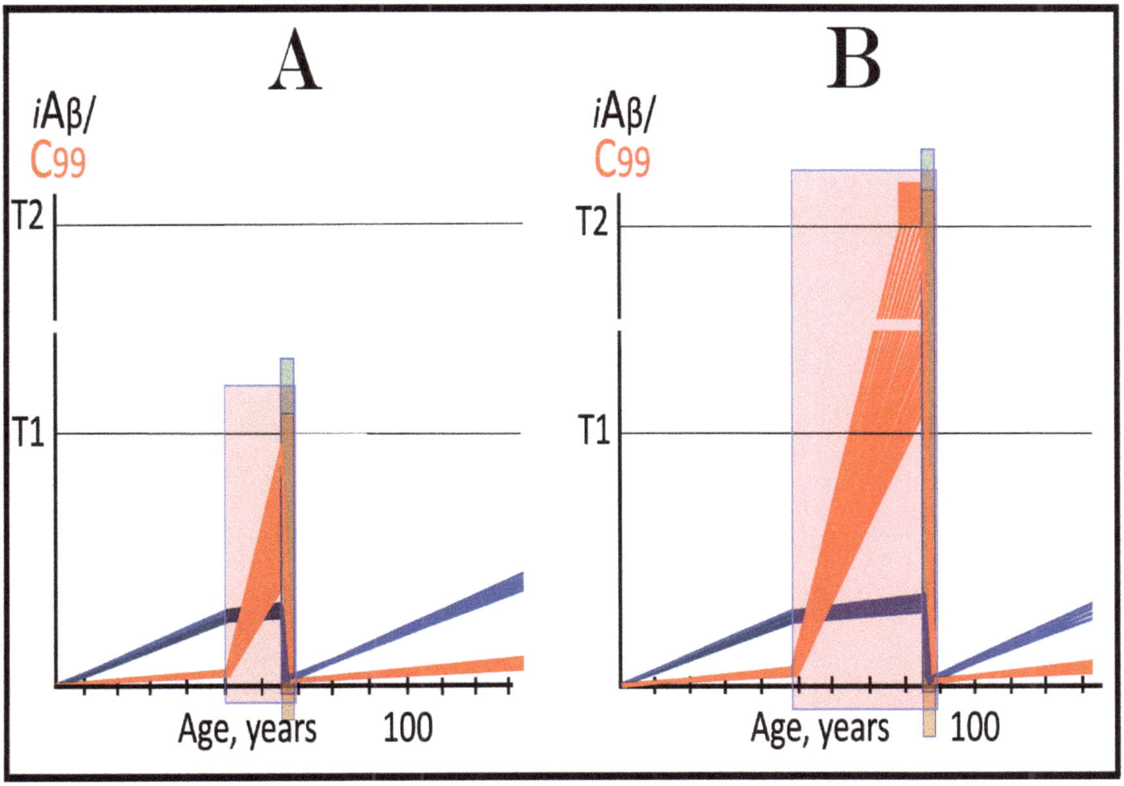

Figure 44. Sustained removal of unconventional stressors converts unconventional AD into conventional one, preventable and treatable by a single transient administration of the composite BACE activation/ISR inhibition therapy. *Blue lines*: Intraneuronal Aβ, iAβ. *Red lines*: The C99 fragment of AβPP. ***T1***: Threshold of cellular concentration of iAβ triggering the activation of PKR and/or HRI kinases, phosphorylation of eIF2α, the elicitation of the neuronal ISR, the initiation of the AβPP-independent C99 production pathway, and the commencement of AD; its extent is the only variable in the present figure. ***T2***: Cellular concentration of iAβ triggering neuronal death via apoptosis or necroptosis. *Red box*: Apoptotic zone, defined as a range of cellular concentrations of iAβ within which cells neuronal cells committed apoptosis or necroptosis or are dead. *Green box*: Duration

of the treatment with the ISR-inhibiting drug. *Orange box*: Duration of the treatment with the BACE1 and/or BACE2-activating drug. *Pink boxes*: The duration of the presence of the unconventional stressor capable of activating one or more of eIF2α kinases and triggering the elicitation of the neuronal ISR. *Panel* (**A**): The anticipated outcome of the transient composite BACE activation/ISR inhibition therapy administered in concert with the sustained removal of unconventional stressors in the prevention of unconventional AD. At the time of the treatment the neuronal ISR has been unconventionally elicited, the AβPP proteolytic pathway has been suppressed and the rate of accumulation of AβPP-derived iAβ reduced. Concurrently, the AβPP-independent C99 production pathway has been unconventionally activated; its C99 product has accumulated but is still below the T1 threshold. The transient administration of the composite BACE activation/ISR inhibition therapy is carried out in concert with the sustained removal of unconventional stressors (or their depletion below levels below the neuronal ISR-eliciting levels). When the transient BACE activation/ISR inhibition treatment is completed, both C99 and AβPP-derived iAβ are substantially depleted and so are unconventional stressors. The neuronal ISR is not re-elicited, and the AβPP-independent C99 production pathway remains inoperative. The accumulation of AβPP-derived iAβ resumes from a low baseline and proceed at a slow rate. It does not reach the T1 threshold and AD does not occur within the lifetime of the treated individual. *Panel* (**B**): The anticipated outcome of the transient composite BACE activation/ISR inhibition therapy administered in concert with the sustained removal of unconventional stressors in the treatment of unconventional AD. At the time of the treatment C99 produced in the unconventionally activated AβPP-independent pathway has crossed the T1 threshold, the pathway became self-sustainable, and AD commenced. C99 has continued to accumulate, crossed the T2 threshold in a fraction of the neurons, and AD symptoms manifested. The transient administration of the composite BACE activation/ISR inhibition therapy is carried out in concert with the sustained removal of unconventional stressors. When BACE-activating and ISR-inhibiting drugs are withdrawn, C99 and AβPP-derived iAβ are substantially depleted. The progression of the disease ceases, and the AβPP-independent C99 generation pathway remains inoperative. The de novo accumulation of AβPP-derived iAβ commences from low baseline and proceeds at a slow rate. It would not reach the T1 threshold, and the disease would not recur within the lifetime of the treated patient. Thus, the sustained removal/depletion of unconventional stressors capable of eliciting the neuronal ISR effectively converts unconventional AD into conventional one, preventable and treatable by a single transient composite ISR inhibition/BACE activation treatment.

In Panel B of Figure 44, C99 produced in the unconventionally activated AβPP-independent pathway has crossed the T1 threshold, the pathway became self-sustainable, and AD commenced. C99 has continued to accumulate, crossed the T2 threshold in a fraction of the neurons, and AD symptoms manifested. At this point the transient composite treatment with activators of BACE1 and/or BACE2 and inhibitors of the ISR is performed in concert with the sustained removal of unconventional stressor or their sustained depletion below the neuronal ISR-eliciting levels. When BACE-activating and ISR-inhibiting drugs are withdrawn, C99 and AβPP-derived iAβ are substantially depleted. The progression of the disease ceases. With unconventional stressors sustainably removed and C99 and AβPP-derived iAβ well below the T1 threshold, the AβPP-independent C99 generation pathway remains inoperative. The de novo accumulation of AβPP-derived iAβ commences from low baseline and proceeds at a slow rate. It would not reach the T1 threshold, and the disease would not recur within the lifetime of the treated patient.

79. Phase Two of the RNA-Dependent Mammalian mRNA Amplification: AβPP Is Potentially Produced in AD-Affected Neurons Unconventionally in an ISR-Driven and ISR-Compatible Process

Sections 10–13 above discussed the "chimeric" phase of RNA-dependent mammalian mRNA amplification in general and that of human AβPP mRNA in particular. It is a powerful process akin to massive gene amplification. Indeed, every molecule of the amplification-eligible mRNA species potentially serves repeatedly as a template (dubbed "progenitor mRNA") for the production of additional mRNA molecules. The first stage in the chimeric phase of mRNA amplification is synthesis of the antisense RNA strand initiating at the 3′poly(A) segment of the mRNA molecule. The presence of the poly(A) segment is, apparently, the only eligibility requirement for serving as a template for antisense RNA. The specificity of the amplification process is introduced at the following stage, which mandates the occurrence of two complementary segments within the antisense strand (the 3′ Terminal Complementary Element, TCE, and the Internal Complementary Element, ICE) that are mutually accessible in the folded configuration of the antisense RNA strand. Consequently, the antisense strand folds into a self-priming structure and is extended into a sense-orientation RNA. This is followed by separation of strands and cleavage of the chimeric RNA intermediate. One of the end products of the chimeric phase of mRNA amplification is a chimeric mRNA molecule consisting of 5′-truncated mRNA containing an antisense segment at its 5′ terminus. It encodes either the original polypeptide or, as in the case of human AβPP mRNA amplification, only a C-terminal fragment, CTF, of the original polypeptide. Another end product of the chimeric pathway is the 3′-truncated antisense RNA molecule. Initially, it was assumed to be a non-functional by-product. However, subsequent observations by Kapranov and coworkers of a class of modified mRNA molecules truncated at the 5′ end and containing 5′-terminal poly(U) segment [289] indicated that 3′-truncated antisense RNA end product of the chimeric phase of mRNA amplification can be utilized as a progenitor in the second, PCR-like phase of the mRNA amplification process.

Both phases of mRNA amplification are illustrated schematically in Figure 45. Stages 1 through 6 (designated "Tier One" in Figure 45) are identical to those discussed in Section 10 and shown in Figure 1. Briefly, the antisense RNA is transcribed from the mRNA progenitor by RNA-dependent RNA polymerase, RdRp. After strands are separated by a helicase complex, antisense RNA, guided by the interaction of its TCE and ICE elements, folds into a self-priming configuration, and its 3′ terminus (TCE) is extended into a sense-orientation molecule. This results in a hairpin-like structure (designated "chimeric RNA intermediate"). Complementary regions of the hairpin structure are then separated by the helicase complex, which mounts the 3′ poly(A) segment and proceeds in the 5′ direction. When it reaches the single-stranded portion of the hairpin structure, it cleaves it. The end product of this phase of mRNA amplification (designated "End Product Tier One" in Figure 45) is the chimeric mRNA described above. In the human AβPP mRNA amplification process it encodes the C100 fragment of AβPP (eventually processed into C99, see Section 23 above).

The cleavage of the chimeric RNA intermediate denotes the end of the first phase and the beginning of the second phase of the mRNA amplification process (designated "Tier Two" in Figure 45) [58,65,66]. In conjunction with the cleavage, the helicase complex or its associated activity polyadenylates the newly generated 3′ terminus of the antisense RNA strand. The resulting antisense RNA molecule contains, therefore, the 3′-terminal poly(A) segment and the 5′-terminal poly(U) segment (transcribed from the 3′ poly(A) of the mRNA progenitor molecule); it constitutes the progenitor molecule in the second phase of mRNA amplification. The presence of the 3′-terminal poly(A) segment makes the antisense RNA molecule an eligible RdRp template. The product of its transcription is the sense-orientation mRNA containing both 3′-terminal polyA) and 5′-terminal poly(U); it is, upon separation of strands by the helicase complex, also an eligible RdRp template. Since both strands are eligible RdRp templates, the entire process constitutes a physiologically occurring intracellular polymerase chain reaction, iPCR (for intracellular PCR).

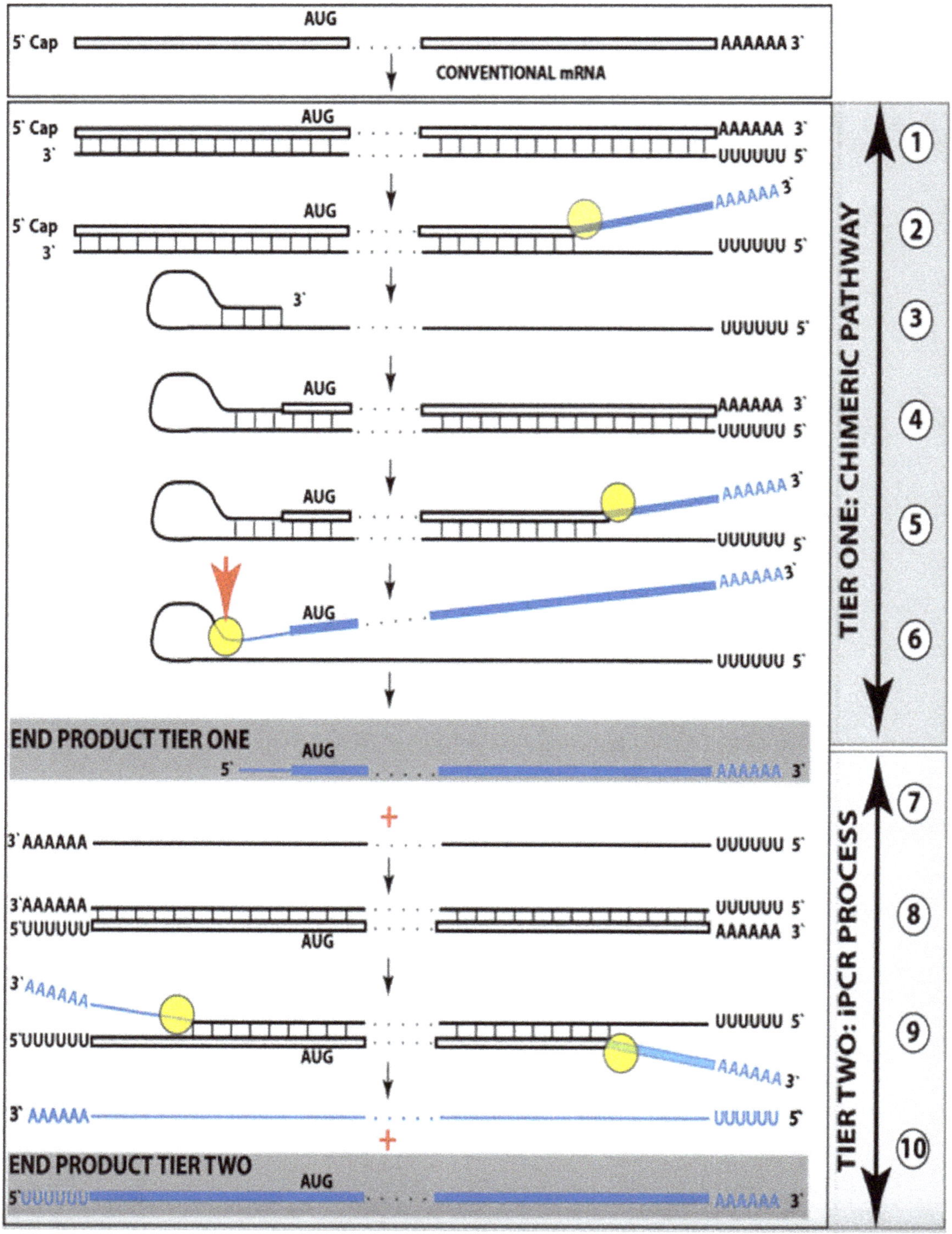

Figure 45. Mammalian RNA-dependent mRNA amplification occurs in two phases: Principal stages. *Single lines*: Antisense RNA. *Boxed lines*: Sense RNA. *Blue boxed lines*: Single-stranded RNA

separated from its complementary RNA strand by helicase. *Yellow circles*: Helicase complex containing helicase strand-separating activity, nucleotide modifying activity and RNA cleaving activity. *Red arrows*: positions of the cleavage of the intermediate generating RNA end products of the amplification process. *RdRp*: RNA-dependent RNA polymerase. *AUG*: Translation initiation codon. *TCE*: 3′ Terminal Complementary Element of the antisense RNA strands. *ICE*: Internal Complementary Element of the antisense RNA strands. ***Top panel***: Conventionally transcribed mRNA molecule referred to as progenitor mRNA. ***Bottom panel***: Stages of the chimeric pathway of the RNA-dependent mRNA amplification; the ICE is located within a portion of antisense RNA corresponding to the 5′ UTR of the progenitor mRNA molecule. (**1**): The progenitor mRNA is transcribed by RdRp into antisense RNA. (**2**): Complementary RNA strands are separated by the helicase activity. Helicase complex mounts the 3′ poly(A) segment and moves along the sense RNA strand modifying, on average, every fifth nucleotide. (**3**): Folding of the antisense RNA into self-priming configuration is guided by the interaction of its TCE and ICE elements. (**4**): RdRp extends the 3′ terminus of self-primed antisense RNA. This produces a hairpin-like molecule containing both sense and antisense RNA components, referred to as the chimeric RNA intermediate. (**5**): Complementary strands of the chimeric RNA intermediate are separated by the helicase activity. Nucleotide modifications introduced during the separation prevent the re-annealing of the sense and antisense RNA strands. (**6**): When the helicase reaches the single-stranded portion of the chimeric intermediate (either the 5′ end of the TCE or a TCE/ICE mismatch) it cleaves the RNA molecule; this cleavage is coupled with polyadenylation of the 3′ terminus of the truncated antisense RNA molecule. (**7**): End products of RNA-dependent mRNA amplification. Antisense RNA is truncated and polyadenylated at its 3′ end and sense portion of the chimeric RNA is truncated at the 5′ end and acquires the cleaved-off antisense RNA fragment. The polyadenylation of the antisense RNA strand denotes the commencement of the second phase of mRNA amplification. (**8**): 3′poly(A)-containing antisense RNA is the initial template (progenitor) in the intracellular PCR (iPCR) amplification process; it is transcribed into the sense RNA strand, initiating at the 3′poly (A), by RdRp. (**9**): The helicase complex separates RNA strands. (**10**): End products of the first iPCR cycle. Each contains 3′poly (A) and 5′poly(U) and each can serve as a PCR template in the RdRp-driven process.

The sense-orientation RNA product of the iPCR phase of mRNA amplification is always the mRNA molecule with the intact protein coding capacity of the progenitor mRNA, even if the chimeric phase of amplification produces mRMA encoding only a CTF of the original polypeptide. This may seem paradoxical but the explanation is simple. Indeed, the protein coding capacity of the chimeric mRNA end product of the first phase of amplification depends on the position of the ICE. If it were located in the region of antisense RNA corresponding to the 5′UTR of the mRNA molecule, the chimeric mRNA end product would contain the entire coding region of the mRNA progenitor. If, however, the ICE is positioned within the antisense RNA segment corresponding to the coding region of the progenitor mRNA molecule, the chimeric RNA end product would be 5′-truncated within the coding region of mRNA progenitor, as occurs during the amplification of human AβPP mRNA (see Sections 11–13 above). In both cases, however, the mRNA end product of the iPCR phase of amplification would contain the entire coding region of the mRNA progenitor. This is because the 5′ truncation of the chimeric mRNA end product is defined by the position of the ICE, which can be anywhere. In contrast, the 5′ truncation of the mRNA produced in the iPCR phase of amplification is no longer than the length of the TCE (because the cleavage of the chimeric RNA intermediate occurs either within the TCE or at its 5′ end), which is much shorter than an average 5′UTR. In human AβPP mRNA 5′UTR is 149 nucleotides long whereas the TCE of its antisense strand is only 33 nucleotides long. Thus, the mRNA end product of the iPCR phase of human AβPP mRNA amplification can be truncated by no more than 33 nucleotides (in contrast, the chimeric mRNA end product of the first phase of human AβPP mRNA amplification is truncated by nearly two thousand nucleotides). Therefore, its translation would result in the entire AβPP.

Both phases of the amplification of human AβPP mRNA can occur only in AD-affected neurons under the integrated stress response conditions, which provides components essential for the operation of RNA-dependent mRNA amplification machinery. Hence, in another apparent paradox, the neuronal ISR suppresses the AβPP proteolytic pathway, including the conventional production of AβPP, but enables the production of the latter in an ISR-driven and ISR-compatible process. Thus, the product of the first, chimeric, phase of human AβPP mRNA amplification, namely C99, apparently drives AD pathology, whereas the product of the second, iPCR, phase of AβPP mRNA amplification, namely AβPP (which cannot be processed proteolytically under the neuronal ISR conditions), potentially facilitates the viability and functionality of the disease-affected neurons.

80. Validation (13): Detection of Predictably 5′-Truncated, 5′ Poly(U)-Containing AβPP mRNA in Human Neuronal Cell-Based AD Models

The notion of the second phase of RNA-dependent mammalian mRNA amplification, as outlined in the preceding section, plausibly explains the origin of a class of human mRNA species that are truncated (in comparison with the original transcription products of their corresponding genes) within their 5′UTRs and possess poly(U) segments appended to their 5′ termini [289]. In this class of mRNAs, all truncations cluster around 8-17 nucleotides, conceivably a typical size of the TCE element (and/or a reflection of the position of a mismatch within the TCE/ICE complex) and are, thus, consistent with the proposed explanation. Moreover, the occurrence of the second phase of the RNA-dependent mRNA amplification was validated in the case of the amplification of alpha-globin and beta-globin mRNA [66]. The occurrence of the first, chimeric, phase of amplification for these mRNA species was proven by the detection of nucleotide sequences containing their "chimeric junctions", i.e., regions containing the antisense RNA segment extended into sense-orientation RNA. The reason that the detection of the chimeric RNA intermediate rather than of chimeric RNA end product (which is much more abundant) was used for validation is that the latter is undetectable by hybridization and, therefore, by cDNA-based sequencing techniques. This is because when the helicase complex moves along the 3′-terminal poly(A)-containing RNA strand, it modifies on average every fifth nucleotide (apparently to preclude the re-annealing of the strands and potentially to enable translation of amplified mRNA under the ISR conditions). The completed chimeric RNA end product is modified throughout. This is not the case for the chimeric intermediate. The chimeric junction region is generated (by extension of the TCE) unmodified and remains such until the extension is completed, and the helicase complex traverses the distance between the junction and the 3′-terminal poly(A). Thus, the pool of chimeric RNA intermediate with yet unmodified chimeric junction region should occur while the amplification process is operational and would provide the detection opportunities.

Such chimeric junction regions were indeed detected for both alpha-globin and beta-globin mRNAs [65,66]. This indicated the positions of anticipated cleavage sites of chimeric RNA intermediates, which, in turn, provided opportunities to validate the occurrence of the second phase of the amplification process. Indeed, the occurrence of mRNA 5′-truncated at the predicted position and containing 5′-terminal poly(U) segments would validate the proposed mechanism. Such a validation was indeed obtained for both alpha and beta-globin mRNAs [65,66]. The same type of validation for the occurrence of the second phase (iPCR) of AβPP mRNA amplification can be obtained in human neuronal cell-based AD models. It would be much easier to detect the predictably 5′-truncated, 5′-terminal poly(U)-containing mRNA segment then 3′-truncated, 3′-terminal poly(A)-containing segment of the antisense RNA. This is because, as shown in Figure 45, the latter would not stay unmodified for long since this is where the strands separation (and modification of the poly(A)-containing strand) begins, whereas the former would remain unmodified for much greater duration of the traverse of the helicase complex along the entire length of the mRNA molecule. Therefore, with the RNA-dependent human AβPP mRNA amplification process operational, the pool of not yet modified 5′ polyuridylated sense RNA molecules (or their 5′ portions) would be much greater than that of 3′ polyadenylated antisense RNA molecules. In Section 15, we identified potential cleavage sites of the chimeric RNA intermediate of the human AβPP mRNA amplification process. This enables us to predict the following validation targets for the occurrence of the second phase of the human AβPP mRNA amplification process in human neuronal cell-based AD models (based on the anticipated cleavage sites of chimeric RNA intermediates shown in Section 15, Figure 3 above):

(1) …UUUUUUUUUUUUUUUUUUUUUUUUUUUUUCCUCGGCAGCGGUAGGCGAGAGCACGCGGAGGAGCGUGCGCGGGGCCCCGGGAGACGGCGGCGGUGGCGGCGCGG…

(2) …UUUUUUUUUUUUUUUUUUUUUUUUUUUUUUGUAGGCGAGAGCACGCGGAGGAGCGUGCGCGGGGCCCCGGGAGACGGCGGCGGUGGCGGCGCGGGCAGAGCAAGGA…

(3) …UUUUUUUUUUUUUUUUUUUUUUUUUUUUUAGGCGAGAGCACGCGGAGGAGCGUGCGCGGGGCCCCGGGAGACGGCGGCGGUGGCGGCGCGGGCAGAGCAAGGACG…

81. Alternative RNA-Based Therapeutic Strategies for AD Targeting AβPP Antisense RNA and 5′ Terminus of AβPP mRNA via: (1) TSS Shift, (2) AβPP Gene Editing, and (3) Anti-Antisense AβPP Oligonucleotides

In terms of the ACH2.0 theory of AD, the C99 fragment of AβPP generated independently of AβPP drives the Alzheimer's disease. Therefore, all therapeutic strategies for AD considered in the present *Perspective* can be distilled to one common point: the prevention or cessation of the activity of the AβPP-independent C99 production pathway. This is, for example, exactly what the reversal of the neuronal integrated stress response (either by direct inhibition or via the removal of stressors capable of its elicitation) does. We do not know yet how to intervene with the AβPP-independent C99 production pathway directly, but long-term systemic interference with this pathway could be no less deleterious than that with integrated stress response. This is definitely the case if this pathway is, in fact, RNA-dependent amplification of mRNA, a scenario we consider overwhelmingly plausible. However, whereas we cannot interfere with the enzymatic machinery of this pathway in general, due to its physiological significance, we certainly can interfere with the amplification of human AβPP mRNA in particular. More specifically, undermining the eligibility of AβPP mRNA to serve as the amplification template is, apparently, a valid therapeutic strategy for AD (in its both, conventional and unconventional forms; this strategy, however, would not impact AACD). This can be accomplished in three ways.

81.1. Shift of Transcription Start Sites, TSSs, of Human AβPP mRNA

The AβPP gene belongs to the category of housekeeping genes that lack the promoter component known as TATA box. The TATA box is typically situated 30 nucleotides upstream from the TSS and defines its position. In genes lacking the TATA box transcription can and usually does initiate at multiple sites, i.e., these genes possess multiple TSSs. This is definitely the case with the human AβPP gene, which has five demonstrated transcription start sites [63]. They are situated at the following positions (in number of nucleotides upstream from the "A" of the AUG translation initiation codon): (-)150, (-)149, (-)146, (-)144,

(-)143. Since nucleotides at these sites are in penultimate 5′-terminal positions within AβPP mRNA (the ultimate 5′-terminal position is occupied by the capG), they are of special importance for the structure of the TCE element of the antisense strand and its ability to form a stable complex with the ICE, without any 3′ overhangs. Of the five TSSs listed above only one, namely (-)149, yields AβPP antisense RNA that satisfies these requirements. The situation is exacerbated by the observation that capG of mRNA is transcribed into 3′-terminal "C" of the antisense RNA molecule and has to be accommodated into the TCE/ICE structure. The TCE/ICE complexes formed by antisense complements of human AβPP mRNA molecules initiated from different TSSs are depicted in Figure 46 (asterisks denote nucleotides corresponding to TSS of mRNA). As stated above, the only AβPP antisense RNA capable of forming stable, 3′ overhang-free TCE/ICE complex is that corresponding to AβPP mRNA initiated from the TSS (-)149. It follows that the frequency of the usage of the TSS (-)149 potentially determines the rate of accumulation of C99 produced independently of AβPP and thus defines the rate of the progression of AD. It also follows that the shift of the TSSs of AβPP mRNA could be a powerful therapeutic tool. Indeed, the modulation of the usage of the known TSSs away from the TSS (-)149 would be beneficial: it would lower the rate of the production of C99 independently of AβPP and reduce the rate of the progression of the disease. Shifting TSSs of human AβPP mRNA more extensively could be far more beneficial. Indeed, shifting it upstream from "regular" TSSs would introduce a substantial 3′ overhang into the TCE/ICE complex; it would, in fact, render the 3′TCE non-terminal and disable its utility in the extension of the 3′ terminus of the antisense RNA molecule. If the TSS is shifted substantially downstream from regularly used sites (but still within the 5′UTR of AβPP mRNA) so that the segment corresponding to the TCE of the antisense molecule is either significantly shortened or eliminated altogether, the TCE/ICE complex would be either destabilized or eliminated. AβPP mRNA amplification would be disabled, and the progression of AD stopped. Importantly, with all manipulations described above, AβPP mRNA would retain its protein coding capacity.

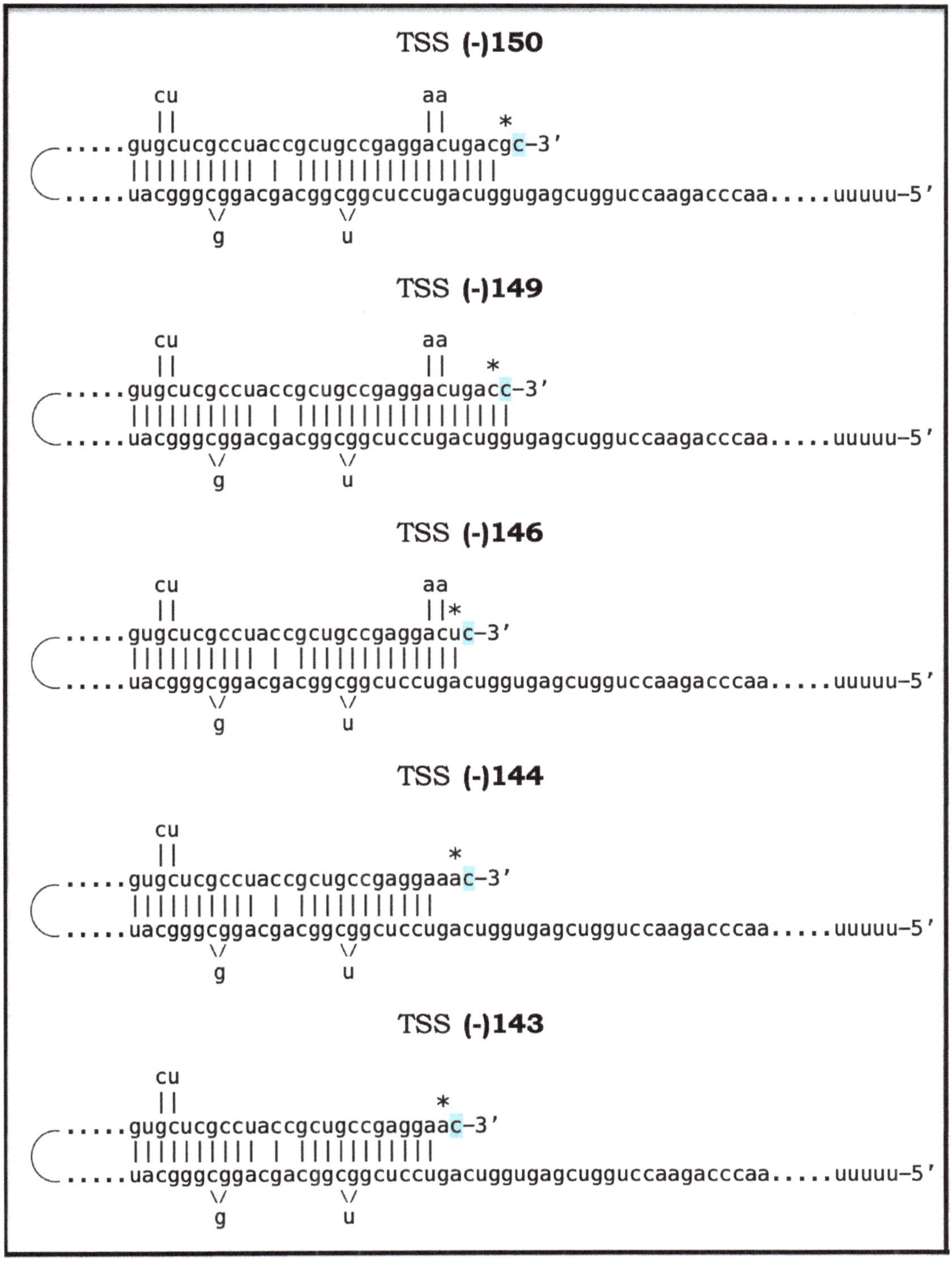

Figure 46. Folding of the antisense RNA molecules corresponding to human AβPP mRNAs initiated at the known transcription start sites. *TSS*: Transcription start site. *(-)150, (-)149, (-)146, (-)144, (-)143*:

Known transcription start sites of human AβPP mRNA. Distances are given in numbers of nucleotides upstream from the translation-initiating AUG codon. *Asterisk*: The nucleotide of the antisense RNA molecule corresponding to the transcription-initiating nucleotide of human AβPP mRNA. Highlighted *in blue*: "C", which is transcribed from the 5′-terminal cap "G" of AβPP mRNA. Only the antisense RNA strand originating from human AβPP mRNA transcribed from the TSS (-)149 is capable of forming self-primed structure without the 3′-overhang.

81.2. Limited AβPP Gene Editing as Therapeutic Strategy for AD

The objective of destabilizing the TCE/ICE complex formed by the antisense AβPP RNA molecule or eliminating it altogether can be accomplished also by a limited editing of the AβPP gene in human neurons. Indeed, the removal of a sufficiently large portion of the gene's segment corresponding to the TCE element of AβPP antisense RNA would destabilize the TCE/ICE complex or even prevent its formation. Alternatively, a portion, if not the entire segment of the AβPP gene corresponding to the TCE element of AβPP antisense RNA can be replaced by a "neutral" nucleotide sequence, i.e., by a sequence, which gives rise to a segment of AβPP antisense RNA (replacing its TCE) that is not complementary to its ICE element. This also would prevent the formation of the TCE/ICE complex. In addition, potentially, changes in the segment of the AβPP gene corresponding to the 5′UTR of AβPP mRNA could be sufficient to alter the folding of AβPP antisense RNA in such a way as to render the TCE and ICE elements mutually inaccessible and thus prevent the formation of the TCE/ICE complex. Such modifications would preclude the RNA-dependent amplification of human AβPP mRNA and, consequently, prevent AD in it both conventional and unconventional forms. Importantly, the proposed alterations would not interfere with the protein coding potential of AβPP mRNA. It should be emphasized that the feasibility of these two strategies (TSS shift and 5′ gene editing) has been validated by the observations that human AβPP mRNA expressed from 5′-modified transgenes in mouse models cannot be amplified (discussed in Section 18).

81.3. Therapeutic Utilization of Anti-Antisense Oligonucleotides Targeting AβPP Antisense RNA in General and Its TCE and ICE Elements in Particular

The two therapeutic strategies discussed above, namely the shift of TSSs of human AβPP mRNA and the alteration of a segment of the AβPP gene encoding 5′UTR of AβPP mRNA, impact AβPP antisense RNA and the interaction of its TCE and ICE elements indirectly, by affecting the structure of either the AβPP gene or AβPP mRNA. This mandates certain restrictions. For example, any alterations must be limited to the 5′UTR of AβPP mRNA (or corresponding portion of the gene) in order not to interfere with the protein coding potential of AβPP mRNA. Even among those, some alterations could interfere significantly with the efficiency of transcription and/or translation of AβPP and thus would be unusable. These limitations are immaterial in the therapeutic strategy utilizing anti-antisense oligonucleotides. In this strategy only AβPP antisense RNA is targeted and affected, and, therefore, there are no restrictions in the choice of particular targets within the RNA molecule. The primary targets are, of course, the TCE and ICE elements of AβPP antisense RNA. The interference with either would affect, possibly prevent, the formation of the TCE/ICE complex. Moreover, potentially, interference with any portion of AβPP antisense RNA may affect the mutual accessibility of the TCE and ICE elements within folded AβPP antisense RNA. Such interference would potentially disable the AβPP mRNA amplification process, and either prevent AD or arrest its progression. Since neither the AβPP gene nor AβPP mRNA are impacted in this approach, translation of the latter would not be hindered in any way. This approach is feasible in light of the recent development of nucleic acid-carrying lipid vehicles capable of crossing the blood–brain barrier [290–293].

82. Validation (14): Prevention of Cellular AD Pathology in Human Neuronal Cell-Based AD Models by Limited Editing of the 5′ Terminus of AβPP mRNA or via the Employment of Anti-Antisense Oligonucleotides

One of the therapeutic approaches described in the preceding section, namely the shift of TSSs of human AβPP mRNA, cannot be currently tested experimentally because techniques to execute such a shift have yet to be developed. In contrast, two other RNA-based therapeutic strategies discussed above can be readily validated in human neuronal cell-based AD models. In these models, cellular AD pathology, including the formation of neurofibrillary tangles, can be induced either conventionally, via the exogenous production of iAβ (which elicits the neuronal integrated stress response when accumulated over the T1 threshold), or unconventionally, through the sustained elicitation of the neuronal integrated stress response by stressors other than iAβ. In both cases, endogenous self-sustainable AβPP-independent production of C99, presumably via RNA-dependent amplification of AβPP mRNA, is activated, and AD pathology is driven by C99 generated independently of AβPP. Such a model, therefore, presents two measurable variables: (1) the extent of AD pathology in general and the formation of NFTs in particular, and (2) the occurrence of C100, i.e., N-terminal Met-C99 (see Section 23 above).

The validation of the therapeutic strategies discussed above can be carried out exactly as proposed in the preceding section. In case of the AβPP gene editing, a segment of the gene corresponding to the TCE element of AβPP antisense RNA can be either removed or replaced by a neutral nucleotide sequence in such a way that in the antisense RNA molecule the TCE is substituted by a nucleotide sequence, which is not complementary to the ICE element. In case of anti-antisense oligonucleotides, they should target either the TCE or the ICE elements of the antisense AβPP RNA. In both cases, gene editing and anti-antisense oligonucleotides, the interference, if successful, would prevent the formation of the TCE/ICE complex. Consequently, the amplification of AβPP mRNA would be precluded. Neither would NFTs be formed nor C100 generated in human neuronal cells following the sustained elicitation of the neuronal ISR.

83. Conclusions: Hope Is in the Air

83.1. The Initial Version of the ACH2.0: Principal Tenets

The present *Perspective* analyses the evolution of the Amyloid Cascade Hypothesis 2.0 (ACH2.0) theory of Alzheimer's disease. The central thread of this evolution is the diminishing role of Aβ in AD. The initial version of the theory marginalized the role of extracellular Aβ, previously believed to be both the causative agent and the driver of the disease and posited that the disease is triggered by intraneuronal Aβ (iAβ) derived from AβPP by the proteolysis of the latter and accumulated over the critical threshold, and is driven by iAβ produced by gamma-secretase cleavage of the C99 fragment of AβPP generated independently from AβPP. In this paradigm, the bridge between differentially produced iAβ is the neuronal integrated stress response. Evidently, AβPP-derived iAβ acts as a neuronal ISR-eliciting stressor. Upon reaching certain cellular concentrations, it activates PKR and/or HRI, both eIF2α kinases. Consequently, eIF2α is phosphorylated at its Ser51 residue, and the elicitation of the neuronal ISR ensues. A pivotal role of the neuronal ISR in AD is enabling the production of components essential for the operation of the AβPP-independent C99 generation pathway. Apparently, iAβ produced by the AβPP proteolysis alone cannot reach the AD pathology-causing range of cellular concentrations. On the other hand, the entire iAβ output of the AβPP-independent C99 production pathway is retained within the neurons, rapidly accumulates, and drives AD pathology. It also propagates the neuronal ISR state, making it independent from the influx of AβPP-derived iAβ, and thus perpetuates the activity of the AβPP-independent C99 generation pathway.

83.2. ACH2.0, Version Two: Conventional and Unconventional AD

A major step in the evolution of the ACH2.0 was the realization that the AβPP-derived iAβ-triggered form of AD is only a special, albeit prevalent, category of the disease, and that Alzheimer's is not a disease of AβPP-derived iAβ but rather a disease of the neuronal integrated stress response. In this paradigm, the sustained elicitation of the neuronal ISR by ANY stressor results in AD. To distinguish between the two categories of AD and their respective triggers, the disease initiated by AβPP-derived iAβ has been referred to as "conventional AD" and its trigger, AβPP-derived iAβ, as the "conventional stressor", whereas the disease triggered by stressors (any stressors) distinct from AβPP-derived iAβ was designated as "unconventional AD" and stressors triggering it have been referred to as "unconventional stressors". Sources of unconventional neuronal ISR-eliciting and, consequently, unconventional AD-triggering stressors are numerous. They can be of mechanical nature, such as traumatic brain injury or chronic traumatic encephalopathy, or can take the form of viral and bacterial infections as well as of various types of inflammation. The common features of all such sources include a decreased cerebral blood flow and compromised blood–brain barrier. The latter may allow unconventional stressors to penetrate the brain and, subsequently, the neurons whereas the former was shown to trigger mitochondrial dysfunction and, consequently, activate the ISR-eliciting HRI kinase in neuronal cells. Multiple forms of dementia classified until now as AD-like dementia (ADLD) and AD-related dementia (ADRD) all, apparently, belong to the unconventional AD category. In unconventional AD the role of Aβ is significantly diminished. AβPP-derived iAβ has little, if any, relevance to the disease but AD pathology is still driven (in the second variant of the ACH2.0) by iAβ produced by gamma-secretase cleavages of C99 generated independently of AβPP.

83.3. ACH2.0, Version Three: AβPP Proteolytic Pathway Is Suppressed in AD-Affected Neurons; The Disease Is Driven by C99 Generated Independently of AβPP

The next milestone in the evolution of the ACH2.0 was the formulation of its Version Three. In this version, the AβPP proteolytic pathway is suppressed in AD-affected neurons following the elicitation of the neuronal ISR. This suppression encompasses the production of AβPP, of BACE enzymes, of components of the gamma-secretase complex, and, consequently, of Aβ and iAβ. Due to the deficiency of gamma-secretase, the processing of C99 generated independently of AβPP into iAβ is also suppressed. Because of the persisting aura of Aβ as the principal player in AD, it took both time and effort to formulate this notion, but this is where the logic and empirical data lead, and it is only prudent for us to follow. This conclusion was derived from a detailed analysis of the current transgenic mouse models of AD. They are rather "attempted models of AD", because, despite vast exogenous overproduction of human AβPP, none of them develops the full spectrum of AD pathology but exhibit only the effects of the neuronal ISR elicited by exogenous AβPP-derived iAβ. The AβPP-independent C99 generation pathway is inoperative in mouse models. Therefore, the analysis left only one plausible explanation for the lack of the massive neurodegeneration, namely that the influx of AβPP-derived iAβ is discontinued or severely suppressed following the elicitation of the neuronal ISR. In retrospect, this is rather a trivial assumption. The global cellular protein synthesis is severely suppressed by the integrated stress response, this is the epitome of the ISR state, and there is no reason why constituents of the AβPP proteolytic pathway would be exempted.

Thus, constituents of the AβPP proteolytic pathway are not exempted but the C99 fragment is produced in the ISR-compatible process. It is both interesting and fitting that tau, a protein forming the major hallmark of AD, neurofibrillary tangles, is also produced under the neuronal ISR conditions (if it were not, the theory would be inconsistent with the data). But in the tau case, it is exempted from the ISR-exerted suppression of cellular protein synthesis. This is because the human tau mRNA contains an internal ribosomal entry site (IRES) [284]. The IRES is positioned within the 5′UTR of tau mRNA; translation, therefore, proceeds in the ISR-compatible manner and yields the complete tau protein

This alteration of the ACH2.0 deprives Aβ from its formerly major role, namely driving AD pathology, thus leaving it with only one function in the disease: triggering the neuronal ISR in the conventional form of AD. It also establishes C99 generated independently of AβPP as the driver of AD pathology and the sustainer of the activity of the AβPP-independent C99 generation pathway. Due to its demonstrated cellular toxicity and the ability to trigger the elicitation of the neuronal ISR, C99 is fully qualified for such a role, which is consistent with the observation of substantial accumulation of C99 in AD-affected neurons but not in the disease-resistant regions of the human brain. The final (to-date) alteration to the ACH2.0 is a notion that, after all, AβPP is produced (though it cannot be processed proteolytically due to the deficiency of gamma-secretase) in the AD-affected neurons not conventionally, but rather in a neuronal ISR-driven and ISR-compatible process. In fact, it is the second phase of the same pathway, which generates C99 independently of AβPP. Thus, interestingly, the same process (although in two distinct phases) produces both the driver of AD pathology (C99) and the substance (AβPP) that potentially facilitates the viability and functionality of AD-affected neurons. The current understanding of AD, in its both conventional and unconventional forms, in the ACH2.0 perspective is summarized graphically in Figure 47.

83.4. ACH-Based AD Drugs in the ACH2.0 Perspective

The current interpretation of the disease in the ACH2.0 perspective allows the evaluation of the potential of currently available as well as future AD drugs. All presently FDA-approved AD drugs belong to the ACH-based category of AD drugs. ACH-based drugs are named so because their design and function have been guided by the ACH. Accordingly, their stated aim is to deplete extracellular Aβ. They achieve this goal by either sequestering their target (e.g., Aβ-specific antibodies) or by reducing its production and thus secretion (e.g., BACE1 inhibitors). Also in this category are gamma-secretase modulators, which shift the production of AβPP-derived Aβ toward shorter, more benign species. ACH-based drugs are legitimate (albeit inefficient) conventional AD-preventive agents in the ACH2.0 perspective. This is because by reducing the pool of extracellular Aβ they also lower the rate of its cellular uptake and, consequently, decrease the rate of iAβ accumulation. In case of BACE inhibitors, the retention of iAβ resulting from gamma-cleavages on intraneuronal membranes is also reduced. The decreased rate of iAβ accumulation translates into delayed or prevented T1 crossing and thus delayed or precluded occurrence of conventional AD. As for gamma-secretase modulators, they elevate the extent of the T1 threshold. As a result, its crossing is delayed or prevented and so is the occurrence of the conventional disease. All these drugs, however, can have no positive effect whatsoever when the self-sustainable AβPP-independent C99 generating pathway is operational. Effects of lecanemab and donanemab in very early AD were marginal because they acted only preventively and only on a marginal neuronal subpopulation that did not yet cross the T1 threshold and where the AβPP-independent C99 generating pathway was not yet activated. It can be fully expected that any ACH-based drug would elicit the same preventive response if administered at similarly early stages of AD. These drugs would have no effect (neither preventive nor curative) of any sort in unconventional AD.

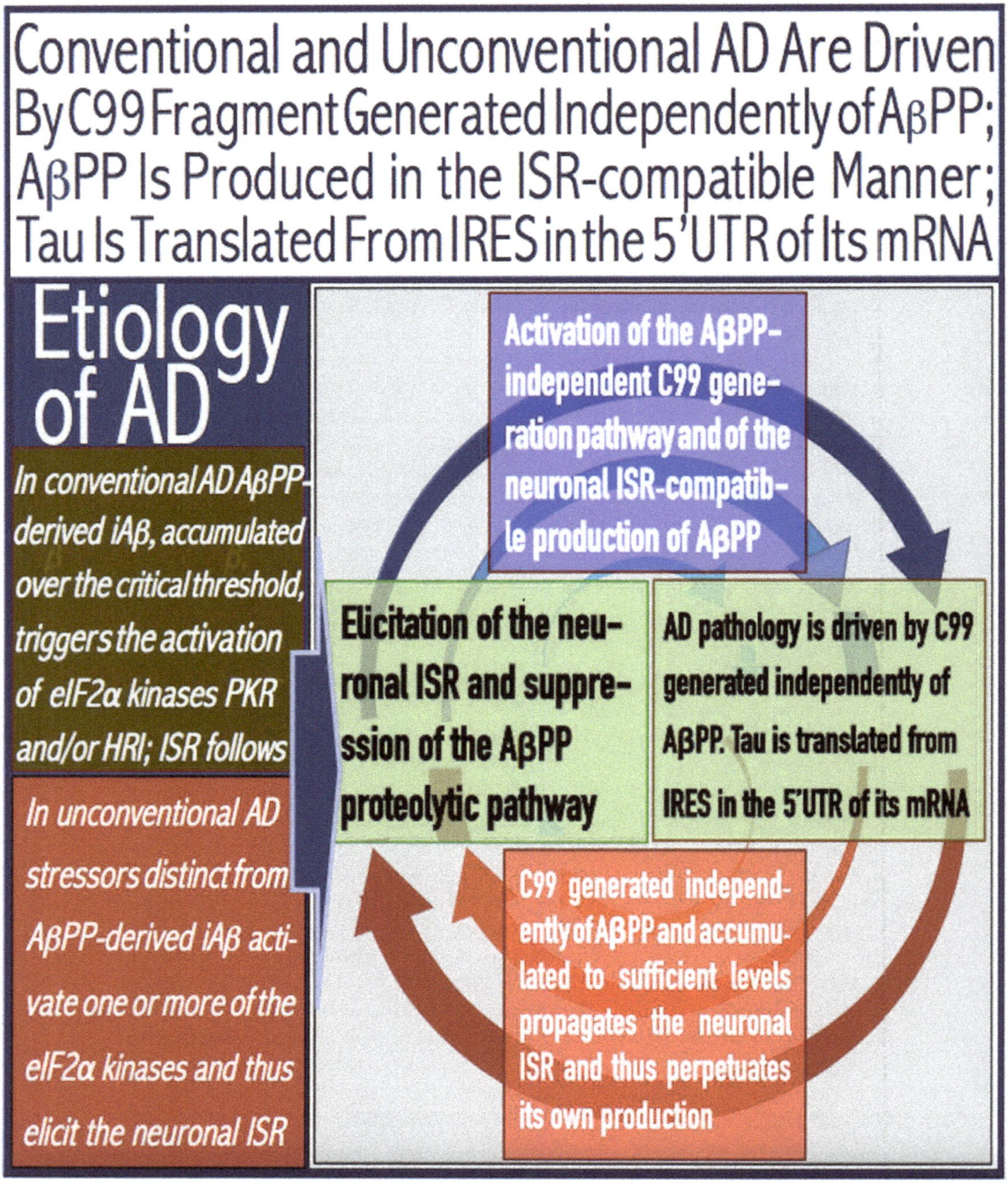

Figure 47. Current understanding of AD: The neuronal ISR suppresses the AβPP proteolytic pathway but enables the AβPP-independent generation of C99, which drives the disease; AβPP is produced in the ISR-compatible process but not processed; Tau is translated from IRES in the 5′UTR of its mRNA. *eIF2α*: eukaryotic translation initiation factor 2α. *PKR and HRI*: kinases that phosphorylate eIF2α at its Ser51 residue. *PACT*: PKR activator. *TNFα*: tumor necrosis factor *α*. *OMA1*: mitochondrial protease activated during mitochondrial dysfunction. *DELE1*: substrate of OMA1 protease. Cleavage of

DELE1 leads to the activation of HRI. Phosphorylation of eIF2α elicits the integrated stress response, which, in turn, enables the generation of components essential for the activity of the AβPP-independent pathway of C99 and iAβ production. iAβ: intraneuronal Aβ. AβPP-derived iAβ accumulates physiologically via the importation of extracellular Aβ and the retention of iAβ produced by gamma-cleavages on intracellular membranes in an exceedingly slow process. In most individuals it does not reach the PKR- and/or HRI-activating, the neuronal ISR-eliciting threshold within their lifetimes and no conventional AD occurs. When this threshold is crossed, the neuronal ISR is elicited and the disease ensues. In unconventional AD, stressors that are distinct from AβPP-derived iAβ and capable of activating one or more of eIF2α kinases trigger the elicitation of the neuronal ISR. Under the ISR conditions the AβPP proteolytic pathway is suppressed; this happens within the framework of the ISR-mediated inhibition of the global cellular protein synthesis and includes AβPP, BACE enzymes, components of the gamma-secretase complex and, consequently, Aβ and iAβ. Concurrently, the neuronal ISR provides the components essential for, and thus enables, the operation of the AβPP-independent C99 generation pathway, and it is this C99, which drives AD pathology. Simultaneously, in the second phase of the same mechanism that underlies AβPP-independent generation of C99, AβPP is produced in the ISR-compatible process but is not processed further. In general, in the AD-affected neurons, any new protein synthesis has to occur in an ISR-compatible manner. In cases of C99 and AβPP this is achieved, apparently, via the nucleotide modifications of their mRNAs. Translation of tau protein, on the other hand, persists under the ISR conditions because its mRNA contains an internal ribosomal entry site in the 5′UTR.

83.5. ACH2.0-Based Drugs in the Prevention of Conventional AD and AACD and in the Treatment of AACD: BACE1 and/or BACE2 Activators

In sharp contrast, the ACH2.0-based drugs hold extraordinary promise for the prevention and treatment of both conventional and unconventional forms of AD, as well as for the prevention and treatment of AACD. By definition, the ACH2.0-based drugs are the agents that either deplete iAβ (and thus remove the trigger of conventional AD when applied preventively) and/or C99 (and thus abolish both the driver of the disease and the sustainer of its own AβPP-independent production) via their selective degradation or terminate the activity of the AβPP-independent C99 generation pathway in any other way. As discussed above, in the current version (Version Three) of ACH2.0, iAβ plays a central role only in conventional AD and in aging-associated cognitive decline, AACD. Its accumulation over the T^0 threshold denotes the commencement of AACD (provided $T^0 < T1$). When iAβ crosses the T1 threshold, it triggers the elicitation of the neuronal integrated response, activates the self-sustainable AβPP-independent C99 generation pathway and initiates conventional AD; AACD morphs into AD at this point. No AACD commences until the T^0 threshold is crossed and no AD occurs until the T1 threshold is reached. This opens up tantalizing preventive opportunities. If iAβ is depleted at mid-life by transient treatment with activators of BACE1 and/or BACE2 (or by any other iAβ–degrading agent), and if the depletion is "deep" enough, de novo accumulating iAβ would not reach the T^0 and T1 thresholds within the lifetime of the individual and neither AACD nor AD would occur. If the treatment is administered following the T^0 but prior to the T1 crossing, AACD would be cured, and neither would it recur nor would AD occur within the individual's lifespan. Thus, a single, one time only transient treatment with BACE1 and/or BACE2 activating drugs can prevent both conventional AD and AACD and cure the latter.

83.6. ACH2.0-Based Drugs in the Treatment of Conventional AD: Concurrent BACE1 and/or BACE2 Activation and Neuronal ISR Inhibition

No less spectacular outcomes can be attained in the treatment of conventional AD. In this case, however, the AβPP-independent C99 generation pathway is operational and, due to the high rate of the influx of C99 produced independently of AβPP, activated BACE1 and/or BACE2 can only reduce but not reverse its accumulation and thus can only slow down but not stop the progression of the disease. The solution is the transient composite C99/iAβ depletion therapy. BACE1 and/or BACE2 activators (or any other iAβ–degrading agent) are administered transiently and concurrently with ISR inhibitors. The latter reverse the neuronal ISR state, restore the production of BACE enzymes, disable the AβPP-independent C99 generation pathway and abrogate the influx of C99 whereas the former activate BACE enzymes, which efficiently deplete both C99 and AβPP-derived iAβ. The progression of the disease ceases and the accumulation of iAβ resumes from a low baseline at a slow pre-T1 crossing rate. It does not reach the T1 threshold, and the disease does not recur within the lifetime of the treated patient. Thus, a single, once in a lifetime transient composite therapy with concurrently administered BACE1 and/or BACE 2 activators and ISR inhibitors can stop for life the progression of conventional Alzheimer's disease and preclude its recurrence.

83.7. ACH2.0-Based Drugs in the Prevention and Treatment of Unconventional AD: Recurrent Composite BACE Activation/ISR Inhibition Therapy; Sustained Removal of Unconventional Stressors as the Ultimate Goal

The prevention and treatment of unconventional presents significantly more challenges. This is because the elicitation of the neuronal ISR and subsequent activation of the AβPP-independent C99 generation pathway (and suppression of AβPP proteolytic pathway) are triggered by unconventional stressors occurring at the levels of AβPP-derived iAβ below the T1 threshold and persistently present thereafter. The disease would commence only when C99 crosses the T1 threshold, and the AβPP-independent pathway of its production becomes self-sustainable. To prevent the disease is to prevent the T1 crossing. C99 (and AβPP-derived iAβ) can be depleted by the transient composite treatment with BACE1 and /or BACE2 activators and ISR inhibitors (transient because long-term administration of ISR inhibitors would be deleterious, and BACE occurrence and, consequently, activation is feasible only under ISR inhibition). The problem is that as soon as drugs are withdrawn, the neuronal ISR would be unconventionally re-elicited, AβPP-independent C99 generation pathway reactivated, and its C99 product would rapidly accumulate. This would only delay the occurrence of AD, possibly for a few years.

Conceptually the same process would happen in the treatment of unconventional AD. Following the composite C99 depletion treatment, the progression of the disease would cease but the AβPP-independent pathway of C99 production would be reactivated; it would rapidly accumulate and cross the T1 threshold, and the disease would recur. A solution, in both the prevention and treatment of unconventional AD, is the recurrent application of the transient BACE activation/ISR inhibition treatment, administered at strategically timed intervals under the guidance of the appropriate biomarkers.

On the other hand, a more fundamental solution, in both the prevention and treatment of unconventional AD, is to administer a single transient composite BACE activation/ISR inhibition treatment but in conjunction with sustained removal (or depletion below neuronal ISR-eliciting levels) of unconventional stressors via the elimination of their sources. Indeed, the removal of unconventional stressors would convert unconventional AD (or perspective unconventional AD) into conventional one. In such a setting, the one time only implementation of the composite treatment with BACE1 and/or BACE2 activators and ISR inhibitors would prevent the occurrence of the disease or its recurrence for the remaining lifetime of the treated individual. In addition, the replacement of BACE-activating drugs with directly acting iAβ-specific degradation agents such as appropriately designed proteolysis-targeting chimeras (PROTACs) and molecular-glue degraders (MGDs) allows long-term treatment and opens further therapeutic opportunities for unconventional AD.

83.8. RNA-Based Approaches: Highly Promising and Eminently Feasible Therapeutic Strategies for AD

The present *Perspective* also introduces a conceptually distinct set of RNA-based therapeutic strategies for AD. These strategies are founded on the overwhelmingly plausible assumption that the AβPP-independent C99 generation pathway is, in fact, the asymmetric RNA-dependent human AβPP mRNA amplification process (asymmetric because only a 3′-terminal portion of AβPP mRNA, encoding only the C99 fragment of AβPP, is amplified in this process). This process occurs, possibly, exclusively in humans, which explains why Alzheimer's disease is, apparently, the exclusively human disorder. The reason for such an exclusivity is that whereas human AβPP mRNA is an eligible template for RNA-dependent amplification, its counterparts in non-human mammals are not; no amplification-eligible AβPP mRNA, no production of C99 independently of AβPP, no disease. The objective of the RNA-based therapeutic strategies for AD is, therefore, to render human AβPP mRNA ineligible for amplification or to interfere on the molecular level at defined stages of the amplification process. The eligibility to serve as amplification template is conferred on human AβPP mRNA by the occurrence on the corresponding antisense RNA strand of two complementary element, one 3′-terminal (the terminal complementary element, TCE) another internal (the internal complementary element, ICE), which are mutually accessible within the folded RNA molecule. These elements are the prime interference targets. Of the two, the TCE is more feasible to interfere with because the necessary modifications of the AβPP gene or AβPP mRNA would occur in the 5′ untranslated region, thus preserving their protein-coding potential. One therapeutic strategy, centered on the AβPP gene, is to shift its transcription start site, TSS. The upstream shift would render the TCE non-terminal whereas the downstream shift would remove either a portion of or the entire TCE; in either case, the amplification process would be disrupted.

Another AβPP gene-centered approach is to remove or to replace, via gene editing in neuronal cells, gene segments encoding a portion of or the entire TCE element. The outcome would be identical to that of a TSS shift: the disruption of the amplification process. The feasibility of these approaches is validated by the observations that human AβPP mRNA expressed from 5′-modified transgenes in mouse models cannot be amplified (discussed in Section 18 above).

The third RNA-based strategy centers on antisense AβPP RNA. In this approach, anti-antisense oligonucleotides targeting either the TCE or the ICE may block their interaction and thus disrupt the amplification process. Moreover, anti-antisense oligonucleotides targeting other regions of AβPP antisense RNA may potentially render the TCE and ICE elements mutually inaccessible within the folded RNA molecule. For two reasons this strategy is more feasible and practical than the other two. First, since the preservation of the coding capacity is not an issue, antisense RNA presents many more potential target sites. Second reason is the feasibility of delivery of oligonucleotides: lipids vehicles have been recently developed that are capable of carrying nucleic acid loads through the blood–brain barrier [290–293]. All three strategies would, depending on the timing of their implementation, either prevent or disrupt the AβPP mRNA amplification process. Accordingly, they would either preclude the occurrence of AD or arrest the progression of the disease equally efficiently in its both conventional and unconventional forms.

In conclusion, the fundamental principles of the ACH2.0 in general and of its latest version in particular are, apparently, solid. Since its inception the theory underwent a remarkable evolution. There is little doubt that it will continue to evolve. However, there is also a firm conviction that the current level of understanding is sufficient to inform and guide the development of potent therapies for AD and AACD. Hope is in the air!

Author Contributions: Conceptualization, V.V. and S.R.-V.; writing-oriinal draft preparation, V.V.; writing-review and editing, V.V. and S.R.-V. All authors have read and agreed to the published version of the manuscript.

Funding: This research was funded by NIH grants numbers NIH R21 GM056179; NIH RO1 AR036819.

Institutional Review Board Statement: Neither human nor animal subjects were involved in the present study.

Acknowledgments: The authors are grateful to Bjorn Olsen for his help, to George Perry for his courage, to Nahum Sonenberg for his encouragement, to Alexander Varshavsky for conveying instrumental information, and to Rachel Rits-Volloch for her logistical support.

Conflicts of Interest: The authors declare no conflicts of interest.

References

1. Volloch, V.; Rits-Volloch, S. The Amyloid Cascade Hypothesis 2.0: On the possibility of once-in-a-lifetime-only treatment for prevention of Alzheimer's disease and for its potential cure at symptomatic stages. *J. Alzheimers Dis. Rep.* **2022**, *6*, 369–399. [CrossRef] [PubMed]
2. Volloch, V.; Rits-Volloch, S. The Amyloid Cascade Hypothesis 2.0: Generalization of the Concept. *J. Alzheimers Dis. Rep.* **2023**, *7*, 21–35. [CrossRef] [PubMed] [PubMed Central]
3. Volloch, V.; Rits-Volloch, S. Effect of Lecanemab and Donanemab in Early Alzheimer's Disease: Mechanistic Interpretation in the Amyloid Cascade Hypothesis 2.0 Perspective. *J. Alzheimers Dis.* **2023**, *93*, 1277–1284. [CrossRef] [PubMed]
4. Volloch, V.; Rits-Volloch, S. The Amyloid Cascade Hypothesis 2.0 for Alzheimer's Disease and Aging-Associated Cognitive Decline: From Molecular Basis to Effective Therapy. *Int. J. Mol. Sci.* **2023**, *24*, 12246. [CrossRef] [PubMed]
5. Volloch, V.; Rits-Volloch, S. Principles of Design of Clinical Trials for Prevention and Treatment of Alzheimer's Disease and Aging-Associated Cognitive Decline in the ACH2.0 Perspective: Potential Outcomes, Challenges and Solutions. *J. Alzheimers Dis. Rep.* **2023**, *7*, 921–955. [CrossRef] [PubMed] [PubMed Central]
6. Volloch, V.; Rits-Volloch, S. Next Generation Therapeutic Strategy for Treatment and Prevention of Alzheimer's Disease and Aging-Associated Cognitive Decline: Transient, Once-in-a-Lifetime-Only Depletion of Intraneuronal Aβ (*i*Aβ) by Its Targeted Degradation via Augmentation of Intra-*i*Aβ-Cleaving Activities of BACE1 and/or BACE2. *Int. J. Mol. Sci.* **2023**, *24*, 17586. [CrossRef] [PubMed] [PubMed Central]
7. Volloch, V.; Rits-Volloch, S. On the Inadequacy of the Current Transgenic Animal Models of Alzheimer's Disease: The Path Forward. *Int. J. Mol. Sci.* **2024**, *25*, 2981. [CrossRef] [PubMed] [PubMed Central]
8. Volloch, V.; Rits-Volloch, S. ACH2.0/E, the Consolidated Theory of Conventional and Unconventional Alzheimer's Disease: Origins, Progression, and Therapeutic Strategies. *Int. J. Mol. Sci.* **2024**, *25*, 6036. [CrossRef] [PubMed] [PubMed Central]
9. Volloch, V.; Rits-Volloch, S. Quintessential Synergy: Concurrent Transient Administration of Integrated Stress Response Inhibitors and BACE1 and/or BACE2 Activators as the Optimal Therapeutic Strategy for Alzheimer's Disease. *Int. J. Mol. Sci.* **2024**, *25*, 9913. [CrossRef] [PubMed] [PubMed Central]
10. Volloch, V.; Rits-Volloch, S. Production of Amyloid-β in the Aβ-Protein-Precursor Proteolytic Pathway Is Discontinued or Severely Suppressed in Alzheimer's Disease-Affected Neurons: Contesting the 'Obvious'. *Genes* **2025**, *16*, 46. [CrossRef] [PubMed] [PubMed Central]
11. Alzheimer, A. Über einen eigenartigen schweren Erkrankungsprozeβ der Hirnrincle. *Neurol. Central.* **1906**, *25*, 1134.
12. Alzheimer, A. Über eine eigenartige Erkrankung der Hirnrinde. *Allg. Zschr. Psychiatr. Psych. Gerichtl. Med.* **1907**, *64*, 146–148.
13. Kraepelin, E. *Psychiatrie*, 8th ed.; Vol I: Allgemeine Psychiatrie; Barth: Leipzig, Germany, 1909.
14. Kraepelin, E. *Psychiatrie*, 8th ed.; Vol II: Klinische Psychiatrie; Barth: Leipzig, Germany, 1910.
15. Alzheimer, A. Die diagnostischen Schwierigkeiten in der Psychiatrie. *Z. Ges. Neurol. Psychiatr.* **1910**, *1*, 1–19. [CrossRef]
16. Alzheimer, A. Über eigenartige Krankheitsfälle des späteren Alters. *Z. Ges. Neurol. Psychiatr.* **1911**, *4*, 356–385. [CrossRef]
17. Blessed, G.; Tomlinson, B.E.; Roth, M. The association between quantitative measures of dementia and of senile change in the cerebral grey matter of elderly subjects. *Br. J. Psychiatry* **1968**, *114*, 797–811. [CrossRef]
18. Glenner, G.G.; Wong, C.W. Alzheimer's disease: Initial report of the purification and characterization of a novel cerebrovascular amyloid protein. *Biochem. Biophys. Res. Commun.* **1984**, *120*, 885–890. [CrossRef]
19. Kang, J.; Lemaire, H.G.; Unterbeck, A.; Salbaum, J.M.; Masters, C.L.; Grzeschik, K.H.; Multhaup, G.; Beyreuther, K.; Müller-Hill, B. The precursor of Alzheimer's disease amyloid A4 protein resembles a cell-surface receptor. *Nature* **1987**, *325*, 733–736. [CrossRef]
20. Robakis, N.K.; Ramakrishna, N.; Wolfe, G.; Wisniewski, H.M. Molecular cloning and characterization of a cDNA encoding the cerebrovascular and the neuritic plaque amyloid peptides. *Proc. Natl. Acad. Sci. USA* **1987**, *84*, 4190–4194. [CrossRef]

21. Tanzi, R.E.; Gusella, J.F.; Watkins, P.C.; Bruns, G.A.; St George-Hyslop, P.; Van Keuren, M.L.; Patterson, D.; Pagan, S.; Kurnit, D.M.; Neve, R.L. Amyloid beta protein gene: cDNA, mRNA distribution, and genetic linkage near the Alzheimer locus. *Science* **1987**, *235*, 880–884. [CrossRef]
22. Goate, A.; Chartier-Harlin, M.C.; Mullan, M.; Brown, J.; Crawford, F.; Fidani, L.; Giuffra, L.; Haynes, A.; Irving, N.; James, L.; et al. Segregation of a missense mutation in the amyloid precursor protein gene with familial Alzheimer's disease. *Nature* **1991**, *349*, 704–706. [CrossRef]
23. Hardy, J.A.; Higgins, G.A. Alzheimer's disease: The amyloid cascade hypothesis. *Science* **1992**, *256*, 184–185. [CrossRef]
24. Keskin, A.D.; Kekuš, M.; Adelsberger, H.; Neumann, U.; Shimshek, D.R.; Song, B.; Zott, B.; Peng, T.; Förstl, H.; Staufenbiel, M.; et al. BACE inhibition-dependent repair of Alzheimer's pathophysiology. *Proc. Natl. Acad. Sci. USA* **2017**, *114*, 8631–8636. [CrossRef] [PubMed]
25. Hu, X.; Das, B.; Hou, H.; He, W.; Yan, R. BACE1 deletion in the adult mouse reverses preformed amyloid deposition and improves cognitive functions. *J. Exp. Med.* **2018**, *10*, 1084. [CrossRef] [PubMed]
26. Selkoe, D.J.; Hardy, J. The amyloid hypothesis of Alzheimer's disease at 25 years. *EMBO Mol. Med.* **2016**, *8*, 595–608. [CrossRef]
27. Egan, M.F.; Kost, J.; Tariot, P.N.; Aisen, P.S.; Cummings, J.L.; Vellas, B.; Sur, C.; Mukai, Y.; Voss, T.; Furtek, C.; et al. Randomized trials of verubecestat for mild-to-moderate Alzheimer's disease. *N. Engl. J. Med.* **2018**, *378*, 1691–1703. [CrossRef]
28. Egan, M.F.; Kost, J.; Voss, T.; Mukai, Y.; Aisen, P.S.; Cummings, J.L.; Tariot, P.N.; Vellas, B.; van Dyck, C.H.; Boada, M.; et al. Randomized trial of verubecestat for prodromal Alzheimer's disease. *N. Engl. J. Med.* **2019**, *380*, 1408–1420. [CrossRef] [PubMed]
29. van Dyck, C.H.; Swanson, C.J.; Aisen, P.; Bateman, R.J.; Chen, C.; Gee, M.; Kanekiyo, M.; Li, D.; Reyderman, L.; Cohen, S.; et al. Lecanemab in Early Alzheimer's Disease. *N. Engl. J. Med.* **2023**, *388*, 9–21. [CrossRef] [PubMed]
30. Mintun, M.A.; Lo, A.C.; Duggan Evans, C.; Wessels, A.M.; Ardayfio, P.A.; Andersen, S.W.; Shcherbinin, S.; Sparks, J.; Sims, J.R.; Brys, M.; et al. Donanemab in Early Alzheimer's Disease. *N. Engl. J. Med.* **2021**, *384*, 1691–1704. [CrossRef]
31. Eli Lilly's Donanemab Significantly Slowed Cognitive and Functional Decline in Phase 3 Study of Early Alzheimer's Disease. Press Release. 2023. Available online: https://investor.lilly.com/news-releases/news-release-details/lillys-donanemab-significantly-slowed-cognitive-and-functional (accessed on 1 January 2025).
32. Sims, J.R.; Zimmer, J.A.; Evans, C.D.; Lu, M.; Ardayfio, P.; Sparks, J.; Wessels, A.M.; Shcherbinin, S.; Wang, H.; Monkul Nery, E.S.; et al. Donanemab in Early Symptomatic Alzheimer Disease: The TRAILBLAZER-ALZ 2 Randomized Clinical Trial. *JAMA* **2023**, *330*, 512–527. [CrossRef]
33. Abbott, A. The next generation of Alzheimer's treatments. *Nature* **2023**, *616*, 26–28. [CrossRef]
34. Katzman, R.; Terry, R.; DeTeresa, R.; Brown, T.; Davies, P.; Fuld, P.; Rebing, X.; Peck, A. Clinical, pathological, and neurochemical changes in dementia: A subgroup with preserved mental status and numerous neocortical plaques. *Ann. Neurol.* **1988**, *23*, 138–144. [CrossRef] [PubMed]
35. Delaere, P.; Duyckaerts, C.; Masters, C.; Beyreuther, K.; Piette, F.; Hauw, J. Large amounts of neocortical beta A4 deposits without neuritic plaques nor tangles in a psychometrically assessed, non-demented person. *Neurosci. Lett.* **1990**, *116*, 87–93. [CrossRef]
36. Dickson, D.; Crystal, H.; Mattiace, L.; Masur, D.; Blau, A.; Davies, P.; Yen, S.; Aronson, M. Identification of normal and pathological aging in prospectively studied nondemented elderly humans. *Neurobiol. Aging.* **1992**, *13*, 178–189. [CrossRef]
37. Aizenstein, H.J.; Nebes, R.D.; Saxton, J.A.; Price, J.C.; Mathis, C.A.; Tsopelas, N.D.; Ziolko, S.K.; James, J.A.; Snitz, B.E.; Houck, P.R.; et al. Frequent amyloid deposition without significant cognitive impairment among the elderly. *Arch. Neurol.* **2008**, *65*, 1509–1517. [CrossRef] [PubMed]
38. Klunk, W.; Mathis, C.; Price, J.; DeKosky, S.; Lopresti, B.; Tsopelas, N.; Saxton, J.; Nebes, R. Amyloid imaging with PET in Alzheimer's disease, mild cognitive impairment, and clinically unimpaired subjects. In *PET in the Evaluation of Alzheimer's Disease and Related Disorders*; Silverman, D., Ed.; Springer: New York, NY, USA, 2009; pp. 119–147.
39. Villemagne, V.L.; Pike, K.E.; Chételat, G.; Ellis, K.A.; Mulligan, R.S.; Bourgeat, P.; Ackermann, U.; Jones, G.; Szoeke, C.; Salvado, O.; et al. Longitudinal assessment of Abeta and cognition in aging and Alzheimer disease. *Ann. Neurol.* **2011**, *69*, 181–192. [CrossRef]
40. Seto, M.; Weiner, R.L.; Dumitrescu, L.; Hohman, T.J. Protective genes and pathways in Alzheimer's disease: Moving towards precision interventions. *Mol. Neurodegener.* **2021**, *16*, 29. [CrossRef] [PubMed]
41. Makin, S. The amyloid hypothesis on trial. *Nature* **2018**, *559*, S4–S7. [CrossRef]
42. Bekris, L.M.; Yu, C.E.; Bird, T.D.; Tsuang, D.W. Genetics of Alzheimer disease. *J. Geriatr. Psychiatry Neurol.* **2010**, *23*, 213–227. [CrossRef]
43. Jonsson, T.; Atwal, J.K.; Steinberg, S.; Snaedal, J.; Jonsson, P.V.; Bjornsson, S.; Stefansson, H.; Sulem, P.; Gudbjartsson, D.; Maloney, J.; et al. A mutation in APP protects against Alzheimer's disease and age-related cognitive decline. *Nature* **2012**, *488*, 96–99. [CrossRef]
44. Harper, A.R.; Nayee, S.; Topol, E.J. Protective alleles and modifier variants in human health and disease. *Nat. Rev. Genet.* **2015**, *16*, 689–701. [CrossRef]

45. Casas, C.; Sergeant, N.; Itier, J.M.; Blanchard, V.; Wirths, O.; van der Kolk, N.; Vingtdeux, V.; van de Steeg, E.; Ret, G.; Canton, T.; et al. Massive CA1/2 neuronal loss with intraneuronal and N-terminal truncated Abeta42 accumulation in a novel Alzheimer transgenic model. *Am. J. Pathol.* **2004**, *165*, 1289–1300. [CrossRef] [PubMed]
46. Bayer, T.A.; Wirths, O. Review on the APP/PS1KI mouse model: Intraneuronal Abeta accumulation triggers axonopathy, neuron loss and working memory impairment. *Genes. Brain Behav.* **2008**, *7* (Suppl. 1), 6–11. [CrossRef]
47. Bayer, T.A.; Breyhan, H.; Duan, K.; Rettig, J.; Wirths, O. Intraneuronal beta-amyloid is a major risk factor--novel evidence from the APP/PS1KI mouse model. *Neurodegener. Dis.* **2008**, *5*, 140–142. [CrossRef] [PubMed]
48. Wirths, O.; Breyhan, H.; Cynis, H.; Schilling, S.; Demuth, H.U.; Bayer, T.A. Intraneuronal pyroglutamate-Abeta 3-42 triggers neurodegeneration and lethal neurological deficits in a transgenic mouse model. *Acta Neuropathol.* **2009**, *118*, 487–496. [CrossRef] [PubMed]
49. Christensen, D.Z.; Bayer, T.A.; Wirths, O. Intracellular Aß triggers neuron loss in the cholinergic system of the APP/PS1KI mouse model of Alzheimer's disease. *Neurobiol. Aging* **2010**, *31*, 1153–1163. [CrossRef]
50. Christensen, D.Z.; Schneider-Axmann, T.; Lucassen, P.J.; Bayer, T.A.; Wirths, O. Accumulation of intraneuronal Abeta correlates with ApoE4 genotype. *Acta Neuropathol.* **2010**, *119*, 555–566. [CrossRef]
51. Bayer, T.A.; Wirths, O. Intracellular accumulation of amyloid-Beta-a predictor for synaptic dysfunction and neuron loss in Alzheimer's disease. *Front. Aging Neurosci.* **2010**, *2*, 8. [CrossRef]
52. Bayer, T.A.; Wirths, O. Intraneuronal Aβ as a trigger for neuron loss: Can this be translated into human pathology? *Biochem. Soc. Trans.* **2011**, *39*, 857–861. [CrossRef]
53. Wirths, O.; Bayer, T.A. Intraneuronal Aβ accumulation and neurodegeneration: Lessons from transgenic models. *Life Sci.* **2012**, *91*, 1148–1152. [CrossRef]
54. Kumar, S.; Wirths, O.; Theil, S.; Gerth, J.; Bayer, T.A.; Walter, J. Early intraneuronal accumulation and increased aggregation of phosphorylated Abeta in a mouse model of Alzheimer's disease. *Acta Neuropathol.* **2013**, *125*, 699–709. [CrossRef]
55. Ripoli, C.; Cocco, S.; Li Puma, D.D.; Piacentini, R.; Mastrodonato, A.; Scala, F.; Puzzo, D.; D'Ascenzo, M.; Grassi, C. Intracellular accumulation of amyloid-β (Aβ) protein plays a major role in Aβ-induced alterations of glutamatergic synaptic transmission and plasticity. *J. Neurosci.* **2014**, *34*, 12893–12903. [CrossRef] [PubMed]
56. Scala, F.; Fusco, S.; Ripoli, C.; Piacentini, R.; Li Puma, D.D.; Spinelli, M.; Laezza, F.; Grassi, C.; D'Ascenzo, M. Intraneuronal Aβ accumulation induces hippocampal neuron hyperexcitability through A-type K(+) current inhibition mediated by activation of caspases and GSK-3. *Neurobiol. Aging* **2015**, *36*, 886–900. [CrossRef]
57. Christensen, D.Z.; Kraus, S.L.; Flohr, A.; Cotel, M.C.; Wirths, O.; Bayer, T.A. Transient intraneuronal Abeta rather than extracellular plaque pathology correlates with neuron loss in the frontal cortex of APP/PS1KI mice. *Acta Neuropathol.* **2008**, *116*, 647–655. [CrossRef]
58. Volloch, V.; Rits-Volloch, S. News from Mars: Two-tier paradox, intracellular PCR, chimeric junction shift, dark matter mRNA and other remarkable features of mammalian RNA-dependent mRNA amplification. Implications for Alzheimer's disease, RNA-based vaccines and mRNA therapeutics. *Ann. Integr. Mol. Med.* **2021**, *2*, 131–173. [CrossRef] [PubMed]
59. Breimer, L.; Denny, P. Alzheimer amyloid aspects. *Nature* **1987**, *326*, 749–750. [CrossRef]
60. Citron, M.; Haass, C.; Selkoe, D. Production of amyloid beta peptide by cultured cells: No evidence for internal initiation of translation at Met596. *Neurobiol. Aging* **1993**, *14*, 571–573. [CrossRef]
61. Macq, A.; Philippe, B.; Octave, N. The amyloid peptide of Alzheimer's disease is not produced by internal initiation of translation generating C-terminal amyloidogenic fragments of its precursor. *Neurosci. Lett.* **1994**, *182*, 227–230. [CrossRef] [PubMed]
62. Mita, S.; Sadlock, J.; Herbert, J.; Schon, E. A cDNA specifying the human amyloid beta precursor protein encodes a 95-kDa polypeptide. *Nucl. Acids Res.* **1988**, *16*, 9351; Erratum in *Nucl. Acids Res.* **1988**, *16*, 11402. [CrossRef]
63. Salbaum, J.; Weidemann, A.; Lemaire, H.; Masters, C.; Beyreuther, K. The promoter of Alzheimer's disease amyloid A4 precursor gene. *EMBO J.* **1988**, *7*, 2807–2813. [CrossRef]
64. Volloch, V.; Schweitzer, B.; Rits, S. Antisense Globin RNA in Murine Erythroid Tissues: Structure, Origin and Possible Function. *Proc. Natl. Acad. Sci. USA* **1996**, *93*, 2476–2481. [CrossRef] [PubMed] [PubMed Central]
65. Volloch, V. Protein-encoding RNA-to-RNA information transfer in mammalian cells: Principles of RNA-dependent mRNA amplification. *Ann. Integr. Mol. Med.* **2019**, *1*, 1002. [CrossRef] [PubMed]
66. Rits, S.; Olsen, B.; Volloch, V. Protein-encoding RNA to RNA information transfer in mammalian cells: RNA-dependent mRNA amplification. Identification of chimeric RNA intermediates and putative RNA end products. *Ann. Integr. Mol. Med.* **2019**, *1*, 1003. [PubMed] [PubMed Central]
67. Volloch, V.; Rits, S.; Olsen, B. RNA-dependent amplification of mammalian mRNA encoding extracellullar matrix proteins: Identification of chimeric RNA intermediates for 1alpha, 1beta, and 1gamma chains of laminin. *Ann. Integr. Mol. Med.* **2019**, *1*, 1004. [PubMed] [PubMed Central]

68. Volloch, V.; Olsen, B.; Rits, S. Precursor-independent overproduction of beta-amyloid in AD: Mitochondrial dysfunction as possible initiator of asymmetric RNA-dependent APP mRNA amplification. An engine that drives Alzheimer's disease. *Ann. Integr. Mol. Med.* **2019**, *1*, 1005. [CrossRef] [PubMed] [PubMed Central]
69. Volloch, V.; Olsen, B.; Rits, S. AD "Statin": Alzheimer's disorder is a "fast" disease preventable by therapeutic intervention initiated even late in life and reversible at the early stages. *Ann. Integr. Mol. Med.* **2020**, *2*, 1006. [CrossRef] [PubMed]
70. Volloch, V.; Olsen, B.; Rits, S. Alzheimer's disease is driven by intraneuronally retained beta-amyloid produced in the AD-specific, APP-independent pathway: Current perspective and experimental models for tomorrow. *Ann. Integr. Mol. Med.* **2020**, *2*, 1007. [PubMed] [PubMed Central]
71. Volloch, V.; Olsen, B.; Rits, S. Alzheimer's disease prevention and treatment: Case for optimism. *Ann. Integr. Mol. Med.* **2020**, *2*, 1008. [CrossRef] [PubMed] [PubMed Central]
72. Volloch, V.; Rits-Volloch, S. Alzheimer's disease is driven by beta-amyloid generated in the amyloid precursor protein-independent pathway and retained intraneuronally: Research and therapeutic strategies in a new AD paradigm. *Ann. Integr. Mol. Med.* **2021**, *2*, 1010. [CrossRef]
73. Volloch, V.; Schwetizer, B.; Rits, S. Evolutionarily Conserved Elements in the 5′-untranslated Region of β Globin mRNA Mediate Site-specific Priming of a Unique Hairpin Structure during cDNA Synthesis. *Nucl. Acids Res.* **1994**, *22*, 5302–5309. [CrossRef] [PubMed] [PubMed Central]
74. Volloch, V. A mechanism for β-amyloid overproduction in Alzheimer's disease: Precursor-independent generation of β-amyloid via antisense RNA-primed mRNA synthesis. *FEBS Lett.* **1996**, *390*, 124–128. [CrossRef] [PubMed]
75. Volloch, V. Mechanism for β-amyloid overproduction in sporadic Alzheimer's Disease: Possible antisense RNA-mediated generation of a 5′-truncated ßAPP mRNA encoding 12 kDa C-terminal fragment of βAPP, the immediate precursor of Aβ. In *Molecular Mechanisms of Dementia*; Wasco, W., Tanzi, R., Eds.; Humana Press: New York, NY, USA, 1997.
76. Volloch, V. Possible mechanism for resistance to Alzheimer's disease (AD) in mice suggests new approach to generate a mouse model for sporadic AD and may explain familial resistance to AD in man. *Exp. Neurobiol.* **1997**, *144*, 214–218. [CrossRef] [PubMed]
77. Volloch, V.; Rits, S. Results of Beta Secretase-Inhibitor Clinical Trials Support Amyloid Precursor Protein-Independent Generation of Beta Amyloid in Sporadic Alzheimer's Disease. *Med. Sci.* **2018**, *6*, 45. [CrossRef] [PubMed]
78. Vassar, R.; Bennett, B.D.; Babu-Khan, S.; Kahn, S.; Mendiaz, E.A.; Denis, P.; Teplow, D.B.; Ross, S.; Amarante, P.; Loeloff, R.; et al. Beta-secretase cleavage of Alheimer's amyloid precursor protein by the transmembrane aspartic protease BACE. *Science* **1999**, *286*, 735–741. [CrossRef] [PubMed]
79. Hussain, I.; Powell, D.; Howlett, D.R.; Tew, D.G.; Meek, T.D.; Chapman, C.; Gloger, I.S.; Murphy, K.E.; Southan, C.D.; Ryan, D.M.; et al. Identification of novel aspartic protease (Asp2) as beta secretase. *Mol. Cell. Neurosci.* **1999**, *14*, 419–427. [CrossRef]
80. Sinha, S.; Anderson, J.; Barbour, R.; Basl, G.; Caccavello, R.; Davis, D.; Doan, M.; Dovey, H.; Frigon, N.; Hong, J.; et al. Purification and cloning of amyloid precursor protein β-secretase from human brain. *Nature* **1999**, *402*, 537–540. [CrossRef]
81. Wang, Y.; Strickland, C.; Voigt, J.; Kennedy, M.; Beyer, B.; Senior, M.; Smith, E.; Nechuta, T.; Madison, V.; Czarniecki, M.; et al. Application of fragment-based NMR screening, X-ray crystallography, structure-based design, and focused chemical library design to identify novel μM leads for the development of nM BACE-1 inhibitors. *J. Med. Chem.* **2010**, *53*, 942–950. [CrossRef] [PubMed]
82. Zhu, Z.; Sun, Z.; Ye, Y.; Voigt, J.; Strickland, C.; Smith, E.; Cumming, J.; Wang, L.; Wong, J.; Wang, Y.; et al. Discovery of cyclic acylguanidines as highly potent and selective β-site amyloid cleaving enzyme (BACE) inhibitors: Part I—Inhibitor design and validation. *J. Med. Chem.* **2010**, *53*, 951–965. [CrossRef]
83. Cumming, J.; Smith, E.; Wang, L.; Misiaszek, J.; Durkin, J.; Pan, J.; Iserloh, U.; Wu, Y.; Zhu, Z.; Strickland, C.; et al. Structure based design of iminohydantoin BACE1 inhibitors: Identification of an orally available, centrally active BACE1 inhibitor. *Bioorg. Med. Chem. Lett.* **2012**, *22*, 2444–2449. [CrossRef]
84. Edwards, P.; Albert, J.; Sylvester, M.; Aharony, D.; Andisik, D.; Callaghan, O.; Campbell, J.; Carr, R.; Chessari, G.; Congreve, M.; et al. Application of fragment-based lead generation to the discovery of novel, cyclic amidine β-secretase inhibitors with nanomolar potency, cellular activity, and high ligand efficiency. *J. Med. Chem.* **2007**, *50*, 5912–5925. [CrossRef]
85. Barrow, J.; Stauffer, S.; Rittle, K.; Ngo, P.; Yang, Z.; Selnick, H.; Graham, S.; Munshi, S.; McGaughey, G.; Holloway, M.; et al. Discovery and X-ray crystallographic analysis of a S-piropiperidine iminohydantoin inhibitor of β-secretase. *J. Med. Chem.* **2008**, *51*, 6259–6262. [CrossRef]
86. Malamas, M.; Erdei, J.; Gunawan, I.; Turner, J.; Hu, Y.; Wagner, E.; Fan, K.; Chopra, R.; Olland, A.; Bard, J.; et al. Design and synthesis of 5,5′-disubstituted aminohydantoins as potent and selective human β-secretase (BACE1) inhibitors. *J. Med. Chem.* **2010**, *53*, 1146–1158. [CrossRef]
87. Rueeger, H.; Rondeau, J.; McCarthy, C.; Moebitz, H.; Tintelnot-Blomley, M.; Neumann, U.; Desrayaud, S. Structure based design, synthesis and SAR of cyclic hydroxyethylamine (HEA) BACE-1 inhibitors. *Bioorg. Med. Chem. Lett.* **2011**, *21*, 1942–1947. [CrossRef]
88. Probst, G.; Xu, Y. Small-Molecule BACE1 Inhibitors: A patent literature review (2006–2011). *Expert. Opin. Ther. Patents* **2012**, *22*, 511–540. [CrossRef]

89. May, P.; Dean, R.; Lowe, S.; Martenyi, F.; Sheehan, S.; Boggs, L.; Monk, S.; Mathes, B.; Mergott, D.; Watson, B.; et al. Robust central reduction of amyloid-β in humans with an orally available, non-peptidic β-secretase inhibitor. *J. Neurosci.* **2011**, *31*, 6507–6516. [CrossRef] [PubMed]
90. Stamford, A.W.; Scott, J.D.; Li, S.W.; Babu, S.; Tadesse, D.; Hunter, R.; Wu, Y.; Misiaszek, J.; Cumming, J.N.; Gilbert, E.J.; et al. Discovery of an orally available, brain penetrant BACE1 inhibitor that affords robust CNS Aβ reduction. *ACS Med. Chem. Lett.* **2012**, *3*, 897–902. [CrossRef] [PubMed]
91. Kennedy, M.E.; Stamford, A.W.; Chen, X.; Cox, K.; Cumming, J.N.; Dockendorf, M.F.; Egan, M.; Ereshefsky, L.; Hodgson, R.A.; Hyde, L.A.; et al. The BACE1 inhibitor verubecestat (MK-8931) reduces CNS β-amyloid in animal models and in Alzheimer's disease patients. *Sci. Transl. Med.* **2016**, *8*, 363ra150. [CrossRef] [PubMed]
92. Pakos-Zebrucka, K.; Koryga, I.; Mnich, K.; Ljujic, M.; Samali, A.; Gorman, A.M. The integrated stress response. *EMBO Rep.* **2016**, *17*, 1374–1395. [CrossRef]
93. Ron, D. Translational control in the endoplasmic reticulum stress response. *J. Clin. Investig.* **2002**, *110*, 1383–1388. [CrossRef]
94. Harding, H.P.; Zhang, Y.; Zeng, H.; Novoa, I.; Lu, P.D.; Calfon, M.; Sadri, N.; Yun, C.; Popko, B.; Paules, R.; et al. An integrated stress response regulates amino acid metabolism and resistance to oxidative stress. *Mol. Cell.* **2003**, *11*, 619–633. [CrossRef]
95. Brostrom, C.O.; Prostko, C.R.; Kaufman, R.J.; Brostrom, M.A. Inhibition of translational initiation by activators of the glucose-regulated stress protein and heat shock protein stress response systems. Role of the interferon-inducible double-stranded RNA-activated eukaryotic initiation factor 2alpha kinase. *J. Biol. Chem.* **1996**, *271*, 24995–25002. [CrossRef]
96. Dever, T.E.; Feng, L.; Wek, R.C.; Hinnebusch, A.G. Phosphorylation of initiation factor 2 alpha by protein kinase GCN2 mediates gene-specific translational control of GCN4 in yeast. *Cell* **1992**, *68*, 585–596. [CrossRef] [PubMed]
97. Wek, R.C.; Jiang, H.Y.; Anthony, T.G. Coping with stress: eIF2 kinases and translational control. *Biochem. Soc. Trans.* **2006**, *34*, 7–11. [CrossRef] [PubMed]
98. Rzymski, T.; Milani, M.; Pike, L.; Buffa, F.; Mellor, H.R.; Winchester, L.; Pires, I.; Hammond, E.; Ragoussis, I.; Harris, A.L. Regulation of autophagy by ATF4 in response to severe hypoxia. *Oncogene* **2010**, *29*, 4424–4435. [CrossRef]
99. Ye, J.; Kumanova, M.; Hart, L.S.; Sloane, K.; Zhang, H.; De Panis, D.N.; Bobrovnikova-Marjon, E.; Diehl, J.A.; Ron, D.; Koumenis, C. The GCN2-ATF4 pathway is critical for tumour cell survival and proliferation in response to nutrient deprivation. *EMBO J.* **2010**, *29*, 2082–2096. [CrossRef]
100. Garcia, M.A.; Meurs, E.F.; Esteban, M. The dsRNA protein kinase PKR: Virus and cell control. *Biochimie* **2007**, *89*, 799–811. [CrossRef] [PubMed]
101. Harding, H.P.; Zhang, Y.; Ron, D. Protein translation and folding are coupled by an endoplasmic-reticulum-resident kinase. *Nature* **1999**, *397*, 271–274. [CrossRef]
102. Cao, A.; Galanello, R. Beta-thalassemia. *Genet. Med.* **2010**, *12*, 61–76. [CrossRef]
103. Costa-Mattioli, M.; Walter, P. The integrated stress response: From mechanism to disease. *Science* **2020**, *368*, eaat5314. [CrossRef]
104. Romero-Ramírez, L.; Nieto-Sampedro, M.; Barreda-Manso, M.A. Integrated Stress Response as a Therapeutic Target for CNS Injuries. *BioMed Res. Int.* **2017**, *2017*, 6953156. [CrossRef]
105. Costa-Mattioli, M.; Gobert, D.; Stern, E.; Gamache, K.; Colina, R.; Cuello, C.; Sossin, W.; Kaufman, R.; Pelletier, J.; Rosenblum, K.; et al. eIF2alpha phosphorylation bidirectionally regulates the switch from short- to long-term synaptic plasticity and memory. *Cell* **2007**, *129*, 195–206. [CrossRef]
106. Rosi, S.; Frias, E.S. The Integrated Stress Response: A Central Memory Switch in Down Syndrome. *Cell Metab.* **2020**, *31*, 15–17. [CrossRef] [PubMed]
107. Kim, H.J.; Raphael, A.R.; LaDow, E.S.; McGurk, L.; Weber, R.A.; Trojanowski, J.Q.; Lee, V.M.; Finkbeiner, S.; Gitler, A.D.; Bonini, N.M. Therapeutic modulation of eIF2α phosphorylation rescues TDP-43 toxicity in amyotrophic lateral sclerosis disease models. *Nat. Genet.* **2014**, *46*, 152–160. [CrossRef]
108. Abisambra, J.F.; Jinwal, U.K.; Blair, L.J.; O'Leary JC 3rd Li, Q.; Brady, S.; Wang, L.; Guidi, C.E.; Zhang, B.; Nordhues, B.A.; Cockman, M.; et al. Tau accumulation activates the unfolded protein response by impairing endoplasmic reticulum-associated degradation. *J. Neurosci.* **2013**, *33*, 9498–9507. [CrossRef] [PubMed]
109. Moreno, J.A.; Radford, H.; Peretti, D.; Steinert, J.R.; Verity, N.; Martin, M.G.; Halliday, M.; Morgan, J.; Dinsdale, D.; Ortori, C.A.; et al. Sustained translational repression by eIF2alpha-P mediates prion neurodegeneration. *Nature* **2012**, *485*, 507–511. [CrossRef]
110. Baleriola, J.; Walker, C.A.; Jean, Y.Y.; Crary, J.F.; Troy, C.M.; Nagy, P.L.; Hengst, U. Axonally synthesized ATF4 transmits a neurodegenerative signal across brain regions. *Cell* **2014**, *158*, 1159–1172. [CrossRef]
111. Halliday, M.; Radford, H.; Sekine, Y.; Moreno, J.; Verity, N.; le Quesne, J.; Ortori, C.A.; Barrett, D.A.; Fromont, C.; Fischer, P.M.; et al. Partial restoration of protein synthesis rates by the small molecule ISRIB prevents neurodegeneration without pancreatic toxicity. *Cell Death Dis.* **2015**, *6*, e1672. [CrossRef] [PubMed]
112. Cheng, W.; Wang, S.; Mestre, A.A.; Fu, C.; Makarem, A.; Xian, F.; Hayes, L.R.; Lopez-Gonzalez, R.; Drenner, K.; Jiang, J.; et al. C9ORF72 GGGGCC repeat-associated non-AUG translation is upregulated by stress through eIF2alpha phosphorylation. *Nat. Commun.* **2018**, *9*, 51. [CrossRef]

113. Martinez, G.; Khatiwada, S.; Costa-Mattioli, M.; Hetz, C. ER Proteostasis Control of Neuronal Physiology and Synaptic Function. *Trends Neurosci.* **2018**, *41*, 610–624. [CrossRef]
114. Wong, Y.L.; LeBon, L.; Basso, A.M.; Kohlhaas, K.L.; Nikkel, A.L.; Robb, H.M.; Donnelly-Roberts, D.L.; Prakash, J.; Swensen, A.M.; Rubinstein, N.D.; et al. eIF2B activator prevents neurological defects caused by a chronic integrated stress response. *eLife* **2019**, *8*, e42940. [CrossRef]
115. Zhu, P.J.; Khatiwada, S.; Cui, Y.; Reineke, L.C.; Dooling, S.W.; Kim, J.J.; Li, W.; Walter, P.; Costa-Mattioli, M. Activation of the ISR mediates the behavioral and neurophysiological abnormalities in Down syndrome. *Science* **2019**, *366*, 843–849. [CrossRef]
116. Page, G.; Rioux Bilan, A.; Ingrand, S.; Lafay-Chebassier, C.; Pain, S.; Perault Pochat, M.C.; Bouras, C.; Bayer, T.; Hugon, J. Activated double-stranded RNA-dependent protein kinase and neuronal death in models of Alzheimer's disease. *Neuroscience* **2006**, *139*, 1343–1354. [CrossRef]
117. Segev, Y.; Barrera, I.; Ounallah-Saad, H.; Wibrand, K.; Sporild, I.; Livne, A.; Rosenberg, T.; David, O.; Mints, M.; Bramham, C.R.; et al. PKR Inhibition Rescues Memory Deficit and ATF4 Overexpression in ApoE epsilon4 Human Replacement Mice. *J. Neurosci.* **2015**, *35*, 12986–12993. [CrossRef] [PubMed]
118. Kernohan, K.D.; Tétreault, M.; Liwak-Muir, U.; Geraghty, M.T.; Qin, W.; Venkateswaran, S.; Davila, J.; Care4Rare Canada Consortium; Holcik, M.; Majewski, J.; et al. Homozygous mutation in the eukaryotic translation initiation factor 2alpha phosphatase gene, PPP1R15B, is associated with severe microcephaly, short stature and intellectual disability. *Hum. Mol. Genet.* **2015**, *24*, 6293–6300. [CrossRef]
119. Hu, Z.; Yu, P.; Zhang, Y.; Yang, Y.; Zhu, M.; Qin, S.; Xu, J.T.; Duan, D.; Wu, Y.; Wang, D.; et al. Inhibition of the ISR abrogates mGluR5-dependent long-term depression and spatial memory deficits in a rat model of Alzheimer's disease. *Transl. Psychiatry* **2022**, *12*, 96. [CrossRef]
120. Oliveira, M.M.; Lourenco, M.V.; Longo, F.; Kasica, N.P.; Yang, W.; Ureta, G.; Ferreira, D.D.P.; Mendonça, P.H.J.; Bernales, S.; Ma, T.; et al. Correction of eIF2-dependent defects in brain protein synthesis, synaptic plasticity, and memory in mouse models of Alzheimer's disease. *Sci. Signal.* **2021**, *14*, eabc5429. [CrossRef] [PubMed]
121. Ma, T.; Trinh, M.A.; Wexler, A.J.; Bourbon, C.; Gatti, E.; Pierre, P.; Cavener, D.R.; Klann, E. Suppression of eIF2α kinases alleviates Alzheimer's disease-related plasticity and memory deficits. *Nat. Neurosci.* **2013**, *16*, 1299–1305. [CrossRef]
122. Chou, A.; Krukowski, K.; Jopson, T.; Zhu, P.J.; Costa-Mattioli, M.; Walter, P.; Rosi, S. Inhibition of the integrated stress response reverses cognitive deficits after traumatic brain injury. *Proc. Natl. Acad. Sci. USA* **2017**, *114*, E6420–E6426. [CrossRef] [PubMed]
123. Zhu, P.J.; Huang, W.; Kalikulov, D.; Yoo, J.W.; Placzek, A.N.; Stoica, L.; Zhou, H.; Bell, J.C.; Friedlander, M.J.; Krnjević, K.; et al. Suppression of PKR promotes network excitability and enhanced cognition by interferon-γ-mediated disinhibition. *Cell* **2011**, *147*, 1384–1396. [CrossRef]
124. Stern, E.; Chinnakkaruppan, A.; David, O.; Sonenberg, N.; Rosenblum, K. Blocking the eIF2α kinase (PKR) enhances positive and negative forms of cortex-dependent taste memory. *J. Neurosci.* **2013**, *33*, 2517–2525. [CrossRef]
125. Ounallah-Saad, H.; Sharma, V.; Edry, E.; Rosenblum, K. Genetic or pharmacological reduction of PERK enhances cortical-dependent taste learning. *J. Neurosci.* **2014**, *34*, 14624–14632. [CrossRef]
126. Batista, G.; Johnson, J.L.; Dominguez, E.; Costa-Mattioli, M.; Pena, J.L. Translational control of auditory imprinting and structural plasticity by eIF2α. *eLife* **2016**, *5*, e17197. [CrossRef]
127. Sidrauski, C.; Acosta-Alvear, D.; Khoutorsky, A.; Vedantham, P.; Hearn, B.R.; Li, H.; Gamache, K.; Gallagher, C.M.; Ang, K.K.; Wilson, C.; et al. Pharmacological brake-release of mRNA translation enhances cognitive memory. *eLife* **2013**, *2*, e00498. [CrossRef]
128. Choi, S.H.; Kim, Y.H.; Hebisch, M.; Sliwinski, C.; Lee, S.; D'Avanzo, C.; Chen, H.; Hooli, B.; Asselin, C.; Muffat, J.; et al. A three-dimensional human neural cell culture model of Alzheimer's disease. *Nature* **2014**, *515*, 274–278. [CrossRef]
129. Meyer, K.D.; Patil, D.P.; Zhou, J.; Zinoviev, A.; Skabkin, M.A.; Elemento, O.; Pestova, T.V.; Qian, S.B.; Jaffrey, S.R. 5′ UTR m(6)A Promotes Cap-Independent Translation. *Cell* **2015**, *163*, 999–1010. [CrossRef] [PubMed]
130. Frottin, F.; Martinez, A.; Peinot, P.; Mitra, S.; Holz, R.; Giglione, C.; Meinnel, T. The proteomics of N-terminal methionine cleavage. *Mol. Cell Proteom. MCP* **2006**, *5*, 2336–2349. [CrossRef]
131. Xiao, Q.; Zhang, F.; Nacev, B.; Liu, J.; Pei, D. Protein N-Terminal Processing: Substrate Specificity of *Escherichia coli* and Human Methionine Aminopeptidases. *Biochemistry* **2010**, *49*, 5588–5599. [CrossRef] [PubMed]
132. Varshavsky, A. The N-end rule pathway and regulation by proteolysis. *Protein Sci.* **2011**, *20*, 1298–1345. [CrossRef] [PubMed]
133. Wingfield, P. N-Terminal Methionine Processing. *Curr. Protoc. Protein Sci.* **2017**, *88*, 6.14.1–6.14.3. [CrossRef]
134. Varland, S.; Osberg, C.; Arnesen, T. N-terminal modifications of cellular proteins: The enzymes involved, their substrate specificities and biological effects. *Proteomics* **2015**, *15*, 2385–2401. [CrossRef]
135. Polevoda, B.; Sherman, F. N-terminal acetyltransferases and sequence requirements for N-terminal acetylation of eukaryotic proteins. *J. Mol. Biol.* **2003**, *325*, 595–622. [CrossRef]
136. Kinoshita, A.; Fukumoto, H.; Shah, T.; Whelan, C.M.; Irizarry, M.C.; Hyman, B.T. Demonstration by FRET of BACE interaction with the amyloid precursor protein at the cell surface and in early endosomes. *J. Cell Sci.* **2003**, *116*, 3339–3346. [CrossRef] [PubMed]

137. Xu, H.; Greengard, P.; Gandy, S. Regulated formation of Golgi secretory vesicles containing Alzheimer β-amyloid precursor protein. *J. Biol. Chem.* **1995**, *270*, 23243–23245. [CrossRef]
138. Mizuguchi, M.; Ikeda, K.; Kim, S. Differential distribution of cellular forms of β-amyloid precursor protein in murine glial cell cultures. *Brain Res.* **1992**, *584*, 219–225. [CrossRef] [PubMed]
139. Cook, D.G.; Forman, M.S.; Sung, J.C.; Leight, S.; Kolson, D.L.; Iwatsubo, T.; Lee, V.M.; Doms, R.W. Alzheimer's Aβ42 is generated in the endoplasmic reticulum/intermediate compartment of NT2N cells. *Nat. Med.* **1997**, *3*, 1021–1023. [CrossRef]
140. Hartmann, T.; Bieger, S.C.; Brühl, B.; Tienari, P.J.; Ida, N.; Allsop, D.; Roberts, G.W.; Masters, C.L.; Dotti, C.G.; Unsicker, K.; et al. Distinct sites of intracellular production for Alzheimer's disease Aβ40/42 amyloid peptides. *Nat. Med.* **1997**, *3*, 1016–1020. [CrossRef] [PubMed]
141. Wild-Bode, C.; Yamazaki, T.; Capell, A.; Leimer, U.; Steiner, H.; Ihara, Y.; Haass, C. Intracellular generation and accumulation of amyloid beta-peptide terminating at amino acid 42. *J. Biol. Chem.* **1997**, *272*, 16085–16088. [CrossRef]
142. Lee, S.J.; Liyanage, U.; Bickel, P.E.; Xia, W.; Lansbury, P.T., Jr.; Kosik, K.S. A detergent-insoluble membrane compartment contains Aβ in vivo. *Nat. Med.* **1998**, *4*, 730–734. [CrossRef]
143. Skovronsky, D.; Doms, R.; Lee, V. Detection of a novel intraneuronal pool of insoluble amyloid βprotein that accumulates with time in culture. *J. Cell Biol.* **1998**, *141*, 1031–1039. [CrossRef]
144. Manczak, M.; Anekonda, T.S.; Henson, E.; Park, B.S.; Quinn, J.; Reddy, P.H. Mitochondria are a direct site of Aβ accumulation in Alzheimer's disease neurons: Implications for free radical generation and oxidative damage in disease progression. *Hum. Mol. Genet.* **2006**, *15*, 1437–1449. [CrossRef]
145. Chafekar, S.; Baas, F.; Scheper, W. Oligomer-specific amyloid-beta toxicity in cell models is mediated by selective uptake. *Biochem. Biophys. Acta* **2008**, *9*, 523–531.
146. Wesen, E.; Jeffries, G.; Dzebo, M.; Esbjorner, M. Endocytic uptake of monomeric amyloid-β peptides is clathrin- and dynamin-independent and results in selective accumulation of Aβ(1–42) compared to Aβ(1–40). *Sci. Rep.* **2017**, *7*, 2021. [CrossRef] [PubMed]
147. Kumar-Singh, S.; Theuns, J.; Van Broeck, B.; Pirici, D.; Vennekens, K.; Corsmit, E.; Cruts, M.; Dermaut, B.; Wang, R.; Van Broeckhoven, C. Mean age-of-onset of familial Alzheimer disease caused by presenilin mutations correlates with both increased Abeta42 and decreased Abeta40. *Hum. Mutat.* **2006**, *27*, 686–695. [CrossRef] [PubMed]
148. Hu, X.; Crick, S.L.; Bu, G.; Frieden, C.; Pappu, R.V.; Lee, J.M. Amyloid seeds formed by cellular uptake, concentration, and aggregation of the amyloid-beta peptide. *Proc. Natl. Acad. Sci. USA* **2009**, *106*, 20324–20329. [CrossRef] [PubMed]
149. Yajima, R.; Tokutake, T.; Koyama, A.; Kasuga, K.; Tezuka, T.; Nishizawa, M.; Ikeuchi, T. ApoE-isoform-dependent cellular uptake of amyloid-β is mediated by lipoprotein receptor LR11/SorLA. *Biochem. Biophys. Res. Commun.* **2015**, *456*, 482–488. [CrossRef]
150. Omtri, R.S.; Davidson, M.W.; Arumugam, B.; Poduslo, J.F.; Kandimalla, K.K. Differences in the cellular uptake and intracellular itineraries of amyloid beta proteins 40 and 42: Ramifications for the Alzheimer's drug discovery. *Mol. Pharm.* **2012**, *9*, 1887. [CrossRef]
151. Bu, G.; Cam, J.; Zerbinatti, C. LRP in amyloid-β production and metabolism. *Ann. N. Y. Acad. Sci.* **2006**, *1086*, 35–53. [CrossRef]
152. Wang, H.Y.; Lee, D.H.; D'Andrea, M.R.; Peterson, P.A.; Shank, R.P.; Reitz, A.B. β-Amyloid1–42 binds to α7 nicotinic acetylcholine receptor with high affinity. Implications for Alzheimer's disease pathology. *J. Biol. Chem.* **2000**, *275*, 5626–5632. [CrossRef]
153. Nagele, R.; D'Andrea, M.; Anderson, W.; Wang, H. Intracellular accumulation of Aβ42 in neurons is facilitated by the α7 nicotinic acetylcholine receptor in Alzheimer's disease. *Neuroscience* **2002**, *110*, 199–211. [CrossRef]
154. Oddo, S.; Caccamo, A.; Green, K.N.; Liang, K.; Tran, L.; Chen, Y.; Leslie, F.M.; LaFerla, F.M. Chronic nicotine administration exacerbates tau pathology in a transgenic model of Alzheimer's disease. *Proc. Natl. Acad. Sci. USA* **2005**, *102*, 3046–3051. [CrossRef]
155. Yan, S.D.; Chen, X.; Fu, J.; Chen, M.; Zhu, H.; Roher, A.; Slattery, T.; Zhao, L.; Nagashima, M.; Morser, J.; et al. RAGE and amyloid-β peptide neurotoxicity in Alzheimer's disease. *Nature* **1996**, *382*, 685–691. [CrossRef]
156. Sasaki, N.; Toki, S.; Chowei, H.; Saito, T.; Nakano, N.; Hayashi, Y.; Takeuchi, M.; Makita, Z. Immunohistochemical distribution of the receptor for advanced glycation end products in neurons and astrocytes in Alzheimer's disease. *Brain Res.* **2001**, *888*, 256–262. [CrossRef] [PubMed]
157. Deane, R.; Du Yan, S.; Submamaryan, R.K.; LaRue, B.; Jovanovic, S.; Hogg, E.; Welch, D.; Manness, L.; Lin, C.; Yu, J.; et al. RAGE mediates amyloid-β peptide transport across the blood–brain barrier and accumulation in brain. *Nat. Med.* **2003**, *9*, 907–913. [CrossRef]
158. Iribarren, P.; Zhou, Y.; Hu, J.; Le, Y.; Wang, J. Role of formyl peptide receptor-like 1 (FPRL1/FPR2) in mononuclear phagocyte responses in Alzheimer disease. *Immunol. Res.* **2005**, *31*, 165–176. [CrossRef] [PubMed]
159. Snyder, E.M.; Nong, Y.; Almeida, C.G.; Paul, S.; Moran, T.; Choi, E.Y.; Nairn, A.C.; Salter, M.W.; Lombroso, P.J.; Gouras, G.K.; et al. Regulation of NMDA receptor trafficking by amyloid-β. *Nat. Neurosci.* **2005**, *8*, 1051–1058. [CrossRef]

160. Chang, R.C.; Suen, K.C.; Ma, C.H.; Elyaman, W.; Ng, H.K.; Hugon, J. Involvement of double-stranded RNA-dependent protein kinase and phosphorylation of eukaryotic initiation factor-2alpha in neuronal degeneration. *J. Neurochem.* **2002**, *83*, 1215–1225. [CrossRef]
161. Peel, A.L. PKR activation in neurodegenerative disease. *J. Neuropathol. Exp. Neurol.* **2004**, *63*, 97–105. [CrossRef]
162. Peel, A.L.; Bredesen, D.E. Activation of the cell stress kinase PKR in Alzheimer's disease and human amyloid precursor protein transgenic mice. *Neurobiol. Dis.* **2003**, *14*, 52–62. [CrossRef] [PubMed]
163. Chang, R.C.; Wong, A.K.; Ng, H.K.; Hugon, J. Phosphorylation of eukaryotic initiation factor-2alpha (eIF2alpha) is associated with neuronal degeneration in Alzheimer's disease. *Neuroreport* **2002**, *13*, 2429–2432. [CrossRef]
164. Onuki, R.; Bando, Y.; Suyama, E.; Katayama, T.; Kawasaki, H.; Baba, T.; Tohyama, M.; Taira, K. An RNA-dependent protein kinase is involved in tunicamycin-induced apoptosis and Alzheimer's disease. *EMBO J.* **2004**, *23*, 959–968. [CrossRef]
165. Paquet, C.; Mouton-Liger, F.; Meurs, E.F.; Mazot, P.; Bouras, C.; Pradier, L.; Gray, F.; Hugon, J. The PKR activator PACT is induced by Abeta: Involvement in Alzheimer's disease. *Brain Pathol.* **2012**, *22*, 219–229. [CrossRef]
166. Lourenco, M.V.; Clarke, J.R.; Frozza, R.L.; Bomfim, T.R.; Forny-Germano, L.; Batista, A.F.; Sathler, L.B.; Brito-Moreira, J.; Amaral, O.B.; Silva, C.A.; et al. TNF-alpha mediates PKR-dependent memory impairment and brain IRS-1 inhibition induced by Alzheimer's beta-amyloid oligomers in mice and monkeys. *Cell Metab.* **2013**, *18*, 831–843. [CrossRef] [PubMed]
167. Zhu, X.; Perry, G.; Moreira, P.I.; Aliev, G.; Cash, A.D.; Hirai, K.; Smith, M.A. Mitochondrial abnormalities and oxidative imbalance in Alzheimer disease. *J. Alzheimers Dis.* **2006**, *9*, 147–153. [CrossRef] [PubMed]
168. Blass, J.P. The mitochondrial spiral. An adequate cause of dementia in the Alzheimer's syndrome. *Ann. N. Y. Acad. Sci.* **2000**, *924*, 170–183. [CrossRef]
169. Manczak, M.; Park, B.S.; Jung, Y.; Reddy, P.H. Differential expression of oxidative phosphorylation genes in patients with Alzheimer's disease: Implications for early mitochondrial dysfunction and oxidative damage. *Neuromol. Med.* **2004**, *5*, 147–162. [CrossRef] [PubMed]
170. Qin, W.; Haroutunian, V.; Katsel, P.; Cardozo, C.P.; Ho, L.; Buxbaum, J.D.; Pasinetti, G.M. PGC-1alpha expression decreases in the Alzheimer disease brain as a function of dementia. *Arch. Neurol.* **2009**, *66*, 352–361. [CrossRef]
171. Du, H.; Guo, L.; Yan, S.; Sosunov, A.A.; McKhann, G.M.; Yan, S.S. Early deficits in synaptic mitochondria in an Alzheimer's disease mouse model. *Proc. Natl. Acad. Sci. USA* **2010**, *107*, 18670–18675. [CrossRef]
172. Lin, M.T.; Simon, D.K.; Ahn, C.H.; Kim, L.M.; Beal, M.F. High aggregate burden of somatic mtDNA point mutations in aging and Alzheimer's disease brain. *Hum. Mol. Genet.* **2002**, *11*, 133–145. [CrossRef]
173. Calkins, M.; Manczak, M.; Mao, P.; Shirendeb, U.; Reddy, P.H. Impaired mitochondrial biogenesis, defective axonal transport of mitochondria, abnormal mitochondrial dynamics and synaptic degeneration in a mouse model of Alzheimer's disease. *Hum. Mol. Genet.* **2011**, *20*, 4515–4529. [CrossRef]
174. Anandatheerthavarada, H.K.; Biswas, G.; Robin, M.A.; Avadhani, N.G. Mitochondrial targeting and a novel transmembrane arrest of Alzheimer's amyloid precursor protein impairs mitochondrial function in neuronal cells. *J. Cell Biol.* **2003**, *161*, 41–54. [CrossRef]
175. Caspersen, C.; Wang, N.; Yao, J.; Sosunov, A.; Chen, X.; Lustbader, J.W.; Xu, H.W.; Stern, D.; McKhann, G.; Yan, S.D. Mitochondrial Abeta: A potential focal point for neuronal metabolic dysfunction in Alzheimer's disease. *FASEB J.* **2005**, *19*, 2040–2041. [CrossRef]
176. Chen, J.X.; Yan, S.S. Role of mitochondrial amyloid-beta in Alzheimer's disease. *J. Alzheimers Dis.* **2010**, *20* (Suppl. 2), S569–S578. [CrossRef] [PubMed]
177. Hansson Petersen, C.A.; Alikhani, N.; Behbahani, H.; Wiehager, B.; Pavlov, P.F.; Alafuzoff, I.; Leinonen, V.; Ito, A.; Winblad, B.; Glaser, E.; et al. The amyloid beta-peptide is imported into mitochondria via the TOM import machinery and localized to mitochondrial cristae. *Proc. Natl. Acad. Sci. USA* **2008**, *105*, 13145–13150. [CrossRef] [PubMed]
178. de la Monte, S.M.; Luong, T.; Neely, T.R.; Robinson, D.; Wands, J.R. Mitochondrial DNA damage as a mechanism of cell loss in Alzheimer's disease. *Lab. Investig.* **2000**, *80*, 1323–1335. [CrossRef]
179. Brooks, W.M.; Lynch, P.J.; Ingle, C.C.; Hatton, A.; Emson, P.C.; Faull, R.L.; Starkey, M.P. Gene expression profiles of metabolic enzyme transcripts in Alzheimer's disease. *Brain Res.* **2007**, *1127*, 127–135. [CrossRef]
180. Wang, X.; Perry, G.; Smith, M.A.; Zhu, X. Amyloid-beta-derived diffusible ligands cause impaired axonal transport of mitochondria in neurons. *Neurodegener. Dis.* **2007**, *7*, 56–59. [CrossRef]
181. Wang, X.; Su, B.; Fujioka, H.; Zhu, X. Dynamin-like protein 1 reduction underlies mitochondrial morphology and distribution abnormalities in fibroblasts from sporadic Alzheimer's disease patients. *Am. J. Pathol.* **2008**, *173*, 470–482. [CrossRef] [PubMed]
182. Wang, X.; Su, B.; Lee, H.G.; Li, X.; Perry, G.; Smith, M.A.; Zhu, X. Impaired balance of mitochondrial fission and fusion in Alzheimer's disease. *J. Neurosci.* **2009**, *29*, 9090–9103. [CrossRef]
183. Wang, X.; Su, B.; Siedlak, S.L.; Moreira, P.I.; Fujioka, H.; Wang, Y.; Casadesus, G.; Zhu, X. Amyloid-beta overproduction causes abnormal mitochondrial dynamics via differential modulation of mitochondrial fission/fusion proteins. *Proc. Natl. Acad. Sci. USA* **2008**, *105*, 19318–19323. [CrossRef]

184. Brewer, G.; Herrera, R.; Philipp, S.; Sosna, J.; Reyes-Ruiz, J.; Glabe, C. Age-Related Intraneuronal Aggregation of Amyloid-β in Endosomes, Mitochondria, Autophagosomes, and Lysosomes. *J. Alzheimers Dis.* **2020**, *73*, 229–246. [CrossRef]
185. Guo, X.; Aviles, G.; Liu, Y.; Tian, R.; Unger, B.A.; Lin, Y.T.; Wiita, A.P.; Xu, K.; Correia, M.A.; Kampmann, M. Mitochondrial stress is relayed to the cytosol by an OMA1-DELE1-HRI pathway. *Nature* **2020**, *579*, 427–432. [CrossRef]
186. Fessler, E.; Eckl, E.M.; Schmitt, S.; Mancilla, I.A.; Meyer-Bender, M.F.; Hanf, M.; Philippou-Massier, J.; Krebs, S.; Zischka, H.; Jae, L.T. A pathway coordinated by DELE1 relays mitochondrial stress to the cytosol. *Nature* **2020**, *579*, 433–437. [CrossRef] [PubMed]
187. Almeida, C.; Takahashi, R.; Gouras, G. β-amyloid accumulation impairs multivesicular body sorting by inhibiting the ubiquitin–proteasome system. *J. Neurosci.* **2006**, *26*, 4277–4288. [CrossRef] [PubMed]
188. Gregori, L.; Fuchs, C.; Figueiredo-Pereira, M.; Van Nostrand, W.; Goldgaber, D. Amyloid β-protein inhibits ubiquitin-dependent protein degradation in vitro. *J. Biol. Chem.* **1995**, *270*, 19702–19708. [CrossRef] [PubMed]
189. Oh, S.; Hong, H.S.; Hwang, E.; Sim, H.J.; Lee, W.; Shin, S.J.; Mook-Jung, I. Amyloid peptide attenuates the proteasome activity in neuronal cells. *Mech. Ageing Dev.* **2005**, *126*, 1292–1299. [CrossRef]
190. Tseng, B.; Green, K.; Chan, J.; Blurton-Jones, M.; LaFerla, F. Aβ inhibits the proteasome and enhances amyloid and tau accumulation. *Neurobiol. Aging* **2008**, *29*, 1607–1610. [CrossRef]
191. Pardossi-Piquard, R.; Checler, F. The physiology of the β-amyloid precursor protein intracellular domain AICD. *J. Neurochem.* **2012**, *120* (Suppl. 1), 109–124. [CrossRef] [PubMed]
192. Szögi, T.; Schuster, I.; Borbély, E.; Gyebrovszki, A.; Bozsó, Z.; Gera, J.; Rajkó, R.; Sántha, M.; Penke, B.; Fülöp, L. Effects of the Pentapeptide P33 on Memory and Synaptic Plasticity in APP/PS1 Transgenic Mice: A Novel Mechanism Presenting the Protein Fe65 as a Target. *Int. J. Mol. Sci.* **2019**, *20*, 3050. [CrossRef]
193. Borg, J.-P.; Yang, Y.; De Taddeo-Borg, M.; Margolis, B.; Turner, R.S. The X11α protein slows cellular amyloid precursor protein processing and reduces Aβ40 and Aβ42 secretion. *J. Biol. Chem.* **1998**, *273*, 14761–14766. [CrossRef]
194. Dunning, C.J.R.; Black, H.L.; Andrews, K.L.; Davenport, E.C.; Conboy, M.; Chawla, S.; Dowle, A.A.; Ashford, D.; Thomas, J.R.; Evans, G.J.O. Multisite tyrosine phosphorylation of the N-terminus of Mint1/X11α by Src kinase regulates the trafficking of amyloid precursor protein. *J. Neurochem.* **2016**, *137*, 518–527. [CrossRef]
195. Miller, C.C.J.; McLoughlin, D.M.; Lau, K.-F.; Tennant, M.E.; Rogelj, B. The X11 proteins, Aβ production and Alzheimer's disease. *Trends Neurosci.* **2006**, *29*, 280–285. [CrossRef]
196. Tamayev, R.; Zhou, D.; D'Adamio, L. The interactome of the amyloid β precursor protein family members is shaped by phosphorylation of their intracellular domains. *Mol. Neurodegener.* **2009**, *4*, 28. [CrossRef] [PubMed]
197. Hoe, H.-S.; Tran, T.S.; Matsuoka, Y.; Howell, B.W.; Rebeck, G.W. DAB1 and Reelin Effects on Amyloid Precursor Protein and ApoE Receptor 2 Trafficking and Processing. *J. Biol. Chem.* **2006**, *281*, 35176–35185. [CrossRef] [PubMed]
198. Russo, C.; Dolcini, V.; Salis, S.; Venezia, V.; Zambrano, N.; Russo, T.; Schettini, G. Signal Transduction through Tyrosine-phosphorylated C-terminal Fragments of Amyloid Precursor Protein via an Enhanced Interaction with Shc/Grb2 Adaptor Proteins in Reactive Astrocytes of Alzheimer's Disease Brain. *J. Biol. Chem.* **2002**, *277*, 35282–35288. [CrossRef] [PubMed]
199. Tarr, P.E.; Roncarati, R.; Pelicci, G.; Pelicci, P.G.; D'Adamio, L. Tyrosine phosphorylation of the β-amyloid precursor protein cytoplasmic tail promotes interaction with Shc. *J. Biol. Chem.* **2002**, *277*, 16798–16804. [CrossRef]
200. Bukhari, H.; Glotzbach, A.; Kolbe, K.; Leonhardt, G.; Loosse, C.; Mueller, T. Small things matter: Implications of APP intracellular domain AICD nuclear signaling in the progression and pathogenesis of Alzheimer's disease. *Prog. Neurobiol.* **2017**, *156*, 189–213. [CrossRef]
201. Chang, K.A.; Kim, H.S.; Ha, T.Y.; Ha, J.W.; Shin, K.Y.; Jeong, Y.H.; Lee, J.P.; Park, C.H.; Kim, S.; Baik, T.K.; et al. Phosphorylation of amyloid precursor protein (APP) at Thr668 regulates the nuclear translocation of the APP intracellular domain and induces neurodegeneration. *Mol. Cell. Biol.* **2006**, *26*, 4327–4338. [CrossRef]
202. Ando, K.; Iijima, K.-I.; Elliott, J.I.; Kirino, Y.; Suzuki, T. Phosphorylation-dependent regulation of the interaction of amyloid precursor protein with Fe65 affects the production of β-amyloid. *J. Biol. Chem.* **2001**, *276*, 40353–40361. [CrossRef]
203. Müller, T.; Meyer, H.E.; Egensperger, R.; Marcus, K. The amyloid precursor protein intracellular domain (AICD) as modulator of gene expression, apoptosis, and cytoskeletal dynamics-relevance for Alzheimer's disease. *Prog. Neurobiol.* **2008**, *85*, 393–406. [CrossRef]
204. Shu, R.; Wong, W.; Ma, Q.H.; Yang, Z.Z.; Zhu, H.; Liu, F.J.; Wang, P.; Ma, J.; Yan, S.; Polo, J.M.; et al. APP intracellular domain acts as a transcriptional regulator of miR-663 suppressing neuronal differentiation. *Cell Death Dis.* **2015**, *6*, e1651. [CrossRef]
205. Grimm, M.O.; Mett, J.; Stahlmann, C.P.; Grösgen, S.; Haupenthal, V.J.; Blümel, T.; Hundsdörfer, B.; Zimmer, V.C.; Mylonas, N.T.; Tanila, H.; et al. APP intracellular domain derived from amyloidogenic β- and γ-secretase cleavage regulates neprilysin expression. *Front. Aging Neurosci.* **2015**, *7*, 77. [CrossRef]
206. Pousinha, P.A.; Mouska, X.; Bianchi, D.; Temido-Ferreira, M.; Rajão-Saraiva, J.; Gomes, R.; Fernandez, S.P.; Salgueiro-Pereira, A.R.; Gandin, C.; Raymond, E.F.; et al. The Amyloid Precursor Protein C-Terminal Domain Alters CA1 Neuron Firing, Modifying Hippocampus Oscillations and Impairing Spatial Memory Encoding. *Cell Rep.* **2019**, *29*, 317–331.e5. [CrossRef] [PubMed]

207. Balusu, S.; Horré, K.; Thrupp, N.; Craessaerts, K.; Snellinx, A.; Serneels, L.; T'Syen, D.; Chrysidou, I.; Arranz, A.M.; Sierksma, A.; et al. MEG3 activates necroptosis in human neuron xenografts modeling Alzheimer's disease. *Science* **2023**, *381*, 1176–1182. [CrossRef]
208. Tucker-Drob, E.M. Cognitive Aging and Dementia: A Life Span Perspective. *Annu. Rev. Dev. Psychol.* **2019**, *1*, 177–196. [CrossRef] [PubMed]
209. Matthews, F.E.; Arthur, A.; Barnes, L.E.; Bond, J.; Jagger, C.; Robinson, L.; Brayne, C.; Medical Research Council Cognitive Function and Ageing Collaboration. A two-decade comparison of prevalence of dementia in individuals aged 65 years and older from three geographical areas of England: Results of the Cognitive Function and Ageing Study I and II. *Lancet* **2013**, *382*, 1405–1412. [CrossRef]
210. De Jonghe, C.; Zehr, C.; Yager, D.; Prada, C.M.; Younkin, S.; Hendriks, L.; Van Broeckhoven, C.; Eckman, C.B. Flemish and Dutch mutations in amyloid beta precursor protein have different effects on amyloid beta secretion. *Neurobiol. Dis.* **1998**, *5*, 281–286. [CrossRef] [PubMed]
211. Martin, B.; Schrader-Fischer, G.; Busciglio, J.; Duke, M.; Paganetti, P.; Yankner, B. Intracellular accumulation of beta-amyloid in cells expressing the Swedish mutant amyloid precursor protein. *J. Biol. Chem.* **1995**, *270*, 26727–26730. [CrossRef]
212. Sannerud, R.; Esselens, C.; Ejsmont, P.; Mattera, R.; Rochin, L.; Tharkeshwar, A.K.; De Baets, G.; De Wever, V.; Habets, R.; Baert, V.; et al. Restricted Location of PSEN2/gamma-secretase determines substrate specificity and generates an intracellular Abeta pool. *Cell* **2016**, *166*, 193–208. [CrossRef]
213. Henley, D.; Sundell, K.; Sethuraman, G.; Dowsett, S.A.; May, P.C. Safety profile of semagacestat, a gamma-secretase inhibitor: IDENTITY trial findings. *Curr. Med. Res.* **2014**, *30*, 2021–2032. [CrossRef]
214. Tagami, S.; Yanagida, K.; Kodama, T.; Takami, M.; Mizuta, N.; Oyama, H.; Nishitomi, K.; Chiu, Y.-W.; Okamoto, T.; Ikeuchi, T.; et al. Semagacestat Is a Pseudo-Inhibitor of γ-Secretase. *Cell Rep.* **2017**, *21*, 259–273. [CrossRef]
215. Imbimbo, B.; Giardina, G. γ-secretase inhibitors and modulators for the treatment of Alzheimer's disease: Disappointments and hopes. *Curr. Top. Med. Chem.* **2011**, *11*, 1555–1570. [CrossRef]
216. Zhao, J.; Liu, X.; Xia, W. Targeting Amyloidogenic Processing of APP in Alzheimer's Disease. *Front. Mol. Neurosci.* **2020**, *13*, 137. [CrossRef] [PubMed]
217. Mekala, S.; Nelson, G.; Li, Y.M. Recent developments of small molecule γ-secretase modulators for Alzheimer's disease. *RSC Med. Chem.* **2020**, *11*, 1003–1022. [CrossRef] [PubMed]
218. Rynearson, K.D.; Ponnusamy, M.; Prikhodko, O.; Xie, Y.; Zhang, C.; Nguyen, P.; Hug, B.; Sawa, M.; Becker, A.; Spencer, B.; et al. Preclinical validation of a potent γ-secretase modulator for Alzheimer's disease prevention. *J. Exp. Med.* **2021**, *218*, e20202560. [CrossRef] [PubMed]
219. Scheuner, D.; Song, B.; McEwen, E.; Liu, C.; Laybutt, R.; Gillespie, P.; Saunders, T.; Bonner-Weir, S.; Kaufman, R.J. Translational control is required for the unfolded protein response and in vivo glucose homeostasis. *Mol. Cell.* **2001**, *7*, 1165–1176. [CrossRef]
220. Huse, J.; Liu, K.; Pijak, D.; Carlin, D.; Lee, V.; Doms, R. β-secretase processing in the trans-Golgi network preferentially generates truncated amyloid species that accumulate in Alzheimer's disease brain. *J. Biol. Chem.* **2002**, *277*, 16278–16284. [CrossRef]
221. Liu, K.; Doms, R.; Lee, M. Glu11 site cleavage and N-terminally truncated A β production upon BACE overexpression. *Biochemistry* **2002**, *41*, 3128–3136. [CrossRef]
222. Lee, E.B.; Zhang, B.; Liu, K.; Greenbaum, E.A.; Doms, R.W.; Trojanowski, J.Q.; Lee, V.M. BACE overexpression alters the subcellular processing of APP and inhibits Aβ deposition in vivo. *J. Cell Biol.* **2005**, *168*, 291–302. [CrossRef]
223. Kimura, A.; Hata, S.; Suzuki, T. Alternative selection of β-site APP-cleaving enzyme 1 (BACE1) cleavage sites in amyloid β-protein precursor (APP) harboring protective and pathogenic mutations within the Aβ sequence. *J. Biol. Chem.* **2016**, *291*, 24041–24053. [CrossRef]
224. Fluhrer, R.; Multhaup, G.; Schlicksupp, A.; Okochi, M.; Takeda, M.; Lammich, S.; Willem, M.; Westmeyer, G.; Bode, W.; Walter, J.; et al. Identification of a beta-secretase activity, which truncates amyloid beta-peptide after its presenilin-dependent generation. *J. Biol. Chem.* **2003**, *278*, 5531–5538. [CrossRef]
225. Shi, X.P.; Tugusheva, K.; Bruce, J.E.; Lucka, A.; Wu, G.X.; Chen-Dodson, E.; Price, E.; Li, Y.; Xu, M.; Huang, Q.; et al. Beta-secretase cleavage at amino acid residue 34 in the amyloid beta peptide is dependent upon gamma-secretase activity. *J. Biol. Chem.* **2003**, *278*, 21286–21294. [CrossRef]
226. Liebsch, F.; Kulic, L.; Teunissen, C.; Shobo, A.; Ulku, I.; Engelschalt, V.; Hancock, M.A.; van der Flier, W.M.; Kunach, P.; Rosa-Neto, P.; et al. Aβ34 is a BACE1-derived degradation intermediate associated with amyloid clearance and Alzheimer's disease progression. *Nat. Commun.* **2019**, *10*, 2240. [CrossRef]
227. Garcia-Gonzales, L.; Pilat, D.; Rivera, S. Emerging Alternative Proteinases in APP Metabolism and Alzheimer's Disease Pathogenesis. *Front. Aging Neurosci.* **2019**, *11*, 244.
228. Farzan, M.; Schnitzler, C.E.; Vasilieva, N.; Leung, D.; Choe, H. BACE2, a β-secretase homolog, cleaves at the β site and within the amyloid-β region of the amyloid-β precursor protein. *Proc. Natl. Acad. Sci. USA* **2000**, *97*, 9712–9717. [CrossRef] [PubMed]

229. Yan, R.; Munzner, J.; Shuck, M.; Bienkowski, M. BACE2 functions as an alternative α-secretase in cells. *J. Biol. Chem.* **2001**, *276*, 34019–34027. [CrossRef]
230. Luo, J.; Zou, H.; Guo, Y.; Huang, K.; Ngan, E.S.; Li, P. BACE2 variant identified from HSCR patient causes AD-like phenotypes in hPSC-derived brain organoids. *Cell Death Discov.* **2022**, *8*, 47. [CrossRef] [PubMed]
231. Basi, G.; Frigon, N.; Barbour, R.; Doan, T.; Gordon, G.; McConlogue, L.; Sinha, S.; Zeller, M. Antagonistic effects of β-site amyloid precursor protein-cleaving enzymes 1 and 2 on β-amyloid peptide production in cells. *J. Biol. Chem.* **2003**, *278*, 31512–31520. [CrossRef]
232. Sellal, F.; Wallon, D.; Martinez-Almoyna, L.; Marelli, C.; Dhar, A.; Oesterlé, H.; Rovelet-Lecrux, A.; Rousseau, S.; Kourkoulis, C.E.; Rosand, J.; et al. APP Mutations in Cerebral Amyloid Angiopathy with or without Cortical Calcifications: Report of Three Families and a Literature Review. *J. Alzheimers Dis.* **2017**, *56*, 37–46. [CrossRef]
233. Hendriks, L.; van Duijn, C.M.; Cras, P.; Cruts, M.; Van Hul, W.; van Harskamp, F.; Warren, A.; McInnis, M.G.; Antonarakis, S.E.; Martin, J.J. Presenile dementia and cerebral haemorrhage linked to a mutation at codon 692 of the beta-amyloid precursor protein gene. *Nat. Genet.* **1992**, *1*, 218–221. [CrossRef]
234. Roks, G.; Van Harskamp, F.; De Koning, I.; Cruts, M.; De Jonghe, C.; Kumar-Singh, S.; Tibben, A.; Tanghe, H.; Niermeijer, M.F.; Hofman, A.; et al. Presentation of amyloidosis in carriers of the codon 692 mutation in the amyloid precursor protein gene (APP692). *Brain* **2000**, *123 Pt 10*, 2130–2140. [CrossRef]
235. Cras, P.; van Harskamp, F.; Hendriks, L.; Ceuterick, C.; van Duijn, C.M.; Stefanko, S.Z.; Hofman, A.; Kros, J.M.; Van Broeckhoven, C.; Martin, J.J. Presenile Alzheimer dementia characterized by amyloid angiopathy and large amyloid core type senile plaques in the APP 692Ala→Gly mutation. *Acta Neuropathol.* **1998**, *96*, 253–260. [CrossRef]
236. Brooks, W.S.; Kwok, J.B.; Halliday, G.M.; Godbolt, A.K.; Rossor, M.N.; Creasey, H.; Jones, A.O.; Schofield, P.R. Hemorrhage is uncommon in new Alzheimer family with Flemish amyloid precursor protein mutation. *Neurology* **2004**, *63*, 1613–1617. [CrossRef]
237. Ryan, N.S.; Biessels, G.J.; Kim, L.; Nicholas, J.M.; Barber, P.A.; Walsh, P.; Gami, P.; Morris, H.R.; Bastos-Leite, A.J.; Schott, J.M.; et al. Genetic determinants of white matter hyperintensities and amyloid angiopathy in familial Alzheimer's disease. *Neurobiol. Aging.* **2015**, *36*, 3140–3151. [CrossRef] [PubMed]
238. Tang, T.C.; Hu, Y.; Kienlen-Campard, P.; El Haylani, L.; Decock, M.; Van Hees, J.; Fu, Z.; Octave, J.N.; Constantinescu, S.N.; Smith, S.O. Conformational changes induced by the A21G Flemish mutation in the amyloid precursor protein lead to increased Aβ production. *Structure* **2014**, *22*, 387–396. [CrossRef]
239. Asai, H.; Ohkawa, N.; Saitoh, Y.; Ghandour, K.; Murayama, E.; Nishizono, H.; Matsuo, M.; Hirayama, T.; Kaneko, R.; Muramatsu, S.I.; et al. Pcdhβ deficiency affects hippocampal CA1 ensemble activity and contextual fear discrimination. *Mol. Brain.* **2020**, *13*, 7. [CrossRef] [PubMed]
240. Rao, Y.L.; Ganaraja, B.; Murlimanju, B.V.; Joy, T.; Krishnamurthy, A.; Agrawal, A. Hippocampus and its involvement in Alzheimer's disease: A review. *3 Biotech* **2022**, *12*, 55. [CrossRef]
241. Gabitto, M.I.; Travaglini, K.J.; Rachleff, V.M.; Kaplan, E.S.; Long, B.; Ariza, J.; Ding, Y.; Mahoney, J.T.; Dee, N.; Goldy, J.; et al. Integrated multimodal cell atlas of Alzheimer's disease. *Nat. Neurosci.* **2024**, *27*, 2366–2383. [CrossRef] [PubMed] [PubMed Central]
242. Ramos-Cejudo, J.; Wisniewski, T.; Marmar, C.; Zetterberg, H.; Blennow, K.; de Leon, M.J.; Fossati, S. Traumatic Brain Injury and Alzheimer's Disease: The Cerebrovascular Link. *eBioMedicine* **2018**, *28*, 21–30. [CrossRef]
243. Mielke, M.M.; Ransom, J.E.; Mandrekar, J.; Turcano, P.; Savica, R.; Brown, A.W. Traumatic Brain Injury and Risk of Alzheimer's Disease and Related Dementias in the Population. *J. Alzheimers Dis.* **2022**, *88*, 1049–1059. [CrossRef] [PubMed]
244. Gu, D.; Ou, S.; Liu, G. Traumatic Brain Injury and Risk of Dementia and Alzheimer's Disease: A Systematic Review and Meta-Analysis. *Neuroepidemiology* **2022**, *56*, 4–16. [CrossRef]
245. Turner, R.C.; Lucke-Wold, B.P.; Robson, M.J.; Lee, J.M.; Bailes, J.E. Alzheimer's disease and chronic traumatic encephalopathy: Distinct but possibly overlapping disease entities. *Brain Inj.* **2016**, *30*, 1279–1292. [CrossRef]
246. Sait, A.; Angeli, C.; Doig, A.J.; Day, P.J.R. Viral Involvement in Alzheimer's Disease. *ACS Chem. Neurosci.* **2021**, *12*, 1049–1060. [CrossRef] [PubMed]
247. Kozlov, M. Massive health-record review links viral illnesses to brain disease. *Nature* **2023**, *614*, 18–19. [CrossRef] [PubMed]
248. Bruno, F.; Abondio, P.; Bruno, R.; Ceraudo, L.; Paparazzo, E.; Citrigno, L.; Luiselli, D.; Bruni, A.C.; Passarino, G.; Colao, R.; et al. Alzheimer's disease as a viral disease: Revisiting the infectious hypothesis. *Ageing Res. Rev.* **2023**, *91*, 102068.
249. Blackhurst, B.M.; Funk, K.E. Viral pathogens increase risk of neurodegenerative disease. *Nat. Rev. Neurol.* **2023**, *19*, 259–260. [CrossRef]
250. Levine, K.S.; Leonard, H.L.; Blauwendraat, C.; Iwaki, H.; Johnson, N.; Bandres-Ciga, S.; Ferrucci, L.; Faghri, F.; Singleton, A.B.; Nalls, M.A. Virus exposure and neurodegenerative disease risk across national biobanks. *Neuron* **2023**, *111*, 1086–1093.e2. [CrossRef]
251. Vigasova, D.; Nemergut, M.; Liskova, B.; Damborsky, J. Multi-pathogen infections and Alzheimer's disease. *Microb. Cell Factories* **2021**, *20*, 25. [CrossRef]

252. Maheshwari, P.; Eslick, G.D. Bacterial infection and Alzheimer's disease: A meta-analysis. *J. Alzheimers Dis.* **2015**, *43*, 957–966. [CrossRef] [PubMed]
253. Catumbela, C.S.G.; Giridharan, V.V.; Barichello, T.; Morales, R. Clinical evidence of human pathogens implicated in Alzheimer's disease pathology and the therapeutic efficacy of antimicrobials: An overview. *Transl. Neurodegener.* **2023**, *12*, 37. [CrossRef]
254. Moné, Y.; Earl, J.P.; Król, J.E.; Ahmed, A.; Sen, B.; Ehrlich, G.D.; Lapides, J.R. Evidence supportive of a bacterial component in the etiology for Alzheimer's disease and for a temporal-spatial development of a pathogenic microbiome in the brain. *Front. Cell. Infect. Microbiol.* **2023**, *13*, 1123228. [CrossRef]
255. Xie, J.; Van Hoecke, L.; Vandenbroucke, R.E. The Impact of Systemic Inflammation on Alzheimer's Disease Pathology. *Front. Immunol.* **2022**, *12*, 796867. [CrossRef]
256. Kinney, J.W.; Bemiller, S.M.; Murtishaw, A.S.; Leisgang, A.M.; Salazar, A.M.; Lamb, B.T. Inflammation as a central mechanism in Alzheimer's disease. *Alzheimers Dement.* **2018**, *4*, 575–590. [CrossRef] [PubMed]
257. Walker, K.A.; Ficek, B.N.; Westbrook, R. Understanding the Role of Systemic Inflammation in Alzheimer's Disease. *ACS Chem. Neurosci.* **2019**, *10*, 3340–3342. [CrossRef] [PubMed]
258. Janbek, J.; Laursen, T.M.; Frimodt-Møller, N.; Magyari, M.; Haas, J.G.; Lathe, R.; Waldemar, G. Hospital-Diagnosed Infections, Autoimmune Diseases, and Subsequent Dementia Incidence. *JAMA Netw. Open* **2023**, *6*, e2332635. [CrossRef]
259. Lindbohm, J.V.; Mars, N.; Sipilä, P.N.; Singh-Manoux, A.; Runz, H.; FinnGen; Livingston, G.; Seshadri, S.; Xavier, R.; Hingorani, A.D.; et al. Immune system-wide Mendelian randomization and triangulation analyses support autoimmunity as a modifiable component in dementia-causing diseases. *Nat. Aging.* **2022**, *2*, 956–972. [CrossRef]
260. Sangha, P.S.; Thakur, M.; Akhtar, Z.; Ramani, S.; Gyamfi, R.S. The Link Between Rheumatoid Arthritis and Dementia: A Review. *Cureus* **2020**, *12*, e7855. [CrossRef]
261. Shmerling, R. Treatments for Rheumatoid Arthritis May Lower Dementia Risk. Available online: https://www.health.harvard.edu/blog/treatments-for-rheumatoid-arthritis-may-lower-dementia-risk-202210032827 (accessed on 1 March 2025).
262. Gupta, D.P.; Lee, Y.S.; Choe, Y.; Kim, K.T.; Song, G.J.; Hwang, S.C. Knee osteoarthritis accelerates amyloid beta deposition and neurodegeneration in a mouse model of Alzheimer's disease. *Mol. Brain.* **2023**, *16*, 1. [CrossRef] [PubMed]
263. Huang, S.W.; Wang, W.T.; Chou, L.C.; Liao, C.D.; Liou, T.H.; Lin, H.W. Osteoarthritis increases the risk of dementia: A nationwide cohort study in Taiwan. *Sci. Rep.* **2015**, *5*, 10145. [CrossRef]
264. Weber, A.; Mak, S.H.; Berenbaum, F.; Sellam, J.; Zheng, Y.P.; Han, Y.; Wen, C. Association between osteoarthritis and increased risk of dementia: A systemic review and meta-analysis. *Medicine* **2019**, *98*, e14355. [CrossRef]
265. Du, J.; Li, A.; Shi, D.; Chen, X.; Wang, Q.; Liu, Z.; Sun, K.; Guo, T.; Alzheimer's Disease Neuroimaging Initiative. Association of *APOE*-ε4, Osteoarthritis, β-Amyloid, and Tau Accumulation in Primary Motor and Somatosensory Regions in Alzheimer Disease. *Neurology* **2023**, *101*, e40–e49. [CrossRef]
266. Ikram, M.; Innes, K.; Sambamoorthi, U. Association of osteoarthritis and pain with Alzheimer's Diseases and Related Dementias among older adults in the United States. *Osteoarthr. Cartil.* **2019**, *27*, 1470–1480. [CrossRef]
267. Cairns, D.M.; Itzhaki, R.F.; Kaplan, D.L. Potential Involvement of Varicella Zoster Virus in Alzheimer's Disease via Reactivation of Quiescent Herpes Simplex Virus Type 1. *J. Alzheimers Dis.* **2022**, *88*, 1189–1200. [CrossRef]
268. Cohen, M.; Austin, E.; Bradu, S.; Jagdeo, J. The Association Between Herpes Simplex Virus and Alzheimer's Disease: A Systematic Review. *J. Drugs Dermatol.* **2023**, *22*, 1046–1052. [CrossRef] [PubMed]
269. Linard, M.; Garrigue, I.; Vellas, B.; Coley, N.; Zetterberg, H.; Blennow, K.; Ashton, N.J.; Payoux, P.; Salabert, A.S.; Dartigues, J.F.; et al. Association between herpes simplex virus infection and Alzheimer's disease biomarkers: Analysis within the MAPT trial. *Sci. Rep.* **2025**, *15*, 2362. [CrossRef] [PubMed]
270. Hyde, V.R.; Zhou, C.; Fernandez, J.R.; Chatterjee, K.; Ramakrishna, P.; Lin, A.; Fisher, G.W.; Çeliker, O.T.; Caldwell, J.; Bender, O.; et al. Anti-herpetic tau preserves neurons via the cGAS-STING-TBK1 pathway in Alzheimer's disease. *Cell Rep.* **2025**, *44*, 115109. [CrossRef]
271. Cairns, D.M.; Smiley, B.M.; Smiley, J.A.; Khorsandian, Y.; Kelly, M.; Itzhaki, R.F.; Kaplan, D.L. Repetitive injury induces phenotypes associated with Alzheimer's disease by reactivating HSV-1 in a human brain tissue model. *Sci. Signal.* **2025**, *18*, eado6430. [CrossRef]
272. Galea, I. The blood-brain barrier in systemic infection and inflammation. *Cell Mol. Immunol.* **2021**, *18*, 2489–2501. [CrossRef]
273. Goldsmith, H.S. Alzheimer's Disease: A Decreased Cerebral Blood Flow to Critical Intraneuronal Elements Is the Cause. *J. Alzheimers Dis.* **2022**, *85*, 1419–1422. [CrossRef]
274. Garsia, J. New Study Shows Lower Cerebral Blood Flow Precedes Brain Tissue Loss in Early Phase of Alzheimer's Disease. Available online: https://news.med.miami.edu/lower-cerebral-blood-flow-precedes-brain-tissue-loss-in-early-phase-of-alzheimers/ (accessed on 1 January 2024).
275. Parodi-Rullán, R.; Sone, J.Y.; Fossati, S. Endothelial Mitochondrial Dysfunction in Cerebral Amyloid Angiopathy and Alzheimer's Disease. *J. Alzheimers Dis.* **2019**, *72*, 1019–1039.

276. Chen, H.L.; Lin, H.C.; Lu, C.H.; Chen, P.C.; Huang, C.C.; Chou, K.H.; Su, M.C.; Friedman, M.; Chen, Y.W.; Lin, W.C. Systemic inflammation and alterations to cerebral blood flow in obstructive sleep apnea. *J. Sleep. Res.* **2017**, *26*, 789–798. [CrossRef]
277. Mun, K.T.; Hinman, J.D. Inflammation and the Link to Vascular Brain Health: Timing Is Brain. *Stroke* **2022**, *53*, 427–436. [CrossRef]
278. Goldsmith, H.S.; Duckett, S.; Chen, W.F. Prevention of cerebral infarction in the dog by intact omentum. *Am. J. Surg.* **1975**, *130*, 317–320. [CrossRef]
279. Goldsmith, H.S.; Duckett, S.; Chen, W.F. Prevention of cerebral infarction in the monkey by omental transposition to the brain. *Stroke* **1978**, *9*, 224–229. [CrossRef]
280. Signoretti, S.; Marmarou, A.; Aygok, G.A.; Fatouros, P.P.; Portella, G.; Bullock, R.M. Assessment of mitochondrial impairment in traumatic brain injury using high-resolution proton magnetic resonance spectroscopy. *J. Neurosurg.* **2008**, *108*, 42–52. [CrossRef] [PubMed]
281. Aygok, G.A.; Marmarou, A.; Fatouros, P.; Kettenmann, B.; Bullock, R.M. Assessment of mitochondrial impairment and cerebral blood flow in severe brain injured patients. *Acta Neurochir. Suppl.* **2008**, *102*, 57–61. [PubMed]
282. Pulina, M.V.; Hopkins, M.; Haroutunian, V.; Greengard, P.; Bustos, V. C99 selectively accumulates in vulnerable neurons in Alzheimer's disease. *Alzheimers Dement.* **2020**, *16*, 273–282. [CrossRef]
283. Takasugi, N.; Komai, M.; Kaneshiro, N.; Ikeda, A.; Kamikubo, Y.; Uehara, T. The Pursuit of the "Inside" of the Amyloid Hypothesis-Is C99 a Promising Therapeutic Target for Alzheimer's Disease? *Cells* **2023**, *12*, 454. [CrossRef] [PubMed]
284. Veo, B.L.; Krushel, L.A. Translation initiation of the human tau mRNA through an internal ribosomal entry site. *J. Alzheimers Dis.* **2009**, *16*, 271–275. [CrossRef]
285. Lee, S.E.; Kwon, D.; Shin, N.; Kong, D.; Kim, N.G.; Kim, H.Y.; Kim, M.J.; Choi, S.W.; Kang, K.S. Accumulation of APP-CTF induces mitophagy dysfunction in the iNSCs model of Alzheimer's disease. *Cell Death Dis.* **2022**, *8*, 1. [CrossRef]
286. Montesinos, J.; Pera, M.; Larrea, D.; Guardia-Laguarta, C.; Agrawal, R.R.; Velasco, K.R.; Yun, T.D.; Stavrovskaya, I.G.; Xu, Y.; Koo, S.Y.; et al. The Alzheimer's disease-associated C99 fragment of APP regulates cellular cholesterol trafficking. *EMBO J.* **2020**, *39*, e103791. [CrossRef]
287. Pera, M.; Montesinos, J.; Larrea, D.; Agrawal, R.R.; Velasco, K.R.; Stavrovskaya, I.G.; Yun, T.D.; Area-Gomez, E. MAM and C99, key players in the pathogenesis of Alzheimer's disease. *Int. Rev. Neurobiol.* **2020**, *154*, 235–278.
288. Lauritzen, I.; Pardossi-Piquard, R.; Bourgeois, A.; Pagnotta, S.; Biferi, M.G.; Barkats, M.; Lacor, P.; Klein, W.; Bauer, C.; Checler, F. Intraneuronal aggregation of the β-CTF fragment of APP (C99) induces Aβ-independent lysosomal-autophagic pathology. *Acta Neuropathol.* **2016**, *132*, 257–276. [CrossRef] [PubMed]
289. Kapranov, P.; Ozsolak, F.; Kim, S.; Foissac, S.; Lipson, D.; Hart, C.; Roels, S.; Borel, C.; Antonarakis, S.; Monaghan, A.; et al. New class of gene-termini-associated human RNAs suggests a novel RNA copying mechanism. *Nature* **2010**, *466*, 642–646. [CrossRef] [PubMed]
290. Khare, P.; Edgecomb, S.X.; Hamadani, C.M.; Tanner, E.E.L.; SManickam, D. Lipid nanoparticle-mediated drug delivery to the brain. *Adv. Drug Deliv. Rev.* **2023**, *197*, 114861. [CrossRef]
291. Cepparulo, P.; Cuomo, O.; Campani, V.; Vinciguerra, A.; Sisalli, M.J.; Nele, V.; Anzilotti, S.; Valsecchi, V.; Casamassa, A.; Brancaccio, P.; et al. Anti-miRNA103/107 encapsulated in transferrin-conjugated lipid nanoparticles crosses blood-brain barrier and reduces brain ischemic damage. *Mol. Ther. Nucleic Acids* **2024**, *35*, 102131. [CrossRef] [PubMed]
292. Wang, C.; Wang, S.; Xue, Y.; Zhong, Y.; Li, H.; Hou, X.; Kang, D.D.; Liu, Z.; Tian, M.; Wang, L.; et al. Intravenous administration of blood-brain barrier-crossing conjugates facilitate biomacromolecule transport into central nervous system. *Nat. Biotechnol.* 2024, *ahead of print*. [CrossRef]
293. Wang, C.; Xue, Y.; Markovic, T.; Li, H.; Wang, S.; Zhong, Y.; Du, S.; Zhang, Y.; Hou, X.; Yu, Y.; et al. Blood-brain-barrier-crossing lipid nanoparticles for mRNA delivery to the central nervous system. *Nat. Mater.* 2025, *ahead of print*. [CrossRef] [PubMed]

MDPI AG
Grosspeteranlage 5
4052 Basel
Switzerland
Tel.: +41 61 683 77 34

International Journal of Molecular Sciences Editorial Office
E-mail: ijms@mdpi.com
www.mdpi.com/journal/ijms